GUIDE TO CULTIVATED PLANTS

GUIDE TO CULTIVATED PLANTS

Ton Elzebroek

and

Koop Wind

Wageningen University,
Crop and Weed Ecology Group,
Plant Production Systems Group

www.cabi.org

CABI is a trading name of CAB International

CABI Head Office
Nosworthy Way
Wallingford
Oxfordshire OX10 8DE
UK

CABI North American Office
875 Massachusetts Avenue
7th Floor
Cambridge, MA 02139
USA

Tel: +44 (0)1491 832111
Fax: +44 (0)1491 833508
E-mail: cabi@cabi.org
Website: www.cabi.org

Tel: +1 617 395 4056
Fax: +1 617 354 6875
E-mail: cabi-nao@cabi.org

A catalogue record for this book is available from the British Library, London, UK.

Library of Congress Cataloging-in-Publication Data

Elzebroek, A. T. G.
  Guide to cultivated plants / A.T.G. Elzebroek and K. Wind.
     p. cm.
  Includes bibliographical references and index.
  ISBN 978-1-84593-356-2 (alk. paper)
1. Plants, Cultivated--Handbooks, manuals, etc. 2. Botany, Economic--Handbooks, manuals, etc. I. Wind, Koop. II. Title.

  SB107.E49 2008
  630--dc22

2007028459

ISBN: 978 1 84593 356 2

Typeset by MRM Graphics, Ltd, Winslow, UK.
Printed and bound in Singapore by Kyodo Press in association with MRM Graphics.

Contents

Foreword

Working in agriculture – whether in education, research, advisory or more practical roles – requires a thorough knowledge of the morphology, botany, ecology, agronomy and use of cultivated crops. This is well recognized, and almost all education programmes in Agriculture, Agronomy, Crop Science, Plant Breeding and related subjects offer mandatory courses to ensure that students acquire such knowledge.

The authors of this book share about 80 years of experience in educating students in these aspects of crop plants. They have often been the first teachers in crop science-related courses that students met in Wageningen, and Koop Wind and Ton Elzebroek have taken their teaching assignments seriously showing both great professionalism and enthusiasm. Their courses were always based on up-to-date written material produced by themselves and they continuously improved the quality of their courses and the hand-outs to the wishes of a dynamic population of students. The quality of their teaching was always clear from the very positive student evaluations of their courses. We often meet former students who have vivid and warm memories of these courses.

When Koop and Ton were approaching retirement we recognized the need to capture their knowledge and experience in the form of a book and we are very glad that they took up the challenge. The product is now in your hands and we – the undersigned – are proud of it.

The book includes 346 crop species, of which 180 are described in some detail. There are 92 main entries comprising 11 commodity groups ranging from vegetables grown in horticulture to forages species and arable crops grown in arable farming systems, to the major fruits and plantation crops. All major crops from temperate, Mediterranean and tropical climates are included. The book contains 370 figures, of which about three-quarters are in full colour.

It is certainly not the first book on this topic and it most likely will not be the last. This book, however, is unique because of the consistency of the descriptions based on a well thought-over outline design, the beautiful illustrations (both pictures and drawings) and the fact that it is obviously written with pleasure, dedication and most of all love for the subject, with an enormous input of time and effort.

We congratulate the authors and the publisher on this product and do hope that readers will enjoy it as much as we do.

P.C. Struik
Professor in Crop Physiology
Wageningen University

K.E. Giller
Professor in Plant Production Systems
Wageningen University

Preface

This book is a standard reference text for students of (agricultural) universities and colleges, extension workers, teachers, farmers, horticulturists and in general for all those who are interested in cultivated plants. Although various monographs of cultivated plant species and multivolume handbooks such as *Tropical Crops* by J.W. Purseglove and the PROSEA (Plant Resources of South-East Asia) handbooks have been published, this book provides concise textual descriptions and attractive illustrations of around 100 entries in a single handbook. Depending on their importance, each entry consists of either a single species, as in the case of potato and rice, or of a group of related species, such as the yams, millets and labiate spices.

The selection of plants is mainly based on their relative importance in terms of cultivated area and economic value. Information on production of the crops is derived from statistical databases concerning 'Primary Crops of Agricultural Production', from the Food and Agriculture Organization of the United Nations. Consistent with this list, commodity groups such as timber trees and forestry plants, ornamental plants, dye- and tannin-producing plants and medicinal plants are not included.

The selected cultivated plants are classified according to their main usage, in the following commodity groups: Beverages and Tobacco; Edible Fruits and Nuts; Elastomers; Fibre Crops; Forages; Oil Crops; Protein Crops; Spices and Flavourings; Starch Crops; Sugar Crops; and Vegetables. The various plants or groups of plants are described conforming to a certain format, including their origin, history and spread; botanical characteristics; cultivars (or varieties), uses and constituents; and ecology and agronomy. Except for specific examples which have major effects on production, the description of diseases and pests and their treatments are not included. Moreover, this subject is more suitable for a separate and detailed description.

To improve the readability of the book, the scientific names of plant species are used without their scientific authorities; these can be found in the index of plant names.

The majority of the described cultivated plants are food plants. Their importance for food supply and food security, now and in the future, is evident.

Acknowledgements

The authors would like to pay special thanks to Professor Rudy Rabbinge (University Professor) who initially proposed the writing of this book, Professor Ken Giller (Plant Production Systems Group), Professor Paul Struik (Crop and Weed Ecology Group) and Emeritus Professor Marius Wessel (Tropical Agriculture), from Wageningen University and Research Centre (WUR), for their advice and suggestions in the preparation of this book. Their judgements concerning the content have proved to be invaluable.

We are grateful to Ir. Dirk L. Schuiling for making available the interactive computer program TROPCROP, which he developed for use in the study of tropical and subtropical cultivated plants. During the realization of this book we often had inspiring conversations with Mr Schuiling, which resulted in improvement of the content.

We thank the Board of the PROSEA Foundation (Plant Resources of South-East Asia) who made available a great number of botanical line drawings originally made for illustrating the series of PROSEA books; Piet Kostense, Harry Wijnhoven and Dick van de Gugten, who contributed to making line drawings; PPO–AGV Lelystad for lending us several slides concerning lettuce and endive; Jan Oude Voshaar from the Wageningse Berg vineyard and Patrick Hendrickx of the Mushroom Research Group of WUR, who contributed to the sections regarding grapes and mushrooms. Several slides were derived from the personal archives of actual and previous staff members of WUR.

As well as using the botanical line drawings of PROSEA, we often consulted their books, which proved to be an important source of information. Also, information regarding cultivated plants in the *Handbook of Energy Crops* (J.A. Duke), published only on the World Wide Web, was very useful to us. Most of the statistics used are derived from statistical databases concerning 'Primary Crops of Agricultural Production', from the Food and Agriculture Organization of the United Nations (FAO).

Photographs and line drawings used without mention of the source are derived from the WUR archive.

1 Beverages and Tobacco

Cacao

Cacao – *Theobroma cacao*; Sterculia family – *Sterculiaceae*

Origin, history and spread

The origin of the cacao tree is thought to have been in the lower tree storey of the evergreen tropical rainforests of the upper Amazon basin. It is assumed that in early times a natural population of cacao was spread throughout the central part of the Amazon, as well as along the Orinoco River towards Venezuela and the Guyanas, and west- and northwards to the south of Mexico. These populations developed into two distinct forms, geographically separated by the Panama isthmus: the Criollo type found in Central America and the Forastero type found in the upper and lower Amazon basin. The Criollo type of cacao was already grown by the Aztecs and Mayas in Central America but there is no evidence that the Forastero cacao was cultivated until the Spaniards started to extend the cultivation of cacao in South America. Until the end of the 18th century cacao production was dominated by the Criollo type.

The continuous increase in cacao cultivation in South America resulted in two separate Forastero types: the Amelonado Forastero type in the Brazilian state of Bahia, originating in the lower Amazon basin, whereas the 'Nacional' Forastero type in Ecuador originated in the upper Amazon basin.

The Trinitario type is a natural hybrid between the Criollo and Forastero types and replaced the earlier Criollo type on Trinidad after 1800. The Portuguese successfully transferred Forastero cacao to West Africa in the 19th century, where it gave rise to the West African Amelonado variety. The first introduction of cacao to South-east Asia can be traced back to the few Criollo seedlings taken from Central America to the Philippines by the Spanish around

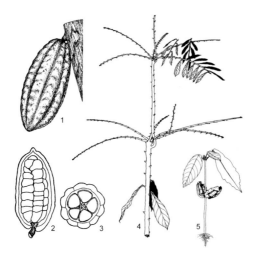

Fig. 1.1. Cacao: 1, fruit ('pod'); 2, fruit in longitudinal section; 3, fruit in cross-section; 4, dimorphic branching habit of young tree; 5, seedling.

the year 1670. The Dutch introduced cacao into Indonesia in the 17th century. Cacao is cultivated now throughout the per-humid and humid tropics, especially in West Africa and South-east Asia.

Botany

Cacao (Fig. 1.1) is an evergreen tree 4–20 m tall, although in cultivation usually 4–6 m. The tap root grows up to 2 m straight down with most of the (up to 6 m long) lateral roots in the upper 20 cm soil layer. Cacao roots are possibly colonized with a mycorrhizal fungus. The main stem of a young cacao plant may grow to 1–1.5 m in height before branching. The stem growth is sympodial, with orthotropic subterminal shoots (chupons) and lateral branching with successive whorls (fan or 'jorquette') of plagiotropic branches. However, the dimorphic branching in cacao is not as absolute as in coffee species; orthotropic shoots can be induced on mature plagiotropic stems.

The leaves are simple, petiolate, usually glabrous, elliptical oblong, 15–50 cm × 4–15 cm. The leaves are produced in flushes; soft and pendulant when young, dark green and leathery at maturity. The petiole is 1–10 cm long with a well-marked pulvinus at both ends. The leaves of the main stem and chupons are spirally arranged, while the leaves on branches are alternate. The stipules at the base of the petiole are shed as the leaves mature.

Inflorescences are borne on leafless older trunk and fan branches (cauliflorous), usually borne on small flower cushions (tubercles) in short, many-flowered fascicles; flowers are five-parted, hermaphrodite, white to reddish and 1–1.5 cm in diameter; pedicels 1–2 cm long. The five sepals are pink or whitish, triangular and rather fleshy; the petals are smaller than the sepals, 3–4 mm long, expanding into a concave, cup-shaped pouch, white with two prominent purple guide lines; upper part spatulate, pale yellow and reflexed. The male part of the flower (androecium) consists of an outer whorl of five long, pointed staminoides, and an inner whorl of five fertile stamens, standing opposite the petals. The ovary is oblong-ovoid, superior, with five carpels. The style is partially divided into five lobes (Fig. 1.2).

A full-grown tree may produce more than 10,000 flowers/year of which perhaps only ten to 50 will develop as mature fruits. Pollination is effected by insects, particularly by specific flying female midges. Unpollinated flowers abscise within 24 h. During the first half of the period of pod development, part of the young fruits (cherelles) stops growing, blackens and shrivels. This fruit thinning process is called 'cherelles wilt'.

The fruit is a berry-like drupe, commonly called a 'pod', with a lignified, smooth or furrowed fruit wall (husk). The 20–60 seeds, usually called 'beans', are embedded in a mucilaginous, whitish and sweet pulp. At maturity the seed consists of two large cotyledons, a small embryo and a leathery skin or shell (testa). The 1000-seed weight, at 30% moisture content, is about 2 kg.

Uses and constituents

At the time the Spaniards conquered Mexico in the early 16th century, they found the cacao tree in the south of the country, where it had already been expertly cultivated for centuries. In Central America the valuable cacao beans were used both as currency and for preparing an invigorating beverage, called 'cacahuatl', by boiling a mix of ground roasted cacao beans with maize, vanilla and chilli peppers.

The Maya Indians considered the plant to be of divine origin. The mythological prophet Quatzalcault took the seeds from paradise and planted them in his garden at Talzitepic. By eating the fruits he acquired universal knowledge and wisdom, which brought him great adoration. The name '*Theobroma*', given by Linnaeus and meaning 'food of the gods', is derived from this legend.

Fig. 1.2. Cacao: cauliflorous inflorescences (left) (photo: J. van Zee) and close-up of a single flower (right) (photo: D.L. Schuiling).

The Spaniards found that cacao beans could be made into a more palatable drink when mixed with sugar and in this form it became popular in Spain, from where it spread to France and England in the 17th century and successively to other European countries. During the 17th and 18th centuries it remained an expensive drink to consume, for instance in the chocolate houses of the aristocracy in London.

The invention of the cacao press, by C.J. van Houten in The Netherlands in 1828, to extract most of the cacao butter, and the process to manufacture milk chocolate, invented by M.D. Peter in Switzerland in 1876, had major impacts on the increase in consumption of cacao products. At the same time the cacao tree was increasingly introduced into many countries.

The manufacture of chocolate consists of various phases. Before the cacao beans (more accurately the seeds) enter international trade, they are fermented and dried in the producing countries. During the process of fermentation, with limited aeration, the mucilage around the seeds breaks down and purple pigment diffuses through the cotyledons. The precursors of the chocolate flavour are produced when the protein and polyphenol compounds react with enzymes, and the astringency disappears. After fermentation the beans are dried in the sun. The beans should be covered during rain and often for the hottest part of the day. The dried beans are normally packed in jute bags and shipped in containers. Once the dried beans have arrived in the chocolate-producing countries, the beans are roasted, after which the full cacao flavour develops. More than 60 aromatic compounds have been found to contribute to cacao flavour.

The roasted beans are ground to cacao mass (also known as 'cacao liquor'). Then cacao butter (fat) is pressed out of part of the cacao mass, lowering the fat content from 55 to 24%; another part of the mass is kept aside. The press cake is ground to powder. Then part of the cacao butter (and sugar) is added to the kept cacao mass to get plain chocolate or, after adding milk, to get milk chocolate. The base of white chocolate is cacao butter only. The cacao powder, when mixed with milk and sugar, makes a very nutritious beverage and the presence of theobromine and caffeine causes a mildly stimulating action.

Chocolate is rich in energy and provides a concentrated food with excellent keeping quality. A typical bar of milk chocolate may contain 15% cacao liquor, 20% cacao butter, 22% milk solids, 40% sugar, small amounts of lecithin emulsifier and vanillin, about 30 mg of caffeine (a cup of coffee contains 100–150 mg of caffeine) and about 44–60 mg of theobromine.

It is suggested that the substantial amount of sugar may partly explain chocolate's supposed addictive properties. On the other hand, scientists reported recently that chocolate may keep high blood pressure down, due to the presence of flavonoids. In general plain chocolate is higher in flavonoids than milk chocolate. During the manufacturing of cacao powder and cacao drinks, most of the flavonoids are removed.

It is possible for a person to be allergic to chocolate, but recent evidence suggests that the assumed allergen property of chocolate is rarer than generally is believed. Chocolate can be toxic, or even fatal, for pets in general but especially for dogs because they like to eat it.

Ecology and agronomy

Cacao is strictly a tropical crop and thrives best in the lowland areas of tropical evergreen and semi-evergreen rainforests. The limits of cultivation are 20°N and S, although the bulk of the crop is grown within 10°N and S. It can be grown at higher altitudes only if other conditions are favourable. Most cacao is grown in areas with an annual rainfall of 1500–2500 mm, with no more than three consecutive months of rainfall less than 100 mm. The optimum temperature ranges between a mean maximum of 30–32°C and a mean minimum of 18–21°C. The absolute minimum temperature should not be below 10°C. Cacao requires well-drained, deep and fertile soils with at least 2% organic carbon and a soil pH of about 6.5. Soils with a large moisture-storing capacity can compensate for periodic drought. Although the wild cacao tree is adapted well to dense shade conditions, it does not necessarily mean that mature cacao, grown as an economic crop, requires such low light intensities (Fig. 1.3).

The cheapest and most practised method of propagating cacao is by seed. Seeds are produced in biclonal gardens by open or controlled pollination to obtain 100% legitimate hybrid seed. Vegetative propagation is used mainly in breeding programmes and for establishing seed gardens. Seedlings are raised in a shaded nursery, usually in polythene bags. Young seedlings sometimes form double stems of which one has to be removed. When the seedlings are 4–6 months old, they are transplanted to the field at densities of 1100–1200 trees/ha. During transplanting damage to the tap root should be avoided. Young cacao seedlings require shade. When the trees grow older and have a sufficiently developed canopy the need for shade decreases. Once the canopies of

Fig. 1.3. Cacao pods hanging from trees under almost closed canopy.

neighbouring trees have met, little or no shade is needed. However, light conditions should be considered in relation to other growth conditions; when these are suboptimal, shade trees are still required. Shade can be provided either by thinning forest or by planting shade trees such as the leguminous species *Gliricidia sepium* and seedless clones of *Leucaena leucocephala*. Cacao is often grown with other crops such as coconut, oil palm and fruit trees.

Weeding is needed during establishment, preferably by slashing without scraping the topsoil in order not to damage the root systems of cacao. The remaining mulch also reduces the evaporation loss from the soil. Once the cacao trees mature, weed growth is suppressed by heavy shade beneath the canopy. Removal of basal chupons is often combined with weeding. Weed control is also done by the use of herbicides, especially when noxious and persistent weed species occur. Cacao is very susceptible to pests and diseases. In particular, sap-sucking mirids feeding on the stem tissue of young twigs may cause progressive deterioration of the canopy and require insecticidal control. Among the diseases, *Phytophtora* pod rot causes large economic losses. Especially in high-rainfall areas, spraying with copper fungicides is needed.

Fertilizer requirements depend to a great extent on soil fertility, age of trees, yields and shade; lightly shaded and unshaded cacao requires more fertilizers, particularly nitrogen. Soil analyses give a better diagnostic value than leaf analyses. Nutrients removed by 1 t of cured cacao beans are 20 kg N, 4 kg P and 10 kg N. As a general guide, per hectare, mature cacao needs 50–100 kg N, 25 kg P, 75 kg K and, sometimes, 15 kg Mg/year.

Young trees need no pruning during the first 2 to 3 years. Later, low drooping branches should be removed to ensure easy access for harvesting and spraying for disease and pest control. The number of fan branches formed at a jorquette is usually four or five. To retain trees at the desired height, chupons should be removed at regular intervals.

In many countries harvesting is extended over the whole year. However, in countries with pronounced wet and dry seasons, the main harvest takes place 5 or 6 months after the start of the wet season. This time interval corresponds with the time required from flower fertilization to fruit ripening. The main harvest lasts for about 3 months, although the crop pattern is also affected by the variety of cacao. Ripe pods are recognized by the colour change. Green-podded Amelonada turn yellow and red pods usually turn an orange shade. The pods remain in a suitable condition for harvesting for 2–3 weeks and have to be cut off the trees with a sharp knife to prevent damage of the flower cushions. Special long-handled tools are used for removing pods which are higher up on the tree.

After harvest, the pods are usually gathered at one or more convenient places, where they are opened and the beans and mucilage are removed. This is preferably done within a few days. The wet beans and mucilage are transferred to wooden boxes for fermentation for 2–4 days for Criollo and Trinitario, or 4–6 days for Forastero cacao (Fig. 1.4). Differences in fermentation and drying methods may result in different qualities of the dried beans. The biochemical effect of fermentation on the quality of the cacao beans is described in the previous section.

Fig. 1.4. Heaps of harvested cacao pods and pulp-enveloped seeds (left) (photo: W. Gerritsma); sun-dried fermented cacao 'beans' (right) (photo: D.L. Schuiling).

The first harvest may take place at the end of the second or third year after planting and increases to a maximum after 10–12 years. Under good growing conditions, a cacao plantation is expected to maintain such yields for 10–15 years. In 2005 the world average yield of dried cacao beans was around 560 kg/ha, with national annual yields per hectare ranging from 490 kg (Ghana), 740 kg (Ivory Coast) to 1250 kg (Indonesia). These data include the yields of smallholder farms and estates. Much higher yields on estates (1500–3000 kg/ha) are no exception. The world production of dried cacao beans was 3.9 million tonnes in 2005. Almost 70% of this total was produced in West Africa, with Ivory Coast, Ghana and Nigeria as main producers. The most important cacao-importing countries are The Netherlands, the USA, Germany and Malaysia.

Coffee

Coffee – *Coffea arabica, Coffea canephora*; Madder family – *Rubiaceae*

Origin, history and spread

Arabica coffee is native to the Ethiopian highlands; wild populations can still be found in the undergrowth of the rainforests in south-western Ethiopia at altitudes between 1400 and 1900 m. Since ancient times, Ethiopian people used to masticate coffee fruits because of the sedative effect of caffeine. Arabica coffee was already cultivated in the 12th century in Yemen. Coffee was brought to Arabia at least in the 15th century, when the discovery of brewing coffee from roasted seeds was made. Arabian traders took it to Europe, India and Ceylon in the 16th and 17th centuries. The Dutch introduced *Coffea arabica* plants, probably originating from Yemen, into Java in 1690. One plant from Java was brought to Amsterdam in 1706. Coffee is self-fertile, so when the plant in Amsterdam flowered, it could subsequently fruit and produce seeds. The Mayor of Amsterdam gave some of the seeds to the French king Louis XIV. In Paris, new coffee plants were bred from this seed. Offspring of these plants were introduced into Martinique in 1720 and subsequently into Jamaica in 1730. Offspring of the plant in the Amsterdam botanic garden were sent to the former Dutch colony of Surinam in 1718; from there plants were taken to French Guyana in 1722 and to Brazil in 1727. From these introductions it was spread widely throughout the Caribbean and Central and South America. Offspring of the Amsterdam plant were also introduced to the Philippines, Hawaii and Africa. Thus, much of the Arabica coffee originated from one plant in Amsterdam, which means that genetic variability was very limited; all the coffee just mentioned belongs to *Coffea arabica* var. *arabica* (syn. var. *typica*). The French took another variety from Yemen, *Coffea arabica* var. *bourbon*, first to the island of La Réunion (formerly called 'Bourbon') in 1715, and later to Latin America and Africa. The variety *bourbon* has a more compact and upright habit than variety *arabica*.

Due to diseases, especially coffee leaf rust (*Hemileia vastatrix*) in Asia, cultivating Arabica coffee became difficult. Therefore, around 1900, much of the Arabica coffee in Asia was replaced by another coffee species, *Coffea canephora*, also called 'Robusta coffee', which was resistant to coffee leaf rust. Robusta coffee is native to the equatorial lowland forests from Guinea to Uganda. Robusta coffee was first introduced into Java. It is now widely grown in the tropical lowlands of South-east Asia, in South America and also in Africa. Arabica coffee is now mainly cultivated in tropical high-altitude areas of South-east Asia, in South America and in Africa. Today about 75% of the world's coffee comes from Arabica coffee.

Botany

Coffee (Fig. 1.5) is an evergreen, glabrous, multi-stemmed shrub or tree, up to 12 m tall; in cultivation pruned to about 2.5 m. The leaves are opposite, simple, dark green, shining, elliptic-ovate, 6–12 cm × 4–8 cm, with prominent veins, and margins somewhat undulate. They have an acuminate apex and a short petiole, about 2 cm; the leaves have connate, subtriangular stipules. The inflorescence consists of an axillary cluster of cymes. There are ten to 30 flowers per node. The flower is bisexual and consists of a small, cup-shaped, green calyx with four to eight lobes; a white, tubular, five- to eight-lobed corolla, 1 cm long and 1–1.5 cm in diameter; the stamens are inserted in the throat; the ovary is inferior, two-celled, with one ovule per cell. Flowers have a jasmine-like scent; they open simultaneously. The fruit is a drupe (in coffee mistakenly often called a 'berry'), about 1.5 cm long, oval-elliptic, the exocarp green when immature, turning yellow first, then red to purple at maturity and black upon drying. Embedded in the fleshy mesocarp there are usually two ellipsoidal, deeply grooved seeds, pressed together and flattened on one side, with horny endocarp. The seed (in coffee mistakenly often called a 'bean') contains mainly endosperm and a small embryo (Fig. 1.6). The 1000-seed weight is 400–500 g.

Fig. 1.5. Arabica coffee: 1, twig with flowers and fruits; 2, flower; 3, stipule; 4, fruit; 5, seed (line drawing: PROSEA volume 16).

The root system is relatively shallow, taproot only up to 1 m deep or less, although some laterals may grow up to 4 m deep for solid anchorage; by far most of the feeding roots are located in the top 30–40 cm of the soil.

Robusta coffee differs from Arabica coffee in some features: it is a more vigorous-growing and taller tree, up to 12 m. Robusta coffee has larger leaves, 15–30 cm × 5–15 cm, and more flowers, up to 80 per node.

Cultivars, uses and constituents

Many cultivars or varieties exist. Popular ones, especially in Brazil, are for example the already mentioned 'Typical' and 'Bourbon'; furthermore 'Mundo Novo' and 'Caturra'. Over 30 mutants are known; breeders have developed many cultivars with resistances such as disease resistance and cold resistance. Some cultivars produce only one instead of two seeds per fruit. Recently, Brazilian scientists identified an Arabica coffee that naturally contains very little caffeine, 0.76 mg instead of the normal 12 mg/g of dry mass; and it does not influence the taste. This discovery is very important for breeding programmes to develop 'healthier' cultivars.

The dried seeds are roasted, ground and brewed to produce one of the most important, stimulating, well-flavoured, non-alcoholic beverages in the world. The stimulating effect is caused by the alkaloid caffeine; the flavour comes from the essential oil caffeol and sugar. The characteristic coffee aroma is only brought out after roasting and grinding the seeds. Coffee is also widely used for flavouring ice cream, desserts, pastries and liqueurs. Coffee is usually marketed

Fig. 1.6. Arabica coffee: 1, flower cluster; 2, clusters of ripe berries; 3, cross-section of berry; 4, dried seeds with silver skin (testa).

unroasted, in sacks. Roasting is mostly carried out by coffee companies in the importing countries. After roasting, especially in Europe, various blends can be made, mainly from Arabica coffee and Robusta coffee. Although the quality of Arabica coffee is better than that of Robusta coffee, the latter is in great demand due to its stronger flavour and the higher yield of soluble solids, which is important for the manufacture of instant coffee. Instant coffee is a soluble powder, prepared by dehydrating extracts of roasted and ground coffee. A third coffee species, *Coffea liberica*, contributes only about 1% to the world coffee production. Its taste is more bitter than that of the other coffee species. It has only local importance.

The coffee consumer can buy coffee as vacuum-sealed packets of whole roast 'beans' or ground 'beans', or several forms of instant coffee. Dried coffee 'beans' contain approximately 10–13% water, 11–16% proteins and free amino acids, 12–14% lipids, 5–9% sugars, 32–48% polysaccharides, 10–15% acids and 4% ash and minerals. The caffeine content ranges from 0.6 to 3.3% and the essential oil content from 10 to 13%. During roasting at recommended temperatures of 200–250°C, the sugars caramelize, the polysaccharides carbonize and the typical aroma is released. The roasting must take place in the consumer country, because roasted coffee quickly loses its aroma.

In Africa ground roasted coffee is sometimes eaten mixed with fat or butter. In the Middle East, fermented coffee pulp is used to prepare a drink. Coffee pulp is also used as manure and mulch, or fed to cattle. In The Netherlands, traditional candies called 'Haagse hopjes' are made from coffee extract.

In folk medicine, coffee is used to treat many different ailments; for example, it is reported to be antidotal, diuretic, hypnotic, stimulant, cardiotonic, and a remedy for asthma, fever, malaria, headache and many more.

Ecology and agronomy

Arabica coffee thrives on deep (2 m), well-drained, slightly acid fertile loams, or clay loams of volcanic origin with pH 5.3–6.6. It is an upland species. The ideal conditions for growth can be found around the equator at altitudes of approximately 1000–2000 m, or lower altitudes from 300 to 1000 m when further from the equator. The average day temperature has to be 18–22°C, with a well-distributed annual rainfall of about 2000 mm. A drier period of 2–3 months is required for initiation of the flower buds. Lower rainfall should be compensated for by irrigation.

Robusta coffee is primarily a tropical humid lowland species. It is adapted to warm equatorial climates with average day temperatures of 22–26°C, and well-distributed annual rainfall of 2000 mm or more, but is also successfully grown in tropical climates with a 2–3-month dry season. Unlike Robusta coffee, Arabica coffee is grown under light shade. As Arabica coffee is self-pollinating, it can be propagated from seed and still keep uniformity. Robusta coffee is also often propagated from seed, but because the species is cross-pollinating, the seed has to be produced in biclonal or polyclonal gardens without other coffee varieties around. Apart from seed, budding, grafting and cuttings have been used for propagation. Today, *in vitro* multiplication, including micro-propagation, is applied as well. Grafts, cuttings and seedlings are usually raised in shaded nurseries. At 6–12 months, when the young plants have four to six leaf pairs, they are transplanted to the field at densities of 1300–2800 trees/ha for Arabica and 1100–1400 for Robusta. Cultivars with a more compact growth habit, like 'Caturra' and certain hybrids, can be planted at high densities of 3000–5000 trees/ha. Planting holes are usually filled with a mixture of topsoil, manure and phosphate. Young plants are very sensitive to weed competition. Weed control methods involve regular manual weeding, application of herbicides, mulching and planting of cover crops.

In Africa coffee is often grown by smallholders, but in countries like Brazil larger plantations are found (Fig. 1.7). Young trees are often intercropped with food crops or cover crops. Fertilizers are commonly used in plantations. Requirement can accurately be determined by foliar analysis and by nutrient removal of the crop. One tonne of dried green 'beans' removes about 35 kg N, 3 kg P, 40 kg K, 3 kg Ca, 2 kg Mg, and some Fe and Mn. Commonly recommended annual fertilizer rates per tree are of the order of 175 g N, 100 g P and 175 g K. Regular pruning is required. This is connected with the specific growth

Fig. 1.7. Three-year-old Arabica coffee plantation with fruiting shrubs, Campinas, Brazil.

and development model of coffee whereby each plagiotropic (horizontal) branch node flowers only once (Fig. 1.8), so that in the course of time fruiting moves towards the tip of older branches and the newly formed branches higher up on the constantly growing orthotropic (vertical) stem. To restrict tree height and maintain vigour in the basal parts, strict rejuvenation pruning is needed, usually

Fig. 1.8. Detail of fruiting, plagiotropic branches of Robusta coffee (photo: W. Gerritsma).

by allowing several vertical shoots to grow and replacing them with new ones every few years. About 3 years after planting, coffee produces the first flowers; it takes about 7 years to come into full bearing. It takes 7–9 months from flowering to maturity of the fruits.

As in non-seasonal climates coffee flowers throughout the year, fruits are harvested the year round. In countries with a distinct seasonal climate, well-defined cropping periods are found. After picking Arabica coffee, fruits are usually pulped, fermented to degrade the mucilage, washed and dried. Robusta fruits, which are often picked in various stages of ripeness, are dried directly. Subsequently the dried coffee is hulled to remove the parchment and remains of the silver skin (Fig. 1.6) and grated in coffee mills, bagged and marketed. The final stage of processing, including blending, roasting and packaging as whole 'beans' or ground coffee, takes place close to the consumer market.

Annual yields per hectare vary widely depending on cultivar, climate, soil fertility and age, from 400 kg up to 6000 kg of green 'beans'. In 2004, the total world production of green coffee 'beans' was almost 8 million t. At present, the major producers of Arabica coffee beans are Brazil and Colombia. Vietnam and Indonesia are the most important producers of Robusta coffee beans.

Tea

Tea – *Camellia sinensis*; Camellia family – *Camelliaceae*

Origin, history and spread

Camellia sinensis has its natural habitat in the lower montane forests from south-western China to north-eastern India. In China it has a long history of use, dating back about 4000 years. Tea has been used in many ceremonies; it is said that it kept Buddhist monks alert during long meditations. There are many references to tea from the time before Christ, as can be found in an old Chinese dictionary. Since ancient times, tea has been valued for its properties as a healthy and refreshing drink. Tea became China's national drink during the Tang Dynasty, AD 618–906; they called it 'ch'a' and 'tay', which is in the English language corrupted into 'tea'. The spread of the plant throughout China and Japan can be attributed to the travelling of Buddhist monks through this region. The Chinese author Lu Yu wrote the first book about tea in about AD 780. Tea cultivation in Japan started in the 9th century. Tea was already known to Arab traders around AD 850; they probably brought tea to Europe via Venetian traders in about 1560. However, it is the Portuguese and the Dutch who claim to have introduced tea and tea drinking into Europe in the 16th and 17th centuries. England entered the trade in the mid- to late 17th century. Tea reached North, Central and South America with European settlers in the 16th century. Tea from Japan and China was introduced into Java in 1690. Subsequently, tea spread into many tropical and subtropical countries. In the 19th century, many plantations were developed, but at first they were often not remunerative. That changed when promising wild tea types were discovered in Assam and Manipur.

Fig. 1.9. Tea: 1, flowering branch; 2, fruiting branch, 3, pluckable shoot (line drawing: PROSEA volume 16).

After 1836, these types were planted in commercial plantations; at first in hilly areas of north-eastern and southern India, next in former Ceylon (Sri Lanka) in the 1870s, and in Java in 1878. Tea was planted in Russia in 1846, although the first successful plantation in this region was developed in Georgia in 1895. From Kew Gardens in England, tea was brought to former Nyasaland (Malawi) in 1886; it was subsequently introduced into Kenya, Tanzania and Uganda in the 1920s and 1930s. In 2004, tea was grown in 45 countries.

Botany

Tea (Fig. 1.9) is an evergreen tree that can be up to 16 m tall in the wild. In cultivation it is usually pruned back to a shrub 0.5–1.5 m high and up to 1.5 m wide. The leaves are alternate, simple, lanceolate to obovate, 4–15 cm × 2–5 cm, leathery, serrate, acute or acuminate. The colour of the older leaves is dark green above and light green below, the lower surface of the leaves is slightly pubescent. The flowers are axillary or subterminal, single or in clusters of two to four. The flower is very fragrant, yellow-white or pinkish, 2–4 cm in diameter. It consists of five to seven sepals and five to seven obovate, concave petals; numerous yellow stamens; petals and outer stamens are united for a short distance at the base; the pistil has a superior three- to five-carpellate ovary, each carpel with four to six ovules. The fruit is a capsule, depressed globose, three-lobed, brownish-green, valvate, up to 2 cm in diameter, thick-walled and woody (Fig. 1.10). The capsule contains one or two subglobose seeds in each lobe. The seeds have a light brown testa, no endosperm but thick cotyledons; they are rich

Fig. 1.10. Tea: flower and flower buds (left) and unripe fruits (right) (photo: D.L. Schuiling).

in oil. The 1000-seed weight is about 450–500 g. Tea has a strong taproot with a dense mat of feeder roots that lack root hairs but have endotrophic mycorrhizae in the top layer of the soil. Some lateral roots grow 3–4 m deep for good anchorage.

Varieties, uses and constituents

Tea cultivars may be classified in two main groups.

1. China teas: slow-growing, multi-stemmed trees, with small, erect, comparatively narrow leaves; flowers borne singly; relatively resistant to cold; and producing a rather low yield. This variety yields some of the most popular teas.
2. Assam teas: quick-growing, single-stemmed, taller trees with large drooping leaves; flowers in clusters of two to four; adapted to tropical lowland conditions; these teas often have larger yields. They give a darker and stronger-tasting beverage. All Assam teas and most Ceylon teas are from this variety.

A large-leaved variety 'Makino' is grown mainly in Japan; characteristic is its bitter extract. Many hybrids between China and Assam types occur. Tea crops can be very heterogeneous due to cross-pollination.

C. sinensis is the plant whose leaves and leaf buds are primarily used to produce tea. For tea production, young light green leaves, usually the tip (bud) and the first two or three leaves beneath, are harvested (Fig. 1.11). The beverage is obtained by infusing the leaves in hot water; it has a stimulant effect due to caffeine. Caffeine content of fresh leaves is about 4%.

Four different tea types can be distinguished.

Fig. 1.11. New shoots ready to be plucked (photo: A.E. Hartemink).

1. Black tea: the leaves are withered, rolled or crushed, fermented, dried, sifted and graded; the processed leaves have a black colour; it is prepared from China and Assam tea.

2. Green tea: the leaves are steamed and dried unfermented. The processed leaves have a green colour; it is prepared from China tea.

3. White tea: the leaves have to be picked before they open fully, the opening leaves and the buds are still covered by soft white hairs; hence the name. Like green tea, the leaves are steamed and dried. The processed leaves have a green colour; it is prepared from China tea.

4. Oolong tea: the leaves are rolled, fermented and dried. While green and white teas are unfermented and black tea is 100% fermented, oolong tea is in between because it is semi-fermented. After processing, the leaves have a deep green colour; it is prepared from China tea.

All four types are harvested from the same species but processed in different ways to obtain different levels of oxidation. Black tea is commercially far the most important type. More than 78% of the present world tea production is consumed as black tea; the largest volume comes from the Assam variety. Several speciality teas are offered to consumers, based on origin ('Darjeeling'), blend ('English breakfast'), form ('pekoe' is cut, 'gunpowder' is rolled) and added flavour like orange, jasmine and many others.

Tea extract is also used for flavouring (alcoholic) beverages, dairy desserts, candies, confectionery and puddings. From the seeds of the tea plant, tea oil can be extracted, which is a sweetish salad and cooking oil. The characteristic flavour and aroma of tea as a beverage are due to the most important constituents, which are polyphenols, caffeine and essential oils. Leaf buds and the youngest leaves have the highest caffeine and polyphenol content and produce the best-quality tea. Dried leaves and leaf buds contain approximately 30–35% polyphenols, 15% protein, 2.5–4.5% caffeine, 22% polysaccharides and carbohydrates, 4% amino acids (including theanine), 5% inorganics, and a few organic acids (mainly ascorbic) and volatile substances. Furthermore, it contains small amounts of vitamins A, B-complex and C.

In folk medicine tea is recommended for the treatment of numerous ailments, especially in China. For example it is regarded as antitoxic, diuretic, carminative, digestive, stimulant and stomachic; tea is reported to be effective in the treatment of bacterial dysentery, gastroenteritis, hepatitis, and many more. Recent studies have indicated that by drinking certain amounts of tea, the average blood cholesterol level may drop; it may cause lower systolic blood pressure level; it increases antioxidant activity; and it may reduce the risk of cardiovascular diseases.

Ecology and agronomy

C. sinensis thrives on a wide range of soils with the texture of sandy loam to clay. They should be well drained, having a good water-holding capacity, be about 2 m deep and with a pH of 4.5–5.6. An annual rainfall of 1700 mm or more is required. Rainfall should not fall below 50 mm per month. The species is cultivated in tropical and subtropical regions. It can be grown from the equator up to latitudes as high as the Black Sea coasts of Russia or northern Japan. It is grown from sea level up to 2300 m altitude. Generally, optimum temperatures for growth are between 18 and 30°C. Climatic conditions have a great influence on the quality of tea. In the tropics fast shoot growth in the lowlands is detrimental to the quality, while at higher altitudes of 1200–1800 m above sea level, high yields combined with excellent quality can be obtained. Hail can cause considerable damage to the leaves; night frost damages the leaves as well but does not kill the tea plant. China cultivars are more cold-tolerant than the Assam cultivars.

Tea is propagated both from seed and by cuttings. Seeds are usually picked from selected, free-growing trees, or from two or more selected clones, the latter yielding biclonal or polyclonal seed. Seed production has to be carried out in special isolated gardens. For germination the seeds are often placed in trays and covered with wet fabric; seeds may also be germinated in sand in beds or trays. After germination the seeds are transferred to shaded nursery beds. After 1.5–2.5 years the young plants can be transplanted to the field. For vegetative propagation, single-noded cuttings are made and placed in small polythene containers, with the leaf and bud just above soil level. The containers are subsequently placed in a polythene tunnel under shade; after 6–9 months the rooted cuttings can be transplanted to the field. Plant densities depend on climate and cultivar, but tea is often planted in rows, at densities of 11,000 to 14,000 bushes/ha. Planting in the row can be 60 cm apart and spacing between the rows 120 cm. Before planting, the soil should be thoroughly cultivated. After planting, suppression of weeds is essential, especially during the period of establishment. Particularly stoloniferous grasses can be very deleterious in Asia.

Sometimes tea is interplanted with green manure crops. Once the plants are established, deep cultivation for weed control has to be avoided, because it will damage the surface feeder roots. At most, a light scraping of the surface can be tolerated. Weed control can also be carried out chemically.

Tea requires regular fertilizer application to produce good yields. Foliar analysis provides information on the nutrient status of the tea bush. Application of fertilizer can also be based on soil analysis data and the nutrient removal of the crop. An annual tea crop of about 1000 kg of processed tea removes approximately 45 kg N, 3 kg P, 17 kg K, 6 kg Ca and 1.5 kg Mg.

In tea cultivation, a naturally small tree has to turn into a low, wide-spreading bush to maintain a convenient height for plucking. The top level of the bushes is called the 'plucking table'. To keep the bushes in good shape, pruning is required; pruning cycles depend on altitude and may vary from once every year up to every 5 years. If possible, pruning should be done during a dormant period. The first time the leaves can be plucked depends on the propagation method and the environment; it varies from 2 to 4 years after planting. Bushes are usually plucked every 7–10 days at lower elevations and every 14 days in colder climates (Fig. 1.12). The economic life of a tea bush is assumed to be 40–50 years; however, many 70–100-year-old bushes are still productive at present.

At harvest, terminal sprouts with two or three leaves are usually hand-plucked (machines for removing leaves have been developed, but until now they are not much used). Subsequently the leaves can be spread thinly on trays and placed in the sun for about 12 h or more, until the leaves are flaccid. As mentioned before, various techniques are used to produce different types of tea; the processes may include withering, rolling, fermentation, steaming, drying, sifting and grading. During these processes, the leaves may become broken up. After the processed tea is dried, the brittle leaves are usually stored in airtight tin boxes or cans and

Fig. 1.12. Harvesting tea leaves, Sumatra, Indonesia (photo: J.D. Ferwerda).

marketed. Ten kilograms of green leaves produce about 2.5 kg of dried tea. World average yield for China teas is about 900 kg of processed tea per hectare per year, ranging from 500 to 1600 kg. The yields of Assam teas range from 600 to 2100 kg/ha. In 2004, the total world production of processed tea was 3.3 million t. The major producers are China, India, Kenya and Sri Lanka.

Tobacco

Tobacco – *Nicotiana tabacum*; Nightshade family – *Solanaceae*

Origin, history and spread

All tobacco varieties belong to the genus *Nicotiana*. The genus contains about 65 species, most of them native to tropical America.

The growing of tobacco began in Central and South America by Indians, over 3000 years ago. For them it was a sacred plant, used for healing practices, used to communicate with the spirits and also used for pleasure. It played a part in all kind of rituals and tribal ceremonies. One way to use tobacco was rolling many leaves to obtain a very large cigar, which they called 'tabaco'. However, tobacco was used in various ways: smoking dried leaves, drinking tobacco juice, chewing the leaves and sniffing tobacco powder. Tobacco played such an important part in everyday life that it was sometimes mythicized: Indians of the Huron tribe told that a long time ago the Great Spirit sent a woman to the earth to help the people. She walked over the world and touched the soil. When she touched the soil with her right hand, there grew potatoes; when she touched the soil with her left hand, there grew corn; and on all places where she sat down to rest, there grew tobacco afterwards.

Tobacco was introduced into Spain and Portugal in the mid-16th century, where it was used as an ornamental plant and a medicine at first, later as a stimulant. In 1566 Jean Nicot de Villemain, France's ambassador to Portugal, introduced tobacco to the French Court. It spread to other European countries and then to Asia and Africa, where its use became general in the 17th century. In North America the first tobacco was planted in Virginia in 1612. In the mid-17th century Linnaeus named the plant genus *Nicotiana*, in homage to Jean Nicot. Linnaeus described two species: *Nicotiana tabacum* and *Nicotiana rustica*. The first tobacco cultivated in Europe and North America was *N. rustica*. Later it was replaced by *N. tabacum*. Both species are believed to be of hybrid origin, because they are not known in the wild. They originate from Peru and Bolivia. The main source of commercial tobacco now is *N. tabacum* (about 90% of the world production). On a much smaller scale, *N. rustica* is grown for use in Oriental tobacco. Today tobacco is grown in 120 countries.

Botany

N. tabacum (Fig. 1.13) is an herbaceous annual growing to 2–3 m high, covered with short, sticky hairs. The erect stem is thick, unbranched, with a taproot.

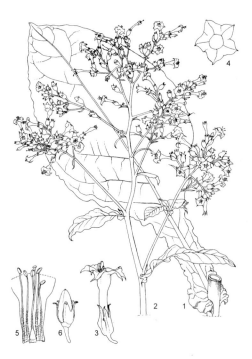

Fig. 1.13. Tobacco: 1, leaf; 2, inflorescence; 3, flower; 4, corolla limb; 5, opened corolla tube showing stamens and pistil; 6, calyx with fruit (line drawing: PROSEA volume 16).

The leaves are more or less stem-clasping, thin, simple, variable in size, up to 30–40 cm long and alternate. The shape is ovate to oblong-lanceolate and the surface is dull. The number of leaves per plant is about 30. The colour of the flowers is usually pink but may be paler or redder. They are borne in terminal panicled racemes with up to 150 flowers per inflorescence. The flowers are about 5 cm long and have a tubular and five-cleft calyx, and a funnel-shaped and five-lobed corolla. Furthermore, there are five stamens, which are attached to the corolla tube and almost as long. The stigma is borne on the end of a long style of about the same length as the stamens. The fruit is a two-celled, many-seeded capsule, 1.5–2 cm long, the greater part enclosed by the calyx (Fig. 1.13). The number of seeds varies from 2000 to 5000 per capsule. The seeds are light to dark brown and very small; 1 g may contain 12,000 seeds.

 N. rustica (Fig. 1.14) resembles *N. tabacum* but there are some differences: the plant is smaller (up to 1.8 m); the corolla is shorter and coloured yellow to green; the leaves are usually petiolate, ovate or cordate in shape with a dark, shiny surface.

Cultivars, uses and constituents

Growers and breeders have developed a wide range of morphologically different types and cultivars, from the small-leaved aromatic tobaccos to the large, broad-leaved cigar tobaccos. The final product determines the needed type or cultivar.

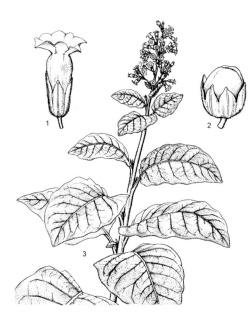

Fig. 1.14. Nicotine tobacco: 1, flower; 2, calyx with fruit; 3, upper part of flowering plant.

Besides taste, the most important reason for using tobacco is the stimulating and light narcotic effects of the alkaloid nicotine. Therefore the leaves are harvested and dried (cured), rolled into cigars or shredded for use in cigarettes and pipes, or processed for chewing or snuff. The final use of tobacco leaves is determined by several factors including cultivar, climate, soil and method of curing. The leaves are cured, fermented and aged to develop aroma. The most important cure methods are air-curing, sun-curing, fire-curing and flue-curing.

1. In air-curing the leaves are hung in well-ventilated barns, or in an open framework in which leaves are protected from wind and sun. It takes 1–2 months to dry. South and Central America are important producers of air-cured tobacco. This group has low sugar content but varies in nicotine content. The colour of the dried leaves is reddish-brown because the tannins present in the leaves oxidize. This tobacco is used mainly for cigars.

2. In fire-curing the leaves are dried in smoke and hot air. It takes about 4 weeks to dry. Fire-cured tobacco has a low sugar and high nicotine content. The colour of the dried leaves is dark brown. Fire-cured tobacco is used mostly for pipe, snuff and chewing tobacco.

3. In flue-curing the leaves are dried by radiant heat from flues connected to an oven, taking 1 week to dry. Flue-cured tobacco has high sugar content and a medium to high nicotine content. The colour of the dried leaves is yellow, orange or mahogany. Most of the flue-cured tobacco is used for cigarettes. Worldwide the most extensive variety, Virginia, is flue-cured.

4. Sun-curing is the drying of uncovered leaves in the sun. The best known are the Oriental tobaccos from Greece, Turkey and other Mediterranean countries. They have very characteristic aromas, a low sugar and nicotine content, and are used in cigarettes.

Fig. 1.15. Close-up of tobacco field, East Java, Indonesia (photo: D.L. Schuiling).

Fresh tobacco leaves (Fig. 1.15) contain 85–90% water, which falls to 12–15% during the curing. On a dry weight basis flue-cured tobaccos contain sugars 18–20%, starch 5–8% and proteins 2%. Air-cured tobaccos contain only a few per cent of sugars but 3–15% proteins. The nicotine content of cured leaves of *N. tabacum* varies from 1 to 4%.

After curing the leaves are graded, bunched and stacked in piles or closed containers for active fermentation and ageing. Most commercial tobaccos are blends of several types. Flavourings such as sugars, fruit juices and spices are often added.

Because of the high nicotine content of *N. rustica* (4–9.5%), this species is also grown for nicotine extraction to produce natural insecticides.

A number of *Nicotiana* species are grown as ornamentals, e.g. *Nicotiana alata, Nicotiana langsdorffii, Nicotiana sylvestris*. Height, width and shape of the leaves are variable. The trumpet-shaped flowers can be pink, red, green or white, often quite fragrant (Fig. 1.16).

Ecology and agronomy

Tobacco is grown under a wide range of climatic conditions (60°N–40°S), but the crop needs a minimum of 120 frost-free days. Tobacco grows on a variety

Fig. 1.16. Inflorescences of tobacco (left) and nicotine tobacco (right) (photos: J. van Zee).

of soils, although preferably on light to medium loams, well-drained and slightly acidic, pH 5–6 (Fig. 1.17).

As tobacco has very small seeds, it is not possible to sow directly in the field, unless the seeds are pelleted. Seedlings are raised in glasshouses and then transplanted to the field. Ten grams of seed is needed for 1 ha. Seeds of most cultivars require light for germination. Young seedlings are planted out by hand or by a mechanical transplanter. Spacing between seedlings and rows varies with the kind of tobacco and with the location. The average plant density is 18,000–25,000/ha for most tobacco types.

The crop needs 400 mm of water during the growing period. Fertilizer rates are 20–60 kg N, 10–20 kg P and 25–60 kg K per hectare. N uptake of tobacco plants must not be too high due to risk of high N content of the leaves, which may cause a decline in the quality of the leaves for the production of tobacco.

Tobacco can be topped or the flower buds can be picked off, as this appears to increase the size and thickness of the leaves. The negative effect is the development of dormant axillary buds into lateral branches (suckers). The suckers have to be removed. To achieve large and tender leaves, shadow-textile is sometimes placed above the growing crop.

Tobacco is susceptible to numerous fungal, bacterial and viral diseases and can be attacked by several species of insects.

The first and lowest leaves are mature about 2 months after transplanting. When the leaf is fully ripe, which means when the colour is yellowish-green and the tip yellow, harvesting can begin. Tobacco is often stalk-cut by machine or harvested by hand, leaf by leaf. The leaves are tied together in pairs on curing

Fig. 1.17. Tobacco field, Pinar del Rio, Cuba (photo: T. Andersson).

sticks or strings. World average yield of cured tobacco leaves is 1.5 t/ha. The world production in 2005 was 6.6 million t of dried tobacco leaves. Asia accounts for 64% with China as main producer (41%), the Americas including the Caribbean for 19%, Europe 8% and Africa 5%.

2 Edible Fruits and Nuts

FRUITS

Apple

Apple – *Malus pumila* (syn. *Malus domestica*); Rose family – *Rosaceae*

Origin, history and spread

Apple has been valued by man for food since very early times as proved by the finds of carbonized apples (*Malus sylvestris*) in the remains of the prehistoric lake dwellings of Switzerland. However, the origin of the cultivated apple is thought to be the area of the Caucasus, Turkestan and Central Asia, where wild apples (*M. sylvestris* and *Malus pumila*) still grow. Woods formed of wild apples and walnuts can still be found in Turkestan. The cultivated *M. pumila* is probably derived from the wild types found in Central Asia. As other wild apple species, such as *Malus sieversii* and *Malus sylvestris*, were present in the area of origin, hybridization between the species took place naturally. Apples then spread throughout the eastern part of the Mediterranean and southern Europe. As the so-called 'Silk Road' passed through the area of origin, it is most likely that nomads and traders contributed to the spreading.

The first written information about an apple orchard is found in Homer's *Odyssey* (900–800 BC). The Romans cultivated apples, as is described by Cato and Pliny. Pliny, a Roman statesman (AD 23–79), described 37 varieties of cultivated apples in his *Historia Naturalis*. In the 2nd century Cato the Elder described the grafting of cuttings on to rootstocks in his work *De Agricultura*. The Romans are believed to have introduced the apple into France and England. As Christianity spread over Europe and religious orders formed settled communities, the apple followed. Monasteries played an important part in the early development of apples.

Colonists took apples in the form of seeds with them to America, Asia, Australia and New Zealand. In the USA there goes a tale that a man named John Chapman contributed highly to the spread of apples. He travelled the country-side with a bag filled with apple seed, and tossed these seeds everywhere to create a country filled with apple trees, earning the nickname 'Johnny Appleseed'. He became the folk hero of apple growers. Today apples are grown throughout the temperate regions and to a small extent in some subtropical countries.

Botany

Apple can be grown as trees, pyramids, dwarf bushes or cordons. When it is grown as a tree it can reach a height of 10 m. Crown width can also be 10 m. The bark colour is reddish-brown to reddish-grey. The shape of the leaves is simple, elliptical-ovate with finely serrated to irregularly toothed or lobed margins. The surface of the leaf is first pubescent, later glabrous. The length of the leaf is 3–9 cm, width 3–7 cm, the colour is dark or olive green at the adaxial side and paler green at the abaxial side. Flowers occur in a truss, the central flower is called the 'king flower' (Fig. 2.1). Because the king flower is the first to open, it often develops a larger fruit than do the others in the cluster. That fruit is called the 'king fruit'.

Apples flower in early spring, just before leaf emergence. Male and female elements are combined in single, five-part flowers. The colour of the five petals can be white or pink. The various stamens are yellow or yellow and red. Bees mostly accomplish pollination. Even when pistils are pollinated, it does not mean that they develop edible apples, because the majority of the young fruits on each tree are aborted. The fruit is in fact a false fruit called a 'pome'. The name is derived from the Latin 'pomum', which means 'fruit'. A pome is formed by the enlarged receptacle. The pome or apple has a paper-like core at the centre in which there are a number of seeds, a fleshy layer around the core, and an outer skin. The fleshy layer or the flesh may be soft and mealy, or crisp, or hard, and may have a variable degree of juiciness. The colour of the outer skin may be green, yellow, red or a combination of red and green or yellow. Red colouring

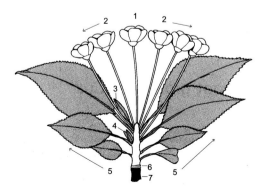

Fig. 2.1. Details of an apple inflorescence: 1, king flower; 2, lateral flowers; 3, bract; 4, bud of secondary leaves; 5, primary leaves; 6, scale scars; 7, spur.

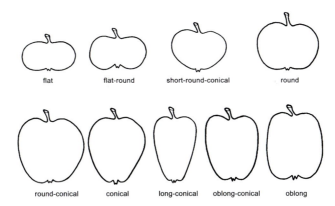

Fig. 2.2. Various fruit shapes of apple.

may be even or striped. The skin may be smooth, ridged or puckered. In the ripe fruit, the remains of the calyx (five sepals) are called the 'eye'. The shape of apples can vary considerably (Fig. 2.2). The diameter is the commercially used dimension for grading purposes. Apple sizes vary from small (diameter 44 mm and below) to very large (diameter 85 mm and above). Apple size depends strongly on climatic conditions. The root system of the apple tree is fibrous and can spread to more than the height of the tree.

Cultivars, uses and constituents

At present there are at least 7500 different varieties of cultivated apples. About 25 varieties are grown commercially such as 'Cox's Orange Pippin', 'Schone

Fig. 2.3. Various apple cultivars (clockwise from left): 'Elstar'; 'Granny Smith'; 'Cripps Pink'; 'Schone van Boskoop' ('Beauty of Boskoop'); 'Alkmene'; 'James Grieve'; and 'Braeburn' (photo: D.L. Schuiling).

van Boskoop', 'Golden Delicious', 'Granny Smith', 'Braeburn', 'Elstar', 'Democrat', 'McIntosh', 'Alkmene', 'James Grieve' and 'Cripps Pink' (Fig. 2.3). Apples are classified into four groups based on flowering seasons: (i) early flowering; (ii) mid-season flowering; (iii) mid-season/late flowering; and (iv) late flowering. This order also indicates when the ripe fruits can be picked. Apples can furthermore be grouped on the basis of utilization: some varieties are best suited for cooking or processing, others as dessert apples, which are mainly consumed fresh, there are varieties for both purposes and others for making cider.

Most apples are eaten fresh. For that purpose, apples need to have the right balance between acidity, sweetness and bitterness, to create richly flavoured fruit. Poor grades of apples are dried or processed into apple juice. Cooking apples are mostly too acidic for consumption as fresh fruit, but they are very suitable for producing jellies, apple sauce, pies and puddings.

Special varieties suitable for cider-making should have the bitterness caused by tannins. Cider is the product of fermentation of apple juice after adding yeast to the juice to produce alcohol. Cider-making has been known for many centuries. One of the earliest references was by Charlemagne at the beginning of the 9th century. An apple brandy or 'Calvados', also called 'applejack', can be obtained by distilling cider. Normandy is well known for growing excellent apple varieties for cider-making. In the USA unfermented apple juice is called 'soft cider' and fermented apple juice 'hard cider'. An apple contains about 85% water and 12% carbohydrates (sucrose and fructose), 0.4–1% acid (the acid content consists of 90% malic acid and 10% citric acid) and 0.2% tannins. Furthermore, apples provide a number of minerals in the diet such as K, Ca, Mg, Fe and P. Located just beneath the skin are vitamins A, B_1, B_2, C and E. The typical apple aroma is a blend of some 250 chemicals, including esters, alcohols and aldehydes. Eating apples is considered to be good for your health as suggested by the once popular quotation: 'an apple a day keeps the doctor away'.

Many varieties of wild apples, also called 'crab apples', are grown as ornamental shrubs because of the attractive flowers and fruits. Some varieties are popular as traditional bonsai.

Ecology and agronomy

Apples are the most important and most widely cultivated fruit of the temperate regions. They can be grown in a wide range of climates, although the average yearly temperature should be above 7.5°C. Apples thrive in a wide range of soils but not in poor acid soils or heavy clay. They prefer well-drained heavy loams with pH 5.0–7.5 and good humus content. Apple trees are best adapted to areas where the average winter temperature is near freezing for at least 2 months, to allow the trees a dormant period. Dormancy is an adaptation to harmful conditions, in which both physical and chemical changes in the tree cause a state of inactivity to protect the tree against damage by dehydration and frost. For good production, apples need at least 600 mm precipitation. They require most of the water in the fruiting period.

An apple tree can be grown from an apple seed, but because cultivated apples are hybrids, the offspring will probably resemble the wild apple. This is the

Fig. 2.4. Modern apple orchard with bushes of cultivar 'Jonagold', The Netherlands.

reason why new apple trees are grown from buds or twigs. The buds or twigs are cut from a desired apple tree and made to grow on strong rootstocks of other young apple trees, known respectively as 'budding' and 'grafting'. In fact, these are cloning techniques. For the different growing types, there are many rootstock varieties. Using the right rootstock is very important because it determines tree size and form. However, apple breeders use seed to obtain new varieties. Once in a while, after selection of countless seedlings, one seedling may develop into a tree that bears better apples than the parent does, and that may be the start of a new variety. Because apples are for the greater part self-infertile, one or more other varieties should be planted nearby for successful pollination. In orchards, apple trees are grown in rows (Figs 2.4 and 2.5). The space between the rows has to leave enough room for tractors and machines, which are used for spraying pesticides, cultivating, mowing grass and harvesting the fruit. The space between the trees in the rows depends on the production system used: trees, pyramids, dwarf bushes or cordons. For the desired development of the latter three, the apples have to be budded or grafted on dwarfing rootstocks. It is important to plant apples in the right way: the fusion of the rootstock and the bud or twig has to be above the soil. In cases where the fusion is buried, the bud or twig will develop roots of its own. If that happens, the effect of the dwarfing rootstock will be lost and the apple develops into a tree.

To get the right number of fruits at each branch and also the right size of apples, it is necessary to prune the tree. In doing so, it is important to distinguish fruiting spurs from leaf buds. Pruning is best done during the dormant period in winter. Even well-pruned trees can bear too many apples and thinning will then be necessary, by hand or chemically. Apart from pruning, trimming may be necessary to keep the trees in shape, particularly in cordons. To obtain sufficient production, fertilizing will be necessary, depending on soil fertility. For fruit pro-

Fig. 2.5. Flowering apple orchard with half-standard trees of cultivar 'Golden Delicious', The Netherlands.

duction fertilizers are often combined with the water supply, called 'fertigation'. Apples are mostly ripe in autumn, and then they can be picked. Not all varieties can be picked at the same time, because there are early- and late-ripening varieties and some in between. To guarantee a continuity of apple supply throughout the year, fruits have to be stored in controlled atmosphere stores. Apples may then be stored possibly for a year, depending on the cultivar.

   In 2005, the world average yield was 12.2 t of fresh apple fruits per hectare. Much higher yields, ranging from 40 to 45 t/ha, are obtained in commercial plantation in north-west Europe and New Zealand. In 2005, the world total production was 63.5 million t of fresh apple fruits, produced in around 90 countries of which China (40%) and the USA (7%) are the leading countries. China and Chile are the largest exporters.

Avocado

Avocado – *Persea americana*; Laurel family – *Lauraceae*

Origin, history and spread

Avocado, also called the 'alligator pear', is native to Central and South America. Three main centres of diversity are distinguished: (i) highland Mexico; (ii) highland Guatemala; and (iii) lowland Guatemala to Costa Rica. The wild species can still be found in mountains and on lower slopes in the forests of those areas. Approximately 9000-year-old avocado seeds have been found at the archaeologically important cave dwellings in the Tehuacán area of Puebla State in

Central Mexico. Because the seed size was similar to the seed size of wild avocados, it may be assumed that the fruits were not cultivated but gathered in the wild. However, archaeological evidence concerning seed size from other caves in the Oaxaca valley in Mexico suggests that there was an indication of selection, which took place from 4000 to 2800 BC. More reliable is the finding of considerably larger seeds at an archaeological site dated to about 500 BC, which proved that avocados were cultivated from selected seeds at that time. The species spread in South America in pre-Columbian times. The Spaniards brought it to the Philippines in the 16th century, and the Dutch brought it to the Dutch East Indies and Mauritius around 1750. It was introduced into Hawaii in 1825, into California in 1833 and into India in 1892. Most of the spread in Asia was in the mid-19th century. The first avocado trees in Israel were planted in 1908.

The name avocado is derived from the Aztec 'ahuácatl', which means 'testicle tree'; the avocado fruit was considered to be an aphrodisiac. To the Spaniards the word *ahuácatl* sounded like their old Spanish word 'avocado', meaning 'lawyer'. The Spanish name was borrowed by the English; the latter named the fruit at first 'avogato pear'. The name 'alligator pear' arose because there is some similarity between the fruit and an alligator based on the green colour, shape and warty appearance of the avocado fruit. At present, avocado is cultivated in many tropical and subtropical countries around the world.

Botany

Avocado (Fig. 2.6) is an evergreen tree of variable form, up to 20 m tall (Fig. 2.7). Budded trees are usually shorter. Dwarf varieties exist. The trunk varies in

Fig. 2.6. Avocado: 1, flowering branch; 2, halved fruit (line drawing: PROSEA volume 2).

Fig. 2.7. Habit of a large, producing avocado tree, Sierra Leone (photo: J.A. Samson).

length, but is often short and 30–60 cm in diameter; spreading with branches that may begin close to the ground. Leaves are spirally arranged, simple and entire, variable in shape and size, elliptic to lanceolate, ovate or obovate, and pointed; the leaf-blade is glaucous beneath, pinnately nerved, 5–30 cm × 3–15 cm, reddish to dark green; petioles 1.5–5 cm long. The growth of the leaves is in flushes. The inflorescences are terminal panicles of cymes, arising usually lateral from continuing vegetative shoots, ending in vegetative buds. The flowers are small, inconspicuous, greenish to cream in colour, 5–10 mm wide and bi-sexual. The flower consists of six sepals, the petals are absent; furthermore there are nine stamens and a one-celled ovary with a slender, hirsute style and a simple stigma. The flowers open twice, first when the stigmas are receptive and again, usually the next day, when the pollen grains are shed (Fig. 2.8). The species is cross-pollinating. Avocado produces an enormous number of flowers, although usually only one fruit sets for every 500 flowers. The fruit is botanically a berry. The shape of the fruits varies from spherical to pear-shaped. The fruits have a smooth or rough to warty skin, which may be green, yellow, reddish or dark purple in colour (Fig. 2.9). The size of the fruits varies considerably, from 7 to 20 cm long; weight varies approximately from 100 to 1000 g. Embedded in the yellow- to green-coloured flesh is a single, large, egg-shaped, brownish-cream to brown seed, 3–5 cm in diameter and 25–50 g in weight.

Fig. 2.8. Close-up of an inflorescence of avocado with flower buds, flowers and young fruits (left); close-up of avocado flower at female anthesis (right).

The root system is shallow, with a low frequency of root hairs; a few primary anchorage roots can penetrate the soil up to 4 m deep.

Cultivars, uses and constituents

Within the species *Persea americana*, three groups or ecological races (races named after the area of origin) can be distinguished: (i) Mexican; (ii) Guatemalan; and (iii) West Indian avocados. The different races can be crossed easily so many hybrids exist. For example, the commercially important cultivar 'Fuerte' is prob-

Fig. 2.9. Various fruit types of avocado.

ably a hybrid between Mexican and Guatamalan types. It is a cultivar with pear-shaped fruits, a smooth green skin and greenish-yellow flesh. Another important variety is the so-called 'hass avocado', with egg-shaped fruits, a dark green or almost black, rough skin, with yellow flesh. Hass avocado belongs to the Guatemalan type. The West Indian type can be distinguished from the other types by the large fruits with lower oil content. Especially in Florida and California, a number of superior cultivars have been developed.

The fruit is eaten fresh, with lemon juice, vinegar, or pepper and salt; it can be a component of fruit salads. The pulp is used as a sandwich spread, in milk shakes, desserts and ice creams. The flesh is slightly nut-like in flavour. Avocado fruits contain much oil, which is extracted from the flesh and used in cosmetics. The oil of avocado is similar in composition to olive oil. The residue after oil extraction can be used as feed for cattle. Avocado is not usually cooked: because of its tannin content the flesh becomes bitter. To preserve the flesh it can be deep-frozen or pickled. Avocado is a main component of the well-known Mexican sauce 'guacamole', which is a blend of puréed avocado flesh, onion, chillies, lemon and Tabasco sauce, and eaten with tortillas.

The contents of the flesh vary widely for the different cultivars; it is approximately 65–85% water, 1–4% protein, 6–23% fat and 3.5–5.5% carbohydrates. The fat is mainly mono-unsaturated. The high fat content gives the flesh a buttery texture. Avocado is rich in Fe and the vitamins A and B. Different parts of the tree have been used in folk medicine to treat many ailments, such as dysentery, neuralgia and cough.

Ecology and agronomy

The species has wide ecological amplitude, due to the differences in evolutionary development of the three ecological races. Cultivars of the West Indian race are adapted to humid tropical conditions, they are cold-sensitive and tolerate soil salinity. Especially during flowering and fruit setting, high atmospheric humidity is required. The Guatemalan race arose in the subtropical highlands of tropical America and therefore its cultivars are somewhat hardier, and are intermediate between the other two. Cultivars of the Mexican race are the hardiest of all avocados; they are adapted to a Mediterranean type of climate and can stand temperatures as low as $-4°C$, but soil salinity is not tolerated.

The optimum temperatures for the three races range from 20 to 35°C for the day and from 10 to 25°C for the night, and the annual rainfall may be up to 2500 mm. Avocado cannot stand high soil moisture content; temporary waterlogging is not tolerated. It can grow in shade but is productive only in full sun. It tolerates a wide range of soils, but it grows best on rich, well-drained soils, with pH 5.0–5.8.

Avocado is propagated by seed, by grafting, or by budding of seedlings. The seeds are planted in nurseries and germinate in 4–6 weeks. The young plants are transplanted into the field when they are 6–12 months old. Seedlings are also grown as rootstocks. Budded and grafted plants are transplanted into the field at 9–16 months old. The spacing depends on cultivar, soil fertility and climate; it ranges from 6 to 12 m. The plants are placed in holes that have to be filled with soil enriched by farmyard manure. Subsequent mulching is advantageous: it suppresses weeds, conserves soil moisture and protects the roots. Weeds may also be controlled by herbicides. In windy areas, young plants should be protected by windbreaks; they can easily be damaged because of the brittleness of the wood. After planting, watering is usually required until the roots are established. Moreover, a continuous moisture supply is required for high yield; however, over-irrigation has to be avoided. Drip irrigation, which can be combined with application of fertilizers, is a good method to keep the soil moist.

Regular fertilizing with a complete nutrient mix or farmyard manure may be required, but only N and sometimes Zn are generally required. Recommended annual N applications vary from 50 to 100 kg/ha. When the avocado trees are young, intercropping with for example pigeonpea occurs, which provides nitrogen through N_2 fixation. Pruning may be required, especially in upright cultivars; the central shoot is usually shortened to develop a spreading type of tree.

Seedling plants come into bearing at 5–6 years; vegetatively propagated plants are usually earlier, although fruits are prematurely removed until the tree is 4 years old.

Fruit development from setting to maturity takes 6–12 months. The fruits will not ripen when still attached to the tree, probably due to an inhibitor in the fruit stem. As avocado shows prolonged flowering, fruits in varying stages of maturity can be found on the tree at the same time. The largest fruits should be picked first. However, it is difficult to determine when the fruit is full-grown. Practical experience is required. After harvesting, full-grown avocados ripen in 1–2 weeks when they are stored at about 20°C.

Annual yields vary widely depending on cultivar, soil condition, climate and age of the tree, from approximately 20 to 100 kg of fresh fruits per tree. In 2005, the world average yield of fresh fruit was 8.2 t/ha. In the same year, the world total production was 3.2 million t. Mexico, accounting for one-third of the world production, is the largest producer. Other important producers are the USA, Indonesia, Brazil, Colombia and Chile.

Banana

Banana – *Musa*; Banana family – *Musaceae*

Origin, history and spread

Although the exact origin of the edible bananas is unknown, it is generally accepted that the Indo-Malaysian region is the main area of genetic diversity, Malaysia probably being the main centre. All modern dessert banana and plantain cultivars are derived from either one or both of the two wild, diploid, seeded *Musa* species: (i) *Musa acuminata* (with genome A, predominantly involved in the dessert type) is native to Malaysia and considered to be the main parent of the edible bananas; and (ii) *Musa balbisiana* (with genome B, predominantly involved in the plantain type), occurring from India to the Philippines and New Guinea.

Further details of the plantain type are described in Chapter 9 on starch crops.

Humans played an important role in the evolution and spread of the edible banana. *M. acuminata* cultivars were taken to areas where the wild *M. balbisiana* is native, after which natural hybridization occurred. Subsequent selection and vegetative propagation resulted in the improvement of cultivars. Bananas were probably taken from Indonesia to Madagascar in the 5th century, and with subsequent dispersal into the heart of the Africa and across the Congo to West Africa. Other sources suggest that banana had already reached Africa in prehistoric times. The early Polynesians brought banana to the Pacific region around the 11th century. The Portuguese took banana from West Africa to the Canary Islands in the 16th century. There is no good evidence for the presence of banana in the Americas before its discovery by Columbus. The first of many introductions into the Caribbean and tropical America was from the Canary Islands to Hispaniola (Haiti) in 1516. Nowadays, banana is widespread in the tropics and subtropics.

Botany

The banana plant (Fig. 2.10), often inaccurately called a banana 'tree', is a large, tree-like perennial herb with a juicy, cylindrical pseudostem consisting of overlapping, tightly rolled leaf-petiole sheaths, 6–8 m tall, 20–50 cm in diameter, arising from a short fleshy rhizome or corm. New shoots (suckers) arise from short rhizomes close to the parent plant, forming a clump of aerial shoots, the eldest shoot

Fig. 2.10. Banana: 1, habit of fruiting plant; 2, cluster of fruits (line drawing: PROSEA volume 2).

replacing the main plant after it fruits and dies. This process of succession continues indefinitely. New leaves, originating from the corm, grow up continuously from the centre of the pseudostem at a rate of one per week in warm weather. The leaf-blades, after unfolding, are oblong to elliptic, 50–400 cm × 70–100 cm, tender and smooth with a pronounced midrib and pinnately arranged parallel veins. They may be green, green with maroon spots, or green on the upper side and red–purplish beneath. The leaves are fleshy stalked and spirally arranged. After about 30–40 leaves have been produced, the apical meristem develops into a stem with a single, terminal inflorescence growing up from the heart of the pseudostem and bending down when exserted. The inflorescence is a terminal, compound spike. The flowers are arranged in clusters along the floral stalk; the female ones in five to ten rows towards the base, above them may be some hermaphrodite or neuter flowers, the male flowers are borne towards the top. Female flowers are about 10 cm long, larger than the male, with an inferior ovary, longer than the perianth; style massive; staminoides five. Male flowers are about 6 cm long, with five stamens, rarely containing pollen. Both male and female flowers produce abundant nectar and are frequently visited by bats, birds and large insects.

Each cluster of flowers is enveloped by a thick, waxy, hood-like bract, purple outside and reddish inside. The bracts will lift from the first female cluster, also called a 'hand', in 3–10 days when the plants are healthy. The male flowers and their bracts are shed in about 1 day after opening of the flowers, leaving most of the upper stalk naked, except at the very tip where an unopened bud remains. However, in some cultivars, such as 'Dwarf Cavendish', the male flowers and bract remain persistent (Fig. 2.11).

Fig. 2.11. Infructescence of 'Dwarf Cavendish' banana with fruits at various development stages (left) (photo: J.F. Wienk); terminal bud of a banana inflorescence, older bract is open and shows male flowers (right) (photo: J. van Zee).

The fruits are berry-like and seedless; young fruits develop from female flowers, at first looking like slender green fingers; when fully grown each cluster of fruits at a node becomes a hand of bananas; individual fruits are called 'fingers'. The stalk droops with increasing weight until the bunch is upside down. The total number of 'hands' per bunch depends mainly on cultivar.

The roots of banana are adventitious, forming a dense mat, mainly in the top 15 cm, and spreading 4–5 m laterally.

Cultivars, uses and constituents

The current, generally accepted classification of dessert banana cultivars is based on the relative contribution of the two wild species to the constitution of the cultivar. Most cultivars are diploid or triploid; cultivars with higher ploidy levels are generally weak and slow-growing. There are numerous cultivars of the dessert banana type. Some of the most important are listed below according their genotype: AA or AB (diploid) and AAA or AAB (triploid).

- *Musa* (AA group): 'Pisang Mas' is important in Papua New Guinea (perhaps 20 cultivars) and Malaysia. The sheaths are dark brown; the leaves yellowish and nearly free of wax. Bunches bear five to nine hands; fruits are small with attractive, golden yellow, thin skin and firm, light orange, aromatic and sweet flesh.
- *Musa* (AB group): 'Lady Finger' is common in tropical countries of South America, the Caribbean and commercial in Queensland and New South Wales. The sheaths have reddish-brown streaks or patches. Bunches bear ten to 14 hands of 12–20 fingers each. The fruit is 10–13 cm long; when ripe they have thin, bright yellow skin and cream-coloured, sweet flesh.
- *Musa* (AAA group): 'Gros Michel' was formerly the leading commercial cultivar in South and Central America, Central Africa and the Caribbean, but has been replaced to a large extent by cultivars less susceptible to diseases. Bunches bear eight to 12 hands. Fruits are medium to large; when ripe they have thick yellow skin and creamy white, fine textured, sweet and aromatic flesh.
- *Musa* (AAA group, Cavendish subgroup): (i) 'Dwarf Cavendish' is widely cultivated in East Africa, South Africa, the Canary Islands and South-east Asia (Fig. 2.12). The plant is 1.5–2.5 m tall. Bunches bear eight to 12 hands. Fruits are medium to large, with light green to greenish-yellow skin. The flesh is white to creamy, soft with fine texture and sweet. The cultivar is easily recognized because the male bracts and flowers are not shed. (ii) 'Giant Cavendish' is widely cultivated for the export market in the Philippines, Taiwan, Colombia, Australia and Hawaii. The plant is 2.5–5 m tall; the pseudostem is dark brown spotted. Bunches are long and cylindrical and bear 14–20 hands. Fruits are larger than those of 'Dwarf Cavendish' and not as delicate. Male bracts and flowers are shed. (iii) 'Pisang masak hijau' is a triploid Cavendish clone of the Philippines, Indonesia and Malaysia. The plant is tall and slender and prone to wind injury. The fruits are commonly used as cooking bananas in Jamaican households.

Fig. 2.12. Drip-irrigated cultivation of 'Dwarf Cavendish' bananas on Tenerife, Spain (photo: J.F. Wienk).

- *Musa* (AAB group): (i) 'Mysore' is the most important dessert banana type (several cultivars) in India and Sri Lanka, where it constitutes about 70% of the total banana crop. Outside India and Sri Lanka it is of importance only in Trinidad where it is cultivated as shade for cacao. The plant is large and vigorous, very hardy and drought-tolerant. Bunches are large and compact. Fruits are medium-sized, plump, bright yellow, thin-skinned and with subacid-flavoured flesh. (ii) 'Silk' is the most popular dessert cultivar in Indonesia and Malaysia. Bunches bear five to nine hands. Fruits are small to medium, yellow-skinned with white, finely textured and slightly subacid-flavoured flesh.

The banana fruit is the main product and is utilized in a multitude of ways; from simply being peeled and consumed fresh, to being sliced and served in various dishes, or being mashed and incorporated in products like ice cream, bread and various dairy products. Unpeeled or peeled, unripe fruits may be cooked or baked. The fruits may be processed into starch, chips, purée (particularly for baby food), vinegar, or they may be dehydrated. In Uganda and parts of Tanzania large quantities of beer are made from bananas, providing an important nutritious drink which is rich in vitamin B due to the yeast content.

Wherever bananas are grown, traditional and typical methods of preparing food from bananas can be found. In Costa Rica, for example, ripe bananas are peeled and boiled slowly for hours to make a thick syrup called 'honey'. Green bananas boiled in the skin are very popular in many Caribbean islands. In the islands of the South Pacific ripe bananas are mashed, mixed with coconut cream, scented with *Citrus* leaves and served as a thick, nice-smelling beverage. In Polynesia people rely on an ancient method of preserving large quantities of bananas for years as emergency food in times of famine. The peeled bananas are wrapped in *Heliconza* leaves and buried in deep pits. Banana leaves are placed on top and covered by soil and rocks. The pits remain unopened until the fermented food, called 'masi', is needed.

The terminal male bud, new shoots of young plants and slices of the pseudostem are sometimes used as vegetables. Clean banana leaves are excellent material for wrapping various kinds of food. The leaves and the pseudostem are also valuable sources of fibre, which is useful in making products like thatching material, fishing lines, rope, fabric, handbags, shoes, mattresses and paper. The peel and pulp of the ripe fruit have antifungal and antibacterial properties.

Rejected green and ripe bananas are used as animal feed, particularly when the cattle range is located close to the banana fields. With dairy cattle, it is recommended that the portion of bananas in the feed should not exceed 20%, due to the low protein content and the somewhat laxative effect.

The nutritional composition may differ between various cultivars. At fruit maturity the edible portion contains approximately 70% water, 1.2% protein, 0.3% fat, 27% carbohydrates and 0.5% fibre. It is rich in K (400 mg/100 g). It is a good source of vitamins C and B_6. Banana has a special place in diets with low fat, cholesterol and salt.

There are numerous ornamental *Musa* species and cultivars, even moderate winter-hardy species like *Musa basjoo*.

Ecology and agronomy

Optimal growing conditions are found in warm humid to per-humid tropical regions, roughly between 30°N and 30°S. However, due to attractiveness of the crop, it is also grown in regions with suboptimal climate conditions. Temperature is the major factor; a mean average of 27°C is the optimum; 15°C is the minimum and 38°C is considered to be the maximum. Chilling injury occurs below 13°C although the corm may survive. Of the previously mentioned cultivars only 'Mysore' and 'Dwarf Cavendish' tolerate temperatures close to freezing.

For optimum growth a mean monthly rainfall of 200–220 mm is required; soil moisture should not be depleted below 60–70% of field capacity. Where waterlogging is likely, bananas are grown on raised beds. Low, perennially wet soils require draining and dry soils require irrigation. The dry season should not exceed 3 months. Deep, friable loam soils are ideal for growing banana because of their good drainage and aeration. High fertility and an organic matter content of over 3% are great advantages. The plant tolerates pH values of 4.5–7.5 although the plant is more susceptible to diseases if the pH is below 5.

Windbreaks are often planted around banana fields. Light winds shred the leaves; stronger winds may twist and distort the crown. Winds to 50 km/h break the petioles; winds of 65 km/h or more will uproot entire plantations.

Bananas are propagated exclusively by vegetative plant parts. Corms or pieces of corms are used in commercial plantations. Most growers prefer corm sections of 1–2 kg. When corms are scarce, smaller pieces are utilized in combination with early fertilization, to compensate for the smaller size. To collect the corms, plants of at least 7 months old, but prior to fruiting, are uprooted and cut off about 10–15 cm above the corm. The corms are usually pared, trimmed and disinfected by immersing for 20 min in hot water (52°C) or in a commercial pesticide solution. Before planting, the corm pieces should be placed in the shade for 48 h, away from all diseased trash.

Propagation by suckers is also utilized, particularly by smallholders and in home gardens, and in case corms are not available in sufficient quantity. The sucker first emerges as a conical shoot with leaves which are reduced to midribs and remnants of leaf-blades. These juvenile leaves are called 'sword' or 'spear' leaves. For propagation, sword leaf suckers are preferred since they bear larger bunches in the first crop. Recently, tissue culture for rapid propagation through disease-free shoot-tip and meristem culture has been developed. Propagation by seed is employed only in breeding programmes.

Planting densities range from 1000 to 3000 plants/ha depending on the ultimate size of the cultivar, soil fertility, planting pattern and other factors. Equidistant planting patterns and row cropping are both practised; double-row planting combines dense populations with good access, for example (3.5 + 1.5) m × 2 m results in 2000 plants/ha and 3.5 m wide alleys. Close planting protects plantations exposed to high winds, but results in fewer suckers and hinders disease control. Planting is best done at the onset of the wet season. Planting holes should be at least 45 cm in diameter and 40 cm deep; planting material is usually set at a depth of 30 cm.

Banana is grown in different cropping systems: home gardens, mixed cropping in farmers' fields, commercial smallholders' orchards and commercial plantations. In home gardens it is grown for home consumption, using minimal inputs and often various cultivars. Surplus produce is sold to market. In mixed cropping, banana is commonly used as a nurse crop for young cacao, coffee, pepper, etc., or as an intercrop in newly established rubber and oil palm plantations, and under mature coconut. When grown as a principal crop, it is usually interplanted with short-term crops such as maize, aubergine, peppers, tomato, okra or sweet potato. A space of at least 1 m in diameter should be kept clear around each banana plant. In smallholder orchards, banana is usually grown as a sole crop in areas ranging from 1.5 to 20 ha. The choice of cultivars is determined by consumers' preference and climatic conditions in the area. Commercial plantations generally produce for the export market only. Cultivars are commonly grown as a single in large areas. It is a capital-intensive system involving high investments in infrastructure and management practices, in order to ensure high yield and good quality to meet the strict requirements of the export market.

Weed control is essential and is required until the plants provide enough shade. Weeding may be done mechanically or by the use of herbicides, preferably pre-emergence or when the plants attain a height of 1.5 m, by contact herbicides.

Banana has a high nutrient requirement, especially N and P, mainly due to the high nutrient removal at harvest. The amount of fertilizer can be adjusted according to soil analysis and expected fruit yield. Adequate amounts may be 80–120 kg N/ha, 100–150 kg P/ha and 100–120 kg K/ha, divided in doses of one-third at planting, one-third after 2 months in a circle of 30 cm diameter around the plants and one-third after another 2 months at double distance around the plants. Broadcasting fertilizer generally gives lower yields.

The time from planting to harvesting the first crop, known as the 'plant crop', is 9–18 months, depending on cultivar, climate and crop management, among other things. Subsequent crops are known as 'ratoon crops'. The stage of maturity at harvest depends on whether it is used locally, when it is cut full-green, or earlier when it is for export. For local consumption, bunches are ripened by hanging them in shady places. Harvest for export generally is done 75 to 80 days after the opening of the first hand, when the fruits are fully de-veloped and 75% mature. Bunches are cut with a curved knife, leaving attached 15–20 cm of the stalk to serve as a handle for carrying. Improved handling methods, like the use of plastic sleeves around bunches, greatly reduce bunch injury. After harvest, the hands are separated from the bunches. The hands are deflowered, washed, sorted, often treated with fungicides, and packed in carton boxes for export. Bananas are usually transported in specially constructed ships in cool storage at temperatures of 11–13°C, depending on the cultivar. On arrival in the importing countries the bananas are placed in ripening rooms at a temperature of 21°C, with relative humidity 90–95%, for 1–2 days, subse-quently for 4–6 days at 18°C. Ethylene is often used to give rapid and uniform ripening (Fig. 2.13).

Fig. 2.13. Cardboard boxes with hands of green bananas, ready for shipment (left); box with artificially ripened bananas, ready for the fresh market (right).

The world average yield of banana was 16.3 t/ha in 2005. Costa Rica (48.6 t/ha) and Guatemala (52.5 t/ha) are among the countries with the highest yields. In 2005, the world production of banana in about 130 countries was estimated at 72.6 million t. About 75% of it was produced in the ten major banana-producing countries, Brazil, China, Ecuador and India being the largest producers. The world exports of banana are highly concentrated. In 2003, a total of 15.4 million t was exported; Ecuador, Costa Rica, Colombia and the Philippines accounted for two-thirds of exported production.

Cherry

Sweet cherry – *Prunus avium*, Sour cherry – *Prunus cerasus*; Rose family – *Rosaceae*

Origin, history and spread

Cherries probably originated in the Caucasus Mountains and other parts of Asia Minor, and spread westwards into Europe due to seed-dropping birds. Cultivation most likely began in Greece, because the first description of a cherry tree is from the Greek Theophrast around 300 BC. At the same time another Greek named Varro wrote a manual for cherry breeding. In the 1st century AD, the Roman author Pliny the Elder mentioned in his writings ten different varieties of cherries. The English word 'cherry' or the Dutch word 'kers' is by many supposed to be derived from the Greek word 'kerasos'. However, others think that the fruit was named after 'Kerasun', which is a town in Turkey. That indicates a connection with the story of Lucullus in 74 BC. Lucullus conquered King Mithridates near Kerasun. When he returned to Rome in his chariot, he brought along from the Kerasun area a branch of a cherry tree full of fruits, while shouting: 'Cherries for Rome!' The Greeks passed it on to the Romans. As the Romans considered cherries to be an essential part of the diet of their soldiers, the troops dispersed cherries throughout Europe. However, there is archaeological evidence that sweet cherries were known in France, Italy, Hungary and England about 4000 BC, which is why it is sometimes assumed that Central Europe was another centre of origin. European colonists brought cherries to their colonies in the 17th century. By that time, many cultivars already existed. The Dutch fruit expert J.H. Knoop (1706–1769) described 26 different cherry varieties in his book *Fructologie*.

Apart from the fruit, cherry is also appreciated for its beautiful blossom. In Japan, a blossoming branch of the cherry tree is a special symbol of spring, and is often used in many pictures and other objects of art.

Nowadays, cherries are broadly distributed around the world and can be found in Europe, North America and Asia.

Botany

The canopy of an adult sweet cherry tree has an erect-pyramidal shape and the tree can be up to 20 m in height. The young foliage is slightly brown in colour.

Fig. 2.14. Cherry orchard, small trees in rows, The Netherlands.

Sour cherries mostly have much smaller trees, at most 10 m in height, with drooping branches and lighter green foliage. At present, both cherry types are in cultivation, often kept less than 4 m tall by pruning, often shrub-like or in hedges (Fig. 2.14). Sweet cherry trees can become very old, up to a maximum of 400 years, whereas sour cherry trees mostly do not grow older than 30 years. In cultivation, the lifespan is mostly much shorter.

The shape of a cherry leaf is oblong ovate or elliptic, ending in a tip, with distinct petioles and veins. The bark is smooth and silvery in colour. The flowers have long pedicels and are borne in clusters on long-lived spurs (Fig. 2.15). The number of flowers can be up to five per cluster. A flower consists of five greenish sepals, five white or pinkish petals, and many stamens. The fruit is a one-seeded, glabrous drupe, 1.5–3 cm diameter, globose to heart-shaped; the seed is called a 'stone'. The fruit's flesh is juicy, soft or firm, and has a thin skin that can be yellow, light red, medium red, dark red, or dark red-black in colour (Fig. 2.16).

Species, uses and constituents

As well as the above-mentioned cherries, there are some minor species, grown mainly in Asia: ground cherry (*Prunus fruticosa*) and Chinese cherry (*Prunus pseudocerasus*). Ground cherry is used in breeding programmes with sour cherry. Furthermore there are hybrids between *P. avium* and *P. cerasus*, called 'duke cherries'.

In previous ages, breeders in Europe, North America and Asia produced many cultivars that have regional importance. Currently, sweet cherries can be

Fig. 2.15. Inflorescences of cherry.

Fig. 2.16. Cherry: detail of a branch with leaves and ripe fruits.

considered as one group whereas sour cherries can be divided into two different groups called 'amarelles' and 'morellos'. Amarelles have reddish, light coloured, low-acid fruits and morellos have dark red, high-acid fruits. The morello group is by far the most important group of the sour cherries. Sour cherries are also called 'tart cherries'. Another species that has importance in cherry culture is *Prunus mahaleb*. It is not grown for the fruits, which are small, almost black and have a very unpleasant flavour, but the species is very suitable for obtaining rootstocks. Rootstocks for sweet cherries are often obtained from seedlings of species like *P. avium*, *Prunus dawyckensis* and *Prunus canescens*.

Sweet cherries are mainly for the fresh market; a smaller part of the production is canned or frozen, and processed into juice, soft drinks and liqueur or brandy. A considerable part of the sweet cherry production is made into maraschino cherries. For the traditional way to produce it, the cherries have to be brined and steeped in the liqueur Marasca, which is distilled juice from fermented wild cherries (*P. cerasus marasca*) from the Dalmatian mountains. However, today almond oil is often used as a substitute for the fermented wild cherry juice. Maraschino cherries are often used in candies and ice creams or covered with chocolate. Sour cherries are mostly frozen, canned or dried, and mainly used for processing into pie fillings, desserts and jams. In Hungary, cold cherry soup is a traditional dish.

Cherries contain 80–84% water, about 1.2% protein, 14–17% carbohydrates, 0.3% fat and 0.3% crude fibre. Furthermore, the fruits contain vitamins A, B_1, B_2, C and minerals. Because cherries are a rich source of antioxidants, they can help prevent cancer and heart disease. Cherries contain cyanogenic glycosides in bark, leaves and stones. These parts are poisonous for humans and livestock and may even cause death.

Ecology and agronomy

Cherry trees grow best in cooler parts of the temperate regions. They have relatively high vernalization requirements, about 1000 h. The trees tolerate frost rather well, although frost during the blooming period may harm fructification. Sour cherries are mostly more cold-tolerant than sweet cherries. They grow on a wide range of soil types, if the soil is well-drained at pH 6–7. However, they do not thrive well on heavy soils and soils which remain wet for extended periods. As far as these factors are concerned, there are differences between the cultivars associated with the different rootstocks of the trees.

Spacing of the trees at planting depends on the sizes of the trees in the different growing systems; the size is also dependent on the rootstock. Nowadays, growers often use dwarfing rootstocks to obtain small cherry trees (Fig. 2.14). The number of trees per hectare also varies considerably, from standard trees to small trained trees, from 120 to 1050. Training trees can be done in upright or V-shaped hedges (vase), the V-shape being realized by removing the main axis or trunk. Pruning will always be necessary but frequency depends on the growing system; standard trees do not need as much pruning as small trained trees. The soil around the trees is mostly left under grass.

Many cherry cultivars possess gametophytic self-incompatibility, which means that it is often necessary to plant several cultivars near each other to realize cross-pollination.

To obtain satisfactory cherry production it is mostly necessary to fertilize the trees. How much fertilizer is needed depends on soil fertility, which means that specific amounts of fertilizer are difficult to recommend. For an orchard that is in full production it can be 120 kg N/ha, 60 kg P/ha and 160 kg K/ha, although on fertile soils often only N is required. To determine an optimal application of fertilizer, soil samples or foliage should be analysed.

Time of harvesting can be determined based on a number of indices, such as the colour and flavour of the fruits, and the force which is needed to remove the fruits. However, cherries for the fresh market have to be firm when they are harvested in order to avoid bruising. Cherries for the fresh market are harvested by hand; the pedicels have to remain on the fruits. Cherries for processing are often shaken from the trees by special equipment. To make this way of harvesting successful, it is recommended to apply a growth regulator, for example Ethephon, to the fruits before harvest to reduce the force which is needed to remove the fruits. Cherries have a shelf-life of at most 2 weeks, which means that cherries for the fresh market have to be cooled and shipped immediately after harvesting. Cherries for processing are often frozen within a day after harvest.

Yields vary considerably. The world average yield in 2005 was 4.6 t/ha. Nowadays 10 t/ha is a high yield. There are indications that yield may increase up to 15 t/ha. In 2005, the world total production was 1.8 million t of fresh fruits. Turkey, the USA and Iran are the main producers of cherries.

Citrus

Orange – *Citrus sinensis*, Lemon – *Citrus limon*, Grapefruit – *Citrus × paradisi*, Mandarin – *Citrus reticulata*; Rue family – *Rutaceae*

Origin, history and spread

The citrus fruits are native to a very large Asiatic area extending from the Himalayan foothills of north-eastern India and north-central China to the Philippines in the east and to Burma, Thailand, Indonesia and New Caledonia in the south-east. The grapefruit (*Citrus × paradisi*) is the only exception, with its origin outside South-east Asia. It appeared on the island of Barbados some time before 1790 as a mutant or possibly as a hybrid of species from the Far East.

The centres of origin of the most important species are difficult to ascertain, because these species are known at present under cultivation only. For thousands of years they have been intercrossed, by man or with natural populations, to such an extent that they can no longer be identified. The earliest records on cultivation of *Citrus* come from China around 2200 BC. The citron (*Citrus medica*) spread from Persia to the eastern Mediterranean, probably following the conquest of Alexander the Great, and ultimately to Italy by the Jewish community some 2000 years ago. Citron was probably the first citrus in Europe.

It is assumed that the Arabs distributed the sour orange (*Citrus aurantium*), lemon (*Citrus limon*), lime (*Citrus aurantifolia*) and pummelo (*Citrus grandis*) throughout the eastern Mediterranean and northern Africa about 800 to 1000 years ago. The sweet orange (*Citrus sinensis*) became important in Europe about 400 years ago when Portuguese sailors brought it from China ('sinensis' meaning 'from China'). The Spaniards introduced sweet orange into South America and Mexico in the 16th century, and probably the French took it to Louisiana, from where it reached California in 1769 and Florida in 1872. The most recent introduction to the Mediterranean basin and later on America was the mandarin (*C. reticulata* or more probably *C. deliciosa*, the so-called 'willow-leaf mandarin'). This took place in 1805 from China to Great Britain and thence to Malta. During the 19th and 20th centuries, many citrus varieties of Asian origin were repeatedly imported by the citrus-growing countries.

Citrus crops are now grown in a belt roughly between latitudes 40°N and 40°S, particularly in areas with a Mediterranean climate, high elevations excepted. Near the equator citrus can be grown up to 2000 m altitude. The Mediterranean area supplies most of the world's fresh citrus fruits, while the US and Brazilian production goes primarily into juice.

The number of valid species in the complicated genus *Citrus* has given rise to endless discussions between specialists as to the best taxonomic classification. The number of species may be nine, 16 or 166; about 800 binomials exist and innumerable cultivars are known. The concept with 16 species, according to Swingle, is generally accepted and is used here.

Four of the commercially most important species worldwide are discussed further below with a main focus on sweet orange. The topics of botany, uses, ecology, agronomy and yields are treated per species.

Orange

Orange (Fig. 2.17), also called 'sweet orange', in contrast with sour orange, is an evergreen tree up to 8 m tall with a rounded, sometimes slightly upright crown, and more or less thorny, especially when young. The leaves are alternately arranged, elliptic to ovate, 5–15 cm × 2–8 cm, rounded at the base, margins undulate to crenate, with a pointed apex; petioles narrowly winged. The flowers are axillary, in few-flowered racemes or singly, 2–3 cm in diameter, bisexual and fragrant; the calyx is five-lobed; the corolla is of five white petals with oil glands; the stamens 20–25 in groups, partially united at the base (Fig. 2.18). The fruit is a subglobose berry known as 'hespiridium', 4–12 cm in diameter, green when unripe and turning greenish-yellow to bright orange at full ripeness (Fig. 2.19). However, when grown in the tropics, in the absence of cold nights, the fruits remain green or turn yellow, instead of becoming bright orange. The leathery peel (rind) is up to 0.5 cm thick, consisting of an orange, outer layer (exocarp) densely dotted with glands containing an essential oil and an inner, dry and white layer (mesocarp). The spongy endocarp consists of a central axis and several segments filled with pulp-vesicles, containing the yellow to orange juicy pulp. Orange trees usually develop a single taproot and horizontally growing

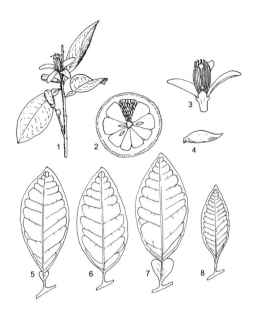

Fig. 2.17. Citrus: 1, flowering branch; 2, fruit in cross-section; 3, flower in longitudinal section; 4, seed. Leaves: 5, orange; 6, lemon; 7, grapefruit; 8, mandarin.

Fig. 2.18. Inflorescence of orange cultivar 'Washington Navel' with open flowers in side-view, one partly opened to show ovary, style and stigma (photo: D.L. Schuiling).

Fig. 2.19. Habit of orange tree (left) (photo: J.D. Ferwerda); detail of fruiting branches (right).

lateral roots with few root hairs. When transplanted, the taproot does not develop.

The numerous cultivars of sweet orange (see examples in Fig. 2.20) are usually divided into three groups.

1. Common oranges with normal fruits: the most important cultivar is 'Valencia'; late maturing (8–9 months from bloom to maturity), medium–large fruits, outstanding in juice quality and quantity, stores very well and is suitable for table and industry. The cultivar 'Pera' is widely used in Brazil. Other important early cultivars in Florida are 'Hamlin' and 'Parson Brown'. The cultivar 'Jaffa', appearing probably prior to 1844 near Jaffa, has large, seedless and juicy fruits with a very thick rind.

2. Navel oranges with a second row of carpels opening at the fruit apex in an umbilicus: the best-known cultivar is 'Washington' with large, seedless fruits which are easy to peel; with crunchy flesh and marked flavour. It is use exclusively as fresh fruit.

3. Blood oranges with red or red-streaked pulp: the distinctive red colour is caused by the pigment anthocyanin. Perhaps the sweetest and most flavourful is cultivar 'Tarocco', which is grown mainly in Italy. The red-skinned cultivar 'Sanguinello' has few seeds and is grown mainly in Spain. The more recent cultivar 'Moro' is grown in California. Blood oranges are used fresh and processed into bottled juice.

Sweet orange has always been mainly used as fresh fruit. Much of the crop is now used for processing juice and frozen concentrated orange juice. The total citrus fruit processing accounts for approximately one-third of the total citrus fruit production. More than 80% of it is orange processing, mostly for orange

Fig. 2.20. Fruits of citrus species and cultivars: 1, grapefruit 'Marsh'; 2, grapefruit 'Ruby Red'; 3, orange 'Sanguinelli'; 4, orange 'Valencia'; 5, orange 'Washington Navel'; 6, lime; 7, lemon; 8, mandarin 'Clementine'; 9, mandarin 'Satsuma'; 10, mandarin 'Ortanique'.

juice production. Roughly 85% of the world market of orange juice is produced in the State of Florida in the USA and the State of Sao Paulo in Brazil.

The edible portion of mature fruits is about 40–50% and contains 80–90% water, 0.7–1.3% protein, 0.1–0.3% fat, 12–13% sugars, 0.5% fibre, 0.5–0.7% ash, 0.5–2.0% citric acid and 45–61 mg ascorbic acid/100 g. The fresh fruits and juice are highly valued because of their reputation as good sources of vitamin C, about 50 mg per 100 ml juice. Vitamin P and vitamin A are also present. The citrus peel contains much pectin and essential oil.

Several other products, such as marmalade, candies and lemonades, are made from the fruit.

The enormous waste disposal problem in the fruit processing industry has resulted in a parallel-running by-product industry. The most important by-products are: essential oils, primarily used in the beverage, perfume and flavouring industries; dried citrus pulp, mainly used as feed for dairy and beef cattle; citrus molasses, used in animal feed and for the production of alcohol by fermentation; and bioflavonoids like hesperidin, which is used as a therapeutic agent in the pharmaceutical industry and as an animal feed supplement.

The sweet orange prefers a clear change of seasons and is therefore a subtropical rather than a tropical species. During the growing period, the temperature should not drop below 13°C. Sweet orange can stand relatively high temperatures of over 38°C. The ideal temperature during winter dormancy is in the range of 2–10°C. Although mature trees can stand brief frosts of up to

−4°C, the fruits are damaged by frost of −1 to −3°C. Young trees may be killed by even mild frosts. An average annual rainfall of at least 1000 mm is required if grown without irrigation. Crop water use is 120–160 mm during a hot month. Excessive rainfall during flowering can be fatal for the crop. A dry period during fruit ripening contributes to a more intense orange colour of the peel. Success in orange culture depends a great deal on the choice of cultivars and their adaptation to distinctive weather conditions. Orange, like all citrus species, is intolerant of high winds and windbreaks should be provided where necessary.

The best soils for orange growing are fertile, light loamy and well-drained soils with a pH range of 5–8. Vegetative growth and fruit yield are greatly reduced on alkaline soils and when the water table is 80 cm or more below the surface.

The most customary method of propagation is by budding on seedling rootstocks. The number of combinations of rootstock type and scion type is virtually innumerable, resulting in a wide range of botanical, physiological and ecological characteristics. Beside the effect of rootstock on fruit yield and quality, citrus growers are particularly interested in the stock–scion interactions which are prominent in citrus; for example, the stock may adapt the scion to heavy soils, the stock may be resistant to soil-borne diseases and/or transfer disease tolerance to the scion.

The most commonly used rootstocks are sour orange (*C. aurantium*), extensively used in the Mediterranean region and Florida, particularly the 'Valencia' orange; trifoliolate orange (*Poncirus trifoliata*), used for satsuma oranges in Japan; rough lemon (*Citrus jambhiri*), used in South Africa, Australia and in Florida for the 'Washington Navel'; and sweet orange (*C. sinensis*), used in California for oranges and lemons. The rootstock seedlings may be budded 10–12 months after germination. The scions (budwood) should be taken from selected, high-yielding trees which are free from diseases and off-type bud mutations. The scions with stems of 8 mm or more in diameter are inserted on the rootstock at a height of 20–30 cm.

Planting density varies from 200 to 400 trees/ha, for example 5 m × 6 m, and planting pattern may be square, rectangular, triangular or hexagonal; space for mechanical cultivation, spraying, irrigation when used and for transport of fruits should be taken into consideration.

The first pruning is done in the nursery and pruning must be continued in the field for 2–3 years. Periodic removal of water sprouts, dead wood and branches that are lower than 30 cm from the ground is essential. Trees that are closely planted in hedges may be pruned mechanically by special equipment.

Mineral nutrition and irrigation in semi-arid regions with subtropical climates are the two basic factors in the husbandry of citrus species and sweet orange in particular. With chemical soil and leaf analyses fertilizer needs can be assessed. Other factors like climatic conditions, quality of irrigation water, physical soil characteristics, previous treatments of the orchard and the presence of a (leguminous) cover crop must also be taken into account. Optimum fruit quality can be obtained when nitrogen is supplied at minimum required levels for optimum production. Excessive N levels can impair fruit growth and quality. Taking all of these factors into consideration, the following annual fertilizer rates can serve as a general guideline: 30–70 kg N/ha to young trees and 100–200 kg N/ha to mature trees, preferably in two applications; about 17 kg P/ha and 60 kg K/ha. Foliar application of potassium nitrate is very useful where soil applications are not effective. Micro-elements, especially Zn, are essential to obtain optimum fruit yield and fruit quality. In areas with too little or unpredictable rainfall, irrigation of 100 mm every 3 weeks will be needed.

Generally, sweet oranges can be harvested 6–9 months after flowering. In both hemispheres most oranges come to maturity in winter, with the main period of harvesting from about October to March in the northern hemisphere and from April to September in the southern hemisphere.

Although various methods of mechanized harvest have been explored, manual harvesting is common. Fruits for the fresh market go through various postharvest treatments like washing, disinfection, drying, colour-adding, waxing, grading, sizing and finally packing. In cool storage sweet orange can be kept for 1 month or longer.

The average yield of sweet orange varies from 40 t/ha in well-managed orchards in Florida, to much lower yields such as 5–15 t/ha in tropical countries. In 2005, the world production of sweet orange was almost 60 million t of fresh fruits, accounting for almost 60% of the total production of all citrus fruits.

Lemon

Lemon (Fig. 2.17) is a rather small, evergreen tree of 3–6 m height. The branches and twigs have sharp thorns. The leaves are ovate, 6–11 cm × 3–6 cm; reddish when young, becoming dark green above and light green below, with a serrated margin. Leaves have a distinctive smell when crushed. Petiole wings are very narrow or absent. Flowers are solitary or in few-flowered clusters, 4–5 cm in diameter; there are five petals, pink in bud, while opened flowers are white above (inside) and purplish below (outside); stamens 20–40 in bundles of three to five or more. The fruit is oval with a characteristic terminal nipple; light yellow when ripe (Fig. 2.21). The peel is rather thick and connected to the pale yellow pulp, gland-dotted and slightly rough. The seeds are ovoid and polyembryonic.

Numerous cultivars are grown in the Mediterranean area, such as 'Monachello' in Italy and 'Verna' in Spain, Algeria and Morocco. 'Eureka' and 'Lisbon' are the principal cultivars in California. The cultivar 'Lisbon' is thought to be a selection from the so-called 'gallego' lemon of Portugal, which in fact constitutes a group of common seedling forms. Cultivars may differ in their suitability for marketing as fresh fruit and for processing, and in their ability to resist decay.

Slices of fresh lemon are served as a garnish on meat, fish and shellfish, or with hot or iced tea, to be squeezed for the flavourful juice; the acidic juice neutralizes the taste of amines in fish and shellfish by converting them to non-volatile ammonium salts. Lemons are widely used in the preparation of lemonade and for culinary and confectionery purposes; in cosmetic and pharmaceutical

Fig. 2.21. Habit of lemon tree (left); fruiting branch (right).

products; for the production of lemon oil, citric acid and pectin. Candied peel, made from the rind, is a very good source of vitamin P.

Fresh lemon juice contains 91% water, 0.5% protein, 0.2% fat and 8.0% carbohydrates; furthermore, vitamin A, ascorbic acid and traces of fibre and minerals are also present. Lemon juice is widely known as an antiscorbutic. Sailing ships of the 18th century carried lemons aboard to prevent scurvy.

Lemon is more sensitive to cold than orange and can be grown in a relatively limited climatic range. The favoured temperature ranges between 15 and 30°C. The best fruit quality is achieved in areas with summers which are too cool for oranges. Lemon can thrive on relatively infertile soils. In California, excellent growth is maintained on silty clay loam of high water-holding capacity. Soil pH should be between 5.5 and 6.5. Annual rainfall requirement ranges from 250 to 1250 mm. In long dry periods the lemon must be irrigated.

Propagation is both from seed and by grafting on to various rootstocks such as sweet orange, mandarin and sour orange (*C. aurantium*). Plant spacing is 7 m × 7 m; the trees must be pruned when young and kept to a height of 3–3.5 m. After 12 years, the trees are cut back severely or replaced. Lemon is very sensitive to herbicides. Fertilizing may be done three times a year at an increasing rate of 2 kg NPK per tree when young to 4.5 kg NPK per tree up to an age of 50 years.

Lemon tree yields vary considerably with cultivar, location and weather conditions. Commercial lemon orchards may yield 15–18 t of fresh fruits per hectare. Lemon and lime (*C. aurantifolia*) are often merged in production statistics, although lime is grown exclusively in tropical climates. In 2005 the world production of lemon and lime was 12.5 million t. Lemons are mainly produced in the USA, Spain, Italy, Argentina and India. Major producers of limes are Mexico and Brazil.

Grapefruit

Grapefruit (Fig. 2.17) is a spreading, evergreen tree of 10–15 m height. The leaf-blade is ovate, 7–15 cm × 4–8 cm, often crenulate; petioles are broadly winged. Flowers are solitary or in small clusters, 4–5 cm in diameter, the calyx is five-lobed, the corolla of five white petals. The fruits are large, usually globose, greenish or pale yellow when ripe (Fig. 2.22). Some cultivars have pink or red flesh (Fig. 2.20).

Grapefruit is mainly used as a breakfast fruit. It is also processed into juice and canned segments. Pulp and molasses are used as cattle feed. The edible portion is around 30–40% of the fruit. Vitamin A is higher than in other citrus fruits, but the concentration of vitamin C is relatively low. Grapefruit does not contain the glucoside hesperidin, like sweet orange, but the bitter glucoside naringin. This gives grapefruit its characteristic flavour: a combination of sweet, acid and mild bitterness.

Grapefruit is mostly propagated by grafting on to rootstocks. Plant spacing of 7 m × 7 m is recommended. It requires less N but more K than sweet orange. Grapefruit needs a warmer climate than sweet orange and appears to be well suited to the tropics. Commercial production started in the USA in the 19th century. The crop is now grown in Florida, California, Texas and Arizona. Important exporting countries are Israel, South Africa, West Indies and Brazil. With good management, a production of 40 t of fresh fruits per hectare is possible. In the subtropics, time of harvest can be delayed and 'stored' fruit on the tree can be marketed later. However, this practice will reduce the next crop.

The world production of grapefruit is not known since it is recorded together with pummelo (*C. grandis*) in FAO statistics. World production in 2005 for the two crops was 3.6 million t, of which grapefruit is by far the most important.

Fig. 2.22. Fruiting branches of mandarin (left) (photo: D.L. Schuiling) and grapefruit (right) (photo: H. Lövenstein).

Mandarin

Mandarin (Fig. 2.17) is a small, evergreen tree of 2–8 m height, sometimes spiny. The leaves are narrow, ovate to elliptic or lanceolate, 4–8 cm × 1.5–4 cm, usually crenate, dark shining green above, yellowish-green below; petioles are usually narrowly winged. Flowers have five petals, 1.5–2.5 cm in diameter, white, and appear singly or in clusters of three or four. Fruits are depressed globose, 5–8 cm in diameter, yellow or orange-red when ripe and with loose peel (Figs 2.20 and 2.22).

In some countries, the terms mandarin and tangerine are used indiscriminately. In agriculture and commerce, often the following classification is used.

- Common mandarins (*C. reticulata*): fruits depressed globose or subglobose, with thin, loose peel, easily separating from the segments, bright orange or scarlet-orange when ripe. Includes many cultivars such as 'Beauty', 'Clementine' (USA), 'Dancy' (USA) and 'Emperor' (Australia). Most of the cultivars do well in the tropics.
- King mandarins or tangors (*Citrus nobilis*): the Indo-China or Cambodia mandarin or kunenbo from Japan. Large fruits with thick peel.
- Satsuma mandarins (*Citrus unshui*): fruits are medium-sized, seedless, orange, with ten to 12 segments. Extensively grown in Japan, they are very hardy and cold-resistant.
- Mediterranean mandarins or willow-leaf mandarin (*C. deliciosa*): fruits are medium-sized, with many seeds; grown throughout the Mediterranean region.

Interspecific hybrids have been made with other citrus species of which the most important is the tangelo (mandarin × grapefruit), with fruits resembling oranges and with flavour intermediate between the parents; an example is 'Ugli', which is grown in and exported from Jamaica.

Mandarins are primarily used as fresh fruit and have long been appreciated for their distinctive and sweet flavour and because they are easy to peel and separate into segments. To a limited extent mandarins are processed, for example as canned segments in syrup.

The mandarin has a wide range of adaptability and is grown under arid, (semi)tropical and subtropical Mediterranean climatic conditions. Although mandarin trees are the most cold-resistant of all commercially grown citrus, their fruits are more sensitive to frost than orange and grapefruit. Mandarin will produce good crops on a wide range of soil types, although light, fertile loams are ideal. It requires close plantings (2 m × 5 m) and a low canopy for ease of harvesting. The ripe fruits are preferably harvested by manual clipping with scissors. Fruits, especially of loose-skinned cultivars, should be harvested with care to avoid damage of the peel. In commercial, well-managed mandarin orchards, mean yields of 25 t/ha are possible. In 2002, the world production of mandarin (including hybrids) was 16.6 million t, which was about 20% of the total citrus production. The major producing countries are China (34%), Spain (10%), Japan (9%) and Brazil (6%).

Grape

Grape – *Vitis vinifera*; Vine family – *Vitaceae*

Origin, history and spread

There are 60 described species in *Vitis*. All of them can be easily inter-crossed and bring forth fertile hybrids, meaning that origin is often difficult to determine. Wild grape is a vine from humid forests and watersides. Cultivation of grape probably began about 6000 years ago in south-west Asia where wild grapes still exist. Also about 6000 years ago, the cultivation of grapes was probably known to the ancient Mesopotamians and Egyptians. The species spread out from south-west Asia to Asia Minor and subsequently to Crete and Greece about 3500–3000 years ago. It reached Rome and next France in around 500 BC. The Greeks and especially the Romans developed viticulture to a high degree and the Romans introduced grapes and wines into their whole domain. Winemaking in China began before the Han Dynasty in 206 BC. Grape is mentioned in the Bible; its later distribution was strongly associated with the spread of Christianity, due to the use of wine in the celebration of the Mass.

Only the Eurasian grape species *Vitis vinifera* is suited for quality wine production. American and East Asian species give inferior wine quality. Grapes from Europe were introduced to the Americas by Portuguese and Spanish settlers in the 16th and 17th centuries. The British brought *V. vinifera* to South Africa in 1655 and to Australia in 1813. *V. vinifera* has no resistance against two fungus diseases (powdery mildew and downy mildew) and the species is also very susceptible to *Phylloxera*, a root aphid. Until about 1850 this caused no problem for Eurasian viticulture as these two fungus diseases and the root aphid were endemic in North America. However, between 1850 and 1880 these fungus diseases and *Phylloxera* reached Europe due to ship transport of American vines. Since that time *V. vinifera* has been cultivated on American rootstocks to prevent *Phylloxera*, because American *Vitis* species have more or less resistance against this root aphid. Fungicides have had to be used since then: sulphur against powdery mildew and copper against downy mildew. During the 20th century wild American *Vitis* species were used in breeding programmes to realize resistance against the two mildew types. However, only just after 1990 were breeders successful in combining resistance and good wine quality.

The grapes that are grown today are often hybrids of unknown identity. Formerly even the Romans used hybridization with native species to introduce *V. vinifera* to new areas. Nowadays, grape is in acreage, production and economic value the most important fruit crop; it is also the most widely grown fruit crop. The importance of this crop is reflected by the specific names for cultivating grapes and winemaking, which are respectively 'viticulture' and 'oenology'. In 2004, grapes were grown in 88 countries, mainly in subtropical climates.

Fig. 2.23. Grape (left to right): leaf; inflorescence with flower buds; bud graft (photos: D.L. Schuiling).

Botany

V. vinifera (Fig. 2.23) is a vigorous, climbing, woody vine, up to 35 m long; however, in cultivation it is usually reduced by annual pruning to 1–3 m tall. The stem or axis is a sympodial vine, which means a stem made up of a series of superposed branches, so as to imitate a simple stem. The bark of older stems tends to peel. Grape has forked tendrils, which are up to 25 cm long; they arise intermittently at two out of three vegetative nodes, opposite the leaves. In fact, each sympodium part terminates in an inflorescence or a tendril. Tendrils can be seen as modified inflorescences. The leaves are simple, circular or circular-ovate in outline, 5–25 cm wide, the leaf-blade is palmately lobed and coarsely toothed; petioles are about 10 cm long. The inflorescence is composed of a panicle with the secondary and ultimate axes cymose (thyrses). The flower is unisexual, and has five sepals that are fused together; five yellow-green petals, joined at the tip, 5 mm long; five stamens; and a superior ovary. The five joined petals form a cap, which falls off at anthesis. The fruit is a berry, ellipsoid to globose, 6–25 mm long, fleshy and juicy, sweet or sour, dark bluish-purple, red, green or yellow, and containing two or three pyriform seeds with rather long beak. *V. vinifera* can have a very long lifespan, up to 100 years is possible.

Grape has an extensive root system that may run very deep; ultimately up to 25 m deep is possible.

Varieties, uses and constituents

Within the commercial grapes, three main groups can be distinguished based on the use of the crop: (i) wine grapes including juice grapes; (ii) dessert grapes; and (iii) dried grapes. There are over 8000 grape cultivars (Figs 2.24 and 2.25); however, relatively few of these have widespread importance. Most of the

Fig. 2.24. Vineyard with cultivar 'Regent', The Netherlands (photo: J. Oude Voshaar).

Fig. 2.25. Fruit bunches of cultivar 'Johanniter', The Netherlands (photo: J. Oude Voshaar).

cultivars are selected for a specific region and purpose. Important wine grapes are the following.

- 'Chardonnay' has small, globose, yellowish-green fruits. This grape is used for the production of the well-known white Burgundy.
- 'Pinot Noir' has somewhat larger, globose to ellipsoid, purple fruits. This grape is used for making red Burgundy.
- 'Cabernet Sauvignon' has rather large, bluish-purple, ellipsoid fruits; it is used for making the famous red wines of Bordeaux.
- 'Merlot' has dark blue globose fruits; it is also used for the red wine of Bordeaux.
- 'Riesling' is a collective name for a group of green grape cultivars with rather small, globose, green fruits, used for producing Rhine wines.

The character of wines depends strongly on ecological factors (Fig. 2.26). When cultivated in different regions with different soil and different climate, the taste of the wine will also be different. Wine character is also strongly affected by the yield per hectare. High yields give thin wines, while low yields give concentrated wines with a long after-taste. Many wines derive their names from the region in which the grapes are grown and the wine is produced.

Most of the world's grapes are made into wine, which is produced by fermenting the juice crushed from grape fruits. Yeast, a fungus that occurs naturally on the skin of the fruits, converts sugar into alcohol and carbon dioxide gas. Artificially cultured yeasts are sometimes added to the sap. The gas is usually allowed to escape; however, in some wines the gas is retained to form sparkling

Fig. 2.26. The Coldstream Hills vineyard in the Yarra Valley, Australia (photo: D.L. Schuiling).

wines like champagne. Wines can be fortified into port or sherry by adding brandy to the wine; and wine can also be distilled to produce brandy or cognac. A part of grape production is processed into (non-alcoholic) juice. For dessert grapes, special cultivars are used, for example 'Muscat of Alexandria', 'Cardinal', 'Kyoho', 'Golden Chasselas' and 'Perlee'. Size, succulence and flavour of dessert grapes have to be of high quality. Some cultivars are seedless. Most of the grapes are grown outdoors, but some of the best table grapes are grown in greenhouses, although on a small scale.

Dried grapes are called 'raisins', 'sultanas' and 'currants'. They can be dried on the vine after removing the leaves; however, usually the ripe bunches are cut and the fruits are spread out to dry on floors exposed to sunlight. Sultanas are seedless raisins. Currants are dried fruits derived from a variety with small black fruits; they are grown mainly in Greece and have been grown for at least 2000 years. Some grape cultivars are adapted for the canning industry.

Grape fruits contain approximately 86% water, 0.5% protein, 0.3% fat, 13% carbohydrates, 1% fibre; furthermore, the minerals Ca, Fe and P; the vitamins A, B_1, B_2 and C; and some acids, malic, tartaric and racemic. Grape seeds contain 6–20% edible oil, which can also be used to produce soaps.

In folk medicine the sap of the vines was used to cure skin and eye diseases; leaves were used to stop bleeding and pain; raisins were given as a treatment for constipation; ripe fruits were used to treat cholera, smallpox, kidney and liver diseases, and more. Recently, modern research on resveratrol began. Resveratrol is a compound of the grape skin, which may be useful for treating cardiovascular diseases.

Ecology and agronomy

Grapes grow on a wide range of soils. However, soil structure is more important than soil fertility; sandy or gravelly clay loams are preferred as they provide good drainage and stimulate root development. Waterlogging is not at all tolerated. The species prefers rather dry soils, but the soils have preferably to be rather deep, due to the extensive and deep-running root system. A soil pH slightly below 7 is considered best. Subtropical or warm temperate climates are most suitable for growing grapes. The species requires warm to hot dry summers and mild, cool winters to rest; it does not tolerate humid, hot climates, because of possible fungus diseases. Grape is reported to tolerate annual precipitations up to 2700 mm, however total water requirement on a sandy soil under cultivation condition is about 650 mm. If rainfall is not sufficient, irrigation will be required. Day temperatures of 25–30°C are optimal for growth. In the winter period, the leafless plant can stand temperatures of up to –20°C; however, in spring young shoots will be killed at –3°C.

Grapes are propagated from cuttings, budding or grafting to resistant rootstocks, which are usually grown in nurseries for about a year. Propagation from seed is used only in breeding programmes. Before planting, the soil has to be ploughed about 25 cm deep; at the same time organic matter has to be added to the soil by turning under farmyard manure or green manure. Spacing varies widely depending on training system and cultivar. The young vines are planted in rows at a spacing of 1.0 to 3.5 m, with the rows 1.2 to 4.0 m apart. Plant densities per hectare range from 700 to 8000. Posts with wires, bowers, pergolas or trellis are needed to train the vines. After planting the vines are usually pruned to a single stem with two or three buds.

Weeds are controlled by shallow cultivation. If erosion can be expected, cultivating has to be sufficient to suppress weeds, so they do not become a problem. Growing a cover crop is another method to suppress weeds. It may have the additional advantage of restricting too vigorous vines, due to competition. However, vineyards are usually clean-cultivated. Fertilizer requirements depend on soil fertility. Soil analysis is required to determine nutrient application; for that, foliar analysis may also be a useful tool as it indicates the nutrient status of the vine. For good yields, annual fertilizer application per hectare may be approximately 100–150 kg N, 25–50 kg P and 0–165 kg K. Minor nutrients may also be needed.

Pruning has to be carried out regularly; it highly stimulates yield, quality and uniformity of the fruits. Especially during the first years after planting, pruning and also training have to be done carefully. In this period, the fundamental framework of the grape vine has to be established. This framework usually will sustain during the vine's entire life.

The duration of the grape vine's crop cycle ranges from 4 to 6 months. Overbearing has to be avoided, in wine grapes as well as in dessert grapes, because it affects bunch and fruit quality. It may lead to inequality of bunches and less sugar accumulation in the fruits. So thinning may be required by removing a number of inflorescences, or after fruit set, by removing a number of bunches or individual fruits.

When grapes can be harvested depends more or less on the use of the harvested fruits: wine, juice, dessert or dried fruits. As grapes do not ripen after harvest, they should not be picked too early. Generally, the longer grapes remain on the vine the higher the sugar content will be. Maturity is indicated by colour and flavour of the fruits, the bunch stem turning to yellowish-brown, high sugar content, and thickening of the juice. The grapes for making wine and juice are crushed as soon as possible after harvesting, to obtain the sap. Dessert grapes are carefully packed and marketed. Fresh grapes can be stored at 1–4°C and a relative humidity of 95% for 3–6 months. Annual yields vary widely from 5 to 30 t/ha; the average of the world is about 9 t/ha. The main grape producers are Italy, France, Spain, the USA, China and Turkey. In 2004 the total world production was about 67 million t.

Mango

Mango – *Mangifera indica*; Cashew family – *Anacardiaceae*

Origin, history and spread

The mango originates in the Indo-Burma region and has been cultivated in India for at least 4000 years. It has since spread throughout the tropics and has

Fig. 2.27. Habit of an undisturbed, solitary mango tree.

become naturalized in many areas. It was introduced into South-east Asia about 1500 years ago and into the east coast of Africa about 1000 years ago. Further expansion to Australia, West Africa and the Americas occurred during the last few hundred years. Although widely distributed in the tropics and subtropics, the mango remains the fruit of India, where it is undoubtedly the most important fruit crop.

Botany

The mango tree is a large, spreading evergreen with a dense rounded or globular crown (Fig. 2.27). It varies greatly in height and habit of growth; 10–15 (20) m tall, 60–120 cm trunk diameter. The leaves are simple and spirally arranged; 12–40 cm long, up to 10 cm wide. The leaves are produced in flushes; young leaves are usually reddish, turning dark green as they mature. The inflorescence is a widely branched, terminal panicle, 10–60 cm long, with 1000–6000 reddish-pink or almost white flowers (Fig. 2.28). Both perfect and staminate flowers occur on the same inflorescence although the sex ratio is a variable component within panicles, trees and among cultivars (Fig. 2.29). The fruit is a fleshy drupe, very variable in shape, size and colour; usually up to 30 cm × 10 cm, yellowish-green to reddish. The edible flesh is yellowish to orange and fibrous or free of fibres.

Cultivars, uses and constituents

The prehistoric existence and the widespread cultivation of mango, combined with the cross-pollinated nature and the propagation by seed for centuries, contributed to the great genetic diversity of mango. In India the total number of cultivars is estimated at over 1000, although only about 200 cultivars are of commercial value (Fig. 2.30). Cultivars can be classified into two groups.

Fig. 2.28. Mango: 1, flowering branch; 2, branchlet with fruit (line drawing: PROSEA volume 2).

1. Mono-embryonic cultivars producing one seedling per seed. The seedlings are not true to type. The large majority of Indian cultivars and their offspring belong to this group. The fruit is often highly coloured and fibreless with a distinct aromatic flavour.

2. Poly-embryonic cultivars producing more than one seedling per seed. These are true to type and in most cultivars tend to suppress the zygotic embryo. Cultivars of this so-called 'Indo-Chinese' type are mostly fibrous with sweet-tasting flesh. The fruits lack attractive coloration.

Fig. 2.29. Inflorescence of mango (left) (photo: J. van Zee); a top view of a single hermaphrodite mango flower (right).

Fig. 2.30. Mango fruits of some indigenous varieties from India: 1, 'Mulgoa'; 2, 'Neelum'; 3, 'Pairi'; 4, 'Bombay yellow'; 5, 'Bombay green'; 6, 'Mundappa' (after S.R. Gangolly).

Mangoes are consumed primarily as fresh, ripe fruit. The ripe fruit is full of juice and has a rich and spicy taste, with a perfect blending of sweetness and acidity. Fruits of inferior varieties, however, are often fibrous and unpleasantly acid and sometimes prejudice the consumer against all mangoes. The fruit is composed of 11–18% skin, 14–22% flesh and 60–75% stone, based on dry matter. Mango is a particularly rich source of carotenoids and vitamin C.

Ripe mangoes are processed into frozen mango products, canned products, ready-to-serve (RTS) beverages and dehydrated products. In India pulp and RTS beverages are the most important processed mango products. Various mango beverages are based on pulp content (10–35%), brix degree (15–72%) and acidity (0.3–1.5%). Unripe mangoes are processed into products like brine stock, pickles, chutneys and dried powder. Peels and stones are the main waste products of processing. Peels are processed into aroma concentrate and pectin. The mango seed kernel is a rich source of starch (47–63%), protein (5–10%), fat (9–16%) and tannins (10–11%) and can be used as feed for cattle and poultry. Young leaves are eaten fresh or cooked as a vegetable.

Ecology and agronomy

Mangoes may be grown in the subtropics at low altitudes and in the tropics at elevations from sea level to 1200 m, although commercial cultivars do best below 600 m. In the subtropics the cold months ensure excellent floral

Fig. 2.31. Immature mangoes on the tree, Cameroon (photo: E. Westphal).

induction. The optimum temperature for growth is 21–27°C and young trees especially cannot tolerate frost. The mango thrives in a wide variety of soils, provided they are not waterlogged, alkaline or rocky; a pH of 5.5 to 7.5 is preferred. The annual rainfall requirement is 700–2500 mm, preferably distributed into distinct wet and dry seasons. Mangoes are notorious for their irregular bearing pattern and fluctuations in yield, with a heavy crop in 'on' years and a poor crop in 'off' years (Fig. 2.31). It is generally thought that periodicity of flowering in mango is due to depletion of food reserves during an 'on' year. Sometimes deblossoming is practised in order to increase flowering the next year.

Promotion of flower induction is sometimes practised by spraying the leaves with a 1% solution of potassium nitrate. In the Philippines, smoky fires below the trees for several days are also used to promote flower induction.

In many mango-growing areas poly-embryonic and mono-embryonic mango selections are grafted on to uniform, nucellar seedling rootstock. Propagation of poly-embryonic cultivars is mainly from seed, but budding and grafting are now becoming the rule, as was already the case for mono-embryonic cultivars. Nursery work takes 1–2 years; trees are preferably planted early in the rainy season. The recommended spacing is in the range of 8–12 m × 8–12 m, that is 69–156 trees/ha.

Trees flower after 5–7 years, whereas clonally propagated, grafted trees are generally smaller and can flower after 3–4 years. The fruits are harvested by hand. For tropical high-quality cultivars 10 t/ha is considered to be a good annual yield. In the subtropics 10–30 t/ha is aimed at, depending on cultivar, in particular its tendency to biennial bearing. In 2005, the world total production of fresh mangoes was almost 28 million t. India (39%) is the largest producer. Mexico is the largest exporter of mangoes.

Papaya

Papaya – *Carica papaya*; Acanthus family – *Acanthaceae*

Origin, history and spread

Papaya originates from the lowland tropics of South America; however, the exact area is unknown as the plant has never been found wild. Papaya probably originated from natural hybridization involving *Carica peltata*, which occurs in Mexico. The species was probably widely cultivated by Indians in Mexico and Central America, prior to 1492. The Spanish brought the plant from tropical America to the Caribbean, the Pacific Islands and South-east Asia in the 16th century. Next, it spread rapidly to India and Africa. It reached Hawaii in the 18th century. Papaya was grown in Uganda in 1874. When Europeans first saw papaya, they called the fruit 'tree melon' because of the similarity of the flavours of papaya and melon. The name 'papaya' is derived from the Indian word 'papayana', which means 'hammering', probably referring to the quality of some papaya components to tenderize meat.

A small papaya industry developed in Florida at the beginning of the 20th century, but the viral disease papaya ring-spot ruined the papaya crop. Scientists of the University of Hawaii succeeded in overcoming the disease by inserting a gene that conferred resistance to the virus, into a papaya cultivar. This cultivar was the first genetically modified fruit crop meant for human consumption. Mainly as result of human selection fruit characteristics and growth habit have changed over time. At present, papaya is grown throughout the tropical and frost-free subtropical regions, and has naturalized in many areas.

Botany

Papaya (Fig. 2.32) is a large, unbranched, perennial, short-lived tree-like herb, up to 10 m high, though in cultivation at most 6 m, usually smaller. The stem appears as a trunk, is light green, brownish or dark purple in colour, bearing many prominent leaf scars. The stem is cylindrical, hollow and has a diameter of 30 cm at most. The leaves emerge directly, spirally, from the upper part of the stem with nearly horizontal, succulent, hollow petioles, which are 35–90 cm long. The leaf-blade is orbicular, deeply palmately five- to nine-lobed; the lobes are deeply and broadly toothed. The leaf is glabrous, up to 75 cm across, having yellowish prominent ribs and veins (Fig. 2.33).

The flowers are male, female, sometimes hermaphrodite, axillary, and usually found on different 'trees'. Flowers have five sepals and five petals; they are fleshy, waxy and slightly fragrant. Female flowers are solitary, or in few-flowered cymes borne on short stalks, 3.5–5 cm long. They consist of a cup-shaped calyx, 3–4 mm long, and a corolla of five almost free, cream-white,

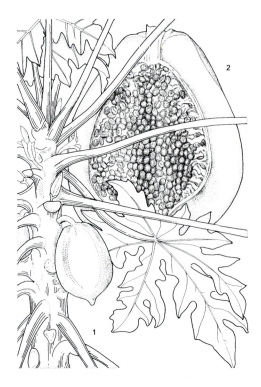

Fig. 2.32. Papaya: 1, part of flowering and fruiting stem; 2, halved fruit (line drawing: PROSEA volume 2).

lanceolate petals; the ovary is ovoid-oblong, 2–3 cm long, with a central cavity in which there are numerous ovules, and five stigmas. Hermaphrodite flowers are more or less similar to female flowers, except they have five or ten stamens with bright yellow anthers. Male flowers are clustered on panicles, which can be up to 100 cm long. Flowers are sessile, clusters of flowers are borne on long stalks; the calyx is small and cup-shaped; the corolla is trumpet-shaped, yellow, 2.5 cm long, and with ten stamens (Fig. 2.33).

Fig. 2.33. Papaya (left to right): leaf-blades (photo: D.L. Schuiling); male inflorescences; detail of branch with solitary female flowers.

Papaya fruits vary in size and form, and in thickness, colour and flavour of the flesh. The fruit is a berry and can be oval to nearly round, slightly pyriform, or elongated club-shaped, 10–30 cm long with diameter of 10–20 cm. The fruits hang on short stalks directly from the stem and shortly beneath the cupola of leaves. When the fruit is ripe, the skin is smooth and yellow; the flesh is yellow, orange or red. Ripe flesh is succulent, aromatic, juicy and sweetish, the flavour somewhat resembling melon. The fruit contains numerous black globose seeds, which are about 5 mm in diameter (Fig. 2.34). Papaya 'trees' may live for 25 years but the yield declines with age.

Development of the root system depends on soil condition. It varies from extensive and dense to small and shallow.

Cultivars, uses and constituents

Despite the great variability in several characteristics of papaya, there were few named cultivars until 1950. Until then, two commercial important cultivars existed: 'Solo' and 'Puna Solo'. Especially in the 1960s, many new cultivars were developed by selection or hybridization. At present there are many good-performing cultivars, some of them are: 'Improved Petersen', 'Sunnybank', 'Hortus Gold', 'Zapote', 'Pusa Majesty', 'Cariflora' and 'Pusa Dwarf'. Next to cultivars, two types can be distinguished: (i) Hawaiian papaya, with a pear-shaped yellow fruit that weighs about 0.5 kg, the plant is relatively small, seldom taller than 3 m; and (ii) Mexican papaya, which has much larger fruits, weighing up to 5 kg, and the plant may be up to 10 m high.

Papaya is mainly cultivated for its ripe fruits; it is used fresh or canned in syrup. It is also made into juice, sauces and jellies; it is an ingredient of several

Fig. 2.34. Papaya (left to right): unripe fruits on tree (photo: J. van Zee); orange-coloured ripe fruits; ripe papaya fruit cut lengthwise showing the numerous seeds inside (photos: D.L. Schuiling).

dishes, pies, ice creams and fruit salads. Young leaves, shoots and fruits are cooked and eaten as a vegetable. In Indonesia, fresh flowers are used as a vegetable. Unripe fruits are sometimes pickled and eaten as summer squash. The bark can be used for making rope, and the leaves are used as a soap substitute. The seeds are edible too, and taste peppery.

Papaya plants produce latex, which is an acrid milky sap. The latex is extracted and dried; it contains proteolytic enzymes: papain, chymopapain and papaya peptidase A. These enzymes are used in several different industries; for example, in the food industry papain is used to tenderize meat and clarify beer, and in the leather industry to make leather flexible and soft. The dried latex is used in manufacture of chewing gum and toothpaste.

In folk medicine, the enzymes are used for the treatment of digestion problems, dyspepsia and other, similar ailments. The leaves and roots of papaya contain cyanogenic glucosides, which develop cyanide. The leaves also contain tannins. Both of these compounds may cause adverse reactions in humans.

The edible part of the fruit contains approximately 87% water, 12% carbohydrates, 0.5% protein, 0.3% fat, 0.7% fibre and 0.5% ash, and also the minerals K, P, Ca and Fe, and vitamins A and C.

Ecology and agronomy

Papaya thrives best in hot, rainy, tropical lowlands. Near the equator it produces good yields up to an elevation of about 1650 m. Optimal day temperatures are 22–32°C. It does not tolerate low temperatures and frost, waterlogging or strong winds. However, papaya is one of the easiest tropical fruits to grow. It has a relatively high water requirement, about 1300 mm, with most water when it is warm and less when it is cool. Papaya needs full sun; shade has a negative effect on the flavour of the fruits. It grows on a wide range of well-drained fertile soils, but preferably on light soils rich in organic matter and with pH 5.5–6.5. The crop is mainly limited to the region between 32°N and 32°S of the equator.

Papaya is mainly propagated by seed, although some cultivars are propagated by cuttings. Seeds germinate, when temperature is about 25°C, 3–4 weeks after sowing. They are grown in containers until the plants are about 20 cm high, which will take about 10 weeks, and subsequently transplanted into the field. Seeds can also be sown in nursery beds before transplanting and seeds may also be planted directly in the field. Since plant sex cannot be determined in seedling stage, most of the male plants have to be removed once they flower, so that the crop mainly consists of female and hermaphrodite plants because they can produce fruits. In commercial cultivation, the spacing is in the range of 1.8–2.4 m × 1.8–4.0 m, which means 1700 to 3000 plants/ha. Weed control can be carried out mechanically, and with herbicides. Plastic mulch is also used to control weeds. For good production, manure or artificial fertilizer is required. In Hawaii, an NPK 8:12:6 mixture at the rate of about 1 kg per mature 'tree' is recommended. Mulching is also profitable.

'Trees' begin to bear when they are 10–14 months old. When too many fruits develop, fruit thinning may be required. As water is the most critical aspect in growing papaya, irrigation will often be needed (Fig. 2.35). Papayas are harvested when the colour of the fruits begin to turn yellow. Mature fruits can be stored at 7°C for about 3 weeks. For papain production, unripe fruits with a diameter of 10 cm are tapped by making vertical cuts and collecting the latex.

Harvesting the fruits is usually done manually by cutting them with a sharp knife. Yields per 'tree' vary from 30–150 fruits per year; in 2005 the world average yield was about 17 t of fresh fruits per hectare. Much higher yields, ranging from 35 to 45 t/ha, are obtained in commercial plantations. In 2005, the total world production of papaya was about 7 million t. Main producers were Brazil (24%), Mexico (15%), Nigeria (12%), India (10%) and Indonesia (9%). Mexico is the largest exporter of fresh papayas.

Fig. 2.35. Drip-irrigated papaya fruit orchard, Barbados (photo: J.F. Wienk).

Peach and Apricot

Peach – *Prunus persica*, Apricot – *Prunus armeniaca*; Rose family – *Rosaceae*

Peach

Origin, history and spread

The origin of peach is China, where the fruit has been cultivated since the early days of Chinese culture. It was already described in the 10th century BC and was the favoured fruit of the old Chinese emperors. The Persians brought the fruit from China to their country and passed it on to the ancient Romans about 2000 years ago, who spread it throughout the Mediterranean region. Old settlers introduced peach into America, Australia and South Africa. Across the whole warm temperate zone from China, through India, Iran and the Mediterranean region, naturalized peaches can be found, often as relicts of ancient cultivations.

Botany

The peach tree is medium-sized, up to 8 m high, with slender branches and brown- to black-coloured bark. The crown is open, spreading and up to 5 m wide (Fig. 2.36). The leaves alternate and are simple, lanceolate, with coarsely serrated margins, 10 cm × 3.5 cm. Flowers are solitary, sessile; the calyx is five-lobed and outside more or less pubescent; the corolla has five pink, rarely

Fig. 2.36. Fruiting peach orchard in New South Wales, Australia (photo: D.L. Schuiling).

Fig. 2.37. Peach: 1, flowering branch; 2, fruit (line drawing: PROSEA volume 2).

white petals; and stamens are numerous. The fruit is a globular to oval drupe (Fig. 2.37). It develops from a superior ovary, and consists of the outer portion or skin (exocarp), the middle portion (mesocarp) of the ovary wall, which becomes fleshy, and the inner portion (endocarp), which becomes a stone that is rough and very hard. The skin of the fruit is velvety and the colour varies from greenish-white to golden yellow, combined with more or less crimson colouring. Some cultivars are completely crimson. The colour of the flesh may be greenish, white or yellow. The size of the fruits varies considerably but may be 7 cm in diameter.

Nectarine is a peach variety with glabrous fruits, a richer flavour and brighter colour. There are white- and yellow-fleshed varieties (Fig. 2.38).

Uses and constituents

When the fruits are ripe, they are soft and juicy with a rich delicious flavour. Fruits are excellent dessert fruits and are eaten raw or cooked, often used in puddings, yoghurts, ice creams, pies, jams, pastries, etc. They are also popular canned in syrup as well as dried.

As well as much moisture the flesh of the fruit contains sugar, protein and minerals, and is rich in vitamin C. Seed kernels are sometimes used as a substitute for almonds; however, they can contain hydrogen cyanide. The toxin causes a bitter taste; very bitter seed kernels should not be eaten. Oil is

Fig. 2.38. Ripe peach fruits (above); whole and halved fruits (below, left to right): peaches; nectarines; and apricots (photos: D.L. Schuiling).

sometimes obtained from the seed kernels and a gum from the stem, which can be used for chewing.

Ecology and agronomy

Most peach varieties require 500 to 1000 h of cold below 7°C to flower normally in spring, so the crop is restricted to the climatic conditions of the warm temperate regions or continental climates with cool but not severe winters. Peaches grow best in well-drained, moist, loamy soil; they do well on limestone, and do not grow in acid soils. The soil pH should be in the range of 6–7. Peaches are very sensitive to soil compaction. Trees growing in loose topsoil underlain by compact subsoil develop a shallow root system. Heavy rains can cause waterlogging that will lead to a shortage of oxygen to the roots. This may result in death of the trees. During dry periods the shallow root system cannot reach enough water, resulting in poor growth and low yields. Peaches flower and bear fruit chiefly on the shoots formed in the previous year; they fruit best when the trees are situated in a sunny position.

The trees are mostly propagated by budding the desired cultivar on to a rootstock; different types of rootstocks are available. It requires about 4–6 years to reach a profitable production and this may continue until trees are 8–18 years old. Planting is generally best done during the dormant period of the trees. The spacing within rows may be 7 to 10 m. To allow full development of the trees, the spacing between the rows should be 7–10 m to permit cultivation and use of machinery without damage to the trees. Pruning of peach trees is required every year for at least four reasons: (i) training the tree to a strong framework; (ii) shaping the tree within a size for convenient harvesting and other operations; (iii) achieving fewer but larger fruits; and (iv) to maintain adequate shoot growth on mature trees.

It is usually necessary to fertilize peach trees annually. N is often the limiting factor in tree growth; peach is an N-demanding tree. How many nutrients should be applied depends on soil fertility and how much is removed in the harvested fruit. Thinning may be necessary to improve the size, grade and finish of the fruits, and to maintain tree vigour. Thinning can be done by hand or chemically. Peaches can be harvested about 105–130 days after flowering. Softening of the flesh by hydrolysis of the pectic materials, development of aromatic constituents and yellowing of the skin are the signs of ripening.

In 2005, the world average yield of fresh fruits of both peach and nectarine was 11.0 t/ha. In commercial plantations yields twice as high are obtained. In 2005, the total world production of peach and nectarine was 11.0 million t of fresh fruits. At present the USA, South America, China, Japan, South Africa, Australia and the Mediterranean region are the major producers of peaches and nectarines.

Apricot

The origin of apricot is western Asia. It grows naturally throughout a very wide area, from western Asia as far east as Beijing. In the Zailinsky Mountains near Alma Ata, wild apricot forests can still be found. The Chinese had probably cultivated apricot by 3000 BC. Alexander the Great brought apricot from Asia to Greece and from there it was taken to Rome. The Romans spread it throughout the Mediterranean and warmer parts of Europe. Early settlers introduced it into North America where it was first grown in California. Apricot is a small deciduous tree growing up to 8 m, but in cultivation generally kept about 3.5 m tall. The species is very closely related to peach. The tree also resembles peach, but leaves are entire, ovate to round ovate with pointed apex, and flowers are mostly white, sometimes pinkish (Fig. 2.39). The shape of the fruit is globose to slightly oblong, glabrous or velvety, diameter 3–6 cm. The colour of the flesh is yellow and less juicy than the flesh of peaches. The skin is yellow to orange, sometimes red-freckled or with a red blush (Fig. 2.38). The apricot fruit was used to symbolize the female genitalia; in medieval France the word 'apricot' was slang for vulva.

Apricots and peaches have more or less similar uses, but only about one-third of apricot production is used for fresh consumption. Like those of the peach, apricot kernels contain constituents similar to those of almond. The

Fig. 2.39. Flowering apricot branches (photo: J.A. Johnson).

kernel contains about 50% oil, which is used in liqueurs, confectionery and as a culinary flavouring. Apricot oil is often used as a substitute for almond oil, which it very closely resembles, in processing cosmetics. Because the oil has a softening effect on the skin, it is used in the manufacture of creams and soaps. The drug 'leatrile' is derived from apricot kernels. It is used to reduce pain and in cancer treatment. At present the latter is controversial because it is thought to be ineffective.

Apricots are grown in a similar way to peaches. They are propagated by budding the desired cultivar on to rootstocks or by seed. Seed needs 2–3 months' cold for stratification. The time needed for germinating can be rather long and can even take up to 18 months. Oats should not be grown between apricots because their roots are considered to have an antagonistic effect on the roots of apricot.

Growing apricots is most successful in mild Mediterranean-type climates, where despite the early flowering of the tree, the danger of spring frost is limited.

Fruits require thinning for optimal size and to prevent biennial bearing. Apricots for fresh consumption have to be picked as ripe as possible, when they are firm-mature, because good flavour never develops in fruit picked prior to physiological maturity. Fully ripe fruits are far too soft to handle without getting bruised. That is not a problem for dried apricots, so they are picked when fully ripe. Apricots have a very short postharvest life of only 1–2 weeks at 0°C and 90% relative humidity.

In 2005, the world average yield of apricot was 6.5 t/ha. High yields of 20–25 t/ha are obtained in the Balkan region. In 2005 the world total production of apricot was 2.8 million t of fresh fruits. Almost half of it was produced in the Mediterranean area. Other important producers are Iran and Pakistan.

Pear

Pear – *Pyrus communis*; Rose family – *Rosaceae*

Origin, history and spread

The genus *Pyrus* is native to south-eastern Europe, Asia Minor, India, north-east and east Asia, and consists of at least 20 species. These have given rise to two main cultivar groups, *Pyrus communis*, mainly in Europe and *Pyrus pyrifolia* in Asia. When the groups arose is not certain; it was probably gradual, first through natural hybridization and later through interspecific hybridization, selection and clonal propagation, carried out by humans. Wild *P. communis* can now be found across Europe, more in the south than in the north. This wild pear is considered to have had a leading role in the origin of the cultivated *P. communis*.

Pears have been known since prehistoric times as proved by remains found in Swiss cave dwellings. In the 9th century BC, Homer wrote in the *Odyssey* that the pear tree was one of the gifts of the gods, grown in the garden of Alcinous. The ancient Romans called the pear 'pyra', from which the name of the genus *Pyrus* is derived. Six hundred years later, pears were certainly cultivated because Theophrastus (370 BC) described the propagation of pears by cuttings, grafts and seeds. Pear was highly valued in antiquity as evidenced by various references of Roman writers including Cato (234 BC), Varro (116 BC) and Virgil (70 BC). Pliny the Elder (AD 23) mentioned at least 35 different cultivated pears. Dioskurides, a Greek medic who lived in Rome in the 1st century, wrote about medicinal herbs and within that framework he referred to winemaking of pears (perry).

The Romans introduced pears into France, Spain, Germany and probably England. The list of fruit trees in Charlemagne's *Capitulare de Villis* (812) contains all the fruit trees that people had to grow in their gardens. About pears Charlemagne writes 'pirarios diversi generi', which means 'several types of pears'. Pears were introduced into North America by French and English settlers. Franciscans took pears to California in the late 1700s; California is still the major production area in the USA. Until the 18th century, most of the pear varieties were either crisp, like apples, or hard-fleshed, fit only for cooking. A Belgian amateur pear breeder named Nicolas Hardenpont succeeded in breeding soft-fleshed, melting dessert pears, such as the cultivar 'Beurré d'Hardenpont'. This was considered to be a revolutionary improvement at that time. Subsequently various cultivars of this type were bred. France, Belgium

Fig. 2.40. Flowers of pear cultivar 'Gieser Wildeman'.

and Germany have been the major centres for pear breeding for a long time. At present, *P. communis* is by far the most important species worldwide.

Botany

Pear trees can reach a height of 20–25 m, although most cultivated pears are dwarf forms. Trees have a tendency to grow very upright and must be trained to develop a spreading growth habit. The growth of the tree is erect or pendulous. The bark is brown to reddish-brown when the tree is young. Older trees have greyish-brown bark with shallow furrows and flat-topped scaly ridges. The twigs are reddish-brown; buds are about 0.5 cm, conical to dome-shaped and sometimes lightly pubescent. The leaves are alternate, simple, ovate with a finely serrated margin, rounded at the base, acuminate at the apex, 2.5 to 10 cm in length, shiny green above, paler and dull below. Flowers appear before or with unfolding of the leaves, and are borne in clusters of six to nine flowers. Flowers are 2.0–3.5 cm in diameter, calyx tubular, with five petals which are ovate and white. The flower has about 20 stamens; the anthers are usually red (Fig. 2.40). The fruit, which in fact is a false fruit, is a large pome (a pome is formed by the enlarged receptacle which surrounds the ovary in the flower); the colour may be green, yellow, russet (a colour formed by combining orange and purple), red or combinations of these. Ripe fruits usually contain four to ten smooth, dark

Fig. 2.41. Pear: entire fruits and cross-sections of the cultivars (left to right): 'Conference'; 'Doyenné du Comice'; 'Gieser Wildeman'; and 'Brederode'.

brown or nearly black seeds in the core (Fig. 2.41). Although transitional forms exist, the shapes of the pomes can be classified into six groups.

1. Round or flattened pears: mostly with a rough skin and russetted.
2. Conical pears: tapering but not waisted, yellow, russetted or red-flushed.
3. Oval pears: mostly russetted.
4. Bergamot or top-shaped pears: mostly with a rough skin and russetted or green when ripe.
5. Pyriform pears with a distinct waist: usually yellow when ripe.
6. Calebasse or long pears: often brown or golden russet, sometimes green.

Others use different classifications because many transitional forms exist in between the forms that are mentioned here.

Cultivars, uses and constituents

The many pear cultivars can be divided into dessert pears and stewing pears. Most cultivars are dessert pears. Dessert pears have to be juicy or buttery, acid yet sweet. They are eaten fresh, in fruit salads, processed in puddings, yoghurt and other desserts, in pastries and in liqueurs. Pears may even be grown in bottles (the bottle is placed around the young fruit on the tree) to become smothered by brandy when they are ripe. Part of the production is canned, or processed into purée and juice.

Stewing pears cannot be eaten fresh because they are hard and lack flavour and juice. Stewed pears form a delicious food when they are cooked slowly in syrup, possibly with cinnamon and a little wine; some cultivars turn red during cooking.

Some of the cultivars are grown for making perry. These pears have a bitter flavour due to the high tannin content. To produce perry, yeast must be added to the juice of the pears to induce alcohol production through fermentation, similar to the making of cider from apples.

A pear, including the skin, contains 83% water, 0.6% protein, 0.3% fat and 14% carbohydrates, as well as minerals, vitamin A, thiamine, riboflavin, niacin and ascorbic acid.

Some well-known pear cultivars are 'Barlett' (syn. 'Williams' Bon Chretien', arose in England before 1770), 'Dr. Jules Guot' (arose in France in 1875), 'Doyenne du Comice' (arose in France in 1849) and 'Conference' (arose in England in 1884). Most of the cultivars grown today are very old.

As pear trees can live up to 300 years, it is possible that very old landraces can still be found near farms and estates. The wood of pear tree is highly valued as timber for making furniture. In north east and east Asia, a number of wild pears exist. Through hybridization, selection and clonal propagation, a number of different cultivated pears arose, called 'Asian pears'. One of them is *P. pyrifolia*. The wild *P. pyrifolia* is the principal progenitor for the cultivated *P. pyrifolia*, which is grown mainly in China and Japan. Cultivars of *P. pyrifolia* range from hard and gritty (with stone cells) pears to delicious, crisp and juicy pears, known as 'nashi' (Fig. 2.42). Nashi is suitable for warmer areas where the winters are mild. It is an erect tree, sometimes spiny, up to 15 m high. Young branches are glabrous, or more often woolly pubescent. The shoots grow in

Fig. 2.42. Oriental pear (nashi): 1, flowering branch; 2, fruit; 3, halved fruit (line drawing: PROSEA volume 2).

flushes. Leaves, inflorescences and flowers resemble *P. communis*. Fruit ripens after 4–5 months. Fruits can be large, with a diameter up to 15 cm. The shape is pyriform to subglobose. The flesh has a crisp and juicy texture and an apple-like flavour; the taste is very sweet. The colour of the skin is golden yellow or silvery green; it turns yellow when mature. The pear is a dessert pear, and has similar properties and uses to *P. communis*. It is the most important fresh fruit in Japan after citrus and apples. Growing nashi is often a very intensive pro-duction system; trees can be trained on a horizontal wire trellis, and branches are tied to the wires. The wires are placed about 1.8 m above the ground. This method requires a lot of pruning.

In the USA, *P. pyrifolia* and *P. communis* were cross-pollinated, resulting in several hybrids. The flesh of these hybrids varies from very hard to very firm with varying levels of grittiness. They are mainly used as processing fruits. A well-known hybrid is 'Kieffer'.

Although *P. pyrifolia* is a tree of warm temperate and subtropical regions, it needs a cold season to break bud dormancy. Without that, the fruit set will be poor. The species cannot withstand frost during flowering and fruit set. *P. pyri-folia* is rather tolerant of drought and thrives on a wide range of soils.

Ecology and agronomy

Pears are grown under a wide range of soil conditions, which vary from deep, well-drained, sandy soils to shallower, heavy and strongly structured soils. It tol-erates a range of soil types if they are moderately fertile; it does not tolerate very alkaline soils. However, pears thrive best on a good, well-drained porous loam, pH 5.5–6.5, and a combination of sufficient moisture and summer warmth. They flower several weeks earlier than apple, so they require warmer conditions earlier in the year. Frost during flowering can cause much damage.

Pear tolerates atmospheric pollution and excessive moisture. Established trees are rather drought-tolerant. Propagation of pears is mainly by budding or grafting on to rootstocks. It is important to use the right rootstock because it de-termines tree size and form. Rootstocks of low vigour are required for dwarfing trees. Today, most pears are grown on quince (*Cydonia oblonga*) rootstocks.

For adequate growth, certain mineral nutrients are required; however too much nitrogen may cause overactive growth of shoots at the cost of fruit pro-duction, so it is important to make an exact fertilization plan. The plan has to be based on several factors such as the expected amount of nutrients that will be removed by the harvested fruits, the expected growth of the trees or bushes, and the amount of nutrients in the soil, found by a soil test. Depending among other things on cropping system and age of the fruit trees, the required amount of nutrients varies widely. Application of nutrients is often split: one application is made at the end of the winter and one at the end of spring. About one-third of the N has to be applied in the postharvest period. Irrigation may be necessary if the required 700 mm of water per year is not achieved. Irrigation may be combined with fertilization, which method is called 'fertigation'.

Pears can be grown as large vigorous trees, as bushes, pyramids, cordons or hedgerows, sometimes with a very high planting density. The number of

Fig. 2.43. Pear orchard (cultivar 'Conference') with V-shaped hedgerows, The Netherlands.

trees per hectare varies from 150 to 5500. Growing V-shaped hedgerows is a development of the last few decades (Fig. 2.43). In orchards, pear trees are grown in rows. The space between the rows has to leave enough room for machinery for cultivating, spraying pesticides, harvesting the fruits and suchlike; it may be 3–3.5 m. The space between the trees within the rows depends on cultivation method. Pruning has to be done in the winter dormant season. How to prune depends on the cultivation method. Pruning is required to keep the trees in the right shape and keep light in the trees; it is also important to get an adequate fruit number, which is important for getting the right size of fruits. Even then, no more than about 5% of the flowers must set fruit for an adequate pear crop. Many flowers do not develop mature fruits because they are not pollinated or young fruits are aborted. If there are still too many fruits on the tree, thinning will be necessary, either by hand or chemically. Most pear cultivars are self-sterile and require cross-pollination by another cultivar to achieve an acceptable yield.

A very serious bacterial disease in North America and Europe is fire blight, which can be most destructive to pears. In the past, fire blight led to big losses in the USA, destroying established pear orchards in one season. At present, good cultivation practices are imperative in fire blight management.

When pears can be harvested depends, among other things, on the cultivar. There are early-maturing and late-maturing cultivars, and a range in between. Pears are one of the few fruits that do not ripen well on the tree and are therefore picked just before the fruits have matured. At picking the fruits can be placed in large bins and brought into cold storage as soon as possible. The average freezing point of a pear is around −2°C; pears kept at −1°C have a long storage life. Use of controlled atmospheres has been found to extend storage life and to improve keeping quality of pears. Pears have to undergo a ripening process after storage; they will ripen when they are allowed to stand at room temperature for several days.

The world average yield of fresh pears was 11.2 t/ha in 2005. In commercial plantation annual yields of over 40 t/ha can be obtained. In 2005, the world production of fresh pears was 19.6 million t. China is by far the largest producer (60%). Other important pear-producing countries are the USA, Spain and Italy. The most important fresh pear-exporting countries are China and The Netherlands.

Pineapple

Pineapple – *Ananas comosus*; Pineapple family – *Bromeliaceae*

Origin, history and spread

Pineapple has its origin in South America, although its exact origin is not known. Colombia and Venezuela are mentioned as places of origin but also Brazil and Paraguay, because wild relatives of pineapple occur in the Parana–Paraguay River area. The maritime tribe of the Carib Indians (other authors mention Tupi Guarani Indians) distributed pineapple throughout tropical America in pre-Columbian times. The Indians, who gave the name of their tribe to the 'Caribbean', brought pineapple to the Caribbean, among others to Guadeloupe. They called the fruit 'anana', which means 'excellent fruit', which explains the name of the genus *Ananas*. Indians cultivated and selected pineapples, they ate the fruit at feasts and used it for tribal rites, and it was also used for making wine. The wide distribution and selection of pineapple types by Indians in pre-Columbian times, also their knowledge of the crop and the cultivation, indicate an ancient domestication, maybe some millennia ago.

Columbus brought the plant in 1493 from Guadeloupe into Europe. Next, Spanish and Portuguese explorers distributed it to India, the Philippines, Malaysia and Africa in the 16th and 17th centuries. In the 18th century, pineapples were grown on a small scale in heated glasshouses in Europe. The first commercial plantation was established on Hawaii in 1885. Hawaii remained the major producer until the 1960s. Pineapple was canned for the first time in 1888 in Malaysia, and canned pineapple was exported for the first time from Singapore to Europe around 1900. The name 'pineapple' is derived from the Spanish word 'piña', which means 'pine' because of the resemblance of the fruit with a pine cone. Carib Indians placed pineapples outside the entrance of their dwellings as symbols of hospitality. Later, the 'fruit' became internationally considered a symbol of welcome; in many locations door frames are decorated with pictures or sculptures of pineapple as a welcome sign.

At present, pineapple is grown throughout the tropics. Pineapple is also grown in greenhouses in the temperate zone.

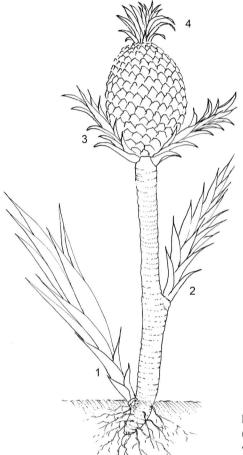

Fig. 2.44. Fruiting pineapple plant (primary leaves left out) showing vegetative parts which are used for propagation: 1, sucker; 2, shoot; 3, slip; 4, crown.

Botany

Pineapple (Fig. 2.44) is a perennial, monocotyledonous, herbaceous plant. The adult plant has a round habit, is 0.8–1.6 m high, with a spread of 1–1.5 m. Pineapple has a short, stout stem, bearing a rosette of waxy, pointed leaves. The leaves are arranged in a tight spiral; the number of functional leaves is 70–80, with shorter, older leaves at the base, gradually increasing in length above, and the youngest leaves at the top decreasing in length. The longest leaves can be 1 m long and 6.5 cm wide. Leaves may be all green, or variously, longitudinally striped with red, yellow or cream. The margins bear usually sharp, upcurved spines. The buds on the stem below ground level produce lateral shoots called 'suckers'; shoots arising from axillary buds are called 'shoots' or 'ratoons', and 'slips' are shoots which are borne on the stalk just below or on the base of the infructescence. These different shoots are used as propagating material for the next crop.

Fig. 2.45. Pineapple plants with immature infructescences, Queensland, Australia (photo: D.L. Schuiling).

At blooming time, the stem elongates and the plant develops a compact inflorescence with numerous small, sessile, purple or red hermaphrodite flowers, each accompanied by a single red, yellow or green bract. The flower consists of three short and fleshy sepals, three petals forming a tube, six stamens, and a pistil with a narrow style and a three-branched stigma. The flower develops parthenocarpically, which means without fertilization, into a fruitlet, which is botanically a berry.

The pineapple 'fruit' is actually a coenocarpium or multiple fruit, which originates from the fusion of many fleshy fruitlets and the extensive thickened axis of the infructescence (Fig. 2.45). The outside of the multiple fruit is a kind of rough, waxy rind formed by the hardened sepals and bracts of the flowers. The rind may be dark green, yellow, orange-yellow or reddish when the 'fruit' is ripe (Fig. 2.46). Commercial cultivars are mainly seedless. The multiple fruit is cylindrical, about 20–30 cm long and 14 cm in diameter, weighing 1–2.5 kg, with yellow flesh.

The apex of the infructescence is vegetative; it consists of a compact tuft of small, short, stiff leaves, called a 'crown', which can be used for propagation. The growth of the plant can be continued after fruiting, by buds in the leaf axils, which can grow into a new plant and come to maturity while still attached to the remnants of the parent plant. Such a plant obtained from an axillary bud

Fig. 2.46. Ripe pineapple infructescences, entire and in longitudinal section.

is called a 'ratoon' plant. In this way, a plant can become more than 50 years old.

The root system of pineapple is shallow and rather limited. At first it consists of adventitious roots. After emergence of the plant, lateral roots are formed; they arise from the axils of the leaf-bases.

Cultivars, uses and constituents

As pineapple has been cultivated for such a long time, a large number of cultivars exist; however, the number of cultivars grown in commercial production is limited. The cultivar 'Cayenne' is far the most widely grown one for commercial production and canning. This cultivar was already cultivated more than 200 years ago in Venezuela by the Maipure Indians. It became the favoured cultivar in Europe because of its well-shaped fruits and fine flavour. Other cultivars which have any importance, often only regionally, are: 'Queen', grown mainly in Australia and South Africa; 'Red Spanish', grown mainly in Mexico, Cuba, Puerto Rico and the West Indies; 'Singapore Spanish', grown in Malaysia; 'Abacaxi', grown only in Brazil; and 'Cabezona', grown in Puerto Rico. *A. comosus* is the only common food plant in the family *Bromeliaceae*, although many ornamentals belong to this family.

The major portion of the grown pineapple crop is canned and exported all over the world. A part of the grown crop is sold fresh and used as dessert fruit.

However, ripe 'fruits' soon deteriorate and must be eaten within 5 days. To extend this period, pineapples are usually marketed half-ripe. The 'fruits' are made into jam, juice and pineapple wine. It is also used for crystallized and glazed fruit. Fresh and/or canned pineapple is used in confectionery, desserts, fruit salads and several dishes; it can excellently be combined with fried chicken. The remnants of the canning process can be used as cattle feed. The organic acids citric, malic and ascorbic can be extracted from the above-ground part of the plant. The stem and the 'fruit' are a source of bromelain, which is an enzyme used to tenderize meat and in pharmaceuticals. However, the acids and bromelain are usually by-products, extracted from the remnants of canning. The leaves of special cultivars yield a strong fibre, used for making fabric and cordage. A variegated pineapple with green, yellow and red stripes is grown as an ornamental, often in a tub.

In folk medicine in Asia, immature fruits have been used as an abortifacient. It is also used to treat warts, abscesses and ulcers. Juice of unripe 'fruits' is said to be purgative.

The edible part of the 'fruit' contains approximately 85% water, 14% sugar, 0.4% protein, 0.1% fat and 0.5% fibre. Other constituents are citric acid, malic acid, ascorbic acid, bromelain, and the vitamins A and C.

Ecology and agronomy

Pineapple is grown in the tropics and the subtropics (Fig. 2.47). The temperature range of the growing areas is 23–32°C; although lower temperatures as low

Fig. 2.47. Pineapple plantation, Queensland, Australia.

as 10°C are tolerated, the plant does not stand frost. A mean annual sunshine duration of 2000 h is achieved in the areas where pineapples are cultivated. The crop will produce 'fruits' under yearly precipitation rates ranging from 650 to 3800 mm, but the ideal annual precipitation should be about 1200 mm and the atmospheric humidity between 70 and 80%. Pineapple is fairly drought-tolerant, due to special water-storage cells. Pineapple is a xerophytic plant, which means that its photosynthetic method is the so-called 'crassulacean acid metabolism'. During the night the plant absorbs carbon dioxide and converts it into acids, which are used in daylight for the synthesis of carbohydrates. This method allows the plant to close the stomata during daytime to minimize water use. The crop does not stand waterlogging. It thrives best on well-drained sandy loams, pH 5.0–6.5, with a high content of organic matter and friable for a depth of at least 60 cm.

The period between planting and harvesting extends further away from the equator and at higher elevations. Pineapples are always propagated vegetatively, except in breeding works. For propagating different material can be used: suckers, shoots or ratoons, slips, crowns and butts. The latter consists of the part of the plant that remains after harvesting the fruit and after removing the base of the stem, roots, leaves and peduncle. For large-scale plantings, slips are preferred. Pineapples are usually planted in double rows, on beds, with a wide path between the beds to allow machinery to carry out field operations. Recommended spacing can be 60 cm between the two rows of the double row and 30 cm between the plants in the row, the path between the beds can be 90–120 cm. It gives a density of 37,000 to 44,000 plants/ha. Yield increases at closer spacing, but fruit size declines. Young shoots are usually planted directly into the field, manually or mechanically.

Before planting the soil has to be prepared thoroughly, weeds have to be removed, the soil should be fertilized and black plastic strips can be laid down on the soil. The shoots can be planted through holes in the plastic. The use of black plastic strips has several advantages: it represses weeds, increases soil temperature and conserves the moisture of the soil.

Manual weeding in pineapple fields is difficult because of the spines on the leaf margins. If strips of plastic or other comparable material are applied in the double rows, weed control is required only in the paths between the beds. Experiments have shown that weeds can reduce yield by as much as 42%. After about 5 months of growth, the crop cover can limit weed growth.

N application, preferably every 4 months, is essential to increase both fruit size and total yield. P is required during the first months of growth, and K is needed for fruit development. How much has to be applied depends on soil fertility. On sandy loam, 9 g N, 1.0 g P and 6 g K per plant is recommended.

Since pineapples flower irregularly, in commercial pineapple plantations the plant hormone ethylene is applied to induce fruiting and to achieve a more uniform time of ripening and harvesting. First harvest varies from 12 to 22 months after planting. In the tropics production is continuous, in the subtropics the 'fruits' are usually harvested in the summer months. The 'fruits' are picked ripe for canning and greener for shipping and the fresh market. For the fresh market, fruits are harvested manually; the rest is mainly harvested mechanically. Per plant, only one fruit is produced. Indicators for the time of harvesting are size, colour and sugar content. After harvest fresh pineapples are washed and waxed prior to trading and shipping. To prevent postharvest diseases, fungicides are usually added to the wax. Pineapples for canning are decrowned, cored, peeled and sliced prior to canning.

Pineapple plants bear for 3–5 years, after that they should be replanted. Yields vary, depending on climate, cultivar and cropping system: it can be up to 75 t of fresh 'fruits' per hectare in the first year; in the succeeding years yields decrease. In 2004, the total world production of fresh pineapples was 15.3 million t. In 2004, Thailand, the Philippines, Brazil, China, India and Nigeria were the main pineapple producers. Costa Rica is the largest exporter of fresh pineapple.

Plum

European plum – *Prunus domestica*, Japanese plum – *Prunus salicina*; Rose family – *Rosaceae*

Origin, history and spread

Numerous species of wild plums grow in the northern hemisphere, from Europe through Asia to North America. Important species for the evolution of modern commercial plums include sloe or blackthorn (*Prunus spinosa*), cherry plum (*Prunus cerasifera*), damson plum (*Prunus damascena*), European plum (*Prunus domestica*), Japanese plum (*Prunus salicina*), chickasaw (*Prunus angustifolia*), Texan plum (*Prunus orthosepala*) and American sloe (*Prunus alleghaniensis*). Sloe is a wild plum of Europe, a small shrub with small blue plums having very acid-flavoured flesh. The name 'sloe' is derived from the Celtic name 'sleha', which means dark blue, referring to the colour of the ripe fruit. It has been proved by archaeological research of Celtic remains, including plumstones, that sloe was eaten by humans in ancient times. Archaeologists measured the lengths of plum-stones and discovered groups of stones that had different average lengths: 7, 9, 12, 15 and 17 mm. This suggests that, as well as sloe, the Celts used other species, probably the wild and/or cultivated cherry plum and damson plum. Cherry plum is native to the Caucasian area. Damson plum is native to Asia Minor. It was first cultivated in ancient times in the region of Damascus. Alexander the Great probably introduced the damson plum into Greece from Persia or Syria. *P. domestica* arose in the Caucasus Mountains near the Caspian Sea as a natural hybrid between sloe and cherry plum. *P. domestica* is the ancestor of many types of European plums. It has been cultivated for at least 2000 years. European settlers introduced European plums in the 16th century into North America. At present damson plums are often considered to be a variety of European plum, *P. domestica* var. *damascena*.

In China, the native *P. salicina* was already cultivated 3000 years ago but it was introduced into Japan only in the 17th century. This so-called 'Japanese plum' spread subsequently throughout the subtropics and the tropical highlands, reaching the USA in 1870. Some examples of native plums of North America include chickasaw, Texan plum and American sloe. Native Americans, prior to the arrival of European settlers, ate wild plums. In recent years, many American plum cultivars have been derived from native species by crossing, often with Japanese plums.

Plums are, next to apples, one of the oldest cultivated fruits. From drawings and literature it is known that the ancient Egyptians, Greeks and Romans ate plums. What species they used is difficult to determine because in the centres of origin so many wild species grew that natural hybridization must have occurred many times.

At present, commercially important plums are divided into two broad categories: European plums and Japanese plums. Globally, European plum is the most important plum.

Botany

The shrub or tree is small to medium-sized, 4–10 m high, crown 4–5 m in diameter, with more erect growth habit than peach (Fig. 2.48). The leaves alternate, and are simple, ovate or elliptic, with acute or obtuse tips and small petioles, 6–10 cm × 2–3 cm. Twigs are glabrous, becoming grey-brown to lustrous red-brown. Flowers are borne mostly in clusters of two or three individuals on short spurs, and solitary or two or three in axils of 1-year-old wood. Flowers are bisexual, white, with a diameter of 1.5–2 cm. They have five sepals, five petals, a single carpel and numerous stamens (Fig. 2.49). Flowers appear before or together with the leaves. The fruit is a drupe, 3–7 cm in diameter, with a

Fig. 2.48. Plum orchard with two cultivars: 'Opal' (left) and 'Reine Victoria' (right), The Netherlands.

Fig. 2.49. Plum: flowering branch of cultivar 'Reine Victoria'.

fleshy yellow to dark red pericarp in which the hard stone is embedded. The stone contains a single seed. The fruits are often known as 'stone fruits', to distinguish them from the 'pip fruits' such as apples and pears. The shape of European plums is oval to round; Japanese plums are round to flattened or conical to heart-shaped and generally larger than European plums. In most Japanese plums the stone is firmly attached to the flesh, in European plums the stone is often free in the flesh. The surface of the fruits is mostly glabrous and waxy. The fruit can range in colour from yellow or red to green, or purplish to dark blue. In general, Japanese plums have rougher bark and more flowers than European plums. They are also more precocious, disease-resistant and vigorous.

Cultivars, uses and constituents

Both commercially and domestically an enormous number of varieties are grown; plums are the most diverse of all stone fruit crops. Both European and Japanese plums are very diverse in colour, internal quality, fruit size and shape.
   European plums are often divided into four groups.

1. Reine Claude or greengage: round, sweet, green, yellow or golden plums, often red-spotted or russetted, high fresh eating quality.
2. Prunes: oval, purple to dark blue plums, very suitable for drying.
3. Yellow egg: small group of round plums, with yellow or red skin and flesh, most often used for canning.
4. Lombard or Victoria: large oval, red blue or pink plums, solely used for fresh eating.

Fig. 2.50. Japanese plum (left to right): entire fruit; longitudinal section without stone; longitudinal section with stone (seed).

Most of the Japanese plums (Fig. 2.50) are grown for the fresh market. Compared with European plums they are usually of inferior flavour. As well as the standard plums, a number of complex hybrids between plums and apricots have been developed including plumcot, aprium and pluot.

Plums are eaten fresh, dried or canned; they are used in jellies, jams, chutneys, sauces, pies, desserts, and for the production of juice. A part of the plum production is used for distilling brandy (slivovitz).

Prune varieties have such firm flesh and such a high sugar content that they are very suitable for drying. They can be dried on the trees or in the sun like raisins, although most of them are dried in hot-air tunnels. These plums are called 'prune-plums' and the dried plums themselves are called 'prunes'. A famous prune is the cultivar 'Prune d'Agen'. The oldest of the prunes is probably the 'Quetsche', a plum from central and southern Europe.

Prunes are eaten raw (back-packers' food), soaked or stewed. Their pulp is used as food for infants.

Fresh plums contain approximately 12% total carbohydrates, 0.7% protein, 0.2% fat and 0.5% fibre; they are a good source of vitamins A, B_1, B_2 and C, and contain minerals including Ca, P, Fe and K. Because of the relatively high fibre content of plums, the juice is often used as help to regulate the functioning of the digestive system. Many *Prunus* species are cultivated as flowering ornamental trees or shrubs. The white, pink, purple or red colour of the blossom is highly valued. In Japan there are famous plum gardens; the blossoms are much used for decoration.

Ecology and agronomy

Different plum varieties are adapted to a wide range of climatic conditions. Generally European plums grow best in the cooler parts of the temperate regions; however, prunes are mostly grown in warmer parts. Japanese plums prefer subtropical climatic conditions. They must be grown in regions where it does not often rain and air humidity is low in the growing season, to prevent diseases.

Both types perform best when grown on deep well-drained soils, ranging in texture from a sandy loam to a sandy clay loam, pH 6.5, although European plums can tolerate heavier clay soils than most tree fruit. Japanese plums are less tolerant of such soils.

European plums are more cold-hardy and have a higher chilling requirement (for breaking dormancy of the buds) than Japanese plums. European plum requires a chilling period of more than 1000 h, Japanese plum 500–800 h. Temperatures below 7°C are believed to be the most effective. Plums need adequate moisture especially during the final growth stage. Plums are mostly propagated by grafting or budding on to rootstocks but they also grow readily from their stones. A number of the best plum varieties have originated as wildings in woods and hedgerows. Among others, clonal selections of *P. domestica* or seedlings of *P. cerasifera* are often used as rootstocks. Most plum cultivars require cross-pollination from a different cultivar. In general fruit set will improve after the addition of a different cultivar as pollinator.

The first years after planting, plums are often pruned into open centre trees. Mature trees should not be pruned heavily. There should be about 25 cm of terminal growth per year on bearing trees. However, trees are kept medium-sized, held to about 6 m by pruning.

Plum trees are often planted at in-row spacing of 3.5–6 m and between-row spacing of 6 m. At present hedgerow planting is becoming more popular. Time to full production for trees is 7–9 years; the yield may then be 18 t/ha. A normal production life of the trees is 25–35 years.

Weed control can be done by mowing, mulching, cultivating and application of herbicides. Grassed orchards or growing a cover crop are good alternatives.

Plums need less nitrogen than other stone fruit, although yield and quality are enhanced by good nutrition. Application of mineral fertilizers depends on soil fertility, but it can be in the form of compound NPK (15:4:12) fertilizer. The rate may be 200–400 g per tree per year, multiplied by tree age in years with a maximum application of 3 kg per tree.

Thinning is often necessary for proper size development, a higher sugar content, and to avoid breaking of branches, especially for Japanese plums, because they often bear crops that are too heavy. Thinning can be done by hand or chemically. Maturing of fruits varies greatly in time; skin colour and firmness are the best indicators. Plums are mostly hand-picked. Plums for canning and prunes for drying are often harvested with tree shakers.

In 2005, the world average yield of fresh plums was 4.0 t/ha. In the same year the world production of fresh plums was around 8.8 million t, which was produced in around 90 countries. The largest producer of fresh plums is China (49%). Other countries with substantial production are Germany, the Balkan region, the USA, France and Chile. Chile is the largest exporter of fresh plums.

Ribes Fruits

Black currant – *Ribes nigrum,* Red currant – *Ribes sativum, Ribes rubrum,* Gooseberry – *Ribes grossularia;* Gooseberry family – *Grossulariaceae*

Introduction

There are about 150 wild *Ribes* species, mainly native to the cool temperate regions of Europe, Asia and North America. Fruits have been gathered since ancient times. Domestication of some species began only about 500 years ago, probably in Germany. European currants were taken to North America by immigrants in the 17th century. However, currants were little grown in North America because they were thought to spread the disease white pine blister rust. Growing currants was even banned by law, although the law was lifted in 1966 in most of the states of the USA. At present, the main commercially grown species are black currant, red currant and gooseberry. The major production takes place in Europe, some in North America. England, Germany, Russia and Poland are important producers.

Ribes spp. form small, deciduous, woody, upright to spreading shrubs, up to 1.5 m high. Stems are usually spineless with the exception of gooseberry. The leaves alternate, and are single, palmately three- to five-lobed. The small greenish or pinkish flowers are borne towards the bases of stems from the previous year and on spurs on older stems. The flower consists of four or five united sepals, four or five united petals, four or five stamens, and an inferior ovary. The stamens are extremely short, arise on the base of the petals and incline slightly towards the style. The flowers are highly attractive to bees. Flowers and fruits are produced on racemes, resulting in loose, drooping clusters of ten to 20 fruits on a 10–15 cm thin stem, called a 'strig'. Individual fruit is a berry and has a thin, smooth, often translucent (with the exception of black currant) skin. Shape of the fruit is often globose in black and red currant and globose to oval in gooseberry. The flesh is very juicy and surrounds several seeds. Dried petals adhere to the fruit at harvest and are difficult to remove. Annual growth is in a single flush in spring. Vegetatively propagated *Ribes* species have shallow roots; the root system is fibrous and can easily be damaged by tillage. Seed-propagated plants form taproots. These *Ribes* species are the hardiest of all fruits from the standpoint of resistance to cold or changing temperatures. Currants should not be confused with the dried currants, which are the dried fruits of a special cultivar of seedless grape.

Black currant

Black currants are grown mainly in Europe. The crop is produced on a woody shrub with shoots that are yellow-brown in the first year, and become black when they age. The leaves are pale green, with numerous aromatic glands beneath. The whitish or greenish flowers are open-campanulate (saucer-like). The fruit varies in colour from black to brown-purple, diameter up to 12 mm.

Fig. 2.51. Inflorescences of red currant (left); infructescences of white, red and black currant (right).

The plant has a strong unpleasant odour. Although most currants have self-fertile flowers, the yield and quality of black currants are improved by cross-pollination by insects. Black currant fruits commonly ripen from the top down. The fruits are usually smaller at the end of the strig and larger at the base where it is attached to the node (Fig. 2.51). Black currant fruit is seldom consumed fresh; the flavour of fresh fruits is generally not appreciated. It is mostly used for making jams, jellies, juice, soft drinks, liqueurs, and for flavouring dairy products. A small quantity is required for confectionery, being used in pastilles and sweets.

The fruit has a high content of citric acid (4%), fructose (2.4–3.7%) and glucose (2.6–3.5%); it is an important source of vitamin C (72–191 mg/100 g), depending on climatic conditions and cultivar.

There are several black currant varieties, some have been grown for many years, and others are new. The varieties differ among other ways in yield, number and size of the fruits, constituents of the fruits, length of the strigs, resistance to frost, resistance to fungi and moment of ripening. Concerning moment of ripening, four groups can be distinguished: (i) early; (ii) midseason; (iii) late; and (iv) very late. Black currant is a perennial crop that takes some years to come into full production but is then expected to remain viable for 10 years or more. Yields may range from 5 to 14 t/ha/year. At present most of the black currant crop is harvested mechanically.

Red currant

Red currants are grown mainly in northern Europe, some in northern America (Fig. 2.52). Cultivated red currants evolved from three species: *Ribes sativum*, *Ribes petraeum* and *Ribes rubrum*. The shrub has a stout main stem and is suitable for growing in standard or cordon form. The leaves are deep blue-green. The inflorescences are mainly carried by buds, which arise around the base of

Fig. 2.52. Field with red currant bushes, The Netherlands.

the shoots that grew the previous year (Fig. 2.51). The greenish flowers are open-campanulate. The colour of the ripe fruits can be red or white, diameter 6–10 mm (Fig. 2.51). Breeders developed varieties of red currant that have the ability to ripen all the fruits on a strig at once. Red currants are eaten fresh or used for jelly, juice, jam, sauces and in pies. They are often used mixed with other fruits that lack sprightliness. For eating fresh, the berries should be left hanging on the shrub for 2–3 weeks after they first turn red, to allow the sugar content to increase. For jelly-making, the fruits should be picked as soon as they turn red. The sugar content is very similar to black currant, the citric acid is 2.5%, vitamin C is 40 mg/100 g. Fruits of red and white currants are rather similar except for the difference in colour and acidity. White currant lacks the red pigment of red currant. The annual yield of red currant can be up to 10 t/ha.

Gooseberry

The most important gooseberry (*Ribes grossularia*) is native to north and central Europe. At present it is grown mainly in northern and central Europe and northern America. The shrub is variable in size and growth habit, widely spreading to almost upright, and most often spiny (Fig. 2.53). The species is sensitive to mildew, however, mildew-resistant cultivars are available. Compared with currants, the raceme is reduced to one to three flowers resulting in one (most often) to three berries. The fruits can be green, yellow, red or purple. Fruit size is 1.5–2.5 cm in diameter (Fig. 2.53). As well as *R. grossularia*, there are more gooseberry species, e.g. *Ribes divaricatum*, which is immune to mildew, making the

Fig. 2.53. Field with gooseberry bushes, The Netherlands (left); fruits of gooseberry (right).

species attractive for hybridization, and *Ribes hirtellum*. Both species are native to North America. Most cultivated gooseberries in North America are hybrids between *R. grossularia* and *R. hirtellum*. In 1984 a new cultivar was selected from a native population of *Ribes oxyacanthoides*, growing in Alberta, Canada. This cultivar produces high-quality, dark red dessert gooseberries with a very good flavour. Moreover, the cultivar is very disease-resistant. Gooseberries are eaten fresh, as dessert fruit, frozen or canned, and processed into pies and preserves. In the past it was also used as an important source of pectin by the confectionery and jam industries. The fruits for processing are often picked while they are still hard, as dessert fruit they may ripen for a longer period. The fresh fruits contain 3.0% glucose and fructose, 0.7% malic acid, 0.7% citric acid and 14–26 mg vitamin C/100 g. There are hybrids between gooseberry and black currant that are larger-fruited and milder-flavoured than black currant: 'Josta' or 'jostaberry' is the best-known hybrid. The life of gooseberry plantations is considerably longer than that of currant plantations. Plantations can still fruit well after 20–30 years. The gooseberry is the first hardy fruit to yield in the year. Yields vary with growth and season, variety and stage of maturity at picking time. Under good conditions an established plantation can yield up to 15 t/ha annually.

Ornamental *Ribes*

The ornamental *Ribes* (*Ribes sanguineum*) resembles the currants but it has more conspicuous flowers. The colour of the flowers can be white, pink or red. Some varieties form wax-coated, bluish-grey berries, in drooping clusters. It is a deciduous, spring-flowering, woody shrub. The shrub can have a height up to 3 m and a similar spread, all depending on pruning. The leaves are lobed and clearly textured. The species is easy to grow on a wide range of soils, and in full sun or partial shade (Fig. 2.54).

Fig. 2.54. Bush (left) and inflorescences (right) of ornamental Ribes (*Ribes sanguineum*).

Ecology and agronomy

Ribes species tolerate a wide range of soils but they grow best in cool, well-drained fertile loam soils with a pH of 6.0–6.5. The plants are sensitive to alkaline or saline soils. Sandy soils are less suitable for *Ribes* species because they dry out too fast, due to their shallow root system. Organic mulch is often desired, both to protect the root system and to keep the soil cool and moist, while also adding humus to the soil.

Ribes species are very hardy and tolerate cold climates with temperatures up to −30°C in midwinter. They do not grow well in hot and dry climates. Temperatures above 30°C may cause leaf injury. The species will grow and produce well in full sun to partial shade. *Ribes* species are usually propagated by hardwood cuttings of 1-year-old wood. The cuttings have to be about 25 cm long; the base should be dipped in rooting hormone and subsequently planted in soil. Seed propagation also occurs. Seeds germinate after being stratified for about 4 months at temperatures just above freezing. Gooseberry can also be grafted on rootstocks of, among others, *Ribes aureum* (syn. buffalo currant, native to the American prairies), which results in a single stem that bears a number of branches at the top, and maintains free of foliage to a height of about 75 cm. This growth habit is also often used in many species of ornamental shrubs.

Ribes species are usually spaced 1–1.25 m apart in rows; the space between the rows depends on mechanical equipment, but can be 2–3.5 m. The species are usually trained as a free-standing bush; these can also be modified into a hedgerow. Sometimes the plants are grown as a single stem (cordon). Once the shrubs are mature, pruning will be necessary. Usually three to five of the oldest branches are pruned each year to achieve rejuvenation of the shrub and subsequently maintain productivity. It also keeps the shrub manageable and healthy.

Pruning should be done in the dormant period. At present a different method of pruning is often used, which is based on a continuous supply of two or three each of 1-, 2- and 3-year-old stems. This method has to start after the first growing season by removing all but two or three stems. This has to be continued after the following growing season by removing all but two or three stems of the current year and the two or three stems that grew the previous year, and so on.

Ribes species have a moderate nutrient requirement. An annual dressing of 8 g N and 3 g K per shrub is usually sufficient. Of course it also depends on actual soil fertility. Phosphorus is needed particularly for root production and it is therefore essential to ensure adequate supplies at planting time. To know how much to supply, soil analysis prior to planting is necessary.

When the supply of water is inadequate it can reduce the current year's crop and even the potential of the crop for the next year. Because of the fibrous, shallow root system, *Ribes* species are ideal for drip irrigation. Water should be supplied until the crop is harvested.

Hoeing between the shrubs and tillage between the rows can control weeds. Placing mulch of organic material also prevents the growth of annual weeds. Besides controlling weeds mechanically, this can be done chemically using a range of herbicides.

About 60–120 days after blooming the crop can be harvested. Harvesting can be done mechanically or by hand. To avoid damaging the fruits, the whole strig should be picked by its stem. After harvesting, fresh fruits can be kept in good condition only up to about 1 week in cold storage. Most commercial production is concentrated in Europe and the former USSR. In 2005, the world production of *Ribes* fruits was 0.6 million t, with Germany and Poland, each accounting for 30%, as major producers.

Rubus Fruits

Raspberry – *Rubus idaeus*, Black raspberry – *Rubus occidentalis*, Wineberry – *Ribes phoenicolasius*, Blackberry – *Rubus × fruticosus*; Rose family – *Rosaceae*

Raspberry

Origin, history and spread

Rubus is a very diverse genus in the rose family. It contains a wide spectrum of wild species, mainly in temperate regions of Europe, Asia and northern America. The species are pioneers of open and disturbed habitats. One of the species is raspberry: *Rubus idaeus* in Europe and *Rubus strigosus* in North America. Some taxonomists regard the two types as subspecies of *R. idaeus*: respectively *R. idaeus* var. *vulgatus* and *R. idaeus* var. *strigosus*.

Seeds of *Rubus* spp. have been found in archaeological sites dating back to the Neolithic period. Pliny the Elder (AD 45) described how the Greeks called raspberries 'ida fruits' after Mount Ida in Asia Minor. The fruits were probably

Fig. 2.55. Field with rows of raspberry canes, The Netherlands.

not cultivated at that time but gathered in the wild. More ancient writers described raspberries, often in a medicinal context. In the 4th century, the Roman Palladius was the first to write about raspberry as a cultivated plant. It was not until the 16th century that raspberry was grown widely in Europe. In North America, raspberry plants were first sold commercially in 1771. Since then, people have been successfully selecting and breeding for types with improved fruits. At present, raspberry is grown throughout the temperate and subtropical regions.

Botany

Raspberry is an erect, semi-erect or trailing, generally thorny shrub, producing renewal shoots from the ground called 'canes' (Fig. 2.55). The canes are reddish-green when young, 1–2 m long, mostly not rooting at the tip when they touch the ground. The bark is brown and scaly on the oldest canes. The canes of *R. strigosus* are more slender and flexible than in *R. idaeus*. The leaves alternate and are pinnated, with three to seven leaflets, 7–25 cm long, with fine prickles or glandular hairs on the petioles. The leaflets are elliptical with toothed margins. The lower surface of the leaves is white.

The flower is greenish, with very small white petals that fall away very quickly. The flower consists of five sepals, five petals, numerous stamens (60 to 90) and a varying number of ovules. Each ovule may develop into a small drupe. The receptacle is elongated into an enlarged cone. The 'fruit' is an aggregate of small drupes attached to the conical receptacle, mostly red, sometimes yellow, soft and not shiny (Fig. 2.56). When the fruits are picked, they separate easily from the receptacle. The receptacle is left on the plant. The root system pene-

Fig. 2.56. Raspberry fruits (left) and blackberry fruits (right).

trates the soil about 1 m; roots have adventitious buds that can develop into suckers, which can be rather invasive.

Uses and constituents

The berries are eaten fresh, cooked, canned, or used in making jam, jelly, syrup, juice, ice cream, bakery products, desserts, wine and liqueur. They are also used in raspberry-scented perfumes and bath products. They are rich in vitamins A, B_1 and C, organic acids (mainly citric, some malic acid), sugar (about 6%, mainly glucose and fructose) and pectin. Raspberry is valued for its high content of antioxidants. Roots and leaves contain tannins and flavonoids. They are claimed to have various medicinal properties; for example, for treating diarrhoea, sore throat and colds.

Ecology and agronomy

Raspberry grows best on moist, well-drained, fertile, deep sandy loam, pH 5.5–6.0, and well supplied with humus. The species is a heavy feeder that will not fruit well on light soils. The subsoil should be loose enough to allow good drainage and permit good root growth.

Raspberry prefers exposure to full sun but is also well adapted to partial shade. Protection from strong wind is necessary. Strong winds can take too much moisture from the canes and the leaves and may dry up the fruits. A structural weak point of the plant is the attachment of the canes to the crown: strong wind can easily cause a breakage.

Different cultivars vary greatly in winter hardiness, although resistance to cold is developed by exposure to cold. It proceeds gradually. Sudden severe cold may kill the buds and the canes.

In the autumn, raspberry canes become dormant (enter a 'rest period'). A cold period is necessary to break the dormancy, when it breaks may vary with region and year.

Raspberry is propagated from suckers, which are taken off the parent crowns and roots and planted in rows about 0.6–1.0 m apart and with 2–3 m between the rows. Posts or trellises to support the canes may be necessary. Planting can best be done in early spring, especially in regions with severe winters.

New canes develop from suckers every year, which bear fruits in the sub-sequent year and then decay. After picking the fruits, the exhausted canes can best be removed. Only about four young canes, which have to be shortened about a third, are kept for the following growing season. Before the summer the surplus suckers should constantly be removed to avoid poor production.

The yield of fruit and size of berries are related to the diameter of the cane. The largest and most vigorous canes produce the best crop. Vigorous growth and development of good foliage have to be promoted by good fertilization, de-pending on soil fertility.

Weed control can be done by tillage or application of herbicides. Tillage should be shallow to avoid damage to the roots. A raspberry planting should be replaced every 6–7 years. In 2005, the world production of raspberries was 0.48 million t of fresh fruit. Poland, Hungary, Russia, Germany, the Balkan region and the Pacific Coast region of North America are the largest producers.

Three more *Rubus* species have some commercial importance.

Black raspberry

Black raspberry is grown mainly in the USA. The species resembles *R. strigosus* in botany and growth habit but has firm black fruits. The fruits are processed into jam and juice, the juice is valuable as a natural colorant. Black raspberry is easy to grow; however, it is a short-lived species, plantings often last only 2–3 years due to diseases. Hybrids have been made between red and black raspberries, giving plants with purple fruit and cane colour.

Wineberry

Wineberry is a native from northern China and Japan. The species resembles raspberry in growth habit but forms much longer canes up to 3 m. Red hairs and many prickles cover the canes. They root at the tips when they reach the ground. The flowers are small with white petals; the relatively large calyx is sticky because of the rather long, red, glandular hairs. The calyx lobes envelop the developing fruits and keep them covered until almost ripe. The colour of the fruit ranges from orange to bright red. The fruits can be used as raspberries.

Blackberry

Wild blackberry species grow mainly in wet areas of the temperate regions. *Rubus ulmifolius* is the most common European species and *Rubus allegha-nensis* is a common North American species. No doubt fruits have been collect-ed from the wild for thousands of years, until today. Real cultivation began in the early 19th century, mainly in North America, where most cultivars were first derived from *R. alleghaniensis*. However, there are many different species and a considerable variation among blackberries. There has been an intense

Fig. 2.57. Fruiting branch of blackberry.

development of new natural hybrid forms, which means that their taxonomy is very complex. This played a part in breeding new cultivars. Because of the complexity, cultivated blackberry is now usually called *Rubus fructicosus* agg.

Blackberry is a climbing plant with long radiating canes up to 4 m long. The canes bear thorns, although there are thornless varieties. Where the cane touches the ground, it will root and form a new plant. The leaves are usually compound, ovate to oblong. The flowers are pinkish white. Older shoots bear the fruits (Fig. 2.57). The fruit is a black cluster of drupes, like raspberry (Fig. 2.56). When the fruit is picked, the receptacle comes away with the fruit and is difficult to remove. The fruits can be used as raspberries. The commercial importance of blackberry is low. Breeders have also developed hybrids between raspberry and blackberry.

Strawberry

Strawberry – *Fragaria × ananassa*; Rose family – *Rosaceae*

Origin, history and spread

Although *Fragaria × ananassa* is now the most important strawberry, some other species played a part in the history of strawberry. *Fragaria vesca* is a wild strawberry, which originates from shady grasslands and forest margins of temperate Eurasia and North America. Archaeologists have found seeds dating from Neolithic, Roman and medieval remains. This species was probably first cultivated by the Romans. It is still widely planted in home gardens. *Fragaria moschata* originates from southern Europe and Russia. It was probably first domesticated in Belgium, France and Germany in the 15th to 17th centuries. The species has an excellent flavour and aroma and is still grown on a small scale because of these qualities. Both *F. vesca* and *F. moschata* can still be found as wild plants in the areas of origin. *Fragaria virginiana* originates from woods and meadows of North America. The early settlers found this species to be grown by Indians. Soon after 1600 the species was introduced into Europe and it is still grown today, especially for making jam. *Fragaria chiloensis* originates from the Pacific coast of North and South America. It was probably first domesticated by the Araucanian Indians in Chile. The plant was introduced into Europe in the 18th century. *Fragaria × ananassa* or garden strawberry was derived from natural crosses between *F. virginiana* and *F. chiloensis* in Europe in the 18th century. In Europe, the two species often grew close together so interbreeding was possible. Modern cultivars are derived from further crossing and selection among the hybrids. At present, the garden strawberry is by far the most important strawberry. It is grown extensively in temperate and subtropical countries. In the tropics strawberries are grown in the highlands.

Botany

Strawberry (Fig. 2.58) is a perennial herb with a very short stem and a rosette of leaves. The plant forms stolons or runners (radiate prostate stems), which bear small leaf rosettes. The leaves are trifoliolate, covered with soft hairs, and glaucous at the lower surface; the margins are serrate. The leaflets are 1.8–7 cm × 1.3–6 cm. The inflorescence is an erect dichasium bearing up to 16 unisexual or bisexual flowers (Fig. 2.59). The flowers have two whorls of five to eight sepals each (5–

Fig. 2.58. Strawberry: 1, flowering plant; 2, 'fruit' (pseudocarp); 3, halved 'fruit' (line drawing: PROSEA volume 2).

Fig. 2.59. Inflorescence and leaves of strawberry.

12 mm × 2.5–4.5 mm) on the margin of a floral cup, also five to eight suborbicular petals 9–12 mm wide, and are white in colour. The number of stamens is 25–37, pistils many. The enlarged receptacle or torus forms a pseudocarp or false fruit (edible part of the strawberry 'fruit'), mostly red, sometimes white in colour. On the surface of the pseudocarp, achenes or seeds are embedded (Fig. 2.60).

Fig. 2.60. Infructescence of strawberry (photo: D.L. Schuiling).

Cultivars, uses and constituents

In many countries new varieties are continually being raised by breeders in the search for more vigorous plants adapted to higher yields, better flavour, resistance to diseases, and local climates. Hybrids have been developed that produce larger yields and larger berries. The wild strawberry *F. vesca* is considered to have the true strawberry flavour; this species is used in breeding programmes to improve the flavour of the fruits of many cultivars. Cultivars are classified in a number of categories based on their response to day length. Most cultivars produce berries seasonally, but there are also ever-bearing cultivars.

Strawberries are often consumed fresh. They are also canned, frozen, and made into jam. A considerable part of the production is preserved as pulp. The pulp is used for strawberry-containing products such as ice cream, pastries, custard and yoghurt. Strawberries contain about 90% water and about 8% carbohydrates; the vitamin C content is about average for fruit. The genus *Fragaria* also contains a few ornamental species.

Ecology and agronomy

Strawberries grow best in full sunlight in mild climates, without extremes of temperature and humidity. Almost any soil can be enriched and improved to create circumstances for good cultivation; however, they prefer fertile, moist and well-drained soil, which is slightly acidic (pH 6.2).

Propagation is vegetative, by taking young leaf rosettes from the stolons. To prevent diseases, leaf rosettes must be taken from certified mother plants, growing at a safe distance from strawberry production fields. Propagation through tissue culture is also possible. Before planting the young leaf rosettes, the soil should be made free from perennial weeds, because it is often very difficult to remove perennial weeds from an established crop. To improve the organic matter content and the fertility of the soil, it is recommended to mix farmyard manure or compost with the soil. At planting, the preferred spacing depends on cultivar and the number of years the crop will be continued, but it can be 22–45 cm apart in the row and 80–90 cm between the rows. For satisfactory development, the leaf rosette should be just level with the soil surface after planting. Plants that are planted too deeply or with roots exposed above the soil often die. Strawberry responds well to nutrients, but if growth is too vigorous, flowering suffers, which will reduce the yield. A yield of 15 t of fresh fruit per hectare removes about 30 kg N, 4 kg P and 60 kg K.

The strawberry field has to be kept weed-free. Weed control can be done either chemically or mechanically, but for a good yield it is important not to disturb the roots, which is hard to avoid because of the shallow root system. Strawberries can also be grown in a mulch of polythene to control weeds (Fig. 2.61). Moreover, this system also conserves moisture in the soil and makes strawing around the plants unnecessary. The purpose of strawing is to keep the fruit clean. During the growing season the stolons must be removed constantly. The fruit ripens within a month after flowering. Harvesting can be done 1 or 2 days after the whole berry turns red, because at that time the berry tastes the best.

Fig. 2.61. A field of strawberry plants grown on plastic-sheet-covered ridges, Queensland, Australia (photo: D.L. Schuiling).

In 2005, the world average yield was 14 t of fresh fruit per hectare. Much higher yields, up to 45 t/ha, are obtained in commercial plantations. Yields of seasonal cultivars are generally lower than those of ever-bearing cultivars. The world production of fresh strawberries was 3.5 million t in 2005. The USA, Spain and the Russian Federation were the largest producers.

NUTS

Almond

Almond – *Prunus dulcis*; Rose family – *Rosaceae*

Origin, history and spread

Although the exact origin of the almond tree is uncertain, it is suggested that almond and its close relative, the peach, probably evolved from the same ancestral species in south-central Asia. With the formation of mountain ranges in the south of Asia, millions of years ago, the peach/almond progenitor evolved into the almond in the arid western part in this region, while peach evolved in the humid eastern areas of south-central China. Almond was already being cultivated in China in the 10th century and in Greece in the 5th century BC. Almond was known in ancient Egypt where it was a valued ingredient in breads served to the pharaohs. Almonds were left in King Tutankhamun's tomb to provide food in his afterlife. Also, the Bible refers several times to almond, such as in the story of Aaron's rod that blossomed and bore almonds, giving the

Fig. 2.62. Almond orchard in Andalucia, Spain.

almond the symbolism of divine approval. The tree is naturalized in western Asia, North Africa and southern Europe, where it is cultivated. It has been introduced successfully into other parts of the world where the climate is suitable, including California, South Australia and South Africa. The almond was brought to California from Spain in the mid-18th century by the Franciscan monks. However, it was not until the late 19th century that the cultivation of almond became successful, due to the development of prominent almond varieties.

Botany

The almond tree is small to medium in size, with an open canopy, usually 3–5 m tall in commercial orchards (Fig. 2.62). Its bark is dark coloured and looks like it has been burned. In cultivation there are relatively few and spineless branches; in the wild, branches are numerous and spiny. Almond and the closely related peach (*P. persica*) are rather similar in appearance. The leaves of both almond and peach are simple and folded in bud, but those of almond are broadest rather below the middle and more minutely toothed, whereas the leaves of peach are broadest about or above the middle; light green, oblong-lanceolate, glabrous, up to 12 cm long and 3.5 cm wide, with rounded teeth. Flowers appear before the leaves, usually in fascicles of one to three, each flower with five pink petals about 2 cm long that gradually fade to white (Fig. 2.63). The almond fruit is an inedible drupe, ovoid-oblong, with a grey-greenish, velvety hull (exocarp), up to 6 cm long, leathery when ripe and splitting open to reveal the single seed (Fig. 2.64). The edible seed is enclosed in the endocarp, which is usually stone-hard, depending on cultivar. The root system of almond is relatively poorly developed.

Fig. 2.63. Almond flowers.

Cultivars, uses and constituents

There are two almond varieties of economic importance: *Prunus dulcis* var. *dulcis*, the sweet almond and *P. dulcis* var. *amara*, the bitter almond. The sweet almond includes probably thousands of cultivars or selections worldwide, although in most regions fewer than ten cultivars contribute substantially to commercial production. The major cultivar in California, occupying around 60% of the almond area, is 'Nonpareil'. Other cultivars, like 'Carmel', 'Mission' and 'Merced', are used as pollinators since almonds are largely self-incompatible.

The various cultivars differ not only in their fruits, colour of flowers, habit, and size and shape of the leaves, but also in their ecological adaptations. For instance, the so-called 'Jordan' almond, a sweet almond variety with a fine flavour, is grown in south-east Spain but is unprofitable elsewhere.

The sweet almond (var. *dulcis*) is the variety grown for its edible seeds, the almond kernels of commerce. Apart from being used whole as a snack food, almonds are much used in bakeries, usually shredded or sliced, and by confectioners (e.g. almond paste and marzipan). The chief components of the edible portion are 4–5% water, 17–21% protein, 48–54% fat, 17–22% total carbohydrate, 2.5–3% fibre and 2–3% ash; it also contains minerals (particularly

Fig. 2.64. Fruiting branches of almond.

Ca and P), β-carotene and vitamins B_1 and B_2. The approximate fatty acid composition of the oil is chiefly oleic (77%), linoleic (17%) and palmitic (5%).

The bitter almond (var. *amara*) is the chief source of almond oil. Almond oil is a non-drying, clear, pale yellow, odourless liquid, with a nutty taste. It is rather similar to the composition of olive oil, but it is devoid of chlorophyll and contains a somewhat larger portion of oleic fatty acid. Almond oil is used: in the food industry; for flavouring soaps and perfumery; in emollient preparations for the skin; and in medicine. The kernel of the nuts and the crude oil contain amagdylin, a bitter-tasting glucoside, which readily breaks down to poisonous prussic acid (HCN). However, the bitterness of the kernel should deter anyone from eating enough to be poisoned. During the refining process of the oil the prussic acid is eliminated. Bitter almond is also used as rootstock for grafting sweet almond.

Outside the production area, almond is also used as an ornamental tree. The wood is used for veneering.

Ecology and agronomy

Almond is adapted to a Mediterranean climate, requiring mild winters and long, dry summers with low humidity. It is usually grown under rainfed conditions, particularly in areas with a winter rainfall of 300–400 mm. Rainfall is deleterious any time from flowering onwards; during bloom it decreases bee activity and therefore fruit set; during fruit development and prior to harvest it induces nut rot and premature fruit opening. Frost during the growing season causes a serious limitation in the feasibility to grow almond: open flowers will be killed below −2°C; young fruits may be killed by temperatures below −0.5°C. Although it will survive midwinter frosts with temperatures as low as −17°C, it has a low chilling requirement during the winter in order to break bud dormancy.

Almond can be grown successfully on a wide range of soil types, preferably on deep, sandy loams with a pH range of 6–8. Less-intensive plantings of the Mediterranean area occur on calcareous, rocky and droughty soil. It is intolerant of wet and poorly drained soils and sensitive to high salinity.

Almond is mainly propagated by budding on to seedling rootstocks. Some areas in the Middle East continue to use seedlings, sometimes grafted with superior plant material.

The trees are planted in rectangular or hexagonal arrangements, with separate rows of pollinators if self-incompatible varieties are grown. Spacings of 7 m × 7 m are commonly used, resulting in about 215 trees/ha. Irrigation after planting is often a necessity. Spacing should be wide enough to permit movement of harvest equipment. Various degrees of pruning are practised after the first year. Pruning at maturity consists of the removal of dead branches, water sprouts and pruning of old canopy branches to provide for the growth of new ones. In practice, trees can be pruned every other year with no or minimal loss of productivity.

Fruiting begins in 3–4-year-old trees, with a maximal production when trees are 6–10 years old. Unlike the relatively short-living, related peach, almond trees can produce for over 50 years and tend to live longer than their planter. Irrigation of almond is preferably done by using micro-irrigation systems such as surface drip, subsurface drip and micro-sprinklers.

Almond trees have high nitrogen and phosphorus requirements, but annual amounts of fertilizer depend highly on soil fertility, presence of a leguminous cover crop, chemical composition of irrigation water and removal of nutrients at harvest. Sometimes leaf tissue analysis is applied to determine the fertilizer requirements.

Fruits are harvested when the flesh starts to split open. In commercial orchards, like in California, the fruits are harvested by mechanical tree shakers, except when trees are still young. In most Mediterranean areas almond harvest is carried out manually.

After harvest the fruits may be dried and de-hulled immediately. The nuts are dried by forced hot air to a moisture content of 5–7%. The nuts are then de-hulled and shelled and sorted for size and appearance. If dry enough, the seeds can be stored for several months. For long-term storage the seeds are frozen.

The average world yield in 2005 was about 1 t of unshelled nuts per hectare. The kernel weight is about 55% of the unshelled nut weight. Today, the cultivation of almond in the USA is concentrated in California's fertile Central Valley, particularly in the areas of Sacramento and San Joaquin. Around 40% of the world total almond production is produced in this region, contributing to around 80% of the total international almond export. Other major producing countries are Spain, Italy and Syria.

Cashew

Cashew – *Anacardium occidentale*; Cashew family – *Anacardiaceae*

Origin, history and spread

Cashew is native to north-eastern Brazil, in the area between the Atlantic rain-forest and the Amazon rainforest. From there the species spread into South and Central America. The Frenchman Thevet, who visited Brazil in 1558, gave the first description of the plant. He observed that the cashew 'apples' and their juice were consumed and that the nuts were roasted and the kernels eaten. In the 16th century, Portuguese colonists introduced it into India and Mozambique. The species was planted in coastal areas, probably mainly for erosion control. Uses for the cashew 'apple' and the nuts were developed later. From India cashew spread to Malaysia and Indonesia. The Spaniards brought it to the Philippines in the 17th century. In most of the areas where cashew was introduced, the species ran wild easily and formed big forests. In Brazil, cashew is known by the Portuguese name 'caju', which is derived from 'acaju', the original name used by the Brazilian Tupi Indians.

The shell of the fruit contains caustic oil, which has irritant and allergenic properties. Touching the fruits may irritate the fingers, it can even cause blisters. South American Indians had discovered that roasting the nuts would remove the caustic oil, allowing the nuts to be cracked and eaten. India developed an improved industrial method to remove the caustic oil, which was the basis for the development of a modern nut industry in this country.

At present cashew is cultivated in many tropical countries.

Botany

The cashew tree is a large, spreading, shrub-like tropical evergreen, 10–13 m high, with a wide dome-shaped crown spreading up to 20 m (Fig. 2.65). In cultivation the tree is smaller and usually only 4–7 m in height. The trunk is mostly single, thick and tortuous. Branches are often winding, lower ones mostly bending towards the ground and often reaching it. The alternate leaves are glabrous, oval to obovate, leathery and thick, 10–20 cm long and 5–10 cm wide. They are distinctly veined and have entire margins (Fig. 2.66). Young leaves often contain red pigment; older leaves are shining dark green. Petioles are 1–2 cm long and swollen at the base. The inflorescences are terminal, drooping panicles of mixed male, female or bisexual flowers. Flowers have five sepals 4–15 mm × 1–2 mm and five petals, which often are striped longitudinally, 7–13 mm × 1–1.5 mm; flowers are fragrant and yellowish-pink in colour. Flowering may occur over several weeks, which means that ripening fruits and flowers can be observed on the tree at the same time (Fig. 2.67). The fruit is a nut, which is borne on a fleshy receptacle. The receptacle becomes extremely big, 5–10 cm long, reddish or yellowish in colour, pear-shaped, and called a cashew 'apple' or pseudo-fruit. Externally and hanging below the 'apple' grows the real fruit, which is a nut, 2.5–4 cm long, grey in colour, and almost

Fig. 2.65. Habit of cashew tree, Seram, Indonesia (photo: D.L. Schuiling).

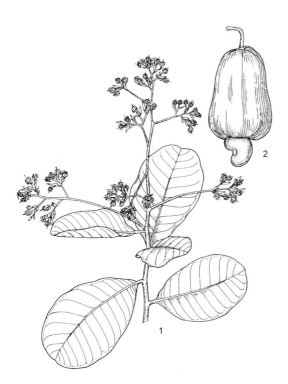

Fig. 2.66. Cashew:
1, flowering branch; 2, fruit
(line drawing: PROSEA
volume 2).

Fig. 2.67. Detail of cashew inflorescence with flowers at anthesis (left) and tip of inflorescence of cashew with open flowers and a fruit with early stage of cashew apple (right) (photos: D.L. Schuiling).

resembling a miniature boxing glove (Fig. 2.68). The nut has a hard double shell: a leathery outer layer and a thin hard inner one, together about 3 mm thick. Between the two layers is a honeycomb structure containing the caustic oil. The nut contains a kidney-shaped seed consisting of a creamy white kernel, covered by a thin reddish-brown testa. The nut develops first; when the nut is more or less full-grown but not yet ripe, the receptacle swells rapidly. Cashew has an extensive root system: the taproot grows up to 3 m deep; the lateral roots are spreading and wide, with sinker roots up to 6 m deep.

Varieties, uses and constituents

Many cashew varieties have been selected in India, based on yield, kernel size and tolerances to drought, fire, low pH and poor soil. A breeding programme in Brazil has produced several high-yielding dwarf clones.

Cashew is a multi-purpose crop; almost all parts of the plant can be used, although it is predominantly cultivated for the nuts. The cashew 'apple' or the enlarged receptacle can be eaten fresh, mixed in fruit salads, preserved as a jelly or preserved in syrup. The juice of the 'apple' can be processed into a beverage or fermented into a wine. The wine can be distilled to obtain a strong alcoholic drink called 'feni'. The kernels of the cashew nuts are roasted and often salted and eaten as (snack) food or used to garnish a variety of sweets and savouries. The kernels contain about 45% edible oil, which can be extracted. The meal remaining after extraction of the seeds is used as animal feed.

The shells yield black (caustic) oil, called 'cashew nut shell liquid' (CNSL), which has many applications. For example, it is used in varnishes as a preservative and waterproofing agent, in oil- and acid-proof cements, in insecticidal

Fig. 2.68. Cashew nut (side view) with young cashew apple (left) (photo: D.L. Schuiling); cashew nuts as marketed (right).

products, in anticorrosion paints, and many more industrial and engineering applications. It is a good natural alternative to petrochemically derived phenol. However, when humans touch the oil it may cause some form of dermatitis. The stem of the cashew tree provides timber that can be used in the manufacture of furniture and boats, and in the production of charcoal. The wounded trunk exudes the so-called 'cashawa gum', a clear gum that is used in pharmaceuticals; because of its insecticidal properties the gum is also used as an adhesive in woodworking and bookbinding to protect against insects. The bark of the tree contains tannins. On a small scale, leaves and young shoots are eaten raw or cooked as a vegetable. In folk medicine in several countries or regions, all parts of the plant are used to treat numerous ailments.

The fresh cashew 'apple' contains approximately 88% water, 0.1% protein, 0.3% fat, 9% carbohydrates, 1% fibre and 0.3% ash. It is a source of P, Ca and Fe. The juice of the cashew 'apple' is rich in vitamins B_2 and C. The seed contains approximately 21% protein, 29% carbohydrates and 35–45% oil; the oil contains 60–74% oleic acid and 8–20% linoleic acid.

Ecology and agronomy

Cashew is primarily a crop for hot tropical lowlands. The optimal average annual day temperature is 24–28°C and the relative humidity 65–80%. Cashew does not tolerate any frost. Annual rainfall of 800–1000 mm will be sufficient; however, in practice it is often 1000–2000 mm. The species is drought-tolerant due to the extensive root system, which gives the plant access to soil moisture. To achieve high yield, a right distribution of rainfall is essential. A relatively dry period during flowering and fruit setting ensures higher yield. Cashew is a hardy crop and can be grown on a wide range of soils, with pH values between 4.5 and 6.5, including poor sandy soils. Cashew does not grow well on heavy clay, waterlogged and saline soils. Soils have to be well-drained and deep.

Cashews are propagated from seed; however, sometimes propagation is performed vegetatively by layering or by softwood grafting. Seeds germinate within 15 to 20 days. Planting selected grafted clones is more expensive, but higher yields can be expected. Before the cashew plants or grafts are planted, a pit about 60 cm deep is usually dug and filled with a mixture of manure and soil. Next, the plants are planted with their roots completely in the manure–soil mixture. If grafts are planted, the graft joint has to remain at least 5 cm above ground level. Spacing depends on cultivar, soil type, cropping system, and especially the age of the trees. It may range from 4 m × 4 m to 20 m × 20 m. Initial spacing can be 4 m × 4 m, which can be maintained for a number of years until the growth of the trees, influenced by proper pruning and training, makes thinning necessary. The final spacing can be, for example, 12 m × 12 m. The ideal time for planting is during the monsoon season, when sufficient rainfall can be expected.

Application of manure and fertilizer can increase the yield considerably. Annual application of 10–15 kg farmyard manure per tree is required, furthermore 500 g N, 55 g P and 100 g K per tree per year. Less fertilizer is required in the first and second year after planting, respectively one-third and two-thirds of the full dose.

A circle of 4 m in diameter around the trunk of the tree has to be kept free from weeds. This can be done by manual or chemical weeding. Mulching the tree basins is also a method to control weeds. A mulch layer also keeps the soil moist, prevents soil erosion and reduces surface evaporation. Intercropping is not common in cultivating cashew; however, annual vegetables like tapioca, pulses, ginger, or coffee and pepper are sometimes grown as intercrops.

To control growth and improve cultivation practices, regular training and pruning is required. The shoots arising below the graft joint, which is the rootstock portion of the tree, have to be removed, especially during the first year of planting. Lower arched branches bending towards the soil, water shoots, disorderly growing branches and dead branches have to be removed as well.

During the first 2 years after planting the inflorescences have to be removed to stimulate vegetative growth. Economic bearing in cashew begins in the third year. Full production is attained by the tenth year and continues until the tree is about 30 years old. The fruiting season can be all year round. From flowering stage to ripe fruit requires 3–4 months.

Ripened cashew fruits fall down to the ground and are collected by hand. Sometimes the fruits are left on the ground for a few days to let the 'apple' dry away. Next, the cashew nuts are usually hand-picked from the 'apples' or the remnants, and dried in the sun. Prior to further processing the nuts are washed. Most of the raw nuts are roasted by the steam roasting process, which means that the nuts are kept under an atmosphere of steam for 30 min. Another method is roasting in hot oil or heating in pans over a fire. During these processes, a part of the CNSL is removed, which makes it safe to crack the nuts without staining the seeds with CNSL. After cooling, the nuts are treated in a cutting machine to separate the shells from the kernels. The cutting machine is usually operated by hand. Shelling in some areas is done entirely manually. Next, the kernels are kept in a hot chamber for about 10 h to be dried. When the kernels are dry, the outer testa is peeled off manually with a knife. After that, the kernels are graded and classified to obtain several varieties. Kernels are usually vacuum-packed or packed in carbon dioxide for export. Vacuum-packed roasted nuts can be stored for about 1 year, and packed in carbon dioxide for about 2 years.

Cashew kernel yields may vary from 0.4 to 2.0 t/ha. In 2004, the total world production of cashew kernels was about 2.3 million t. Since 1994 the world production has more than doubled and is still increasing. Vietnam is the major producer, accounting for 28% of the world production. Furthermore, India, Nigeria, Brazil and Tanzania are important producers of cashew nuts.

Hazel

Hazel – *Corylus avellana*; Birch family – *Betulaceae*

Origin, history and spread

Hazel is native to almost all of Europe and Asia Minor. After the retreat of the ice of the glacial period 18,000–17,000 BC, hazel moved to the north. Pollen analysis has revealed a large amount of hazelnut pollen in north European Boreal strata, dated to 6000 BC. Man has eaten the nuts since the earliest times. The nuts were collected for food by Mesolithic people and were well known by the ancient Greeks and Chinese. In ancient Roman and Celtic cultures, people believed in mystical powers of hazelnut. Cultivation of hazel started mainly in the 19th century. As well as *Corylus avellana* or hazel, there is also cultivation of *Corylus maxima* or filbert. Filbert is a native of south-east Europe. Hazelnuts have been grown in the USA since the late 1800s. The name '*Corylus*' comes from the Greek word 'korys', which means 'helmet' and refers to the calyx covering a large part of the nut. At present, hazel and filbert are grown in temperate and Mediterranean regions.

Botany

Hazel is a small, much branched shrub, up to 5 m high, the branches growing directly from the base with a spreading habit. The shrub maintains an overall upright shape with a rounded head (Fig. 2.69). The leaves are roundish oval, glabrous above, somewhat pubescent on the lower veins, and have a point at the distal end. The leaves are 5–10 cm long and 4–6 cm wide; the margin is double-serrated. Hazel is monoecious and male and female flowers are unisexual. The male flowers are borne in catkins at nodes on 1-year-old wood. The catkins are 3–7 cm long, drooping, and have bright yellow anthers. The small female flowers appear in short, bud-like spikes with crimson styles (Fig. 2.70). At anthesis the styles can remain receptive for up to 3 months. The fertilized flowers develop into clusters of nuts, which turn brown when mature. The nuts are globose or ovoid, 1.5–2 cm long, with a hard shell and partially enclosed by a deeply lobed husk. The kernel is free inside the shell and separates freely when cracked (Fig. 2.71). Filbert resembles hazel but is more robust and about

Fig. 2.69. Young hazel plantation in Kent, England (photo: D.L. Schuiling).

Fig. 2.70. Hazel: female inflorescence with exserted red stigmas (left) and male inflorescences at anthesis (right).

Fig. 2.71. Hazel: 1, young fruits on the tree; 2, ripe hazelnuts, shell halves and kernel; 3, ripe filbert nuts, shell halves and kernel.

7 m high. The nuts are larger and oblong ovoid in shape (Fig. 2.71). The husk covers a greater part of the nut, or the entire nut. Leaves are mostly green but may be red.

Cultivars, uses and constituents

The economically most important cultivars throughout the world have been derived from the species *C. avellana* by selection or cross-breeding. The kernels of hazelnuts have a delicious taste and are eaten as a snack, sometimes roasted or salted. They are used for desserts and confectionery, for example in candy, chocolate, ice cream, nougat and pie. The kernels are also processed into sandwich spreads, often combined with chocolate. The fat content of the kernel can be about 70%, and consists of linoleic, oleic, palmitic and stearic acids. The nuts contain the vitamins E, B_1 and B_2.

Ecology and agronomy

Hazelnuts are grown in regions with a large amount of water, mild, humid winters and cool summers. Cultivation of hazelnut resembles that of fruit in an orchard. They prefer a well-drained soil with good water-holding capacity. The soil pH should be slightly acidic. Propagation may be by seed or vegetative by layering of the suckers which arise from the base of the shrub. Seed can best be germinating in greenhouses. Spacing for seedlings is about 6 m in and between the rows. Hazelnut shrubs begin to bear from 4 to 6 years after planting. Commercial production begins by the sixth year. For a good production it is necessary to remove the suckers regularly. When the nuts begin to drop, they are left on the ground until final harvest. Then the nuts are swept in windrows and collected by a machine. Subsequently, the nuts are cleaned, washed and dried.

In 2005, the world average yield of hazelnut, including filbert, was 1.4 t/ha. Yields of around 2.5 t/ha are obtained in commercial orchards. In the same year, the world total production was 0.75 million t. Turkey (67%) and Italy (17%) are the main producers.

Pistachio

Pistachio – *Pistacia vera*; Cashew family – *Anacardiaceae*

Origin, history and spread

Pistachio is native to the east of the Mediterranean region and Asia Minor. It has been a part of the human diet at least since the late Palaeolithic era. One of the places where pistachio grew in abundance in ancient times was central Persia, nowadays called Iran. The species was first cultivated about 3000 years ago in that area. From there the plant spread in Roman times throughout the rest of the Mediterranean region. More recently, around 1900, it was introduced into Australia and the USA. Pistachios have always been highly valued. For example, it is said that the Queen of Sheba always recognized quality and precious things, and one of these things was pistachio. She liked the 'nuts' so much that she claimed the whole production of her realm for herself and her court. Another example is a Muslim legend which says that pistachio was one of the foods brought to Earth by Adam. In the Bible is said that Jacob's sons brought pistachios to their brother Joseph in Egypt. The name 'pistachio' is probably derived from the Persian word 'pesteh', the Persian name of the plant. At present, pistachio is grown in the Middle East, the Mediterranean region and the USA.

Botany

Pistachio (Fig. 2.72) is a deciduous tree, up to 8 m tall; however, in cultivation it is usually pruned to a smaller size to make harvesting easier. Pistachio is a dioecious tree, so male and female flowers are borne on different trees. Unpruned, female trees are usually smaller than male trees. The bark of the tree is reddish-brown. The leaves are pinnately compound, 10–20 cm long, and consist of three to seven ovate leaflets, which are each 5–10 cm long. The colour of the leaves is greyish-green; at the end of the growing season, the colour changes to yellow. Inflorescences of many small greenish flowers are borne laterally on 1-year-old wood. The inflorescence is a panicle, usually with 13 primary branches, bearing one terminal and five to 20 lateral flowers each (Fig. 2.73).

Flowers have no petals, they have five sepals; male flowers have five stamens and female flowers have a single, superior ovary. The pistachio fruit is an ovoid drupe. It consists of a fleshy outer layer formed by exocarp and mesocarp, which surround the hard endocarp that is beige in colour. Inside the endocarp there is one single green seed or kernel. When the fruit ripens and the outer layer removes, the hard endocarp splits open partially. When the heavy clusters of fruits are ripening, they show a certain similarity to grapes.

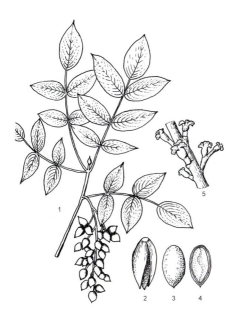

Fig. 2.72. Pistachio: 1, fruiting branch; 2, ripe fruit; 3, seed; 4, seed in longitudinal section; 5, detail of female inflorescence.

Fig. 2.73. Close-up of newly sprouted pistachio twig, showing emerging panicles below terminal leafy bud (photo: D.L. Schuiling).

Species, uses and constituents

There are 11 species in the genus *Pistacia*, of which only *Pistacia vera* is edible. A related species *Pistacia lentiscus* produces resin globules, which are the source of gum mastic, an oleoresin used among other things in chewing gum, perfumes, pharmaceuticals and varnishes. When the resin globules fall from the tree and become buried in the soil, they gradually convert into amber.

Within the species *P. vera*, there are more than 50 male and female cultivars. For the world's pistachio production, the female cultivar 'Kerman' plays a mayor part in cultivation due to the large size of the 'nuts'. The male cultivars 'Peters' and 'Chico' are known as excellent pollinators. In Iran, wild *P. vera* trees growing in forests or as solitary trees are also used for harvesting the fruits. The pistachio kernels are highly valued, both for their green colour and pleasant flavour. The kernels can be eaten roasted and salted as a snack food; they are also used for decorating and flavouring ice cream, nougat and cakes, used in meat dishes, processed into a spread, and used in confectionery such as the well-known Arab/Greek 'baklava'.

The pistachio kernel consists of 5% water, 19% protein, 52% fat, 19% carbohydrates and 2% fibre; it also contains the vitamins A and C, thiamine and riboflavin, and the minerals Fe, Ca, P and K. Pistachio has also been used medicinally; the seeds are thought to be digestive and sedative, and were used as a remedy for sclerosis of the liver, bruises, chest ailments and other problems. Arab people consider pistachio to be an aphrodisiac. Some *Pistacia* species are grown as ornamentals, often in tubs.

Ecology and agronomy

Pistachio (Fig. 2.74) is frost-hardy and withstands temperatures of up to −24°C, although the flowers may freeze at temperatures slightly below 0°C. To break dormancy, a period with low temperatures, at least below 7°C, is necessary. For obtaining good fruit production, pistachio trees require about 1000 chill hours for vernalization. The species is relatively drought-tolerant and thrives well in hot, almost desert-like conditions. However, to produce abundantly, sufficient water is required; therefore irrigation may increase production considerably. Pistachio needs a well-drained soil, which may be slightly alkaline or saline with high lime content and pH.

Pistachio can be propagated by seed, but grafting is the most used propagating technique. The trees are often planted in orchards, sometimes solitary. As pistachio trees come in male and female cultivars, a mixture of the two has to be planted. One male can pollinate about 12 female trees. In orchards, a variety of spacings are used from 4 m to 10 m. In narrow spacings, every other row can be removed about 15 years after planting.

Pistachio trees show a very strong apical dominance, so pruning is required to stimulate branching. Training and pruning is also necessary to obtain a well-shaped canopy of fruiting wood. When the growing conditions are good, a tree requires about 10 years to reach full bearing. The soil underneath the trees can be kept free from weeds by cultivating the soil, regularly mulching the weed

Fig. 2.74. Pistachio trees, still leafless but with sprouting buds, on the slopes of Mount Etna, Sicily, Italy (photo: D.L. Schuiling).

vegetation, or using herbicides. Application of fertilizers varies widely and depends among other factors on cropping system and expected yield. Since pistachio trees grow slowly, they do not require large amounts of nitrogen fertilizer. One single, balanced NPK fertilizer, applied in spring, is mostly sufficient. In the USA, the recommended fertilizer application rate is 40 kg NPK (20:10:10) per hectare for an established orchard.

The time between flowering and harvesting is about 20 weeks. When the mesocarp of the fruits becomes loose from the hard endocarp, the fruits can be harvested. The fruits are harvested by shaking them from the trees with so-called trunk shakers, or by knocking them down with poles. Young trees may become damaged by the trunk shaker and have to be harvested by hand. The fruits are sometimes sun-dried, but mostly dried in forced-air driers to a moisture content of 5%. After that, the 'nuts' are salted and roasted (Fig. 2.75).

The world average yield is about 1.2 t of 'nuts' per hectare. However, yields may vary from 0.2 to 4 t/ha. In 2004, the total world production was about 0.5 million t. Iran and the USA, especially California, are the main producers.

Fig. 2.75. Pistachio nuts as marketed. The edible green seed is inside the semi-split endocarp (photo: D.L. Schuiling).

Sweet chestnut

Sweet chestnut – *Castanea sativa*; Beech family – *Fagaceae*

Origin, history and spread

Sweet chestnut is native to southern Europe, western Asia and northern Africa. In these areas chestnut has been grown for many centuries; perhaps even more than 3000 years. The Greek Xenophon (444 BC) wrote in one of his works that the children of the Persian nobility were fattened on chestnuts. From the writings of Virgil (30 BC) it is clear that chestnuts were abundant in Italy at his time. The Romans probably introduced the tree to the northern countries in Europe. Charlemagne commended the propagation of chestnuts to his people in 812. In his list of cultivated plants *Capitulare de Villis*, the species 'Castaniarios' is mentioned. European colonists brought *Castanea sativa* to the New World. At present, most of the commercially grown chestnuts can be found in temperate and Mediterranean regions.

Botany

Sweet chestnut is a large tree growing up to 35 m high (Fig. 2.76). The leaves are oblong lanceolate (20 cm long). They have serrated edges with forward-pointed teeth, and 15–20 pairs of straight, parallel and prominent veins. The upper surface of the leaf is shiny while the underside is paler and pubescent. The crown of the tree spreads to about half its height. The bark of older trees is dark brown with a network of longitudinal and spiral ridges.

Fig. 2.76. Silhouettes of old, solitary sweet chestnut trees: left with foliage, right leafless (line drawing after John Wilkinson).

Fig. 2.77. Inflorescences of sweet chestnut, male catkins around and female flowers in the centre.

The male and female flowers are borne in separate inflorescences (Fig. 2.77). The male flowers are spike-like, long, slender, yellow-white, and grow upright in catkins. The catkins are 10–20 cm long and conspicuous because of the yellow anthers. The female flowers are green and rounded, growing at the base of male catkins. Female flowers are usually borne with three together; they

Fig. 2.78. Sweet chestnut: 1, branchlet with leaves; 2, fruits; 3, opened husk; 4, seeds (nuts).

are surrounded by green bracts and bear seven to nine red styles. The fruits are nuts, enclosed in green prickly burrs, which turn yellow-brown and become hard when they mature. Nuts are dark brown in colour, often lightly striped (Fig. 2.78).

Species, uses and constituents

Apart from sweet chestnut (*C. sativa*) there are 12 other species. Natural hybrids often occur when more species grow in the same area. Two other well-known chestnuts with edible nuts are Chinese chestnut (*Castanea molissima*) and Japanese chestnut (*Castenea crenata*). Chinese chestnut has dense hairs on the young leaves; the mature leaves are broader than those of other species; and the relatively large nuts have a small hilum. Japanese chestnuts have narrow leaves with glandular scales on the undersides; they produce the largest nuts of all, which have a large spreading hilum. Japanese chestnuts are poorer in quality than Chinese chestnuts. Both species form a smaller tree than *C. sativa*. In southern Europe, chestnuts have been considered a food for poor people for many centuries. The nuts were ground into flour and used in soups, porridges and stews. They can be boiled and eaten as potato, or roasted and eaten as snack food. In the southern countries of Europe, roasted chestnuts are often sold in the streets, and chestnuts are also used as feed for animals such as pigs and poultry. In France, chestnuts are sometimes preserved in sugar to produce the delicacy known as 'maron glacé'. Chestnuts are sometimes eaten as purée combined with roast game.

Chestnut contains approximately 50% starch, 5% oil and 5% protein (on a dry weight basis), as well as vitamins B1, B_2 and C, which means that it forms a rather well-balanced food. Composition of chestnut more resembles wheat rather than any of the other edible nuts such as hazel, almond and walnut. As well as its use for nut production, *C. sativa* is also a magnificent tree for parks and avenues. It also delivers a good-quality timber, resembling oak.

Ecology and agronomy

Sweet chestnut can be grown almost throughout the temperate regions but thrives best in the warmer parts and subtropical regions, where it also yields more nuts. When the tree is established it is tolerant of drought, although lack of water in the juvenile stage will create a shrub and prevent growth to tree size. The tree prefers to grow in full sun. Sweet chestnut grows best in sandy, well-drained soils and is highly tolerant of acidity.

Cultivated chestnuts are generally propagated by budding or grafting on seedling stocks. The seedlings are produced by sowing fresh seeds. Dried seeds may not germinate. Chestnuts have both male and female flowers on the same tree but they are largely self-incompatible, so for achieving an acceptable yield of nuts more trees have to grow together. It is hardly possible to grow sweet chestnuts in the USA because of a fungal disease called 'chestnut blight'. This disease is not a problem in Europe. *Castanea dentata* is a native of North America, which disappeared almost completely because of chestnut blight. To grow chestnuts in North America, it is advisable to use blight-resistant varieties of Japanese or Chinese chestnuts.

In 2005, the world average yield of chestnut was 3.4 t/ha. In the same year, the total world production was about 1.1 million t. China, accounting for 75% of it, is by far the most important chestnut producer at a yield average of 6.6 t/ha.

Walnut

Walnut – *Juglans regia*; Walnut family – *Juglandaceae*

Origin, history and spread

Walnut, also called English walnut, European walnut or Persian walnut, is native from south-eastern Europe to west and central Asia and China. It has been grown since ancient times. The Greeks threw walnuts at weddings as a fertility symbol. The Roman Varro (116 BC) described the presence of walnut in Italy at that time. Probably influenced by the Romans and the drifting tribes in Europe, walnuts spread westwards and became adopted into the culture of the Celts in ancient France. Walnuts also spread eastwards to the Orient. The old caravan routes most likely played an important part in this process. In the 9th century, Charlemagne decided that a walnut tree should be planted in all gardens in his empire. In the 18th century, Spanish missionaries brought walnuts to Chile and California. Also in the 18th century, German emigrants introduced walnuts into Pennsylvania. In 1932, a Canadian, Reverend Paul C. Crath, took a different, more winter-hardy type of walnut from the Carpathian Mountains to Canada.

The scientific name '*Juglans*' means 'Jupiter's nut' and '*regia*' means 'royal'. In an old story it was told that in the ancient days when men lived upon acorns, the gods lived upon walnuts. Thus walnuts were considered food for the gods, demonstrating that walnut was already highly valued a long time ago.

Gaul is the ancient name of France; walnut became established there under the name 'Gaul nut', which later was corrupted in English as 'walnut'. However, in another interpretation it is suggested that the name walnut arose from the Germanic word 'Walhhnutu', which became 'walnut' in the old German language. It means nuts of the 'Walchen'. Walchen is a Germanic word for Celts.

In Europe, there is renewed interest in growing walnut in agroforestry systems. English walnut is grown mainly in temperate and Mediterranean regions.

Botany

English walnut is a tall, broad-headed tree, up to 30 m high (Fig. 2.79). At present dwarf varieties are also grown. The bark is grey and smooth on young

Fig. 2.79. Habit of a young walnut tree.

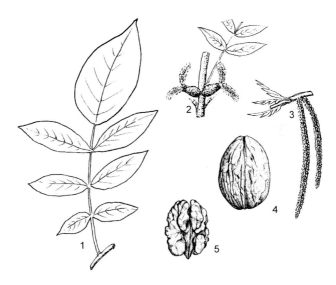

Fig. 2.80. Walnut: 1, leaf; 2, female inflorescences; 3, male inflorescences; 4, seed (nut); 5, endosperm.

trees and fissures with age. The pinnate leaves are alternately arranged. Leaves have five to 11, but generally seven, obovate or elliptic leaflets, 8–15 cm long, which are paired, except at the ends, margins irregularly serrate. The colour of the leaves is red-brown in the young stage, becoming brownish-green when they mature. Leaves are very aromatic. Glandular hairs on the lower side of the leaves produce essential oil, especially in warmer regions. The oil is more or less intoxicating. Male flowers develop from dormant buds of the previous year. They have a small lobed perianth and three to 40 stamens, which are borne in long, pendulous, greenish catkins, 6–12 cm long. The female flowers develop in clusters of two or three, sometimes more, as terminal spikes on the current year's twigs. The fruit is a green drupe, containing a wrinkled stone (nut). The stone contains the edible parts, two crumpled seed-lobes (Figs 2.80 and 2.81).

Species, uses and constituents

There are at least 15 different species in the genus *Juglans*. They have in common that they all produce edible nuts, but only one is of any commercial importance, the English walnut (*Juglans regia*). English walnut is grown in many countries, mainly in Europe, Asia and North America. Three other species are only regionally important. These are black walnut (*Juglans nigra*) and butternut (*Juglans cinerea*) in North America, and Japanese walnut (*Juglans ailanthifolia*) in Japan.

Walnuts are consumed fresh, roasted or salted; they can be used in salads, confectionery, and for making liqueur. Walnut oil, pressed or extracted from the nuts, can be used as edible oil or in making soap. Young green fruits can be eaten pickled in vinegar. The kernel contains 50–65% fat, 18–25% protein,

Fig. 2.81. Unripe walnut fruits.

vitamins B_1, B_2, B_6 and C, and a number of minerals, especially K and P. Fats contain 38% linoleic and linolenic acid. Extract of shells can be used for tanning leather and staining furniture. Shells can also be used as antiskid agents for tyres, blasting grit, and in the preparation of activated carbon. From the outer fleshy part of the fruit a yellow dye can be produced. Walnut wood is high-quality, it is hard and durable and valued very much as timber, for the production of furniture, rifle butts and musical instruments. The tree is often grown as an ornamental in avenues, parks and gardens.

The regionally important species differ as follows.

1. Black walnut – *J. nigra*. Black walnut is a native of North America. The nuts are larger than English walnuts, about 4 cm long, with a thick shell, which can hardly be cracked. Special nutcrackers are made for this purpose. The nuts occur singly or in clusters of two or three. The kernel is rather strong in flavour and is used in pastries, confectionery, ice cream, and suchlike. The tree can grow up to 40 m in height and the trunk can be 2 m in diameter. The trunk has dark brown, almost black bark, with rough ridges separated by narrow furrows. The branches are pubescent. The leaves are dark green and up to 60 cm long, with 11–23 lanceolate, serrate leaflets, which are about 8 cm long. The tree is valuable for timber.

2. Butternut – *J. cinerea*. Butternut is also native to North America. The fruit is 5–8 cm long and 2.5–4 cm across, sticky with glandular hairs. Fruits are produced singly or in clusters of two to five. The four-ribbed nuts are carved with deep furrows. The shell is hard but easy to crack. The kernel has a good flavour. The tree may grow up to 35 m, and the trunk has a diameter of up to 1 m. The bark is grey, old bark separated by furrows, and the branches are pubescent. The compound leaves are 40–70 cm long and have seven to 17 irregularly serrate

leaflets, 5–13 cm long, and glandular hairs. The leaves are pubescent when young, becoming smooth when they mature. Butternut is very winter-hardy. Butternut produces valuable timber; the wood is often used for wood carving.
3. Japanese walnut – *J. ailanthifolia*. The Japanese walnut is native to Japan. It is also called 'heart-nut' because the fruit, the nut, and kernel resemble a heart in shape. The fruits are produced in clusters of ten or more. The nuts are smaller than black walnut and have a rather thick shell; however, they can be cracked rather easily. Japanese walnut resembles butternut in productivity, flavour and uses. The tree has a wide branching crown and can reach a height of about 20 m. The compound leaves have about 13 leaflets and can become 1 m long. The leaf axis and the branches are glandular hairy.

Ecology and agronomy

Walnuts are adapted to northern latitudes ranging from approximately 35°N to 45°N; however for the colder parts of the region, winter-hardy varieties have to be chosen. They thrive best in warm temperate climates, although the numerous walnut varieties differ widely regarding cold and heat resistance. High summer temperatures above 35°C can cause damage to the kernels. Walnut prefers deep, fertile, well-drained, sandy or loamy soils, with a soil pH ranging from 6.5 to 7.0, and with good water availability.

Propagation is mostly by grafting or budding on seedling rootstocks. Planting distance between trees depends on cultivation methods, soil fertility and cultivar, but may be from 8 m × 8 m to 20 m × 20 m. However, to achieve a heavy production quickly, the trees are often planted at higher densities. About 10 years after planting, thinning can start, followed by gradual removal of excess trees in the years to come. After planting the young trees, the tops are often cut back at about 1.5 m from ground level. This stimulates lower buds to grow and form branches, which is necessary to create the desired framework of the tree. Regular pruning has to be done to keep the framework in good shape. Walnuts respond well to cultivation and fertilization. Depending on soil fertility, 100–150 kg N/ha is required for nut production, and in heavily bearing orchards 200–250 kg N/ha may be applied. Other elements should be supplied in proportion.

When the nuts are ripe, the husks crack open and allow nuts to fall to the ground. Trees are often shaken to hasten the process. New cultivars begin producing nuts in 5–6 years. Well-cultivated, older orchards may yield up to 7.5 t/ha. In 2005, China, the USA, Iran and Turkey were the most important producers; the total world production of walnuts was 1.5 million t.

3 Elastomers

Rubber

Rubber – *Hevea brasiliensis*; Spurge family – *Euphorbiaceae*

Origin, history and spread

Hevea brasiliensis originated in the Brazilian Amazon basin; the latex of this tree was already known to ancient Indian tribes in Ecuador and Brazil who harvested the latex from wild trees. The Indians used it to produce torches, boots and bottles. For the Aztecs and Mayas, rubber was religiously and socially an important material. They often burned rubber during ceremonies of human sacrifice. They also produced bouncing rubber balls for a ball game called 'tlachtli'. Indian tribes also tapped rubber from other plant species.

Early explorers, including Columbus, noticed the use of rubber by the Indians, but for more than three centuries it was considered a curiosity and more or less neglected. The French scientist C.M. de la Condamine 'rediscovered' rubber during an expedition in 1736. The French called rubber 'caoutchouc', after 'ca-o-chu', the name given to the rubber tree by the South American Indians, meaning 'weeping tree'.

The name 'rubber' was given to the latex product by the British. They found out that a text, written with a pencil on paper, could be removed by rubbing it with the latex product and subsequently called it 'rubber'. In the 18th century, domestication of hevea rubber began and plantations were developed. That was relatively late in history; hevea rubber can be seen as one of our youngest domesticated major crops.

Gradually, the cultivation of hevea rubber in South America was seriously impeded by a fungal disease caused by *Microcyclus ulei*. Subsequently, the British introduced *H. brasiliensis* seedlings via Kew Gardens into South-east

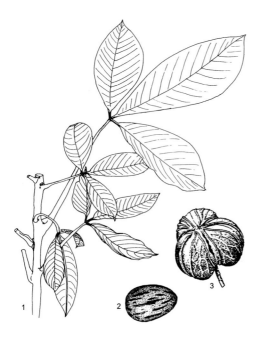

Fig. 3.1. Rubber: 1, detail of a young tree with leaves; 2, seed; 3, fruit.

Asia, where subsequently many plantations were established. Due to technical innovations, the rubber industry began to develop in the 19th century. The invention of a machine called a 'masticator' in 1820 made it possible to soften, mix, and shape natural rubber. In 1839, Hayward and Goodyear discovered that the elastic properties of rubber could be improved considerably by treating the rubber with sulphur and heat, which process is called 'vulcanization'. After the invention of the combustion engine in the 19th century, the demand for rubber increased greatly due to the expanding need for tyres for motor vehicles and aircraft, and also for the production of other goods such as bicycle tyres and raincoats.

Botany

Hevea rubber (Fig. 3.1) is a tree up to 40 m tall, although only 15–20 m in cultivation. The greyish-brown trunk is straight and tapering, up to 50 cm in diameter, and unbranched for most of its height, terminating in a much-branched conical canopy. The bark is smooth to slightly corky. The species contains milky sap (latex) in the inner bark. The leaves are arranged spirally, alternate and are trifoliolate, stipulate and have petioles up to 10 cm long; the leaflets are entire, obovate, pinnately veined, narrowed-acute at both ends, usually about 5 cm × 15 cm, and elliptic-lanceolate in outline (Fig. 3.2). The inflorescences are axillary, pubescent panicles on the basal part of a new flush, half as long as the leaf; the numerous, sweet-scenting flowers are small, monoecious, and have a creamy or greenish-yellow five-lobed calyx, the petals are absent. The female flowers are apical and the male flowers lateral on each branch of the inflorescence. The female flower contains a compound ovary with three locules,

Fig. 3.2. A flush (new leaves) in a rubber tree (photo: J. van Zee).

and topped by three sessile stigmas. The male flower has two rings of five stamens. The fruit is a three-lobed, subglobose, woody capsule, 3–5 cm in diameter, with three carpels; each carpel contains one seed. The seeds are ellipsoidal, 2.5–3 cm long, mottled brown, with 1000-seed weight of 2–4 kg. The species has a well-developed taproot, up to 2 m long with laterals spreading up to 10 m.

Species, uses and constituents

As well as *H. brasiliensis*, at least eight other *Hevea* species occur that yield latex which can be used for producing rubber. However, the yield and quality of the latex of *H. brasiliensis* are far superior to all other species, which is why about 99% of the world's natural rubber production is obtained from *H. brasiliensis*. Since 1979 latex allergy has evolved in Europe and North America, probably due to the use of inferior hevea rubber. For several people it means that touching hevea rubber may cause an allergic reaction or even a severe allergic shock. Therefore a number of 'rubber' products are made from the milky sap obtained from the totally different species *Parthenium argentatum* (*Compositae*), a herb grown in Mexico, also called 'guayule'. The advantage of this crop is also that harvesting can be fully mechanized, contrary to hevea rubber.

There are many different plants that exude milky sap or latex from an injury; the function may be that, after the latex is solidified, it protects the plant against the penetration of fungi and bacteria through the injury.

The main users of hevea rubber are tyre and tube manufacturers; they account for about 60% of production. The remaining part is divided among manufacturers of rubber car components (e.g. joint rings and weather strips); engineering components (e.g. building mounts, anti-vibration mounts and flooring); and many consumer products (e.g. various medical devices, rubber gloves, pacifiers, squash balls, elastic bands, diving equipment, adhesive tape, air beds, mattresses, erasers, condoms, shoes, boots, and many more).

When a plantation has to be replanted, the old trees yield excellent timber for furniture, parquet and other wood products. From the seeds, semi-drying oil can be extracted, which can be used in producing soap and paint.

Fresh latex consists of a colloidal suspension of rubber particles in an aqueous serum. The content of rubber hydrocarbon varies from 25 to 40% and is synthesized in the plant from carbohydrates. Other constituents present in the latex are proteins, resins, tannins, glucosides, sugars, alkaloids and mineral salts. When oil is pressed from the seeds, the remaining press cake or meal is used as feed for livestock.

Ecology and agronomy

As wild hevea rubber is native to the tropical rainforest of the Amazon basin, the species is adapted to tropical lowland conditions with a hot and humid climate. It thrives best at a well-distributed annual rainfall of 2000–2500 mm or more. As wet bark cannot be tapped, rain should preferably fall in the late afternoon and during the night. The plant is sensitive to strong winds. Optimum day temperatures range from 25 to 35°C. Rubber tolerates a wide range of soils but thrives best in deep well-drained, moist, loamy soils, with an adequate moisture storage capacity, and pH 5–6. The plant can stand a short period of waterlogging, provided that the water is not stagnant.

Hevea rubber is propagated by seeds or vegetatively by budding. Seeds have to be sown fresh because they may lose viability in 10–20 days. This period can be extended to about 5 weeks if seeds are stored in moist charcoal or sawdust in perforated polythene bags. To avoid too much variability between seedling trees and seedling rootstocks, so-called 'clonals' are used, which are seeds from clonal rubber plantings. Smallholders, the main producers of rubber, plant mainly seedling trees while commercial plantations use budded rootstocks as planting material. Successful bud-grafting was developed by Van Helten and Maas on Java in 1916 and on Sumatra in 1917, respectively. As a source of bud-wood the highest-yielding mother trees are chosen. Seeds are at first germinated on shaded beds and subsequently transferred to a nursery and planted in the ground. Bud-grafting takes place when seedlings are 12 to 18 months old. Prior to transplanting to the field, plants are uprooted and both the stem and the taproot are cut back to a length of about 50 cm. The bare root stumps are planted at the beginning of the rainy period in large planting holes with a mixture of surface soil, phosphate and manure. A more recent development is raising nursery seedlings in polythene bags and so-called 'green budding' of 4- to 6-month-old rootstocks, which shortens the nursery period but requires greater skill.

In the field the young trees are pruned to restrict development to one single stem, free from any branches, up to 3 m to ensure enough bark that can be tapped (Fig. 3.3). Plant densities may vary at first, but after thinning, the final number of trees is 250–300/ha, which means spacing 6–7 m apart. In some countries, wider spacing is used to enable intercropping, for example with coffee or cacao. In new rubber plantations, the natural ground cover is sometimes maintained and controlled by periodic slashing. Weeding is usually only required in a circle around the young rubber tree, until the weed is shaded out. Instead of maintaining the natural cover, leguminous cover crops such as *Centrosema*, *Calopogonium* and *Pueraria* are often used. Ground cover is profitable because it prevents loss of soil structure and soil erosion. These leguminous cover crops are particularly advantageous because they fix nitrogen, although inoculating the seed or the soil with the proper *Rhizobium* bacteria may be required. Due to N_2 fixation, no N fertilizer may be needed. However, on some soil types, P, Mg and K may be required at rates (per hectare) of 20 kg P, 60 kg K and 20 kg Mg. If no leguminous cover crop is used, 50 kg N may also be given. In general, hevea rubber is less demanding in terms of soil fertility than most other tree crops.

Harvesting begins when the trees are 5–8 years old, while maximum production is reached when the tree is about 15 years old. Commercial latex production is sustained for about 25 years.

Fig. 3.3. Rubber plantation, showing tapping panels on the trunks, Indonesia (photo: J. Ferwerda).

Fig. 3.4. Tapping on a second panel, Indonesia (photo: J. Ferwerda).

Tapping rubber trees is done by hand (Fig. 3.4). The latex is located in the phloem of the inner bark. To harvest the latex, a tapping cut is made at an angle of 30° from the horizontal, from high left to low right. Usually only the basal part of the tree (1.5 m) is tapped. Special knives are used to cut to the correct depth. A common practice is to use a half-spiral cut and to tap on alternate days. The bark above the cut is renewed from the cambium and can be retapped after about 10 years. Latex yields can be stimulated by applying ethylene-releasing chemicals on the bark of the tapped tree. This is mainly a labour-saving practice for obtaining reasonably high yields at a lower tapping frequency. The latex is collected in a cup below the incision. When the latex arrives at the factory it is filtered and then either coagulated with formic acid for the production of sheet rubber, crepe rubber or block rubber, or just concentrated by centrifugation.

It is marketed as natural raw rubber. In South-east Asia the average annual estate latex yield is about 1.5 t/ha. On smallholdings yields are about half this amount. In 2004, the total world production of natural rubber was about 9 million t. This is about 40% of the total world consumption of rubber. The remaining 60% is synthetic rubber. At present, Thailand, Indonesia and Malaysia are the major producers of natural rubber, together accounting for about 80% of the world total production, while the production from South and Central America is at present less than 3%.

4 Fibre Crops

Cotton

Sea island cotton – *Gossypium barbadense,* Upland cotton – *Gossypium hirsutum;*
Mallow family – *Malvaceae*

Origin, history and spread

The genus *Gossypium* contains about 34 species, from which only two are at
present commercially important: 'upland cotton' and 'sea island cotton'. These
species probably originated in the southern part of North America, the northern
part of South America and Central America. Discoveries on archaeological sites
in Mexico and Peru suggest that domestication began around 5000 BC. In
ancient times, all sorts of different cotton species spread to many countries.
Around 3000 BC cotton was already cultivated in Africa, India and Pakistan. In
Africa, cotton was first spun and woven in the ancient Nubian Kingdom. Two
of the species used were *Gossypium harbaceum* and *Gossypium arboretum.*
In the 16th century, Spanish and Portuguese settlers recognized the superiority
of the cotton species in the Americas. At first they brought perennial types to
many countries, but in the long run they were replaced by the annual types
Gossypium hirsutum and *Gossypium barbadense,* the former becoming by
far the most important species.

One of the steps in processing cotton is separating the seeds from the fibre,
called 'ginning'. Doing this by hand is very time-consuming. In 1794 Eli Whitney
invented the 'saw gin'. This machine removed the seed from the fibre with a
spiked cylinder that pulled the fibre through wooden slots which were too narrow
for seeds to pass through. This invention was revolutionary; processing cotton
using a saw gin could be done 50 to 100 times faster than processing by hand.
It made mass production of cotton possible, so many new plantations were

Fig. 4.1. Upland cotton: 1, flowering branch; 2, flower in longitudinal section; 3, fruit; 4, opened fruit (line drawing: PROSEA volume 17).

developed and the demand for cheap labour grew, especially in the south of the present-day USA. A negative aspect is that the expansion of cotton cultivation is inextricably associated with the slave trade from Africa. The abolition of slavery caused the Civil War in North America and cotton production declined dramatically. The result was that cotton supplies from North America were cut off, which stimulated the cultivation of cotton in many tropical and subtropical countries, and the warmer countries of the temperate regions.

At present about 80% of the world cotton production is 'upland cotton' (*G. hirsutum*) and about 15% is 'sea island cotton' (*G. barbadense*).

Botany

Cotton (Fig. 4.1) is originally a perennial, but nowadays mainly an annual, much branching, shrub-like herb, covered with glandular hairs, and up to 3 m high. The species usually has a single stem with dimorph branches, which means both vegetative and fertile fruiting branches. The fertile branches are in fact sympodial, with a succession of short lateral branches. The leaves are spirally arranged, they are hirsute and palmately three- to five-lobed or parted; the leaf-blade is cordate, as broad as long and 7–15 cm across; the margins are entire. The petiole is 3–8 cm long. On each fertile branch, six to eight solitary flowers may arise, although with intervals. The three bracteoles around the flower, called the 'epicalyx', are leafy, pointed, highly serrated and toothed; they persist beyond maturity. The creamy-white to yellow flowers are large and showy (Fig. 4.2). Flowers consist of a reduced five-dentate or five-lobed calyx; a corolla of five large joined petals; petals 4–8 cm long, sometimes with a red spot at the base; more than 100 stamens joined into a staminal column, which surrounds the

Fig. 4.2. Flowers of sea island cotton (left) (photo: B.P. Schuiling) and upland cotton (right).

pistil; the ovary is superior. The fruit is a capsule (often called a 'boll'), light green to dark green, globular to avoid, opening loculicidally, subtended by the epicalyx, 4–6 cm long, having three to five valves with a smooth surface and many black gland dots (Fig. 4.3). The fruit contains usually 36 seeds, which are about 1 cm long, ovoid and dark brown; 1000-seed weight is 100–130 g.

Fig. 4.3. Fruits of sea island cotton (left) (photo: J. van Zee) and upland cotton (right).

Fig. 4.4. Upland cotton: shrub (left); close-up of mature fruit opened in four, exposing the packages of seed fibre ready to be picked (right) (photo: D.L. Schuiling).

On the epidermis of the seeds, two kinds of hairs arise: (i) long hairs called 'lint', up to 5 cm in length; and (ii) very short hairs which are strongly attached to the seed, called 'fuzz' (2–7 mm) (Fig. 4.4).

Cotton develops a long taproot with numerous lateral roots; the root system may penetrate the soil as deep as 3 m.

Cultivars, uses and constituents

There are many cotton cultivars, mainly classified on the length of lint. In the textile industry the length of the fibre is expressed in the term 'staple'. Five different categories can be distinguished: (i) short staple; (ii) medium staple; (iii) medium long staple; (iv) long staple; and (v) extra long staple. The bulk of cotton lint is used for textile manufacture; it also supplies yarn, twine, tyre cord, and cordage. Fabric made of cotton has several good characteristics. It is strong, well washable, durable, can be printed easily, pleasant to wear (cool), and it can be mixed readily with synthetic fibres. Therefore, cotton is economically the most important fibre crop. Cotton fibre contains 88–96% α-cellulose, 3–6% hemicelluloses and 1–2% lignin. Cotton seed contains 18–24% oil and 16–20% protein; the oil is semi-drying, used as a salad and cooking oil and in the manufacture of margarine, soap and paints. The main constituents of the oil are linoleic, palmitic and oleic acids. The oil is extracted from the seed by pressing. The pressed cake is used as an important protein concentrate for livestock. However, the seed contains 1–2 % of the poisonous gossypol, which is toxic to pigs and poultry but not deleterious to ruminants. Gossypol is a polyphenol that brings about a dark colour and a strong scent to the crude oil. Plant breeders are developing new cultivars that have reduced gossypol content. The seed hulls can be used as roughage or bedding for livestock. Fuzz is used in the manufacture of felts, upholstery, mattresses, wicks, carpets, surgical cottons, rayons, paper and plastics. Cotton stalks can be manufactured into paper and paper board; when dried they can be used as household fuel.

In folk medicine, cotton seed has been used to treat nasal polyps and uterine fibroids; gossypol has shown anticancer activities; tea made of seeds is used against bronchitis, diarrhoea and dysentery; flowers are considered diuretic and emollient. Gossypol is used in China as a male contraceptive.

Ecology and agronomy

Cotton has a relatively long growing season, and needs about 200 frost-free days. Optimum day temperature for growth is about 32°C, with a minimum temperature of about 12°C and a maximum temperature of about 39°C. The species cannot withstand frost and does not tolerate shade. As the species evolved in the hotter and drier part of the world, production is highest in irrigated desert areas. Cotton grows from sea level up to 1200 m altitude. It is grown mainly in zones with much sunshine during the growing and ripening season of the crop. Cloudy regions are less suitable because the vegetative phase of the cotton plants may be prolonged and the yield reduced. The average rainfall during the growing season should be 500–1500 mm. When annual rainfall is below 500 mm, irrigation is required. Cotton is sensitive to waterlogging. During ripening, the weather should be dry because rainfall during the period when the fruits are open decreases the fibre quality.

Cotton is adapted to a wide range of soils but prefers a deep, friable, moisture-holding sandy, loamy or clayey soil, with high organic matter content and a pH of 5–7. A suitable cotton soil should be deep enough to allow taproot growth to a depth of about 3 m. Cotton is intolerant of a shallow soil.

Cotton is propagated by seed; using high-quality seed is essential. For unhindered sowing, the remaining fuzz has to be removed from the seed mechanically or by acid treatment. Before sowing the soil has to be cultivated deep and thoroughly to allow proper development of the taproot, and to reduce weed competition during the early development stage of the seedlings.

Because of the ability of the plant to develop many branches, it can be grown successfully over a range of plant densities. Densities used depend on cultivar, climate, soil and cropping system. Final spacings are 50–120 cm between rows and 20–60 cm in the row. Seed rate is 10–20 kg/ha. Cotton seedlings do not compete well with weeds; good weed control remains essential until the plants have formed a closed canopy. Weed control can be done by harrowing, hoeing or chemically. Nutrient requirements of cotton are moderate. The uptake for a yield of 2000 kg of seed cotton per hectare is approximately 125 kg N, 21 kg P and 79 kg K. The seed cotton itself removes 48 kg N, 8 kg P and 17 kg K. The nutrients in the crop residues can be left on or returned to the field as manure. Half the N and all the P and K should be applied just before sowing, which stimulates the vegetative development of the plants. The second N application is carried out about 2 months later to stimulate fruiting.

Harvesting can be done 4–6 months after sowing; the interval between fertilization and opening of the fruit (or boll) ranges from 50 to 70 days. For smallholders, and on plantations in countries where the labour costs are low, harvesting the bolls is usually done by hand. Hand-picking is considered to produce the best-quality cotton; usually three pickings are carried out. When labour costs are high (e.g. in the USA) the bolls are picked in one go, by machines. To avoid the harvested product being contaminated by leaves and weeds, chemical defoliation is often applied prior to mechanical harvest. After harvesting, the lint is removed from the seed, cleaned and baled. Subsequently it is usually sent away for spinning. The fuzz may stay on the seedcoat, or can be removed by special machines or by an acid treatment. Finally, oil is usually extracted from the seeds.

In 2004, yields of seed cotton (seed cotton is the combination of lint and seeds) varied considerably, from 620 kg/ha (Algeria) to 4850 kg/ha (Australia). In the same year, the world seed cotton production was about 70 million t. The total world production of cotton lint was about 24 million t. The main producing countries are China, the USA, India, Pakistan, Uzbekistan and Brazil.

Flax

Linen flax – linseed – *Linum usitatissimum*; Flax family – *Linaceae*

Origin, history and spread

There are strong indications that flax originated in India because of the large diversity of different forms of flax in the Indian subcontinent. From there it was spread to the north and the west by nomads and traders. The cultivation of the crop was initiated in the Mediterranean region and Asia Minor. In Babylon, paintings in burial chambers of flax cultivation date from around 3000 BC. The ancient Egyptians used linen for clothing and for the wrapping of mummies around 2300 BC. The Phoenicians probably brought the plant to Europe, where it became widespread during the next centuries. Flax is mentioned many times in classic Greek and Roman writings, and also in the Bible. In the Stone Age, flax was already grown in Europe. In Europe and the Mediterranean region, varieties developed into linen flax, while linseed varieties were developed in warmer climates of Asia and India. Although the main purpose of growing flax in Europe was the production of fibre, the seeds were also used for oil production or were eaten. For example, in Denmark, linseed was found in the stomach of the mummified Tollundman. In 812, flax was mentioned in Charlemagne's *Capitulare de Villis*. The early colonists took flax to North America and later to New Zealand and Australia. In the 19th and 20th centuries the cultivation of flax declined due to the rise of cotton cultivation and the appearance of synthetic fibres. Today, linen flax is grown in cool temperate regions and linseed is grown in warmer climates.

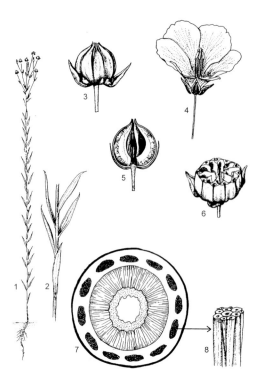

Fig. 4.5. Linen flax: 1, plant habit; 2, part of a stem with leaves; 3, fruit (capsule); 4, flower; 5, longitudinal section of the fruit; 6, cross-section of the fruit; 7, cross-section of the stem with fibre bundles; 8, fibre bundle.

Botany

Flax (Fig. 4.5) is an annual herb that grows to a height of 40–120 cm, depending on cultivar, climate, plant density, soil structure, fertility and available moisture. The stem is thin, erect and wiry, and approximately 30 groups of flexible fibre bundles are embedded within the stem cortex. Leaves are small and alternate, simple, lanceolate, glabrous, 2–4 cm long and greyish-green in colour. There is little branching, except in the upper 12–24 cm of the plant; linseed has more branches than linen flax. The branches are terminated by flowers, forming cymose inflorescences. The flowers are bisexual and measure about 2 cm across. Each flower has five small lanceolate sepals, five blue or white petals, five stamens, and an ovary with five erect styles (Fig. 4.6). The flowers are usually self-pollinating although cross-fertilization is possible. The fruit is a rounded capsule, yellow-brown in colour, containing about ten seeds. The seeds are oval in shape, flat, very shiny, about 4–5 mm long, and brown in colour; the 1000-seed weight is 5–6 g. Flax has a branched taproot, which may extend to a depth of about 1 m.

Cultivars, uses and constituents

Flax cultivars can be classified into two main groups: (i) cultivars grown for the stem fibres (linen flax); and (ii) cultivars grown for the oil of the seeds (linseed) and on a smaller scale for the high mucilage potential of the

Fig. 4.6. Flowering field of linen flax, The Netherlands.

seedcoat. A small group of dual-purpose cultivars is grown for both the fibre and seed. Short-stemmed varieties with large flowers are used as ornamentals.

Flax products have many applications; for example, the fibre is used for manufacturing textile and geo-textiles, painting canvas, paper and various types of twine. There is an increasing interest in applying linen to fashion.

Linseed oil is a drying oil used in the manufacturing of linoleum, paint, printing inks, varnish and cosmetics. Cold-pressed oil can be used in foods. The fatty acid composition of linseed oil is stearic and palmitic acids 6–16%; oleic acid 13–36%; linoleic acid 10–25%; and linolenic acid 30–60%. The residue left after oil extraction is used as a protein concentrate for livestock. It is important to be aware of the presence of the glucoside linamarin in the residue, because in the presence of the enzyme linolase it may form the poisonous hydrogen cyanide. Heating the residue prevents this problem. The seedcoat absorbs water easily and produces sticky mucilage. This mucilage is used in products for treating digestive complaints.

Ecology and agronomy

Flax thrives best in the cool temperate climate zone. It requires at least 150–200 mm rainfall during the growing season, preferably evenly distributed. The total crop duration is normally 90–120 days. Flax grows best on heavy loam soils that retain moisture well and with a pH of 5 to 7.

Flax is a long-day plant. Flax must be seeded in moist soil, 2 to 3 cm deep, in rows 15 to 20 cm apart. A seeding rate of 30–50 kg/ha is

Fig. 4.7. Almost mature linen flax in the field.

recommended for linseed, to obtain about 300 plants/m². Linen flax needs a higher plant density to avoid branching of the stem and thus obtain good fibre quality and yield (Fig. 4.7). The recommended seeding rate is 90–120 kg/ha. Extreme high seeding rates should be avoided because they may promote lodging. Early sowing produces the best yields. Seedlings can withstand temperatures down to −3°C. Since flax is not a very competitive crop, weed control is important and can be achieved through pre- and post-emergent herbicides.

Flax requires relatively small amounts of nutrients; too much nutrient can decrease fibre quality. Nutrient application depends on a range of factors but it may be 35–80 kg N/ha, 20–40 kg P/ha and 50–75 kg K/ha; using organic manure is better than using artificial fertilizers.

Linseed is harvested when fully ripe. Seed yields vary but can be up to 2 t/ha. From 10 kg of seed, about 2 l of oil can be obtained. The fibre of the straw can be used for the manufacture of paper and cardboard.

Linen flax is harvested before the seed is ripe, preferably by pulling the stems. After deseeding the straw, it may be dried in sheaves. Next starts a process to separate the fibres from the rest of the stem. First, retting (rotting) is needed, in tanks of warm water or by exposing the stems in the field to the prolonged action of dew (Fig. 4.8). In retting, the middle lamella of the cell walls breaks down. To obtain the fibres, retting is followed by drying, scutching, beating, hackling and combing.

Yield of fibre varies but may be 0.7 t/ha. Today, Canada is the world leader in the production and export of linseed; most linen flax is grown in Europe and China. In 2004, the total flax fibre production was about 0.79 million t; linseed production was about 2 million t.

Fig. 4.8. Pulled linen flax in the field, exposed to the action of dew for retting.

Hemp

Hemp – *Cannabis sativa*; Hemp family – *Cannabaceae*

Origin, history and spread

Since prehistoric times hemp has been an important plant for its fibre, oil, medicine and as a recreational and religious drug. The exact origin of hemp and its cultivation are not very clear; however, both historians and archaeologists agree that hemp was one of the first non-food crops to be cultivated. Hemp probably originates from Central Asia. There is strong archaeological evidence of the widespread use of hemp in China as an economic crop by around 4500 BC. The vast majority of hemp fibre remains have been recovered from archaeological sites in China. The oldest written record of the use of hemp is a Chinese herbal from the third millennium BC, in which the medicinal use of hemp was described. In a tomb in the Shaanxi Province of China in 1957, archaeologists discovered a piece of paper containing hemp fibre, dated to 140–87 BC. This is considered to be the oldest piece of paper ever recovered. According to Vavilov, domestication of hemp probably occurred independently in several centres in northeast Asia around six millennia ago. By around 1000 BC, hemp had probably migrated west and south with nomads and traders and spread over India, the Middle East, Asia Minor, Africa and Europe. In Egypt, the presence of hashish was found in the body tissues of mummies, dating back to about 1000 BC. The Greeks and Romans used hemp for the production of rope and coarse fabric. They probably did not use it as a drug. In Persia and Arabia they certainly used hemp as a drug because the term 'hashish' is Arabian and taken from 'hashish

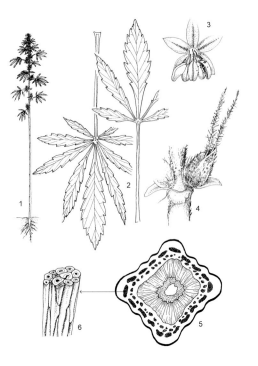

Fig. 4.9. Hemp: 1, plant habit; 2, leaves; 3, male flower; 4, female flower; 5, cross-section of the stem with fibre bundles; 6, fibre bundle.

al kief', which means 'dried herb of pleasure'. There is considerable evidence for the diffusion of hemp in Europe and the Middle East. In Europe the plant was grown during many centuries for the production of ropes, fabric and rigging of sailing vessels. From the 16th to the 18th centuries, hemp was an important fibre crop in Europe and North America. Hemp production declined in the 19th and 20th centuries due to the large-scale cultivation of cotton, the advent of synthetic fibres and because the growing of hemp has been made illegal in many countries.

Botany

Hemp is an annual herb and can grow to 1–5 m tall (Figs 4.9–4.11). The stem is grooved, with resinous pubescence, angular and hollow. The stem contains long, thick-walled fibres in the phloem and short, thin-walled fibres in the xylem. The basal leaves are opposite, the upper leaves alternate. The second pair of true leaves has three serrate leaflets, radiating from the tip of a long petiole. The following pair of leaves has five leaflets and so on, up to possibly 13 leaflets. The leaflets are lanceolate, up to 10 cm long and 1.5 cm wide with serrated edges. Hemp is mostly dioecious but monoecious cultivars have been bred. The male flowers are in axillary and terminal panicles, apetalous, with five yellowish petals and five poricidal stamens. The female flowers are both axillary and terminal; they have one single-ovulate ovary. The fruit is a greyish-brown, shining achene, variously marked or plain, tightly embracing the seed. The fruit is ellipsoid,

Fig. 4.10. Young fibre hemp plants with elongated stems, just before flowering.

Fig. 4.11. Fibre hemp can be up to 4 m high.

2–6 mm long and 2–4 mm in diameter. In a dioecious crop, male and female plants are usually present in more or less similar numbers. Male plants die 3–5 weeks earlier than female plants. Hemp has a shallow root system consisting of a taproot and laterals.

Cultivars, uses and constituents

Over the years, many varieties and cultivars have been selected or bred for the main purposes: fibre, oil or drugs. Fibre hemp is tall, branches little and yields small quantities of seed. Oil hemp is short, matures early and produces large quantities of seed. Hemp grown for drugs is short, much-branched with dark green leaves. In between these main groups there are many varieties, which differ in one or more characteristics. The psychoactive effect of hemp products, used as drugs, is mainly caused by the presence of tetrahydrocannabinol (THC). Two types of drug are produced: (i) hashish, pure resin, which is scraped from the flowering tops of the plant, the oil obtained from resin is known as 'hashish oil'; and (ii) marijuana, dried unfertilized inflorescences of female plants, or the dried leaves and flowers of the male and female plants. More resin is produced in tropical than in temperate climates. At present, hemp is still grown commercially in some Eastern European countries for the production of fabric and rope. In Western Europe, the growing of fibre hemp has practically disappeared, but now there is renewed interest. In the last decades a large multidisciplinary research programme was initiated in the European Union. The goal was to develop hemp as an alternative fibre crop for pulp for the paper industry. Whether that will be successful depends among others things on the success in breeding cultivars with increased stem yield, better bark quality and low levels of THC (THC is the narcotic component of the drugs that can be achieved from *Cannabis*; high concentrations can prevent the cultivation of hemp, due to legal provisions, in many countries).

Modern medicine uses hemp products in alleviating the pain of cancer and other diseases and as a mood stabilizer. The growing of hemp can have a dual purpose: fibre and oil production. The oil from the seeds can be used in varnish and food. The oil has a relatively high level of polyunsaturates and contains γ-linoleic acid and antioxidants.

Ecology and agronomy

Hemp for fibre requires a mild temperate climate and about 700 mm of annual rainfall. Although hemp for drugs grows best in a tropical climate, it tolerates annual temperatures of 6 to 27°C. It grows well on fertile, light soils at a pH of at least 5. It also tolerates a wide range of rainfall and alkalinity. To sow a crop of fibre hemp, about 20 kg of seeds is required per hectare; plant density has to be approximately 900,000/ha. If the plant density is very high, the crop is self-thinning in temperate climates. Seed must be sown as early in spring as possible in temperate climates, because it can germinate at low temperatures. The seeds must be drilled at a depth of 3.5 cm. The recommended fertilization for fibre hemp is 12 kg N, 12 kg K and

2 kg P per tonne of dry matter. The use of herbicides is mostly not needed because the crop can suppress weeds well. The crop achieves its full potential when it is not limited by shortage of water and nutrients, by pest or disease attack, or by other stresses, and depending on the interception of the amount of light, stem dry matter yield may be up to 14 t/ha. The crop should be harvested about 1 month after flowering, which means about 4 to 5 months after sowing. Traditionally, harvesting hemp involves a period of field drying. A new option is chopping the stems mechanically for ensiling, as an alternative way to preserve.

   In 2004, the world total production of hemp fibre was about 67,000 t. The main producers are China, Spain and South Korea. In 2004, the world total production of hemp seed was about 31,500 t. The main producing nation is China.

Jute

Jute – *Corchorus* spp.; Lime-tree family – *Tiliaceae*

Origin, history and spread

There are more than 30 *Corchorus* species, but only two are cultivated on a larger scale: (i) *Corchorus capsularis* (white jute), which probably originates from southern China; and (ii) *Corchorus olitorius* (tossa jute), which probably originates from Africa, although some authors consider India as the origin. *C. olitorius* occurs in the wild in Africa and India. The greater genetic diversity within the species in Africa justifies the supposition that Africa is the first centre of origin, with possibly a secondary centre in India.

   Jute has been cultivated from ancient times in Africa and the Indian subcontinent. At first the plant was grown mainly for its leaves, which were used as a vegetable or medicine. That still applies in Africa and the Middle East today. Valuation of jute as a fibre plant, especially in the Indian subcontinent, came much later. First the fibre crop had only regional importance. Jute has been part of Bengali culture for many centuries. There is evidence of trade of jute cloth in 16th-century Bengal. Jute came first to Europe in 1828 where it was valued as a good and cheaper substitute for hemp. Initially the fibre was mainly processed into cordage for ships. In 1850 there was a substantial export of raw jute fibre from Calcutta to England, especially to Dundee where jute mills were concentrated and the fibre could be processed. Because of the introduction of spinning and weaving machines, the processing of jute fibre could expand considerably. Dundee became an important centre for jute processing. The first jute mill in India was established near Calcutta in 1854. Gradually, jute fibre could be spun into good-quality yarn because of newer technologies. The yarn was woven into jute cloth, which could be used to process sacking bags and handbags at first. Later, the number of jute products increased. At present, India and Bangladesh, especially the Ganges–Brahmaputra delta, and China are the main jute fibre producers, accounting together for about 95% of the world's

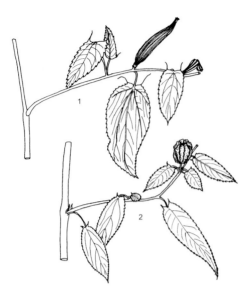

Fig. 4.12. Fruiting branches of jute: 1, tossa jute; 2, white jute.

production. The production of jute in Bangladesh is so important for the economy of the country that people call jute fibre 'golden fibre'.

Botany

Both jute species (Fig. 4.12) are erect herbaceous annuals, 1–4 m high. Fibre types are taller and less branching than vegetable types. The stem is slender, cylindrical, and 1–4 cm in diameter at the base. When maturing, the stem becomes woody at the base. The stem contains fibres (bast fibre), which are composed primarily of cellulose and lignin. The leaves alternate, they are simple, narrowly ovate to obovate or elliptical, 3–14 cm × 1–6 cm; the base is rounded; the leaf margin is serrated with two lower teeth prolonged into fine-pointed auricles, up to 1.5 cm long. The petiole is 0.5–6 cm long. The leaves have stipules, 5–12 mm long. The inflorescences are borne solitary at nodes; they have one to three flowers. The flower consists of five linear-obovate sepals, 3–8 mm × 1–2 mm; five obovate or narrowly obovate, yellow petals, 4–7 mm × 2–2.5 mm; and 20–50 stamens (Fig. 4.13). The ovary is obovoid or cylindrical and sometimes covered with small stiff hairs. The ovary comprises five to ten cells with ten to 42 ovules per cell. The style is 1–2 mm long.

The fruit of *C. capsularis* is a capsule, depressed globose, 1–1.5 cm in diameter, longitudinally grooved, ten-valved, with 35–50 seeds. The seeds are rhomboid to obovoid, dark brown and 2–3 mm long. The fruit of *C. olitorius* (Fig. 4.14) is a capsule, cylindrical, 3–8 cm × 3–6 mm, straight, longitudinally ten- to 12-ribbed, five- or six-valved, with 130–200 seeds. The seeds are rhomboid, 1–2 mm long, and green to black in colour; the 1000-seed weight is 2.0–3.3 g. Both species have a taproot, up to 60 cm deep, with many lateral roots.

Fig. 4.13. Leaves, flower buds and flower of tossa jute (photo: D.L. Schuiling).

Fig. 4.14. Fruits of tossa jute (photo J. van Zee).

Varieties, uses and constituents

Within the two species there are many cultivars. *C. olitorius* cultivars can be divided into two main groups, a group grown as a vegetable and a group grown as fibre plants. *C. capsularis* cultivars are grown mainly as a fibre crop. *C. olitorius* cultivars are usually higher yielding, with fibre of better quality, although *C. capsularis* cultivars are more adapted to lowland conditions with temporary waterlogging (not in the juvenile stage).

Jute fibres, composed primarily of cellulose and lignin, are mainly used for manufacturing products for packaging goods such as grain, potatoes, coffee, onions, cotton, etc. At present, this use accounts for about 75% of the world's jute production. Other uses include jute fabrics as carpet backing, wall coverings, carpets, paper, cordage, canvas, backing for linoleum, and felts. The fibre is also used to produce geo-textiles, to control erosion on hillsides. The fibre is valued because of its strength and durability. However, jute has partly lost its past importance due to synthetic fibres. At present, jute fibres are blended with synthetic fibres.

The leaves, tops and tender shoots of tossa jute are eaten as vegetables in Asia, Africa and the Middle East. The cooked leaves and shoots form a mucilaginous sauce, which can be very well combined with starchy dishes made from cassava, yam, millet or sweet potato. The leaves are also consumed with various other dishes. Tossa jute in Africa and the Middle East is grown mainly for this purpose. In the Arab world, the dried leaves are known as 'molukhyia' and used in soups.

The leaves of white jute are applied medicinally, for example to treat dysentery, cough and headache. The seeds of both jute species are toxic; however, ground into a powder they have been used as a carminative (white jute) and as a purgative (tossa jute).

The fresh leaves of tossa jute contain approximately 80% water, 5% protein, 12.5% carbohydrate, 0.3% fat and 2% fibre, and also Ca, P, Fe, β-carotene, thiamine, riboflavin, niacin and ascorbic acid.

Ecology and agronomy

Jute is a rainy-season crop. It grows best in areas with a daily temperature range of 22–35°C, a relative humidity of 65–90% and an annual rainfall of 1000–2000 mm. Jute can be grown on a wide range of soils, from clays to sandy loams, with pH between 5 and 8.6, although optimum pH is 6.6–7.0. Soils should be deep, well-drained and fertile, and high in organic matter content. *C. capsularis* is to some extent tolerant to saline conditions.

Jute is propagated from seed. In India and Bangladesh seeds are sown in a well-prepared soil, which is ploughed and cross-ploughed up to six times. Heavy soils have to be ploughed more often than light soils. Next, manure is added to the soil. Manure may be animal dung, ashes, composted plants, or rotted water plants. Seed is usually broadcast behind the plough, although sometimes line sowing is carried out. Seed rating is about 10–12 kg/ha for broadcasting and slightly less for line sowing. In good soil conditions, seeds may germinate in 2 or 3 days. When the plants are about 20 cm tall, they are thinned out to reach spacings of 10–15 cm × 15 cm; the relatively high plant density avoids branching of the plants. Weed control is usually done by harrowing and weeding several times, but it can also be done chemically as herbicides are available.

In Bangladesh farmers usually do not use artificial fertilizers, but in other countries some NPK fertilizers are applied. Nutrient requirements depend on soil fertility and expected yield. A jute crop of 34 t of green plants per hectare (about 2 t of retted fibre) removes on average about 63 kg N, 14 kg P and 132 kg K.

The crop can be harvested 4 or 5 months after sowing, when the crop is in small fruit stage; that is when 50% of the plants are in pods. Harvesting before that stage means that the fibre can be too weak. Harvesting when the fruits are mature and the seeds are ripe is also not acceptable because then, although the fibre is stronger, it is too coarse.

At harvest, the plants are cut close to the ground and left in the field for several days to allow the leaves to dry up and drop off. Before the fibre can be obtained from the bast or the phloem layer of the stem, the plants have to undergo a process called 'retting'. The stems have to be laid in water with a temperature of about 35°C for 8 to 15 days. Enzymes, which are the result of bacterial actions, decompose the cambium and the cortex, which makes it possible to extract the fibre of the stem. Various chemicals can be used to improve the process. After retting, the fibres can be stripped from the woody core. As jute fibre is bast fibre, just like the fibre of flax and hemp, the whole process is more or less similar to those of flax and hemp (see Flax, this chapter). In lowlands, where the crop can be standing in water at harvest-time, the retting starts immediately after harvest, which means that the leaves do not drop off. If leaves remain on the plants during retting, the colour of the fibre will become darker. Jute goes wild easily so, in many countries where the plant is grown, it may become a serious weed.

Fibre represents about 6% of the green weight. Fibre yields range from 0.8 to 3 t/ha. The genetic potential is 4 t/ha.

A number of factors determine jute as a suitable crop for smallholders in developing countries: it is a rainfed crop, which can be grown without irrigation; it is a labour-intensive crop; and it requires small quantities of fertilizer and pesticides.

In 2004 about 1.4 million ha were covered by jute, and the world's jute fibre production was about 2.9 million t. Nowadays, jute is second in the world production of natural textile fibres.

Sisal

Sisal – *Agave sisalana*; Agave family – *Agavaceae*

Origin, history and spread

Sisal, also called 'sisal hemp', is probably native to south Mexico; however, that is not entirely certain because wild forms are unknown. Agaves were already cultivated about 1000 years ago. For the Aztecs it was an important plant. Buds and flowers of the plant were eaten raw or cooked. *Agave* species have been

used for producing rope, clothing, paper, medicines and needles. There is archaeological proof that *Agave* species have been used as food for about 9000
years.

  Because the fibre was originally exported from Mexico via the port of Sisal
in Yucatan, the plant received the name *Agave sisalana*. The species is also
called 'century plant', because people mistakenly believed that the species flowered only once in a century. Sisal was exported from Mexico in the 19th century
to Indonesia and the Philippines, and from there it spread in the 19th and 20th
centuries throughout South-east Asia, India, the Pacific Islands and Australia. Dr
Henry Perrine, who was consul of the USA at Campeche in Yucatan, introduced
the first sisal plants to the USA, in Florida in the year 1826. In 1893, 1000 sisal
bulbils were sent from Florida to Germany. The Germans brought bulbils to
East Africa. A few years later, bulbils were sent from Kew Gardens in England
to Kenya. These imports and later imports of bulbils laid the basis for the current
plantations in Tanzania, Kenya, Uganda and Mozambique. The first commercial
plantation in Brazil dates back to the 1930s and the first fibre export was made
in 1948. In the 1960s cultivation of sisal extended considerably.

  At present sisal is grown mainly in South America and Africa.

Botany

Sisal is a monocarpic, perennial herb (Fig. 4.15). When the plant is not flowering, the stem is short and thick, up to 1 m. When flowering, the stem or pole

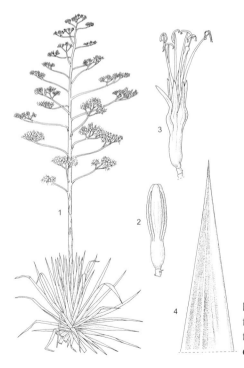

Fig. 4.15. Sisal: 1, habit of
flowering plant; 2, flower bud; 3,
flower; 4, apical part of leaf (line
drawing: PROSEA volume 17).

can reach a height of 9 m. The plant consists of a rosette of numerous, linear-lanceolate (sword-shaped), succulent leaves, which may be 1–2 m tall, up to 15 cm wide and 2–5 cm thick. At first, the leaves are glaucous with a bluish tinge, becoming green when maturing.

Young leaves may have small spines along their margins; they disappear when the plant matures. Leaves have a terminal, dark brown, rigid, very sharp spine, 2–3 cm long. The cross-section at the base of the leaf resembles a flattened triangle. Leaves are hard and contain bundles of fibres which are strong and long and run the length of the leaves. The average fibre content of the leaves is about 4%. Leaves are arranged in an ascending spiral. Leaves may live for 5 years and still be photosynthetically active. Cultivated sisal plants may produce about 200 leaves before the emergence of the inflorescence.

Sisal has a paniculate inflorescence. The flowers are perfect and tube-like, consisting of six yellowish-green petals, 5–6 cm long, and six stamens. The fruit is a capsule; however, it develops rarely. Flowering occurs from the bottom to the top of the inflorescence and may last for several months. The plant flowers only once, in commercial cultivation after 5 to 12 years, the number of years depending on cultivar and environment. After flowering the plant dies.

Reproduction is usually vegetative by means of bulbils (small plantlets). Bulbils are borne in the axils of the bracteoles of the inflorescence after flowering (Fig. 4.16). Sisal produces subterraneous rhizomes from buds in the axils of the lower leaves. Along the rhizomes there are buds that may grow into new plants, forming colonies. Sisal has a relatively small root system; roots are fibrous and originate from the base of the leaf scars at the bottom of the stem. Most of the roots are concentrated in the upper 40 cm of the soil, where they spread

Fig. 4.16. Bulbils in the inflorescence of sisal.

horizontally up to 5 m. A number of roots grow deeper than 40 cm, which results in good anchorage.

Species, uses and constituents

A. sisalana is only one of about 300 species of the *Agave* genus, although it is by far the most important. It is grown for its very strong fibres, which are mainly used for producing rope. Because of the higher content of lignin in the fibre bundles, sisal fibres are more rigid, stiffer and have greater tenacity than other bast and leaf fibres.

Next to rope production, sisal is locally also used to produce many different products such as nets, carpet pads, baskets, sandals, clothing, paper pulp and construction material. In the Western world, much sisal was used as binding twine for grain-harvesting machines. However, at present it is partly replaced by synthetic fibre.

As well as *A. sisalana*, a number of other *Agave* species have importance for fibre production. One such is 'Henequen' (*Agave fourcroydes*), which is grown mainly in Mexico, especially in the Yucatan region. The fibre is also known as 'sisal' and used for the production of rope, sacks and carpets, mainly in Mexico. Henequen was domesticated by the Mayas in pre-Hispanic times. The term 'sisal' is used for both the plants and the fibres of several species, although mainly *A. sisalana* and *A. fourcroydes* are involved; other species include 'maguey', also called 'cantala' (*Agave cantala* and *Agave americana*), grown in the Philippines, India and Indonesia, and *Agave letonae*, a fibre crop that is grown in El Salvador. Some cultivars of *A. americana* are variegated, the leaves having white or yellow, marginal or central stripes from the base to the apex. These cultivars are grown as ornamentals, and grow well in tubs.

In Mexico, the sap from several *Agave* species is collected and fermented to make the beverage 'pulque', and after distilling a spirit called 'mezcal'. The spirit 'tequila' is derived from the fermented and distilled sap of *Agave tequilana*. Two other species, *Agave lecheguilla* and *Agave funkiana*, produce fibres that are used only to make brushes.

The components of the dry weight of sisal fibre are approximately 55–65% α-cellulose, 11–18% hemicelluloses, 7–15% lignin, 1% pectin and 1–8% ash.

Ecology and agronomy

Sisal grows best in the sub-humid tropics, up to 1800 m altitude, with an annual rainfall of 1000–1500 mm. However, it is often grown with less rainfall. It needs full sunlight; maximum temperature should not exceed 32°C and minimum temperature not below 16°C. It does not tolerate frost, hail and waterlogging. The species is drought-resistant; it is morphologically adapted to manage water scarcity by its extensive root system and the arrangement and shape of the leaves, which, like a funnel, concentrate rainwater on a small area. Moreover, it is a xerophytic plant, which means that its photosynthetic pathway is the crassulacean acid metabolism. During the night the plant absorbs carbon dioxide and converts it into acids, which are used in daylight for the synthesis of

Fig. 4.17. Sisal plantation with ground cover of puero (*Pueraria phaseoloides*) between the double rows, Tanzania (photo: J. Wienk).

carbohydrates. This allows the plant to close the stomata during daytime to minimize water use. Sisal grows best on well-drained, non-saline, sandy-loam soils at pH 5.5–7.5.

Sisal is mainly propagated with bulbils. The bulbils are first planted in nurseries. After 12–18 months the plantlets can be transplanted into the field. Suckers can also been used for propagating. At present, tissue culture is also used for obtaining excellent planting material.

Before transplanting, weeds have to be removed from the field, usually by mechanical or manual hoeing, by ploughing or by using herbicides. This has to be repeated regularly during the first 3 years after planting. Later, weeds can be allowed to grow for some time, and after slashing left in the field as a mulch layer which reduces further weed growth and preserves soil moisture.

Plant densities range widely, from 4000 to 7000 plants/ha, depending on cropping system, climate, soil conditions and the end use of the crop. For obtaining the perfect crop for manufacturing paper, the highest plant densities are used.

When plants are cultivated in rows, a much used density of 5000 plants/ha can be obtained by a spacing of 2.5 m between the rows and 0.8 m between plants within the row (Fig. 4.17).

The required nutrients can partially be obtained from the waste material after fibre extraction and from mineralization of soil components. This may be sufficient; however, applying some extra nutrients is usually profitable. It can be based on the removal of nutrients, which per tonne of fibre is about 27–33 kg N, 5–7 kg P, 60–80 kg K, 42–70 kg Ca and 34–40 kg Mg.

Harvesting leaves may start about 2–4 years after planting; it can be repeated yearly for many years, until the emergence of the inflorescence. By

Fig. 4.18. Harvesting the last leaves from sisal plants.

removing leaves regularly, the emergence of the inflorescence will be postponed (Fig. 4.18). The leaves grow in circles of about 13 leaves and, each harvest, usually five circles are cut off. Immediately after harvesting, the tips with the sharp spines are removed. Next, the fibres have to be extracted from the leaves. The order of work of fibre extraction is as follows: (i) fresh leaves; (ii) decorticating; (iii) washing with water; (iv) binding; (v) cleaning; and (vi) drying. Decorticating is carried out with a special machine as follows: fresh leaves are hand-fed into the machine; in the machine the leaves are beaten heavily, which involves crushing and scraping; the broken fibres, pulp and liquid are washed away. The remaining long fibre is removed from the machine and can be washed and dried; the washed-away waste of the decorticating process can be used as fertilizer or animal feed.

Next, the fibres are usually graded, based on length of the fibre, bundle strength, presence of impurities, and colour. Dried fibres are transported to the mills to be converted into yarns, twines, ropes, etc. In India, the fibres are also extracted from the leaves by retting (see Flax, this chapter). The annual yield of dried fibre ranges from 0.5 to 2.8 t/ha. In China, yields up to 4.5 t/ha are obtained. In 2004, the total world production of sisal fibre was around 0.3 million t. Brazil is by far the world's major producer of sisal fibre; other important producers are Tanzania, Kenya, Mexico and Haiti.

5 Forages

© CAB International 2008. *Guide to Cultivated Plants* (A.T.G. Elzebroek and K. Wind)

FORAGE GRASSES

Grass family – *Gramineae*

Introduction

The grass family is the most important plant family with regard to the production of food for man and animals and as a part of natural vegetation. It comprises over 600 genera and about 10,000 species. Cereals are the best known domesticated grasses; the most important cereals are described in Chapter 9.

Domestication of forage grasses is a relatively recent development, initially in the cool temperate climate zone and subsequently in tropical zones. Meadow and forage crops were important in the agriculture of classical history. Pliny the Elder (AD 23–79) mentioned in his *Naturalis Historia* sowing of 'hayseed' along with vetch, lupine, lucerne and other soil builders and fodder crops. The ryegrasses (*Lolium* spp.), cock's foot (*Dactylis glomerata*), timothy (*Phleum pratense*) and other grasses were well known and appreciated in Europe before history records them. If sown, the seed was taken from local meadows, often as hay in seed, or from seed which was left in the barn after hay had been fed. Seeds from different regions performed differently and were in fact different landraces that had evolved over long periods of time. Grass-breeding companies continue to utilize elite landrace populations to this day. During the last century, the number of new grass cultivars increased greatly, especially in the temperate climate zone. In 2006 for example, the number of cultivars of perennial rye-grass (*Lolium perenne*), for forage purposes only, exceeded 60 in the official List of Recommended Varieties of Field Crops of The Netherlands. However, the worldwide number of genera and species which are used in breeding programmes of forage grasses and in grassland improvement is astonishingly low, certainly if compared with the total number of species of the grass family. Grassland improvement in the tropics is still occurring on a relatively small scale. Large-scale utilization of natural and semi-natural grassland, as well as utilization like grazing of the understorey in tree crop plantations, are still common practices. However, grassland improvement by means of the introduction of legumes appears often to be successful, especially when applied by smallholders.

Taxonomy

The earliest family name, *Gramineae*, is sanctioned by the International Code of Botanical Nomenclature. The later name, *Poaceae*, based upon a legitimate included genus, is an allowable alternative. Various specialists have worked on the systematic classification of the grass family during the past two centuries, resulting in some 20 different classifications. Three major groups can be distinguished with increasing clarity, and deserve the status of subfamily, of which some of the most important characteristics are summarized below.

1. Subfamily *Panicoideae*: main distribution in tropical and subtropical regions; photosynthetic pathway predominantly C4; number of florets per spikelet, two; spikelet disarticulation mostly below the glumes; sterile or reduced florets basal; first seedling leaf broad and often procumbent; number of veins of lemma, irregular. Includes genera such as *Zea*, *Setaria*, *Pennisetum*, *Paspalum*, *Brachiaria*, *Panicum* and *Sorghum*.
2. Subfamily *Chloridoideae*: main distribution in arid and semi-arid habitats; photosynthetic pathway predominantly C4; number of florets per spikelet, one to many; spikelet disarticulation mostly above the glumes; sterile or reduced florets terminal; first seedling leaf broad and often curved; number of veins of lemma, three. Includes genera such as *Chloris*, *Eragrostis*, *Cynodon* and *Sporobolus*.
3. Subfamily *Pooideae*: main distribution in temperate climate zones; photosynthetic pathway predominantly C3; number of florets per spikelet, one to many; spikelet disarticulation mostly above the glumes; sterile or reduced florets terminal, sometimes basal; first seedling leaf narrow and erect or non-erect; number of veins of lemma, mostly five or more. Includes genera such as *Lolium*, *Dactylis*, *Poa*, *Festuca*, *Triticum*, *Secale*, *Hordeum*, *Avena* and *Phleum*.

Nowadays, two further small groups are usually accorded the status of subfamily:

4. Subfamily *Oryzoideae*: well-adapted to aquatic habitats, including the important genus *Oryza* (rice).
5. Subfamily *Bambusoideae*: well-adapted to forest habitats, including woody and semi-woody bamboo species.

Morphology of grasses

The typical morphology of the grass plant is an upright, cylindrical stem, anchored in the soil by its roots, and jointed by transverse nodes bearing distichously arranged alternate leaves. Leaves consist of the upper part, the leaf-blade, which is free from the stem, and the lower part enveloping the stem, the leaf-sheath. The inflorescence usually is comprised of an arrangement of many spikelets. The spikelets are composed of one or more florets which contain the flower parts. The spikelets are subtended by, usually two, oppositely arranged glumes. Many modifications, such as the absence of one or both glumes, occur and are often species-specific.

The structure of the inflorescence and the number of florets per spikelet are the most important characteristics used in the identification of generative grass species.

The different inflorescence structures can be distinguished by the branching of the main axis and the presence or absence of pedicels. Although many modifications occur, the basic forms of solitary inflorescences are as follows.

- Spike: the spikelets are attached directly – without a pedicel – to the un-branched main axis.
- Raceme: the spikelets are pedicelled and attached to the unbranched main axis.
- Panicle: whorled or individually arranged lateral branches radiate from a central main axis and branch further to terminate in pedicelled spikelets.
- Spike-like panicle: in fact a panicle but with very short branches, resulting in a cylindrically shaped inflorescence; also called a 'false spike'.
- Spike-like raceme: a jointed axis carries pairs of spikelets, one pedicelled and the other sessile, at each node; mostly present in compound inflorescences.

Compound inflorescences exist of more than one unit of one of the basic forms of solitary inflorescences; arrangements can be paired, digitate, subdigitate or racemose.

Although the taxonomy of the grass family is almost entirely based on floral characteristics, there are numerous identification keys based on vegetative characteristics. Features used in such keys are, for example, youngest leaf rolled or folded, shape and size of the ligule, presence of auricles, the degree of ribbing of the leaf-blade, presence of rhizomes, stolons, hairiness, etc.

The most important morphological features, as well as brief information about the ecology and agronomy of a selected number of genera and species of cultivated forage grasses, are discussed below.

Forage Grasses of Temperate Climate Zones

Lolium spp. – Rye-grasses

> *Lolium perenne* – Perennial rye-grass. A loosely to densely tufted perennial with culms up to 90 cm tall. Leaf-sheaths are red-violet at the base; leaf-blades are folded when young, glossy below, smooth or slightly rough above, with small auricles. Ligule is up to 2 mm long. Inflorescence is a spike (Fig. 5.1); spikelets alternating on opposite sides of the axis, 7–20 mm long, four- to 14-flowered,

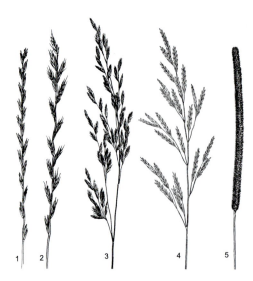

Fig. 5.1. Inflorescences of some important temperate forage grasses: 1, perennial rye-grass (*Lolium perenne*); 2, Italian rye-grass (*Lolium multiflorum*); 3, tall fescue (*Festuca arundinacea*); 4, meadow fescue (*Festuca pratensis*); 5, timothy grass (*Phleum pratense*).

without awns. Probably the most important forage grass in the cool, temperate climate zone. It is prominent in old pastures and meadows, especially on rich heavy soils. It is extensively sown in most parts of north-west Europe (Fig. 5.2), but not persistent in areas with severe winter frosts. Annual dry matter yields of 10–14 t/ha are obtained.

Lolium multiflorum – Italian rye-grass. Annual or biennial with tufted or solitary culms up to 100 cm tall. Leaf-sheaths are red-violet at the base; leaf-

Fig. 5.2. Pasture with perennial rye-grass, Flevopolder, The Netherlands.

blades are rolled when young, with narrow spreading auricles. Ligule is 1–2 mm long. Inflorescence is a spike (Fig. 5.1); spikelets alternating on opposite sides of the axis, 8–25 mm long, five- to 15-flowered, lemmas with fine straight awn up to 10 mm long. It is a native of the Mediterranean region and a valuable fodder grass, often sown for silage or hay. Annual dry matter yields of pure stands of 10–16 t/ha are obtained.

Festuca spp. – Fescue

Festuca arundinacea – Tall fescue. A tufted perennial with culms up to 200 cm tall, sometimes forming large dense tussocks. Leaf-sheaths are smooth or rough; leaf-blades are rolled when young, with small, minutely haired auricles. Ligule is up to 2 mm long. Inflorescence is a panicle (Fig. 5.1); spikelets 10–18 mm long, three- to ten-flowered. Lemmas have a fine, rough awn 1–4 mm long. The taller robust types grow on heavy soils and low-lying meadows. The shorter types are found in grazed pastures in drier areas. It occurs throughout Europe, north-west Africa and temperate Asia, and was introduced into the USA.

    Festuca pratensis – Meadow fescue. A loosely tufted perennial with culms up to 120 cm tall. Leaf-sheaths are smooth and rounded; leaf-blades are rolled when young, bright green, with narrow, hairless auricles. Ligule is up to 1 mm long. Inflorescence is a panicle (Fig. 5.1); spikelets are 10–20 mm long, five- to 15-flowered, usually awnless. A valuable grass for grazing, hay and silage, it is often abundant in low-lying, wet meadows on loamy or heavy soils. Occurs throughout Europe and south-west Asia, and was introduced into the USA.

Phleum – Cat's-tail

Phleum pratense – Timothy grass. A loosely to densely tufted perennial with culms up to 150 cm tall. Leaf-sheaths are smooth and rounded, becoming dark brown at the base; leaf-blade is rolled when young, greyish-green, without auricles. Ligule is blunt, up to 6 mm long. Inflorescence is a spike-like panicle (Fig. 5.1); spikelets flattened, 3–4 mm long, with a single flower. Glumes have a rigid, rough awn 1–2 mm long. A valuable fodder grass, it is often sown in mixtures with perennial rye-grass. Found in most parts of Europe, from Scandinavia to Portugal and the Balkan region, it is very winter-hardy and well adapted to mowing, somewhat less to grazing.

Forage Grasses of Tropical Climate Zones

Brachiaria spp.

Brachiaria ruzizienzis – Congo signal grass (Africa), Ruzi grass (Australia). A spreading perennial with short rhizomes and with culms up to 1 m tall. Leaf-blades are hairy on both sides, lanceolate, 100–250 mm × 10–12 mm. Ligule is fringed with hairs. Inflorescence has three to six racemosely arranged (spike-like) racemes, rachis of racemes broadly winged (Fig. 5.3); spikelets hairy,

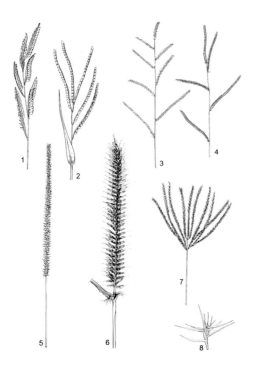

Fig. 5.3. Inflorescences of some important tropical forage grasses: 1, Congo signal grass (*Brachiaria ruzizienzis*); 2, signal grass (*Brachiaria decumbens*); 3, brownseed paspalum (*Paspalum plicatulum*); 4, Dallis grass (*Paspalum dilatatum*); 5, setaria (*Setaria sphacelata*); 6, elephant grass (*Pennisetum purpureum*); 7, Rhodes grass (*Chloris gayana*); 8, kikuyu grass (*Pennisetum clandestinum*) (line drawings: PROSEA volume 4).

3 mm × 5 mm. Length of lower glume is half of the spikelet length. Native of the Ruzizi plains in Congo, Rwanda and Burundi, it is now widely distributed in the tropics. It forms a dense mat under grazing, which withstands grazing well. Ruzi grass combines well with legumes. When fertilized, annual dry matter yields above 20 t/ha are obtained.

Brachiaria decumbens – Signal grass. A trailing perennial with culms up to 60 (100) cm tall. Prostrate stems root at the nodes. Leaf-blades are hairy at the base, narrow lanceolate, 40–140 mm × 8–12 mm. Ligule is 3 mm, membranous. Inflorescence has two to five racemosely arranged (spike-like) racemes, rachis of racemes flat (Fig. 5.3). Spikelets are hairy, 4–5 mm long. Length of lower glume is one-third to half the spikelet length. Native to open grasslands and partial shade on the Great Lakes Plateau in Uganda and adjoining countries, now widespread in the tropics and subtropics. A valuable high-rainfall pasture grass, but will survive a dry season of 4–5 months. It also withstands heavy grazing and trampling. When fertilized, dry matter yields above 30 t/ha are obtained.

Paspalum spp.

Paspalum plicatulum – Brownseed paspalum (USA), Plicatulum (Australia). A tufted perennial with sub-erect culms up to 120 cm tall. Leaf-sheaths are glabrous. Leaf-blades are about 400 mm × 10 mm, folded and fringed with hairs at the base. Ligule is 1.5 mm long. Inflorescence has ten to 13 racemosely arranged racemes (Fig. 5.3). Racemes are 2–6 cm long, 1.5–2 mm wide. Spikelets are in pairs, usually one of a pair undeveloped towards the base of raceme, 2–3 mm × 1.5–2 mm. Common name is derived from the dark brown caryopsis. Native of South and Central America, it is now distributed throughout the world from Africa to Australia, the USA and South-east Asia. It is adapted to tropical and subtropical climates with over 750 mm of annual rainfall. Used as pasture grass for grazing, silage or hay-making, as understorey in coconut plantations and in ley pastures. Annual dry matter yields vary from 8.5 to 24 t/ha. It is persistent on infertile soils and combines well with legumes.

Paspalum dilatatum – Dallis grass (USA). A tufted perennial with short creeping rhizomes and ascending to erect culms up to 150 cm tall. Leaf-sheaths are slightly keeled, often pilose. Leaf-blades are linear, 5–40 cm × 3–13 mm. Ligule is up to 6 mm long. Inflorescence has three to five (nine) alternately arranged racemes (Fig. 5.3). Spikelets are paired, overlapping, ovate, 3–4 mm long. Native to the humid subtropics of Brazil, Argentina and Uruguay, now widely distributed in coastal subtropical Australia, the south-eastern part of the USA, India, Malaysia and some African countries. Well adapted to humid subtropical climates with over 1000 mm of annual rainfall. It is preferably used for grazing, but is also suited for making hay and silage. Dallis grass is very palatable when young but this decreases rapidly with age. Annual dry matter yields range from 3 to 15 t/ha, depending on soil fertility or fertilization.

Pennisetum spp.

Pennisetum purpureum – Elephant grass, Napier grass. A robust, deep-rooting perennial, with short rhizomes and culms 2–3.5 (7) m tall and 3 cm in diameter. Leaf-sheaths are glabrous to short bristly haired. Leaf-blades have prominent midrib, up to 120 cm × 5 cm, glabrous and hairy at the base. Ligule is fringed with hairs. Inflorescence is a bristly spike-like panicle (false spike) up to 30 cm long, usually yellow-brown (Fig. 5.3). Spikelets are 5–7 mm long, solitary or in clusters of up to five. There is little or no seed formation. Elephant grass is of tropical African origin and has been introduced into all tropical and frost-free subtropical regions of the world. It is naturalized in areas of South-east Asia where annual rainfall exceeds 1000 mm. Elephant grass is commonly used in a cut-and-carry system, feeding in stables, or is made into silage. Grazing at 6–9-week intervals at a height of about 90 cm gives good utilization. Annual dry matter yields range from 2 to 10 t/ha when unfertilized and from 6 to 40 t/ha when fertilized. In 1959 Vicente-Chandler and colleagues established, with heavy fertilizing, a world record of 84.8 t of dry matter per hectare.

Pennisetum clandestinum – Kikuyu grass. A prostrate perennial, with short culms up to 45 cm tall when ungrazed, spreading vigorously from rhizomes and stolons. Leaves arise alternately from multi-branched stolons. Leaf-sheaths are 1–2 cm long, pale green and densely hairy. Leaf-blade is tightly folded when young, linear, 10–150 mm × 1–5 mm. Ligule has a rim of short hairs. Inflorescence is a spike with two to four clustered spikelets which are partly enclosed within the uppermost leaf-sheath (Fig. 5.3). The florets are protogenous and the stamens are rapidly exserted on long filaments, usually in the early morning. Kikuyu grass is of East and Central African origin, particularly from highland grasslands at forest margins at altitudes of 2000–2700 m, receiving 1000–1600 mm of annual rainfall. It has been introduced widely into humid, (sub) tropical areas between latitudes 0° and 35°S. It is not only valued as pasture grass for ruminants, but also as grass for lawns and recreational areas. It is potentially a weed on arable land and in irrigation channels. Once established and with suitable soil and moisture conditions, kikuyu grass will dominate a pasture within 3–9 months. It responds well to intensive grazing. It can be combined with legumes like white and red clover or *Desmodium* species, but management (close grazing) and (P-containing) fertilizer management must be good to maintain the mixture. Annual dry matter yields of kikuyu grass vary from 9 to 30 t/ha.

Chloris gayana – Rhodes grass

A glabrous, usually stoloniferous perennial with fine, leafy culms up to 2 m tall, but very variable. Leaf-sheaths are glabrous except the uppermost part. Leaf-blades are flat, 25–50 cm × 3–9 mm, tapering towards the apex, margins rough, hairy at the base. Ligule is about 1 mm long, fringed with short hairs. The inflorescence consists usually of six to 15 digitately arranged spikes, each 4–15 cm in length (Fig. 5.3). Spikelets are green or purplish, about 3.5 mm long, with awned lemmas. Rhodes grass is native to Central Africa and adjacent regions. It was first cultivated in South Africa and at the beginning of the 20th century was introduced into the USA, South and Central America, parts of Asia and Italy. Now it is widely grown in (sub) tropical countries. Rhodes grass is used for grazing and making hay. After flowering, the feeding value declines rapidly. Silage-making generally gives variable results. It has excellent ability to spread naturally. It requires 600–1000 mm of annual rainfall but can withstand a dry season of up to 6 months. Optimum temperature for growth is 35°C but it is more tolerant to low temperatures than most other (sub)tropical grasses. Examples of recorded annual dry matter yields of Rhodes grass are: North Sumatra 38 t/ha, Zambia 58 t/ha and Texas 16 t/ha.

Setaria sphacelata – Setaria (Australia), Golden timothy (Zambia)

A tufted perennial with more or less erect culms up to 3 m tall. Leaf-sheaths are prominently keeled, often red-pigmented, sometimes hairy. Leaf-blades are 10–70 cm × 11–20 mm. Inflorescence is an elongated spike-like panicle, 10–50 cm long (Fig. 5.3). Young tillers are strongly flattened. Spikelets are borne in clusters of two or three on short branches, each spikelet 2.5–3 mm long and subtended by stiff bristles. Various cultivars may differ in morphological characteristics. Setaria is native to and widely distributed in tropical and subtropical Africa. In was first cultivated in Kenya and has since been widely

Fig. 5.4. Pasture improved with setaria, Serdang, Malaysia.

planted in Africa, Asia (Fig. 5.4) and Australia. It is an important forage, very palatable when young and commonly used under grazing and cut-and-carry systems. Annual dry matter yields of pure stands of 19–31 t/ha are obtained.

FORAGE LEGUMES – TEMPERATE

Clover species – *Trifolium* spp., White clover – *Trifolium repens*, Red clover – *Trifolium pratense*, Subterranean clover – *Trifolium subterraneum*, Berseem or Egyptian clover – *Trifolium alexandrinum*, Persian clover – *Trifolium resupinatum*, Lucerne – *Medicago sativa*; Pea family – *Leguminosae*

Clover species

The eastern Mediterranean is the centre of origin of the clovers. They have been cultivated in southern Europe for many centuries. The name *Trifolium* is derived from the compound leaves of these species, which are trifoliolate. The stipules at the petioles of the leaves of the different species vary in size, shape and colour, and can be used in identification of the species (Fig. 5.5).

The flowers occur crowded together in racemes; the number of flowers is variable. The flowers are irregular with only one plane of symmetry. Five, mostly green, joined sepals form the calyx. The corolla is the conspicuous part of the flower, especially one large petal called the 'standard'. Underneath the standard

Fig. 5.5. Leaves and stipules of some temperate forage legumes: 1, white clover; 2, red clover; 3, berseem or Egyptian clover; 4, Persian clover; 5, lucerne.

there are four petals, the two small petals in the centre are joined together and are known as the 'keel', because of their boat-shaped appearance. On both sides of the keel there is a petal that is broad above and tapering towards the base, which is called the 'wing'. Within the keel lie ten stamens forming a tube that surrounds the ovary. The colour of the flower is variable. The fruit is a pod, containing a variable number of ovules.

That growing clovers can increase soil fertility has been known for centuries. Clovers have the ability to fix atmospheric N_2 in their roots in a symbiotic association with bacteria of the genus *Rhizobium*. Bacteria penetrate the roots and proliferate within plant cells, colonizing nodules on the roots. Whether nodules are active is shown by the pink colour of the cross-section of the nodules. The pink colour is caused by 'leghaemoglobin', a protein with high affinity for oxygen. It is asserted that a clover crop can fix up to 600 kg N/ha/year, although 250 kg is probably normal.

Because of the ability to fix N_2, clovers are valuable in agriculture, particularly grassland farming. The relatively high protein content and digestibility of clovers means that clovers are highly valued as feed for livestock. The strive for more sustainable agriculture, and the desire to reduce the dependence on mineral fertilizers, will lead to increased usage of legumes.

Clovers are often grown in mixtures with grasses in a form of ley cropping. In some areas no fertilizers are applied to pastures including grasses and clover. Bloat can be a problem for animals that eat too large a proportion of clover in their diet. Use of grain supplements can prevent it.

White clover

White clover is a perennial herb. It has a prostrate growth habit (Fig. 5.6) with a branching network of stolons. This makes white clover well adapted to grazing and mowing; only the leaves are removed without damage to stolons

Fig. 5.6. Habits of flowering plants of white clover (left) and red clover (right) (drawing: Misset BV, Doetinchem, The Netherlands).

and the growing point. The plant is entirely glabrous; leaflets often have a light green V-shaped leaf mark. Flowers are white and the number of flowers in the raceme can be up to 100. There are five or six seeds per pod, each seed being heart-shaped; 1000-seed weight is 0.7 g. White clover can be found throughout the temperate regions and is limited in distribution only by extreme cold, heat and drought. The plant grows best on well-drained silt loams and clays, pH 6–7. It is the most important pasture legume in temperate regions, almost always grown in mixture with grasses. It improves the feeding value of a pasture due to its higher intake, high digestibility and crude protein levels. Mixtures of grasses and white clover may be used for high-quality hay or silage. White clover cultivars contain variable quantities of a cyanogenic gluco-side, which does not affect most livestock but can cause photo-sensitization in horses.

Red clover

Red clover is much-branched, semi-erect biennial herb (Fig. 5.6). The whole plant is pubescent. The upper surface of the leaves usually has fewer hairs than the underside. The leaflets have a light green V-shaped leaf mark. The stipules are often marked with purple-coloured veins. Inflorescences arise in the axils of the leaves and at the tips of the shoots. The inflorescence is a globular raceme, containing about 30 pink to purple flowers, and almost cupped by two nearly sessile leaves. The pods are short and terminate in a long point, and contain a single seed. The seed is asymmetrically heart-shaped; 1000-seed weight is 1.8 g. Red clover is not well adapted to regular grazing or mowing because the axillary buds and growing points are removed. For some varieties regrowth from basal buds is possible and three or four cuts can be taken, although that will usually weaken the plants, so survival in the second year is difficult. Other varieties have only one wave of shoot elongation. After removing the shoots, the basal buds stay dormant until the next growing season. This means that red clover is not suitable for intensively used pastures. Red clover can probably best be grown in pure stand as a hay or silage crop, although some cultivars are grown in a mixed grass/clover sward for conserving as hay or silage. Red clover is sometimes undersown into cereal crops such as spring barley. The cereal crop (cover crop) has to be harvested early and conserved (e.g. silage) before full maturity to give red clover enough time for good development before the autumn. Cover crops minimize weed problems. There are several red clover varieties which vary in time of flowering. Like all clovers, red clover is valued as a fertility-restoring crop. It grows best on clay soils, with a pH of 6.

Subterranean clover

Subterranean clover is an annual herb, usually grown in pastures. It is a prostrate plant with long and branched creeping, not rooted, stems. The above-ground portion of the plant is pubescent. The leaflets are heart-shaped and have a light green leaf mark. Stipules are broad with a short point, often with purple-coloured veins. The inflorescences are small axillary racemes bearing three to six fertile flowers and a number of upper sterile flowers. The petals are white or cream, sometimes pinkish. The calyx is hairy and may have one or two red rings around it. The pods contain a single seed. After fertilization the stalk of the flower head bends towards the soil and the head is subsequently pushed into the soil. The pods and the seeds mature when they are buried. In summer when the seeds have set, the plants dry off and die. The buried seeds make it possible for the species to re-establish in the autumn. To make re-establishment possible the residual top growth of mainly grasses must be removed before the autumn rains begin. This can be achieved by forced grazing or burning.

Subterranean clover is especially well suited to areas that have warm, moist winters and dry summers. Through its prostrate growth habit, subterranean clover is capable of withstanding close grazing by sheep. Subterranean clover is highly nutritious and is valued as feed for livestock. However, some varieties contain phyto-oestrogens. The effect of grazing subterranean clover-dominant pastures by sheep may be that ewes have a reduced ovulation and thus a reduced lambing percentage. Lambs may show enlarged teat and mammary gland development. Dairy and beef cattle are usually not affected. To avoid these problems ewes should not be allowed to graze subterranean clover pastures before and during the period of mating. Breeders have now developed cultivars with low oestrogenic activity.

Subterranean clover tolerates acid soil conditions and medium to poor drainage; it grows best on soils which are well supplied with P and S.

Berseem or Egyptian clover

Berseem clover is an erect annual herb. The leaflets are oblong and do not bear a leaf mark. The stipules are lanceolate. The shape of the inflorescence is ovate and the flowers are white. The species has a deep root system.

Berseem is a vigorous clover that is used for pasture, green fodder or silage. It is less suitable for hay-making because it is difficult to dry the succulent stems and the leaves drop off easily when the plant is dried. Under favourable conditions it is possible to take six to eight cuts per year; recovery after cutting is usually good. The highest yield of protein and the relatively lowest fibre content are obtained by cutting the plants when they are 40–50 cm tall. Berseem is well adapted to climatic conditions of the Mediterranean area, although it is also grown in India, Pakistan and North and South America. It needs at least 250 mm rainfall in the growing period and cannot withstand severe frost. Berseem is often grown under irrigation to profit maximally of its vigour and its capability to recover rapidly after grazing or cutting.

Berseem tolerates rather high salt concentrations in the soil and a pH on the alkaline side of neutrality.

Persian clover

Persian clover is an erect annual herb, which can be 70 cm tall. The plant is glabrous; the leaflets do not have a leaf mark. It has only a single stem. The purple or pink flowers form a small head. It can be distinguished from other clovers by the flowers that are borne upside down, the largest petal (standard) is situated at the bottom of the flower, and the pea-shaped flower pods.

Two distinct types of Persian clover are grown: soft-seeded types and hard-seeded types.

It is grown by itself or mixed with grasses. Once established with grasses, it can reseed easily. Persian clover may produce a high forage yield and has a good regrowth potential after cutting or grazing. It provides highly palatable and digestible feed. When used as green feed, it has to be mixed with some dry roughage to prevent bloat.

Persian clover is adapted to areas with an average annual rainfall from 450 to 650 mm. It prefers heavy moist soils; it is not suited to acid sandy soils.

Lucerne

Origin, history and spread

Lucerne is probably native to an area south of the Black and Caspian Seas, and the Caucasus. It still occurs naturally in Iran and eastern Anatolia. Lucerne has been cultivated since antiquity. It was considered to be excellent feed for horses. The first written reference to lucerne dates from the 8th century BC. At that

Fig. 5.7. Lucerne: 1, flowering and fruiting branch; 2, leaf; 3, taproot with lateral roots and a few nodules; 4, flower, view from below; 5, flower, side-view; 6, fruit (legume); 7, seeds.

time the plant was called 'aspasti' in Persia, which became 'alfalfa' in the Arabic language. During the Persian wars in the 5th century BC, lucerne was brought to Greece. The Greeks passed it on to the ancient Romans, who subsequently spread it throughout Europe. The Spanish introduced it into South America in the 16th century; from there it spread northwards into California and the rest of North America. Colonists subsequently introduced lucerne to Australia, New Zealand and South Africa.

At present, lucerne is grown throughout the warm temperate regions and the subtropics.

Botany

Lucerne (Fig. 5.7) is a perennial herb, new growth starting each year from the crown. The crown is a plant section with buds near soil level. The stem can grow up to 1 m and up to 25 axillary shoots may rise from one single crown. The stem is four-angled, much-branched, glabrous or the upper part pubescent. The pinnately trifoliolate leaves are arranged alternately on the stems. Each leaf has two pointed stipules (Fig. 5.5) and a pedicelled terminal leaflet. Leaflets are obovate-oblong, ovate or linear, tapering to the base, crenate above, 10–45 mm long. The inflorescence is a roughly conically shaped raceme consisting of five to 40 flowers (Fig. 5.8). The colour of the flowers may be yellow, mauve or blue. The calyx is tubular with teeth longer than the tube. The corolla

Fig. 5.8. Inflorescences of lucerne.

is 6–15 mm long and consists of five petals, the two keel petals fitting tightly together. The fruit is a pod, 3–9 mm in diameter, with two or three spirals. There are six to eight seeds per pod; seeds are yellow or brown, ovoid irregularly, cordate or reniform; 1000-seed weight is 2.0 g. The taproot can penetrate the soil 7–9 m.

Varieties, uses and constituents

In the course of history, many varieties arose by selecting and breeding which are adapted to completely different local conditions. Varieties may also differ in disease resistance and growth characters such as spreading growth, either by underground rhizomes or by adventitious shoots rising from roots.

Lucerne is highly valued legume forage; it has the highest protein production per hectare of all forages. To balance the high protein level of the forage, it is mostly fed combined with a high-energy crop, often a grain such as wheat or maize. Ruminants fed on lucerne alone may suffer from bloat; feeding maize or other grains along with lucerne can prevent this. Lucerne can be used fresh by grazing or feeding immediately after harvesting, it can be made into hay or silage. Tolerance to grazing is sometimes improved by cross-breeding with *Medicago falcata* (*M. falcata* resembles lucerne but is more prostrate and has yellow flowers). Lucerne and mixtures of lucerne and grass make good silage. When lucerne is used alone, fermentation is improved by the addition of molasses. Digestibility and nutritive value decline with maturity, so it has to be harvested while it is young and leafy. Dried lucerne meal is added to poultry feed mainly because of the presence of carotene, which may colour the egg yolk dark orange.

Seedlings of lucerne are used for human consumption, in salads or for garnishing dishes. In China and Russia lucerne leaves are used as a vegetable.

Green forage of lucerne contains approximately 80% water, 5% protein, 1% fat, 9.5% total carbohydrates, 3.5% fibre and 2.5% ash.

As well as lucerne a number of annual species belong to the genus *Medicago*, referred to as bur-clovers. They are native to the Mediterranean region but reached North America. The species have rather weak stems, trifoliolate leaves with heart-shaped leaflets and pods that are spiny and coiled. Bur-clovers are valued in pastures or as green-manure crops.

Ecology and agronomy

Lucerne grows best on friable, well-drained, loamy lime-rich soil with loose topsoil. It is somewhat tolerant to salt. It can withstand high temperatures of about 40°C as well as low temperatures; it tolerates frost. Annual rainfall of 500–600 mm is required in temperate regions, but lucerne will survive on less. The crop is rather drought-resistant because of the deep-penetrating root system.

Lucerne has the ability to fix atmospheric N_2 in a symbiotic association with rhizobia, which penetrate the roots and colonize indeterminate nodules formed on the roots.

Establishment of lucerne after sowing can succeed only when the land is well-ploughed, weeds are absent and effective nodulation occurs. Lucerne seed has to be inoculated with *Sinorhizobium meliloti* bacteria, which are relatively specific for lucerne. When the soil is more or less acidic, seed can be inoculated and subsequently coated with lime. Lucerne may be sown pure or in a mixture with grasses. Seed rates vary from 10 to 20 kg/ha. Before the plants are established, irrigation may be necessary. Weed control in the seedling stage is very important because at this stage the plants are very sensitive to competition. When the crop is well established, frequent cultivation or spraying with herbicides to control weeds is usually required. To maintain soil fertility, application of farmyard manure and phosphate-containing fertilizer at sowing and after each cut is needed.

Lucerne can be harvested when the first flowers open. The number of possible cuts per year varies from one to ten, depending on water availability, although three or four cuts are normal. All processes from mowing to storage, both for hay-making and ensiling, can be completely mechanized. Hay-making has to be done very carefully because when leaves are dry they are extremely brittle during the day. To avoid large losses the hay is often baled early in the morning when there is some dew on the hay. Lucerne is therefore sometimes artificially dried and pelleted.

Average annual dry matter yield is about 15 t/ha. About half the total world acreage of lucerne is cultivated in North America.

FORAGE LEGUMES – TROPICAL

Stylo – *Stylosanthes guianensis,* Pinto peanut – *Arachis pintoi,* Desmodium – *Desmodium intortum;* Pea family – *Leguminosae*

Introduction

The three described legumes have several characteristics in common; they are all members of the family *Leguminosae,* which is one of the most important families in agriculture. They are all often grown as forage legumes, often in association with grasses. The species are also grown as cover crops or green-manure crops. Forage crops are grazed, cut or mown for fresh stall-feeding, dried or conserved by other methods. Forage legumes provide an N-rich component of an animal's diet; they improve soil fertility and stimulate the growth of associate species. On the roots of these legumes developing nodules are inhabited by rhizobia. The nodules are visible to the naked eye, but vary in their size and shape. The soil usually contains sufficient bacteria, but if these legumes have not been cultivated before, the soil, or the seed, may require inoculation. Part of the fixed N becomes available in the soil after decay of leaves, roots and other parts of plants; another part is ingested by animals, subsequently becoming recycled in manure. Both of these pathways lead to more N being available for associate plants, usually grasses, and succeeding crops.

These legumes are usually propagated from seed. Pinto peanut is also sometimes multiplied using cuttings. As the seed testa is sometimes impenetrable to moisture, seeds often need pretreatment for germination. Common methods are mechanical scarification, a hot-water treatment, or a treatment with acid.

The three species can be oversown in natural pastures, or drilled in cultivated land, often in mixtures with grasses. In pastures, the seeds are easily spread by passage trough-grazing animals. Due to N_2 fixation, N fertilizers are usually not required. However, to correct known soil deficiencies, application of P and K may be necessary. Weed control may be required, especially in pure stands. It is carried out by slashing, weeding or by chemicals.

Stylo

Stylo is native to Central and South America, especially central Brazil; the species spread naturally to the north and the south. There is also one African species called *Stylosanthes fruticosa.* Since the 20th century, stylo has been cultivated and found naturalized in many humid tropical regions. The species is an erect or semi-erect, herbaceous, perennial small shrub (Fig. 5.9). The stems can be glabrous or pubescent and up to 1.8 m tall. When the plant matures, the stems become woody at the base. The leaves are trifoliolate; the leaflets are lanceolate to elliptic, acute at the tip, 1.0–6.0 cm long, glabrous or pubescent, petioles 6–15 mm long. The inflorescence consists of several small sessile spikes, containing two to 40 flowers. The flowers are yellow or orange, often with black stripes, and subtended by leaf-like bracts (Fig. 5.10). The flower consists of a five-

Fig. 5.9. Stylo: 1, habit of
flowering and fruiting branch; 2,
fruit (line drawing: PROSEA
volume 4).

pointed glabrous or sparsely pubescent calyx tube; the standard is 4–8 mm ×
3–5 mm; the wings and the keel are 3.5–5 mm long. The fruit is a pod, 2–3
mm × 1.5–2.5 mm, usually glabrous and single-seeded; the fruit is indistinctly
veined, with a minute beak and strongly inflexed. The seeds are kidney-shaped
and yellowish-brown or purple. Stylo is adaptable to a wide range of soils,

Fig. 5.10. Stylo: flowers (left) and close-up of foliage (right) (photos: D.L. Schuiling).

including those low in phosphate content. Its climate range has been extended by selection of different cultivars, at present it is grown in tropical and subtropical climates. The species is propagated by seed, seeding rate is 2–6 kg/ha. Dry matter digestibility of young plants is 60–70%, but it will reduce at maturity. The palatability of stylo is not very high, which protects the species from overgrazing. A pure stand of stylo may yield up to 10 t of dry mass per hectare; its contribution in mixed pastures can be 2–6 t/ha.

Other forage stylos are *Stylosanthes hamata*, a species that is adapted to dry environmental conditions and low phosphate content of the soil; and *Stylosanthes scabra*, which is extremely drought-resistant.

Pinto peanut

Pinto peanut is native to the valleys of central Brazil. It was first collected and domesticated in 1954. Since then it has been distributed to other countries in South America, Central America, the USA, South-east Asia, Australia and the Pacific.

Pinto peanut (Fig. 5.11) is a stoloniferous perennial herb. The stems are prostrate at first, becoming ascendant, up to 25 cm high, especially in dense swards. The leaves are tetrafoliolate, margins entire; the distal leaflets are obovate, and the proximal leaflets oblong-obovate, obtuse at the tip and slightly cordate at the base. Leaflets may be 4.5 cm × 3.5 cm, but smaller in regularly mown stands. Flowers arise individually from the axillary buds; the flower is a characteristic pea-family flower (see Pea, Chapter 7), standard 12–17 mm wide, and yellow in colour. The ovary is borne on a gynophore, which elongates after

Fig. 5.11. Pinto peanut: flowering plants (left) and detail of mixed crop of pinto peanut and palisade signal grass (*Brachiaria brizantha*) (photos: M. Ibrahim).

pollination and pushes the ovary into the soil. The fruit is a pod, reticulated, and 10–14 mm × 6–8 mm. The terminal pod on the gynophore usually contains one seed, sometimes two. Accordingly, the pods, including the seeds, mature subterraneously. The seed is 8–11 mm × 4–6 mm, light brown; 1000-seed weight is 110–200 g. The species develops a strong taproot and many feeding roots.

Pinto peanut is adapted to a wide range of soils with low to neutral pH and low to high fertility. The climate has to be preferably humid (sub) tropical, with an annual rainfall of 2000–2200 mm. It is essentially a lowland species. The species is capable of growing under shaded conditions; therefore, it is suitable for growing in tree plantations. The plant is tolerant of periodic flooding; it cannot withstand frost.

Pinto peanut is often propagated from seeds; seeding rate is 10–15 kg in pods per hectare. If seed is not available, it can be propagated from cuttings. Once established in a pasture, the species is highly persistent due to its subterranean seed production. Pinto peanut is very tolerant of heavy grazing. It is a very promising forage; for example, smallholder farmers in Colombia proved that after introducing pinto peanut into the ration of cattle, milk production per lactation increased by 31% and cow fertility by 5%.

The plant is well accepted by cattle in all stages of growth. The dry matter digestibility varies from 60 to 76%. The dry matter yield of pinto peanut in mixtures can range from 5 to 10 t/ha. In contrast with the other leguminous species described here, pinto peanut is also grown as an ornamental.

Desmodium

Desmodium or greenleaf desmodium originates from the region extending from southern Mexico to southern Brazil. It was domesticated as a pasture plant in the 20th century. At present, it can be found in pastures in the humid regions of the subtropics and elevated tropics.

Greenleaf desmodium (Fig. 5.12) is a perennial herb with pubescent, trailing or climbing, square, grooved, glandular, reddish-brown stems up to 5 m long. The hairs are hooked, which make the stems adhere to clothing and animals' fur. The stems have a diameter of 1.5 to 4 mm, the internodes are 3–11 cm long, and the plant roots at the nodes if in contact with moist soil. Leaves are alternate and trifoliolate; leaflets are dark green, usually with reddish-brown marks on the upper surface, ovate-acuminate, softly pubescent on both sides, 3–12 cm × 1.5–7 cm; petioles up to 5 cm long. The terminal leaflet is usually larger than the other two. The inflorescence is a compact axillary or terminal panicle, up to 30 cm long. The flowers are borne in pairs, about 8 mm long, and with 30 to 50 deep lilac to pink-coloured flowers. They are characteristic pea-family flowers (see Pea, Chapter 7). The fruit is a small, narrow pod, 15–50 mm × 3–4 mm, curved to the main axis and covered with hooked hairs. The pod is up to 12-articulate, indented, and breaks up into pieces each containing one seed at maturity. Seeds are brown to green, kidney-shaped, 2 mm × 1.3 mm. The root system consists of a taproot and adventitious roots.

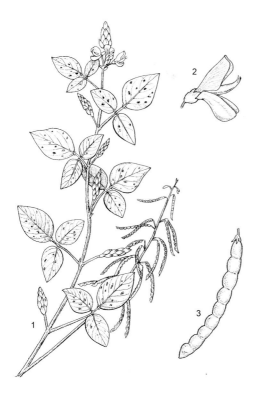

Fig. 5.12. Greenleaf desmodium: 1, flowering and fruiting branch; 2, flower; 3, fruit (legume) (line drawing: PROSEA volume 4).

Desmodium grows on a wide range of soils with pH not less than 5.0 (preferably 5.5) and not saline. It is grown at sea level in the subtropics but the species prefers elevated areas from 500 to 1000 m in the tropics. The species requires at least 1100 mm annual rainfall, preferably more. Desmodium cannot stand heavy frost; it is tolerant of some shade.

The plant is usually propagated from seed, although rarely also by rooted cuttings; seeding rates are 1–2 kg/ha. Desmodium is usually grazed by cattle, although sometimes it is cut and brought to the stable or cowshed for fresh feeding. Dry matter digestibility is about 55–60%. In pure stands, the annual dry matter yields of greenleaf desmodium may range from 12 to 19 t/ha. Next to greenleaf desmodium, the closely related silverleaf desmodium (*Desmodium uncinatum*) is also grown in pastures. Silverleaf desmodium differs from greenleaf desmodium by its dark green leaves with a silvery shiny spot, and the lack of the reddish-brown marks on the upper surface of the leaf.

6   Oil Crops

Castor bean

Castor bean – *Ricinis communis*; Spurge family – *Euphorbiaceae*

Origin, history and spread

It is assumed that castor bean is native to north-eastern Africa and the Middle East. It is a very old crop, which is proved by several indications in tombs of ancient Egyptians, and was probably domesticated some 6000 years ago. The Egyptians used the oil, which can be obtained from the seed, for their lamps, but the oil was probably also used medicinally. At an early date, the crop spread through the Mediterranean and parts of Asia, and, since about AD 800, in many other areas with a rather dry and hot climate. The name '*Ricinis*' is the Latin name for 'tick', because the appearance of the seed shows similarity with visual characteristics of a number of tick species. Castor bean is sometimes named 'Palma Christi'.

Nowadays the crop is grown in many tropical and subtropical regions, or in temperate regions with a hot summer. In these regions, the species often escapes cultivation and becomes a ruderal plant or a weed.

Botany

Castor bean (Fig. 6.1) is a perennial, monoecious, soft-woody shrub or small tree, which in the tropics can reach a height of 13 m and a trunk diameter of up to 35 cm. In cultivation, the species is usually annual and 1–5 m tall (Fig. 6.2). The stem may then be 5–15 cm in diameter, and is herbaceous and succulent. Stem and branches are often glaucous and have conspicuous nodes that often bear glands. The glabrous leaves alternate; they are orbicular, palmately

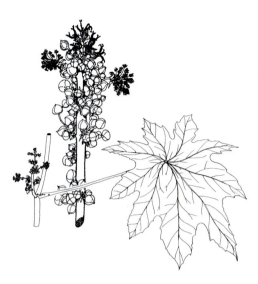

Fig. 6.1. Detail of inflorescence and leaf of castor.

compound, 10–50 cm broad, with six to 11 toothed lobes, margins are slightly serrated. Leaves have long petioles and prominent central veins. Stem, branches and leaves can be green, reddish or purple. The inflorescence is an erect terminal panicle, up to 40 cm long. The flowers are numerous; male flowers are borne at the base and female flowers are borne at the top of the

Fig. 6.2. A field of castor, Kenya (photo: J. Ferwerda).

Fig. 6.3. Castor: top of inflorescence with spiny fruits (left); smooth fruits and male flowers in bud and at anthesis (right) (photo: J. van Zee).

inflorescence (Fig. 6.3). Flowers have no petals; they have three to five sepals. Male flowers have numerous whitish-yellow stamens; female flowers have superior ovaries with three single-ovule cells and three styles with red, star-shaped stigmas. The fruit is an ellipsoid to subglobose capsule, at first green but turning brown when ripening, up to 30 mm long, often spiny, sometimes smooth. The fruit or capsule contains three carpels, each one containing a single seed. At maturity, the carpels may split open and eject the seeds. These characteristics occur mainly in wild types; spontaneous opening and ejecting seeds are usually not found in cultivars. Seeds are ellipsoid and bean-like, 10–20 mm long, and with a conspicuous caruncle at the base (Fig. 6.4). Seeds are often named 'beans' but they are not true beans as with the *Leguminosae*. The seedcoat is shiny and mottled, often grey-brown and black. The 1000-seed weight ranges from 250 to 1000 g. Castor bean has a strong taproot with prominent lateral roots.

Varieties, uses and constituents

A number of varieties can be distinguished based on different characteristics such as height of the plant, size of the leaves, colour of the plant, and several seed characteristics. Some varieties are selected as ornamentals. In temperate regions, the species is suitable for creating a tropical effect in a garden.

Castor bean is primarily grown for the oil from the seed. The oil is not edible, and for many ages it was used only for illumination and medicines. Nowadays it has

Fig. 6.4. Seeds of castor (photo: G. Guiking).

many different applications. Crude castor oil does not dry easily but dehydration of the oil improves the drying quality and makes it suitable for manufacturing paints, lacquers, inks and varnishes. Castor oil is also used as a high-grade lubricant and in hydraulic fluids. Important characteristics are the high lubricity and the high viscosity, which remains constant over a wide range of temperatures. The oil is also used in the manufacture of other products such as medicines (primarily purgatives), plastics, waxes, synthetic resins, cosmetics, nylon, and many more.

The meal or press cake, which is the residues of the oil extraction or pressing, is highly toxic. The toxicity is caused by the protein ricin, which is one of the deadliest natural poisons. A few sources suggest that even one single seed contains enough ricin to kill a child. The estimated lethal dose in an adult is 1 mg/kg of body weight. A few hours after consuming castor beans, symptoms of poisoning can be noticed. After several days there can be severe dehydration and a decrease in urine and blood pressure, which can lead to death. In the past, even in the recent past, ricin was regularly used to murder people. The toxicity can be destroyed by heat. The leaves of castor bean are also toxic but to a much lesser degree. At present ricin is used in cancer research, because it has been shown to have an antitumour effect. The seeds are sometimes used in jewellery, such as necklaces.

The 'dry' seeds contain approximately 5% water, 45–55% oil, 15–30% protein, 15–25% crude fibre, 5–10% carbohydrates and 2–4% ash. The oil consists of about 90% ricinoleic acid, and furthermore oleic acid, linoleic acid, palmitic acid and stearic acid.

Ecology and agronomy

Castor bean grows best at average day temperatures of 20–25°C, with a minimum of 15°C and a maximum of 38°C. For good growth the crop requires about 500 mm of rain between sowing and flowering. The species prefers full sun and low air humidity. When the plants are established, they can tolerate drought because of their deep root system. The crop requires a growing season of 140–180 days, and a day length of 12 h or more. Castor bean thrives best on well-drained, fertile, loamy soils with fine or medium texture, at pH 5–6.5.

Castor beans are propagated by seed. Before sowing, deep tillage of the soil is desirable to encourage development and deep penetration of the taproot. The seeds should be planted when the soil is warm enough, about 15°C. The seeds are planted usually in rows at a depth of 4–7 cm. Emergence of the seedlings may take up to 14 days. Depending on germination rate (which is usually low), 1000-seed weight, plant size and cropping system, 15–40 kg of seeds are needed per hectare. The number of plants per hectare may vary from 25,000 to 40,000. In-row spacing ranges from 25 to 40 cm, spacing between the rows is very variable and can be up to 1 m. In some tropical countries, castor bean is often intercropped or grown along field borders.

As germination is slow and seedlings are poor competitors, weed control is very important. The culture has to start free of weeds, which can be achieved by a pre-emergence treatment, either by herbicides or mechanically. About 6 weeks after sowing, weed control has to be carried out. It is often done mechanically, by hoeing or earthing up, sometimes by hand-weeding. If necessary, weed control has to be repeated.

As castor bean exhausts the soil quickly, fertilizers are needed. The rates depend on soil fertility and expected yield. Plant residues left on the field can serve as a green manure. One tonne of fruits removes about 25 kg N, 2.5 kg P, 8 kg K, 2.5 kg Ca and 1.8 kg Mg.

The crop can be harvested when all the capsules are dry and most of the leaves have fallen from the plant. Harvesting of large fields is fully mechanized. After harvesting and during storage, the seeds have to be handled with care because of their thin and fragile outer coat (testa), which is easily broken. Yields of seeds per year vary widely from 0.5 to 3 t/ha. In 2004 the world production of castor beans was about 1.3 million t . The most important castor bean producers are India, China, Brazil and the former USSR. The USA is the world's largest consumer of ricinus oil.

Coconut

Coconut – *Cocos nucifera*; Palm family – *Arecaceae*

Origin, history and spread

Coconut is native to the Indo-Pacific region, where the greatest genetic diversity occurs. However, its primary centre of origin and early distribution are uncertain. Coconuts probably spread spontaneously throughout many tropical areas in ancient times. The palms often grew, and still grow, near to the shore and ripe fruits fall to the ground. At high tide, the fruits float and drift away. Other coconut palms grow along riversides; their fruits may fall into the river and float out to sea. The fruits of wild coconuts have thick husks and germinate slowly. Therefore, the fruits can float long distances at sea without losing vigour. After being washed ashore, new trees can develop. It has been shown that some fruits are capable of producing a new plant after floating in the sea for 110 days. In that period, a distance of about 5000 km can be traversed. Fossil coconuts have been found in, among other regions, India, Melanesia and New Zealand. The plant was first described in AD 545 by an Egyptian monk named Cosmos Indicopleustes, who visited India and Ceylon. Coconut tree was probably first cultivated in India and Southeast Asia. Marco Polo visited Sumatra and India in 1280. He also described the local coconuts, which he called 'Indian nut' ('nux indica'). In ancient times, Austronesians probably spread coconut through the Pacific. The first European explorers in Asia and the Pacific found coconuts in almost all coastal areas. Subsequently, they took coconuts to the Caribbean, the Atlantic coast of tropical America and West Africa, making the species truly pan-tropical.

The botanical species name '*nucifera*' is derived from the Latin words 'nux' and 'ferre', meaning respectively 'nut' and 'to bear' (nut-bearing). Today, there are plantings of coconuts spread over tropical Asia, Central and South America, and Africa.

Botany

The coconut is a monocotyledon; it is composed of a series of joints, each having a node, a leaf and a short internode. It is an evergreen palm tree, which has after a number of years an erect, often slightly curved, smooth, greyish-brown trunk, marked with rings of scars left by fallen leaves, and terminated by a crown of leaves (Fig. 6.5). The length of the trunk can be up to 30 m, dwarf types about 10 m, 30–45 cm in diameter. The spirally arranged leaves are

Fig. 6.5. Plantation of coconut palms, Tidore, Indonesia.

Fig. 6.6. Coconut: inflorescence with numerous male flowers and few female flower buds (left) (photo: J. van Zee); flowering and fruiting apex of a coconut palm (right) (photo: D.L. Schuiling).

pinnately compound, feather-shaped, up to 6 m long and 2 m wide. Leaflets are linear-lanceolate, 0.5–1 m long, narrow, tapering, rigid and with entire margins. The leaf stalks are 1–2 m long and without thorns. The inflorescence is borne in the axil of each leaf of a bearing palm. Developing flowers are protected by two sheaths which form a spathe. When full grown, the spathe is 1–1.5 m long. The pressure of the inflorescence inside causes the spathe to rupture near the tip. The spathe may fall off when the inflorescence emerges. The inflorescence is 1–2 m long and consists of a central axis with many lateral branches (Fig. 6.6). Flowers are monoecious; female flowers are borne at the base of the branches, male flowers above. The number of female flowers per inflorescence varies from 20 to 40, male flowers are numerous. Female flowers are 2–3 cm in diameter; male flowers 1–1.5 cm. Flowers bear three sepals and three petals, and are yellowish in colour, male flowers contain six stamens, female flowers an ovary consisting of three carpels. The fruit is a drupe, roughly ovoid, up to 40 cm long and 30 cm wide, consisting of a smooth exocarp, a fibrous mesocarp (husk) and a stone-hard endocarp (shell). Within the endocarp is a single seed with a thin brown testa; white, oily, 1–2 cm thick endosperm (the 'meat') and a small, 0.5–1 cm long embryo lying embedded in the endosperm. The endosperm is soft at first but becomes firm at maturity. In addition to the soft or firm endosperm, the partially hollow seed also contains fluid endosperm, a liquid called 'coconut water', which is absorbed completely as the fruit matures (Fig. 6.7). The nut, consisting of endocarp and seed, is 25–30 cm long and has

Fig. 6.7. Longitudinal section of a young coconut fruit: 1, smooth exocarp; 2, fibrous mesocarp; 3, endocarp (stone-hard when ripe); 4, solid, white endosperm. Liquid endosperm (coconut milk) not represented (photo: D.L. Schuiling).

a diameter of 15–20 cm. At one end of the nut are three sunken holes ('eyes'), one of which is of softer tissue to enable the young shoot and root to emerge at germination. The exocarp of the fruit is green at first and will be greyish-brown when mature.

The swollen base of the trunk is surrounded by a mass of up to 4000 adventitious roots. Roots are mostly located in the top 1.5 m of the soil.

Varieties, uses and constituents

Cocus nucifera is the only species of the genus *Cocos*. Within the species two major groups of coconuts can be recognized: tall and dwarf types. The dwarf type probably originated as a mutation of the tall type. More than 95% of the cultivated coconuts are tall types. Tall types first flower 6–10 years after planting and they have a lifespan of 60 to 100 years. Dwarf types first flower 3 or 4 years after planting and have a lifespan of about 30 years. Within the two groups there are many varieties. In isolated regions and through selection, varieties have developed which are adapted to the different localities, having well-defined phenotypic characters, for example: Indian Tall, Ceylon Tall, Malayan Tall, Java Tall, Jamaica Tall, Dwarf Green, Dwarf Orange and Malayan Dwarf.

The coconut is a palm of many uses. It has been called 'the tree of life'. In Sanskrit it is called 'kalpa vriksha', which means 'the tree that provides the necessities of life'. This indicates that the palm and its many uses have been valued for a long time; indeed, today humans will use almost every part of the coconut palm. The liquid or soft-solid endosperm of the unripe coconut is drunk or eaten fresh or used in many different dishes, sauces and beverages; an alcoholic beverage that contains coconut is the well-known 'Pina Colada'. The liquid or coconut water is a refreshing and nutritious drink. It contains mainly sugar and a little oil. When the nut matures, the endosperm becomes more solid, the sugar content decreases and the oil content increases. The endosperm has a nice, nutty fragrance and a typical, slightly sweet taste. The solid endosperm, also called 'meat', is sometimes directly eaten as a food, although most of the meat is dried to produce 'copra', which is the main form in which coconut is traded. Most of the copra is used for oil production. Copra contains about 60–70% coconut oil. Coconut oil can be extracted from copra and used in a variety of ways including cooking, processing margarine, and manufacturing soaps. The oil is composed mainly of triglycerides of saturated fatty acids: 40–55% lauric acid and 15–20% myristic acid dominate. Several other fatty acids are found at lower concentrations, such as caprylic acid, capric acid, palmitic acid and oleic acid; the latter is the only unsaturated fatty acid. Shredded meat is often used in confectionery. The taste of meat can be combined very well with chocolate. By mixing grated meat with hot water, a milky-white liquid called 'coconut milk' will be obtained. It contains oil and aromatic substances and is used in a wide variety of Asian dishes.

The remaining nutshells after extracting copra from the nuts can be used to produce charcoal as cooking fuel. Coconut cake, the residue after the extraction of the oil from copra, is a valuable animal feed. Leaves are used for thatching roofs or for constructing shelters; leaflets are used to make baskets, hats, mats and bags. The hard wood of the trunk is called 'porcupine wood' and can be used as timber for furniture or buildings. The hollowed swollen base of the trunk is used as a drum in Hawaiian music-making. The fibrous husk of the coconut (mesocarp), known as 'coir', consists of short coarse fibres, and is used for making ropes, building materials and coconut matting. Coir dust is a component of potting mixtures for houseplants and ornamentals, because of the high water-holding capacity.

Apical buds of the palms can be cut off and eaten as 'palm-hearts'. It is an expensive vegetable because the palm will die after cutting. The end of the unopened spathe, surrounding the young inflorescence, can be cut for tapping palm juice. The juice has high sugar content and can be fermented to produce 'palm wine'. Distilled palm wine produces the well-known spirit 'arrack'. Evaporated palm juice produces a crude sugar called 'jaggery' (this name is also used for unrefined cane sugar).

The composition of fresh endosperm is 35–50% water, 35–45% oil, 3–4% protein, 9–11% carbohydrates, 2–3% fibre and 1–2% ash.

Root extracts, milk, oil and other coconut products have been used to treat various ailments, for example venereal diseases and dysentery; some coconut components are considered to be antipyretic, diuretic, laxative and antidiarrhoeic. The palm is also highly valued as an ornamental, for example along boulevards and on lawns. Finally, the coconut palm is much used as an inviting symbol of the tropics. Travel agencies like to use the image of the palm, combined with a blue sea and white beaches, in their tour guides and brochures.

Ecology and agronomy

For good growth, the palm requires an average minimum temperature of 22°C; optimum temperature is 27°C. Temperatures below 7°C may cause damage to young palms. Coconut requires full sunlight; annual sunlight has to be about 2000 h. Annual precipitation should be 1000–2000 mm, evenly distributed throughout the year. The palm prefers high relative humidity. Coconut palms will grow on most well-drained soils. The soils may have a wide pH range but preferably 5.5–7. Because the species is tolerant of salty soils, it can occupy a strip of land at the top of the beach. Because of that, the palm can be found along many tropical shorelines. However, salt is not required for growth, and coconut can be grown successfully inland. Most commercial plantings are confined to the tropical lowlands; however, coconut can also be cultivated successfully in a few warmer subtropical areas.

Propagation is entirely from seed. Large well-matured fruits are collected and usually placed in a nursery (Fig. 6.8), partly covered by soil. Several months later, the seedlings can be transplanted into the field by placing them in large dug holes, about 1 m in diameter and 1 m deep. The soil removed from the hole can be mixed with farmyard manure or other organic material before it is replaced, to provide nutrients and to retain moisture. After planting, the seedling has to be watered regularly until the plant is well established. A mulch layer on the soil around the plant will also help to retain soil moisture and repress weed growth. In commercial plantings the palms are planted at a spacing of about 10 m × 10 m. Although many coconut palms are never fertilized, they respond well to fertilization. Application of nutrients can be based on leaf analyses, which

Fig. 6.8. Germinating coconuts in a nursery.

is a quick method to determine nutrient requirement of the palm. It can also be based on the annual removal of crop nutrients per hectare: with yield of 7000 nuts/ha, the nutrient removal is approximately 49 kg N, 7 kg P, 95 kg K, 5 kg Ca, 8 kg Mg, 11 kg Na and 4 kg S.

Covering with a green-manure crop is often practised in young plantations; it provides N, represses weeds and prevents erosion. Weeding will sometimes be necessary, especially for young palms. Coconut is often intercropped with several different perennials, such as cacao, coffee, banana and pineapple.

The tall coconut starts fruiting 6–10 years after germinating; the species reaches full production at 15–20 years of age, and continues to fruit until the age of about 80 years. Fruits require about a year to mature, and harvesting can be done regularly throughout the year. Harvesting coconut fruits can be a risky undertaking. If fruits must not be entirely ripe, they have to be removed from the often very tall trees. It may be done by climbers who use the roughness left by old leaf-bases on the stem, to climb to the top. They are usually attached to the trunk with a rope. If the length of the palm allows it, the harvester can stay on the ground and cut the fruit with a knife on the end of a long bamboo cane. In some countries, trained monkeys are sent up the trees to throw the fruits down. If fruits are required entirely ripe, the fruits will be left to fall naturally.

To obtain the endosperm (coconut meat), the fruits are split open, usually on a sharpened iron stake, which is stuck into the ground. Next, the exocarp and the husk (fibrous mesocarp) are removed. The nuts are then broken open with a cutlass and the meat taken out of the shells. Coconut is mainly a small-holder's crop and less than 10% of the total area consists of larger plantations. Yield of coconuts varies widely, from 7000 to 16,000 nuts/ha/year. An average number of 6000 nuts is required for 1 t of copra.

Major producers are the Philippines, Indonesia, India, Sri Lanka, Mexico, Brazil, Mozambique, Tanzania and Ghana. The annual world production of copra, which is the main coconut product, is about 5 million t. The world area of coconut palm is about 11 million ha.

Groundnut

Groundnut – *Arachis hypogaea*; Pea family – *Leguminosae*

Origin, history and spread

The genus *Arachis* contains 40–70 species in seven sections and is of South American origin, from the region east of the Andes, south of the Amazon and north of the Río de la Plata. Most of the species are diploid but the cultivated groundnut, *Arachis hypogaea*, is an allotetraploid and was probably derived from the wild tetraploid *Arachis monticola*, which is native to north-western Argentina. The most likely diploid progenitors were *Arachis cardenasii* and *Arachis batizocoi*. The earliest archaeological records of groundnut in cultivation are from Peru, 4000–3000 BC. Cultivated groundnuts were

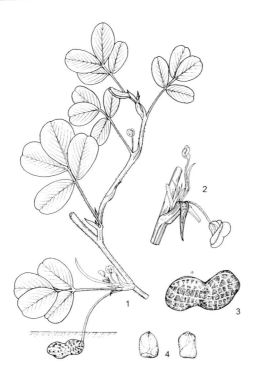

Fig. 6.9. Groundnut: 1, branch with flowers and fruit; 2, inflorescence; 3, fruit; 4, seeds (line drawing: PROSEA volume 1).

widely dispersed through South and Central America by the time Europeans reached the continent. After European contact, several major genotypes were dispersed around the world. The Portuguese apparently took groundnut from Brazil to West Africa and later to India in the 16th century. At the same time the Spanish introduced groundnut from Mexico into the western Pacific, whence it spread to China, Indonesia and Madagascar. Groundnuts apparently reached North America from Africa with the slave trade in the 17th century. It is remarkable that a relatively young protein and oil crop like groundnut became of such great importance to humanity. Within a century the world production of groundnut (in shell) increased from 0.13 million t in 1910 to over 36 million t in 2005. The main reason for this is the excellent nutritive value of groundnut. In contrast with the area of origin, where it is little used, the cultivation of groundnut has spread extensively in Asia, Africa and North America. Africa and Asia are now even regarded as secondary centres of diversity.

Botany

Groundnut (Fig. 6.9) is an annual, prostrate to erect herb, usually 15–60 cm high. Branching habit, branch length and hairiness distinguish the main botanical varieties from each other. The leaves are four-foliolate, with two opposite pairs of obovate leaflets of 3–7 cm × 2–3 cm and a petiole 3–7 cm long. Most of the root system is generally concentrated at a depth of 5–35 cm. The taproot can grow to a length of 90–130 cm in well-drained soils. The lateral roots are

Fig. 6.10. Flowering stem part in bunch-type groundnut (photo: J. van Zee).

shorter but basically similar to taproots; root hairs are present only in axillary branches, and N_2-fixing nodules are present. The spreading types usually have a more vigorous root system than the bunch types.

The flower is sessile but appears stalked after the growth of a 4–6 cm long tubular hypanthium (fused lower parts of calyx, corolla and staminal tube) just before anthesis (Fig. 6.10). The calyx is five-toothed and the typical papilionoid corolla is inserted on top of the hypanthium. The standard petal is generally orange, the wings yellow and the keel almost hyaline and faintly yellow. Groundnut is mostly self-pollinating because the stamens and stigma ripen before the flower opens. The pollination occurs in the early morning and fertilization around noon. The lifespan of a single flower is around 12 h. After pollination, the base of the ovary, called a 'peg' or 'gynophore', elongates and carries the ovary, located at its tip, into the soil. Elongation of the peg stops after it has penetrated the soil to about 5 cm. The apical region then swells and enlarges into a fruit (Fig. 6.11). The fruit is a cylindrical pod, 1–8 cm × 0.5–2 cm, containing one to six seeds (Fig. 6.12). The 1000-seed weight is 300–800 g.

Cultivars, uses and constituents

The gene pool of cultivated groundnuts has been divided by Kaprovickas (1973) into two, generally accepted subspecies, each with two main botanical varieties, as follows:

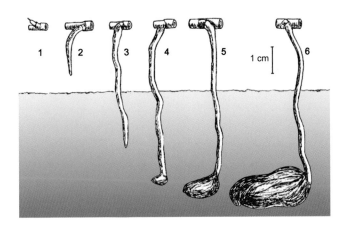

Fig. 6.11. Successive stages of fruit development in groundnut (after Smith, 1950).

1. Subspecies *hypogaea* (Virginia cultivar group), no floral axis on main stem, alternating pairs of vegetative and floral axes on laterals:
(i) Variety *hypogaea* – less hairy, branches short (Virginia type);
(ii) Variety *hirsuta* – more hairy, branches long (Peruvian runner type).
2. Subspecies *fastigata* (Spanish-Valencia cultivar group), floral axis on main stem, continuous runs of floral axis on laterals:
(i) Variety *fastigata* – little branched (Valencia type);
(ii) Variety *vulgaris* – more branched (Spanish type).

As a consequence of differences in growing habit and flowering phenology, the period of pod ripening is more or less extended. Runner types may have over-ripe, ripe and immature pods at the same time. Erect bunch types show a more synchronized ripening of the pods.

Breeding programmes have crossed cultivars from the two subspecies, and some subspecific traits such as branching pattern are no longer distinctive in cultivars developed from such crosses. Recent breeding programmes are focused not only on to yield, agroecological adaptation, tolerance to environmental stresses and resistance to pest and diseases, but also on improvement of flavour and quality desired by processors and consumers.

About 75% of the world production of groundnuts is used for the manufacture of edible oil. The press cake is a protein-rich livestock feed but is also used to produce groundnut flour, which is used in many human foods. Unfortunately groundnut protein can cause immediate hypersensitivity reactions such as angio-oedema and asthma in some individuals.

The seeds – raw, boiled or roasted – are also consumed directly, especially in the tropics where it is one of the most important food crops in some countries. The seeds are roasted for use in confectioneries, ground to make peanut butter, and eaten whole as snacks. The vegetative residues from the crop are excellent forage. Groundnut-in-shell consists of about 70% seeds, which are rich in both oil (43–55%) and protein (22–30%). Cultivars of the Virginia type tend

Fig. 6.12. Seeds of three different groundnut varieties, Bolivia (photo: D.L. Schuiling).

to have lower oil content than Spanish types. The protein content varies with type, cultivar, locality, year and physiological maturity of the seed. Groundnut protein, once used only as animal feed after oil extraction, is gaining creditability as a readily available source of protein to meet world demand. The protein quality is of a sufficient standard when added to a staple cereal diet. Groundnut seeds contain 10–15% starch, are rich in P, and are a good source of vitamins B and E.

Aflotoxin contamination is a major problem in many groundnut-producing countries. Aflotoxins are highly toxic, cancer-causing substances produced by the fungus *Aspergillus flavus*, which often infects pods and seeds.

Pinto peanut, *Arachis pintoi*, is a related, perennial species which is valued in the humid tropics as a pasture legume (see Chapter 5). Apios, *Apios americana*, commonly also called 'groundnut', is a leguminous native of North America which produces small, starchy, edible tubers in strings. Because of its fragrant, purplish-brown flowers it is also valued as an ornamental.

Ecology and agronomy

Groundnut is now cultivated between latitudes 40°N and 40°S in the warm tropics and subtropics up to an altitude of about 1000 m. Groundnut can be grown in temperate, humid regions with sufficiently long warm summers.

The optimum temperature for vegetative growth is between 20 and 30°C depending on cultivar. The lower critical temperature for growth is about 15°C. Approximately 1600 growing degree-days and 800 h of bright sunshine are required to harvest a good crop of groundnut. Groundnut is a species which is insensitive to day length, although opening of flower buds and total numbers of flowers are highly dependent on light intensity. Low light intensities can cause abortion of flowers.

Between 500 and 600 mm of water, reasonably well-distributed through the growing season, allows satisfactory production. Because pods develop underground and must be recovered at harvest, a friable, well-drained soil and a dry period for ripening and harvesting are prerequisites for cultivation. In wet conditions, harvesting and subsequent collection, trashing and drying of the crop are difficult.

Although groundnut is considered to be tolerant of acid soil and some cultivars grow well in alkaline soils to pH 8.5, optimal growth, nodulation and N_2 fixation are best at a pH in the range of 5.5–6.5. Groundnuts are particularly sensitive to low concentration levels of available Ca. Sandy or acid soils often need gypsum adding to prevent Ca deficiency. A shortage of Ca in the podding zone will result in empty pods or 'pops'. The response to N fertilizer is often small and erratic, even on N-deficient soils. This is in spite of the fact that, at high yield levels, the N requirement of nodulated groundnuts cannot be met entirely from symbiotic N. Inoculation of groundnut seed with efficient strains of *Bradyrhizobium* sp. of the cowpea group is necessary for areas where groundnut has not grown before. The efficiency of N_2 fixation by rhizobia depends on the genotype, rhizobial strain, soil moisture, temperature and nutrients. A groundnut crop may fix as much as 250 kg N/ha or more. Soil P concentrations required for groundnut are often lower than those required for other crops.

In most countries, groundnut is a rainfed and at least a partly mechanized row-crop (Fig. 6.13). Commercial crops are grown from seed at plant densities of up to 250,000/ha for Spanish short-season cultivars and up to 125,000/ha for Virginia long-season cultivars. Plant spacing between rows ranges from 20 to 40 cm; plant spacing within the row is about 20 cm. Although groundnut can emerge even from a depth of 20 cm, sowing deeper than 7.5 cm results in reduced emergence and yield. In heavier soils a sowing depth of 4–5 cm is recommended. Mechanical sowing is by precision planter, usually with a picker wheel rather than a plate, to minimize mechanical damage. Pre-emergence and post-emergence weed control is essential to achieve maximum economic yields,

Fig. 6.13. A monocrop of groundnut, grown on ridges (photo: J. Ferwerda).

either with herbicides or by hand weeding. Weeding cannot be done once peg penetration and pod development of the crop commence. A typical groundnut plant produces about 500 flowers throughout the growing season while only about 40 pods per plant eventually reach the desired size and are suitable for harvest. Harvesting should be done in bright sunshine by lifting the plants with pods intact so that the pods and vines can be dried thoroughly. When the pods are dried to about 10% moisture, they are separated from the vines mechanically or by hand. To minimize the development of the fungus *A. flavus*, the groundnuts should be dried to less than 14% moisture.

As a legume, the groundnut occupies an important position in cropping patterns based on cereals or root crops, for example, with pearl millet and sorghum in West Africa and India, or cassava and rice in South-east Asia. The world average yield of groundnut-in-shell is 1.4 t/ha. In marginal areas, yield is about 0.7 t/ha in sole cropping. In commercial cultivations, under optimum growth conditions, yields of more than 5.0 t/ha are possible. The largest recorded commercial yield of groundnut-in-shell, 9.6 t/ha, was harvested in Zimbabwe. China is by far the largest producer and counts for around 40% of the total world groundnut production. India, Nigeria and the USA are also important producers.

Oil palm

Oil palm – *Elaeis guineensis*; Palm family – *Arecaceae*

Origin, history and spread

Oil palm, also called 'African oil palm', originates probably from the Niger delta in West Africa. The natural habitat of this palm is likely to be at the edges of swamps and along riversides of the tropical rainforest region. The oldest indications of the use of palm oil are residues found in Egyptian pyramids dating back to about 5000 BC. In West Africa, more or less cultivated, wild oil palms have occurred in semi-wild groves for many centuries; palm oil has been (and still is) the major cooking oil of local populations. Written records of oil palm are available from Portuguese explorers of West Africa in the 15th century. In the 17th century oil palm reached Brazil through the slave trade. From there it spread to the rest of tropical America. The Dutch brought oil palm in 1848 from Mauritius to the Bogor Botanic Gardens in Indonesia. From Bogor it gradually spread to other countries in South-east Asia. Larger-scale exports of palm oil from Africa to Europe began at the end of the 18th century. Around 1900, West Africa was the centre of palm oil production, but that changed in the course of the 20th century.

The oil palm plantation industry started late, not before the first decade of the 20th century, in Indonesia, mainly in Sumatra. Next, commercial plantations were planted in Malaysia. In Africa, the first commercial plantations were established after 1920. The name *Elaeis* is derived from the Greek 'elaia', which means 'olive', it refers to the oil richness of the fruit. *Elaeis guineensis* means 'olive from Guinea'. Oil palm is one of the world's most important sources of vegetable oil and is extensively cultivated throughout the humid tropics.

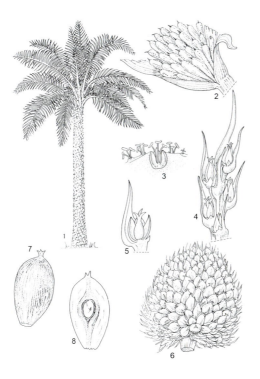

Fig. 6.14. Oil palm: 1, habit of fruiting tree; 2, male inflorescence; 3, detail of male flowers; 4, detail of female inflorescence; 5, female flower; 6, infructescence; 7, fruit; 8, fruit in longitudinal section (line drawing: PROSEA volume 14).

Botany

African oil palm (Fig. 6.14) is an erect monoecious palm that can be 8–25 m tall, however in cultivation usually no more than 15 m tall, with a terminal crown of about 50 leaves (Fig. 6.15). The trunk is unbranched, stout, 30–75 cm in

Fig. 6.15. Habit of young oil palm, Surinam.

Fig. 6.16. Male inflorescence and young fruit bunch of oil palm (photo: D.L. Schuiling).

diameter, ringed by scars of the leaf-bases, and covered by the persistent leaf-bases above. The basal part of the trunk is swollen. The leaves are spirally arranged, compound, juvenile leaves lanceolate, later becoming pinnate to paripinnate, consisting of 200–300 leaflets, and up to 8 m long. The leaflets are linear, 60–120 cm long, 2–5 cm wide and irregularly inserted on the rachis of the leaf. The leaves have sheaths which are tubular at first, later disintegrating into a mass of fibre. The fibres stay attached to the base of the petiole. The petiole is angled, bearing spines, and up to 2 m long. The inflorescence is axillary, solitary and unisexual, male and female flowers are borne in separate inflorescences on the same tree (Fig. 6.16). Several adjacent axils producing inflorescences of one sex are followed by several axils producing the other sex. The inflorescence is tightly enclosed in an ovate spathe before anthesis. Later, the spathe ruptures due to the pressure inside from the growing inflorescence. The central rachis has 100–200 spirally arranged spikes. The male inflorescence is ovoid, 20–25 cm long, with cylindrical spikes with numerous closely packed flowers. The female inflorescence is subglobose, 25–35 cm long, with a thick rachis; spikes are thick and fleshy, each in the axil of a spiny bract, with ten to 25 spirally arranged flowers and a terminal spine. The female inflorescence develops a large, tightly packed fruit bunch called an 'infructescence', which is about 50 cm long and 30 cm wide, weighing 4–15 kg, and consisting of 500

Fig. 6.17. Close-up of mature oil palm fruit bunch (photo: E. Westphal).

fruits or more (Fig. 6.17). The palm may produce two to six infructescences a year. The fruit is a globose to elongated sessile drupe, 2–5 cm long, weighing up to 30 g, and usually violet-black in colour. The exocarp is smooth, shiny, and orange-red when mature. The mesocarp is yellow-orange, fibrous, oleiferous, and encloses the dark brown stony shell or endocarp, which contains the seed.

   The root system is adventitious, forming a dense mat in the upper 0.5 m layer of the soil and spreading up to 5 m. Several thicker primary roots grow vertically, more than 1.5 m deep, giving the palm solidity. A palm tree can reach the age of about 50 years, although the economic lifespan is much shorter, usually 20–30 years.

Types, uses and constituents

African oil palms are usually classified into three types based on fruit character-istics, particularly the 'shell thickness'. The three types are: (i) 'Dura', having a thick endocarp of 2–8 mm; (ii) 'Pisifera', which has an endocarp that is thinner than 0.5 mm or does not have a lignified endocarp at all, is usually unproductive, but is used as the male parent in breeding programmes; and (iii) 'Tenera', having a relatively thin endocarp of 0.5–4 mm but more oil-bearing mesocarp than 'Dura'. 'Tenera' is the offspring of the cross-breeding of female 'Dura' flowers and 'Pisifera' pollen. Most of the cultivated oil palms belong to the 'Tenera' type. Most of the wild oil palms of West Africa belong to the 'Dura' type. The oil palm fruit yields two kinds of oil: palm oil from the fibrous, fleshy mesocarp; and palm kernel oil from the seeds. The mesocarp contains 45–55% oil, which is yellow to orange in colour. For manufacturing edible products, the oil is usually bleached.

Palm oil is used in the manufacture of margarines, cooking fats, candles and soaps; it is also used as a lubricant in the textile and rubber industry. Epoxidized palm oil is a plasticizer and stabilizer in polyvinyl chloride plastics. Crude oil can be used as bio-fuel for diesel engines.

The endosperm of the kernel contains about 50% oil, which is almost colourless, white or slightly yellow, and solid at lower temperatures. Palm kernel oil is used for the manufacture of edible fats, mayonnaise, ice cream, confectioneries, soaps and detergents.

Palm press fibre, palm oil sludge, and palm kernel cake are by-products of extraction and purification of oil, and can be fed to livestock, usually mixed with other feeds. By tapping the unopened male inflorescence, sugary sap can be obtained for making palm wine by fermenting the sap. Sap-tapping is profitable: 1 ha of 150 palm trees may yield 4000 l of sap, which is much more valuable than the oil yield of the same trees. The sap contains about 4.3 g sucrose and 3.4 g glucose per 100 ml. Sap is an important source of vitamin B-complex.

The central shoot of the palm or palm heart can be cut off and eaten as a vegetable, although it means that the palm will die. The leaves of oil palm can be used for thatching roofs. The leaflets are sometimes used to weave baskets and mats; leaflets also produce a strong fibre for fishing lines, etc. Oil palm is, like coconut, sometimes planted as an ornamental in lawns or along avenues.

The mesocarp of mature fruits contains 40–55% oil and 15–18% fibre. The endosperm of the kernel contains 6–8% water, 48–52% oil, 7–9% protein, 30–32% carbohydrates and 4–5% fibre. In folk medicine, the oil is used as a diuretic and anodyne; it is also used against rheumatism. It was considered to be an aphrodisiac.

The fresh fruit contains approximately 25% water, 2% protein, 59% fat, 13% carbohydrate, 3% fibre and 1% ash, and also Ca, P, Fe, carotene, thiamine, riboflavin, niacin and ascorbic acid. Palm oil (mesocarp) contains unsaturated fatty acids, mainly palmitic acid, oleic acid and linoleic acid. Palm kernel oil (from the endosperm) has a high content of saturated fatty acids, mainly lauric acid. Palm kernel oil is similar to coconut oil.

Ecology and agronomy

Oil palm shows the highest bunch and oil production in areas with an evenly distributed annual rainfall of about 2000 mm, a mean maximum temperature of about 29–33°C, a mean minimum temperature between 21 and 24°C, and at least 5 h of sunshine daily. These conditions are for instance found in Sumatra and Malaysia. Temporary flooding, as often happens along rivers, is tolerated but prolonged waterlogging is unfavourable. The species is found in agroecologies from savannah to rainforest, but grows best in the tropical lowlands. Oil palm is adapted to a wide range of soils, ranging from sandy to clayey soils, and pH values from 4 to 7. Soils should be deep and well-drained. Adequate water-holding capacity within the rooting depth is of great importance.

Oil palm is propagated from seeds. Rapid and even germination of seeds within 80 days can be obtained with dry-heat and wet-heat treatments.

Pre-germinated seeds are usually sown in polythene bags and transplanted from the nursery into the field when the plants are 12–18 months old.

In plantation agriculture land preparation usually involves planting of leguminous cover crops, to suppress weed growth and to maintain and improve soil fertility. Planting procedures involve digging of planting holes and circular weeding around the young plants. Usually a triangular spacing of 9 m is used, which gives about 125 palms/ha. At this spacing a closed palm canopy is reached within 5 years after planting. Smallholders usually do not use cover crops but tempo rarily cultivate the inter-rows with food crops. The semi-wild groves of oil palms in West Africa are often cleared to let 75–150 palms stand per hectare. Plants grow slowly at first; it takes 6–8 years before newly developed leaves reach their full size.

The annual uptake of nutrients by adult oil palm trees that yield 25 t of bunches per hectare is 1.4 kg N, 0.2 kg P, 1.8 kg K, 0.4 kg Mg and 0.6 kg Ca per tree. About 30–40% of the nutrients is removed by the harvested bunches, the rest is returned to the soil by the organic material that falls from the tree or is immobilized in the palm. Limited additional fertilizer is usually required if remains and wastes of oil extraction are returned to the field. In Malaysia, 2–4 kg N fertilizer and 1.5–3 kg K fertilizer per palm are commonly applied annually.

First harvesting of bunches begins when the palms are 3.5–4.5 years old, and reaches a maximum at about 10 years. At an age of about 25 years yields are so low and palms are so tall (10–12 m) that harvesting becomes a major problem and the palms are usually replaced by young ones or another perennial crop such as rubber.

Bunches ripen throughout the year, so harvesting also has to be done throughout the year, usually with an interval of about 10 days. Harvested bunches are brought to palm-oil mills for extraction of the oil. This is done by pressing or centrifuging the pulp of the mesocarp, or by macerating the pulp and boiling in water, after which the floating oil can be skimmed off. The kernel oil is extracted by shelling and grinding the seeds, then pressing them, seldom by using chemical solvents. The cultivation of oil palm is often criticized because it may happen at the cost of rainforest. Annual yields of fruits vary greatly from 12 to 32 t/ha, and palm oil yields from 2.5 to 7 t/ha. The main oil palm fruit-growing countries are Indonesia and Malaysia, together counting for 77% of the total world production. Nigeria is an important producer in Africa. The total annual world production of oil palm fruits has increased exponentially over the last 40 years, from 13.8 million t in 1965 to 173.3 million t in 2005.

Olive

Olive – *Olea europea*; Olive family – *Oleaceae*

Origin, history and spread

Olive and grapevine are considered to be the first cultivated tree crops. The domestication of the olive tree was not possible until man realized that it was very

Fig. 6.18. Olive orchard in Andalusia, Spain.

easy to propagate the olive tree vegetatively. Its cultivation is believed to date back about 6000 years in the ancient Mesopotamian region, from where it spread throughout the Mediterranean basin with the spread of Phoenician, Greek, Roman and Arab civilizations. Eventually every human settlement selected and propagated the best trees, mainly based on fruit size and oil content, which gave rise to many varieties. Today the world's most important region of olive cultivation can still be found in the Mediterranean countries. In the 17th century the Spaniards and Portuguese took the olive from Europe to America. More recently the olive tree has also been introduced into countries such as Australia, Chile and South Africa.

Botany

The olive tree is a large shrub in its native state. In cultivation it is a small to medium-sized, sometimes up to 8 m tall, evergreen tree, with a round, spreading crown (Fig. 6.18). Olive trees are the longest-lived trees of all fruit crops. Some trees in Europe are claimed to be more than 1000 years old (Fig. 6.19).

The leathery, oppositely arranged lanceolate leaves, 2–7 cm long, are green above and silvery-scaly underneath. The small, fragrant, cream-coloured flowers (ten to 40) are borne on panicles rising from leaf axils. The flowers are perfect or staminate and largely wind-pollinated, with most varieties being self-pollinating. Fruit set is usually improved by cross-pollination with other varieties. Incompatibility and self-incompatibility do occur. Mature olive trees produce large numbers of flowers, but fruit set is usually below 5%. The fruit is a drupe,

Fig. 6.19. A centuries-old olive tree with a characteristic gnarled trunk, Greece (photo: D.L. Schuiling).

green at first and dark blue or purplish when ripe, containing a single hard seed (Fig. 6.20). The olive never bears fruit on the same place twice and usually bears on the previous year's growth.

Cultivars, uses and constituents

The olive areas in the Mediterranean basin traditionally have their own restricted number of cultivars. The International Olive Oil Council developed a methodology for the identification of olive cultivars. However, the structure and composition of olive genetic resources are not yet fully known. Olive cultivars usually fall into one of two commercial uses: 'Oil' and 'Table'. 'Oil' cultivars predominate. The most famous are 'Picual', 'Arbequina', 'Cornicabra', 'Hojiblanca' and 'Empeltre' in Spain; 'Frantoio', 'Moraiolo', 'Leccino' and 'Pendolino' in Italy; 'Koroneiki' in Greece; 'Chemlali' in Tunisia; and 'Ayvalik' in Turkey. The 'Table' olive cultivars include 'Manzanilla' and 'Gorda' from Spain; 'Kalamata' from Greece; 'Ascolano' from Italy; and 'Barouni' from Tunisia. Throughout the world more than 70 varieties of olive are grown.

In the Mediterranean, 90% of the olive trees are grown for the oil. On average 5 kg of ripe olives are needed to produce 1 l of olive oil. Like wine, no two olive oils are alike. Each is a product of soil, climate, olive cultivar, age and

Fig. 6.20. Close-up of tip of olive inflorescence (left) and green olive fruit (right).

processing method. European Union directives have defined the different types of olive oil as follows.

1. Extra virgin olive oil: the fruits are only pressed once and the oil has an acidity, expressed in oleic acid, of not more than 1% and it should be of the purest quality. The best oils have acidity below 0.5%.
2. Virgin olive oil: this oil is also taken from the first pressing, but may have some defects in flavour. The acidity of this oil is 2% or less.
3. Olive oil: this is made up from a mixture of refined oils and a small amount of extra virgin oil. The acidity is 5%.
4. Sansa oil: this oil is made from the pressed residues of the olives by means of chemical extraction. It can be used for frying, but has a poor quality.

Olive oil, unlike wine, does not improve with age, and should be stored in dark glass bottles and used within 2 years of production. Extra virgin olive oil is a precious food, rich in chlorophyll, carotene, antioxidants (lecithin, polyphenols) and vitamins A, E and D. For use as table olives, the fresh, inedible olives are steeped in water to reduce their bitterness and bottled in a mild solution of salt. Olive oil is also used for manufacturing soap, cosmetics and medicines. Fruitless varieties are used for ornamental purposes. The wood is valued for small cabinet-work.

Ecology and agronomy

Olive is a drought-resistant perennial, surviving on only 200–300 mm of annual rainfall. For good yields irrigation is a necessity. Olive is sensitive to frost but requires a cold period in order to flower. It will grow on almost any well-drained soil up to pH 8.5 and is tolerant of mild saline conditions. Modern orchards are planted at 7 m × 7 m, 8 m × 8 m or 8 m × 6 m, allowing one stem per tree. Planting systems involve the planting of cuttings of different sizes or small trees. Grafting is also practised. Pruning is very essential for the olive and takes place in winter to obtain the basic shape, including removal of suckers, and in early summer to reduce the number of bearing branches. In southern Spain and northern Africa, inter-row ploughing or harrowing of olive orchards is sometimes practised for weed control and to improve infiltration of rainwater and reduce evaporation. Fertilizer requirement depends on soil fertility, age of the trees and fruit production per tree. Both soil and leaf analyses may be needed. Fertilizers are often applied in a circle around the tree which corresponds with the diameter of the crown. Application rates per tree may be 0.5–1.5 kg N, 4–8 kg K and 2–3 kg P.

The timing of harvest is very critical. Most growers prefer as much oil of good quality as possible. This means that the olives have to be of optimal ripeness. On the other hand, if the olives tend to be overripe, the fruits will oxidize immediately after harvest, which results in olive oil of inferior quality. In intensively managed orchards the olives are picked by hand. The pickers stand on ladders and strip the fruiting branches using a wooden comb. The fruits are then caught in nets that are stretched closely above ground level. In poorer regions the olives are collected from time to time after the fruits have dropped spontaneously into the nets. Because the time between harvest and pressing is critical, this latter method sometimes leads to problems. In cooler climates the harvested olives sometimes are left for a day or two near the press. During this time the fruits will warm up, while in the meantime the oil content will increase. If this is carried out carefully it will not influence the oil quality.

An average olive tree produces 15–20 kg of olives per year during 150 years or more. The world average yield of olives is 1.9 t/ha. In commercial orchards the yield may be twice as high. Alternate bearing is a problem in some cultivars, especially in semi-arid zones and in unirrigated groves. In the growing season 2004/05 the world olive oil production was 3.0 million t, of which 87% was produced in Europe and 12% in countries in North Africa and the Middle East. In these countries 90% of the olive trees are grown for the oil. In the same growing season, the world total production of table olives was 1.7 million t, chiefly grown in the same countries which produce olive oil. Spain, Italy and Turkey are the largest producers of both olive oil and table olives. About 6% of canned olives are produced in California, USA.

Rapeseed

Rapeseed – *Brassica napus, Brassica rapa;* Wallflower family – *Brassicaceae*

Origin, history and spread

Within the *Brassica* genus three diploid species are known: (i) *Brassica nigra* ($n=8$); (ii) *Brassica oleracea* ($n=9$); and (iii) *Brassica rapa* (syn. *Brassica campestris*, $n=10$). From these species three allotetraploids are achieved

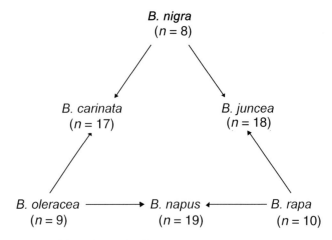

Fig. 6.21. Diagram of the relationship between important *Brassica* species.

through natural interspecific hybridization: (i) *Brassica carinata* (n=17); (ii) *Brassica juncea* (n=18); and (iii) *Brassica napus* (n=19). The two most important oilseed crops are *B. napus*, restricted to Europe and North Africa, and *B. rapa*, which has a much wider distribution (Fig. 6.21).

The primary centre of origin for *B. napus* is uncertain as the wild form has never been found. However, in the Mediterranean region there is an area of overlap in distribution of *B. oleracea* and *B. rapa*, which makes hybridization possible. So if a wild form of *B. napus* exists, its most likely origin is the area of overlap. The primary centre of origin for *B. rapa* comprises the Mediterranean, Afghanistan and Pakistan; in these areas the wild form can still be found.

Domestication of *B. rapa* and *B. juncea* started more than 4000 years ago in south and east Asia. The Romans grew *B. rapa* but they probably used the turnips as vegetables. The name 'rape' is derived from the Latin word 'rapum', which means 'turnip'.

In Charlemagne's *Capitulare De Villis* from AD 800, three *Brassica* species are mentioned: 'Sinape' (*B. nigra*), 'Ravacaulos' (*B. napus*) and 'Caulos' (*B. napus*). The latter two were grown as vegetables, probably for the swollen root and the leaves. In Dutch documents from 1360 and 1421 the word 'Raepsaet' (rapeseed) is mentioned, which suggests that rapeseed was already cultivated in The Netherlands. According to Toxopeus (2001), *B. rapa* had become a major crop in The Netherlands by the 14th century and subsequently in other countries of north-western Europe. Rape oil became the most important lamp oil in northern Europe. In the 17th century *B. napus* was developed in The Netherlands and spread from there into Europe. Rapeseed is basically a temperate-region crop, but breeding and selection have considerably increased its range.

Botany

The characteristic inflorescence of the *Brassica* spp. is racemose, with numerous yellow flowers, each with four sepals, four petals and six stamens, four long and

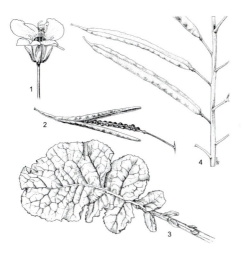

Fig. 6.22. Oilseed rape (*Brassica napus*): 1, flower; 2, dehisced fruit (siliqua); 3, leaf; 4, detail of stem with mature fruits.

two short ones. The fruit is a yellow-brown siliqua with a long, seedless, tapering beak, containing black, reddish-brown or yellow, more or less globose seeds. The seeds are 1–3 mm in diameter and attached to a thin false septum. The seedcoat is finely reticulate; the 1000-seed weight ranges from 2 to 5 g. *Brassica* spp. have a firm, 60–80 cm deep, taproot with numerous laterals.

B. napus (Fig. 6.22) is an annual or biennial herb; the plant has leaves on a short stem in the juvenile stage; when maturing it has an erect branching stem, up to 2 m high. The shape of the leaves is lanceolate. The lower leaves are deeply lobed, grass-green and bristly; the upper leaves are less deeply lobed, bluish-green and glaucous. The uppermost leaves are unlobed and clasp the stem partly with their deeply cordate base. Flowers have pale yellow to bright yellow petals, 11–15 mm long (Fig. 6.23), producing a fruit (siliqua) of 5–11 cm length, each containing 20–40 seeds (Fig. 6.24). The roots are slender.

Fig. 6.23. Detail of inflorescence of oilseed rape.

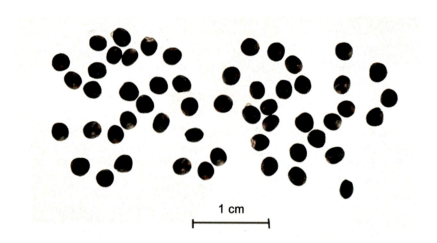

Fig. 6.24. Seeds of oilseed rape.

B. rapa (Fig. 6.25) has a rosette of leaves in the juvenile stage, and on maturing an erect branching stem. *B. rapa* has a lighter, shorter, more erect growth habit than *B. napus*. The shape of the leaves is lanceolate. The basal leaves of the flowering plant are petiolated, the top leaves are sessile and fully clasping the stem. All the leaves are bright green. Compared with *B. napus*, *B. rapa* has paler, broader-based leaves, which can be hairy. The flowers have bright yellow petals, 6–10 mm long, producing a siliqua 4–7 cm long with ten to 20 seeds. The roots are tuberous.

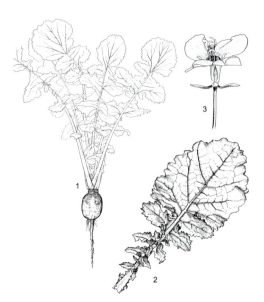

Fig. 6.25. Turnip (*Brassica rapa*): 1, vegetative plant with rosette of leaves and swollen root; 2, leaf; 3, flower.

Cultivars, uses and constituents

For a long period rapeseed was grown for producing lamp oil and later also for industrial lubrication. It was not popular as edible oil because of its pungency. It was also not suitable as edible oil because it contained glucosinolates and erucic acid. In oil processing or after eating in digestion, glucosinolates are broken down by the enzyme myrosinase into glucose and isothiocyanates or nitriles, which may cause serious illness in humans and monogastric animals. Erucic acid has toxic properties and is thought to increase the risk of cardiovascular ailments. In the 1960s a number of higher-yielding cultivars were developed of both *B. napus* and *B. rapa*, collectively known as 'canola'. These cultivars have a low content of glucosinolates and erucic acid and are therefore called 'double-zero'. Canola oil is not pungent and is fully interchangeable with other edible oils.

The average composition of rapeseed, in percentage of dry matter, is as follows: 45% oil, 25% carbohydrates, 25% protein, 2% fibre and 5% ash. The most important fatty acids in oil composition are 8–33% oleic acid, 12–21% linoleic acid, 8–14% linolenic acid and 25–55% erucic acid. In the 'double-zero' cultivars the erucic acid content is reduced to 0.2–1.5%, while oleic acid content is much higher, 55–63%, and the linoleic acid content, 20–24%. The latest developments are 'triple-zero' cultivars, which have also a low linolenic acid content. Linolenic acid causes the bad odour of the oil in cooking. A low content will improve the odour. Nowadays the oil of canola is used as cooking oil and in the manufacture of margarine and other food products. After oil extraction, the remnant has high protein content and is used as feed for livestock.

As well as rapeseed production, special varieties of *B. napus* and *B. rapa* can also been grown for their swollen root and hypocotyl, respectively known as 'swede' and 'turnip'. These plants are harvested in a vegetative stage. Both swedes and turnips are grown for human food and as feed for livestock. In folk medicine rapeseed oil is believed to strengthen the skin and keep it healthy. It is also used for massage.

Ecology and agronomy

As rapeseed is basically a temperate-region crop (Fig. 6.26), it prefers moderate temperatures during growth, day temperature below 25°C; higher temperatures during flowering and seed setting may decrease yield and oil content. Rapeseed is resistant to frost down to –10°C. In temperate regions rapeseed is grown between 60°N and 40°S, in the tropics at 1500–2000 m altitude and in the subtropics during the cool season of the year. Depending on vernalization requirement, there are winter and spring cultivars. *B. rapa* is more winter-hardy than *B. napus*. Approximately 450–500 mm rainfall is needed, especially during vegetative growth and the flowering period. However, to achieve maximum seed production a total of 700 mm is required. More rainfall may cause damage by fungal infections and waterlogging. Photoperiod requirements of rapeseed cultivars vary greatly. Some cultivars have been grown within the Polar Circle with 24 h of daylight; other cultivars were grown in India with 10 h of daylight.

Fig. 6.26. Field with flowering oilseed rape in Niedersachsen, Germany (photo: D.L. Schuiling).

Cultivars can be locally adapted and may not reach maturity outside their normal latitude. Rapeseed can be grown on a wide range of soils, from heavy clay to light sandy or volcanic ash soils. However, soils have to be well-drained to avoid damage by waterlogging. A pH of 5.5–8 is suitable for optimum growth; however, the crop is more tolerant to acidity than to alkalinity. Rapeseed tolerates some salinity; in The Netherlands, *B. napus* was mostly the first grown crop on land recently reclaimed from the sea.

Rapeseed is propagated by seed and is sown directly in the field, mostly by machine drilling in rows 20–45 cm apart; however, row widths depend strongly on machinery for pest and weed control. As the seed is rather small, the seedbed must be well prepared. This preparation depends on type of soil: for example, on heavy clay soils a very fine seedbed can result in compaction or crusting; over-cultivation of sandy soils can dry out the seedbed. The seedbed has to be weed-free. Seeds can be drilled 0.5–2.0 cm deep, depending on the soil type. In tropical areas it may be a little bit deeper (up to 5 cm) because the topsoil can dry out quickly; however, under these circumstances the soil has to remain loose and higher seed rates are used.

Seed rates can vary widely from 3 to 12 kg/ha, depending on cultivar used and sowing technique. In general, 7–9 kg/ha is sufficient for *B. napus* and 5–7 kg/ha for *B. rapa*. Rape seedlings compete poorly with vigorous weeds, so weed control in early growth is needed. Once stems begin to elongate, the crop is able to repress most of the weeds. In general, two mechanical or chemical weedings are sufficient. To achieve an acceptable seed production, irrigation may be necessary shortly before flowering and during seed filling; irrigation combined with supply of fertilizers can have a great effect.

Rapeseed responds well to fertilizers, particularly to N. A typical fertilizer for a summer rapeseed crop with a seed yield of 2 t/ha or more is 40–50 kg N, 20–25 kg P and 20–25 kg K per hectare in the seedbed, and another 40–50 kg N before flowering; however, it all depends on soil fertility.

Rapeseed usually will reach maturity about 180–240 days after sowing winter cultivars and 85–125 days with spring-sown cultivars. It can be harvested when the stems and the fruits are yellow and the fruits are rattling when shaken. Harvesting is best done before the whole crop is fully mature to avoid yield loss due to seed shattering. Therefore further drying to 6–8% moisture content is necessary to avoid decline of quality during storage. World average yield of rapeseed is about 1.5 t/ha, varying from 0.5 t/ha in India, 1.5 t/ha in Canada and up to 4.0 t/ha in Europe. In 2004, the world total rapeseed production was about 46 million t. China, Canada, India, Germany and France are the main producers.

Safflower

Safflower – *Carthamus tinctorius*; Daisy family – *Asteraceae*

Origin, history and spread

Safflower is not known as a wild species but it probably originated in the Middle East. Secondary centres of origin may have been Afghanistan, Ethiopia and India. The species was domesticated a long time ago. It was known as a cultivated plant in ancient Egypt, around 2000 BC. The crop was initially grown for the yellow or red dye that can be obtained from the fresh florets, later also for its edible oil. Safflower was probably introduced into China by traders who used the so-called 'Silk Road' around 200 BC, and from there into Japan in the 5th century. Both in China and Japan it was highly valued for dyeing silk. At the Imperial Courts of China and Japan, colours were used for ranking; different colours were associated with different ranks. For example, at the Japanese court only women of high rank could wear the colour 'kurenai', which is the red colour obtained from safflower. Kurenai was said to be more valuable than gold. Arab traders probably introduced safflower into India, where they used its dye for colouring textiles and food. In food it was often used as a substitute for the very expensive spice saffron, which is why safflower is also called 'false saffron'. From the Middle East safflower spread westwards into Europe and North and South America. The genus name is derived from the Arabic 'kurthum' and/or the Hebrew 'kartami', both meaning 'dye'. The species name comes from the Latin 'tinctor', which means 'a person who dyes'. Safflower means 'yellow flower'; it is partly derived from the Arabic word 'asfar', which means 'yellow'.

Botany

Safflower (Fig. 6.27) is a herbaceous, thistle-like annual that may grow to a height of 0.5–2.0 m, depending on sowing date and plant spacing. The stem

Fig. 6.27. Safflower: 1, growth habit (schematic); 2, flowering branch; 3, head cut lengthwise; 4, detail of a head; 5, apical part of a floret slit open; 6, ovary with pappus; 7, fruit (achene) (line drawing: PROSEA volume 14).

is whitish, erect, cylindrical, and solid with soft pith. The plant can produce many branches, each with a terminal flower head or capitulum (Fig. 6.28). The elliptic, entire, sharp-pointed leaves usually bear many long sharp spines, the base sessile or half-clasping the stem. The leaf-blades are 4–20 cm × 1–5 cm,

Fig. 6.28. Field of flowering safflower seen from above (photo: J. van Zee).

Fig. 6.29. Inflorescence (left) and fruits (right) of safflower.

the margins are bristly-toothed. Leaves are spirally arranged. The flower heads are globular and consist of 20–100 white, yellow, orange or red flowers, surrounded by bracts (modified leaves), which have scattered spines on the margins (Fig. 6.29). The flowers are tubular, about 4 cm long, with a five-lobed corolla, five stamens, and an ellipsoid ovary. They are borne on a flattened receptacle, with bristles interspersing the flowers. There are only disc flowers present. The fruit is an achene, which is smooth, four-sided and generally lacks a pappus. The colour of the hull can be white to cream or light brown, sometimes striped (Fig. 6.29). The 1000-seed weight is 40–80 g.

Safflower has a well-developed root system with a thick, fleshy and long taproot that can grow to 3 m, which makes the species suitable for dry climates.

Uses and constituents

Traditionally, safflower was grown for its florets, used for colouring and flavouring foods and dyeing textiles. The florets contain two major pigments: yellow carthamidin and orange-red carthamin. The florets were also used in medicines. Obtaining dye from the florets is not much used nowadays, due to replacement by chemical dyes. However, there is still a minor dye production for traditional purposes and also for the cosmetics industry.

Today, safflower is still widely cultivated commercially, mainly for the oil in the seed, and also for birdseed, although that includes different types of safflower. The seed may contain 35–60% oil. For oil production, a linoleic type of safflower is often used. The oil is very high in polyunsaturated acids; it contains 70–80% linoleic acid, which is considerably higher than the percentage in the oils of maize, soybean, olive and groundnut. Safflower oil is therefore highly valued as edible oil. It is used as a salad oil, cooking oil, and in producing soft margarine. Furthermore, the seed contains 10–15% oleic acid, 5–7% palmitic acid and 2–3% stearic acid. However, cultivars differ greatly in oil composition: some have high linoleic and low oleic acid content, while others have high oleic and low linoleic content. The meal that remains after oil extraction contains about 25% protein when the hulls are included, without the hulls it is about 40%. The meal can be used as a protein supplement for livestock. As safflower oil is a drying or semi-drying oil and does not yellow with age, it can be used for paint and varnish production. Young leaves of the plant are sometimes eaten as a vegetable. On a small scale, immature plants are used as forage.

Ecology and agronomy

Safflower is basically a crop of semi-arid, subtropical regions, although breeders have succeeded in expanding the range by selecting and breeding new cultivars. It is grown now within latitudes 20°S and 40°N. Average temperatures of 17–20°C are desirable for vegetative growth and 24–32°C for flowering and seed development. Water requirement depends on climatic conditions, but annual rainfall has to be at least 600–1000 mm. To achieve high yields, 1500–2500 mm is required. The crop requires full sunlight. Safflower is grown on a wide range of soils and tolerates a pH range of 5.5–8. However, it thrives best in deep, well-drained sandy loam of neutral reaction, with a high water-holding capacity and high level of stored moisture.

Safflower is propagated by seed. The seeds can be drilled by a grain drill in rows, 5–6.5 cm deep. The rows may be 15–30 cm apart and the spacing in the row may be 3–8 cm. However, in practice it varies greatly because it depends on availability of water and planting date. The planting date can be early because seedlings of safflower tolerate temperatures as low as −7°C.

Application of fertilizer depends on many factors such as yield expectations, moisture availability, previous cropping or soil fertility, and planting date. In practice the amount of applied N ranges from 80 to 200 kg/ha, with P and K in proportion.

Weed control is extremely important to optimize seed production; especially broadleaf weeds can reduce the seed production by about 70%. It is therefore very important to control weeds, mechanically or chemically. Safflower planted in rows can be harrowed shallow or tilled several times until just before flowering. What methods can be used depends mainly on the distance between the rows. Growing safflower has a rotational benefit: the species provides a disease-break crop especially for cereals, sugarbeet, cotton and grain legumes; by its deep-penetrating root system it can reach water and nutrients that have moved below the rooting depth of shallow-rooting crops.

Safflower should be harvested when the leaves become dry and brown and the moisture content of the seed is 8% or less in order to ensure a safe and long-term storage of the seed. This makes direct combine-harvesting possible. If the moisture content is too high, further drying will be necessary. Windrowing in the field or artificial drying can achieve this. The average seed yield of safflower grown under rainfed, intensive cultivation is 1.5 t/ha and about 3.0 t/ha under irrigation. In 2005, the total world seed production was 0.8 million t. At present the crop is grown mainly in Mexico, India and the USA.

Sesame

Sesame – *Sesamum indicum*; Sesame family – *Pedaliaceae*

Origin, history and spread

Sesame (*Sesamum indicum*, syn. *Sesamum orientale*) is probably native to Africa, although others suggest that the species originated on the Indian subcontinent. There are 4000-year-old archaeological relics of sesame, found in the Harappa valley in Pakistan. Sesame is probably one of the oldest cultivated crops known to man. In ancient times it was already a highly valued crop in Babylon, Assyria and Egypt. In 2300 BC, Babylonian women ate the seed, mixed with honey, to improve their fertility. Assyrian tablets from 2300 BC describe the drinking of sesame wine by the gods. Ancient Egyptians ate the seeds and used sesame oil for their oil lamps. Sesame seeds were also found in the tomb of Tutankhamun. Sesame was on the 3500-year-old 'Medical Papyrus' of Thebes in Egypt about herbal remedies. The name 'sesame' is probably derived from the Greek word 'sesamon', which has probably a Semitic origin, meaning 'oil' or 'fat'. The well-known phrase 'Open sesame!' from the famous fairy tale 'Ali Baba and The Forty Thieves' was meant to open a cave with a treasure, and it refers to the splitting capsule of the ripening sesame plant that also gives access to riches (seeds and oil).

In 600 BC in China, sesame was considered such a valuable crop that the seeds were sometimes used as currency. The Chinese used sesame oil as lamp oil; the soot of the burned oil was often used as stamp-pad ink. The Portuguese introduced sesame into South America, and African slaves brought sesame to North America.

Botany

Sesame (Figs 6.30 and 6.31) is an erect, branched or unbranched, annual herb, usually 1–1.5 m tall, occasionally up to 2 m tall. The stem is firm, square, pubescent and about 1 cm in diameter. The leaves are opposite or alternate; shape varies from ovate to lanceolate, lower leaves trilobed, upper leaves undivided irregularly serrate, often pointed. Leaf size varies considerably, leaf-blade length 12–31 cm, blade width 2–24 cm, length of the petiole 5–20 cm. Higher leaves are shorter-petioled than lower ones. Older cultivars have smooth and flat leaves; new cultivars have often cupped leaves with leaf-like outgrowth on their lower side. Flowers are borne, single or more, on short glandular pedicels in axils of leaves (Fig. 6.32); they are campanulate, two-lipped, white, pink or mauve in colour and 2–3.5 cm long. The flower consists of a five-lobed slender calyx, a five-lobed campanulate corolla that is horizontally flattened, four stamens, the upper two shorter than the lower two, and a superior ovary which is rectangular longitudinally in shape with a 1 cm long style bearing a two-lobed stigma. The fruit is an oblong, grooved capsule, 1.3–7 cm long, with a short triangular beak, with two to four chambers, containing 30–120 white, yellow, brown, grey or black seeds. Seeds are small, obovate and flattened, 2–3 mm in diameter and 0.5–1 mm thick (Fig. 6.33); 1000-seed weight ranges from 1.2 to 4.5 g. The

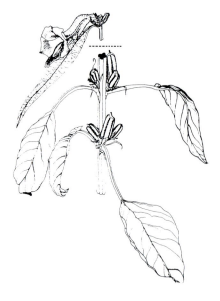

Fig. 6.30. Details of sesame plant: flower in axil above; fruits in axils below.

root system is extensive; the taproot can be 100 cm long, bearing many laterals.

Species, uses and constituents

The genus *Sesamum* comprises about 35 different species. As many of these species can be crossed successfully, there are many hybrids. This is one of the reasons why there is so much variability in the sesame cultivars. The cultivars

Fig. 6.31. Field with flowering sesame, Mali (photo: D.L. Schuiling).

Fig. 6.32. Close-up of sesame flowers with pubescent corolla (photo: D.L. Schuiling).

Fig. 6.33. Detail of a sesame stem with closed and dehisced fruits (left) and fruit in longitudinal and cross-section (right) (photo: B.P. Schuiling).

are usually divided into two types: (i) the older and seed shattering cultivars; and (ii) the relatively recently bred, mostly non-shattering cultivars.

About 65% of the world's sesame seed production is processed into oil and meal. Sesame oil is semi-drying, pale yellow in colour and almost odourless. The oil has good stability due to the presence of antioxidants. It is an excellent salad oil, but is also used for cooking. The oil is also used in manufacturing many products such as margarine, soap, skin softeners, paints and drugs. Part of the sesame oil is used in the manufacture of insecticides. The meal or cake that is left after the oil is pressed or extracted from the seed is an excellent high-protein feed for poultry and livestock. Sesame seeds are available hulled, un-hulled, and also ground into flour.

Whole seeds are used on sesame-seed buns, on bread, on crackers, and in cakes and confectionery. The seeds are valued because of the nutty taste after roasting. In Africa, sesame seeds are used in soups. In India, roasted and salted sesame seeds are consumed as a snack. In China, Japan and Korea, roasted seeds are also used as a spice in several dishes. In the Middle East, sesame seeds are manufactured into a paste called 'tahini', which is used as a spread.

Sesame seed contains about 5% water, 19–30% protein, 34–60% oil, 15% carbohydrates and 5% fibre. It also contains vitamin A, thiamine, riboflavin, niacin and some ascorbic acid. The composition of the oil varies considerably; the oil consists of glycerides, on average with about 47% oleic acid, about 39% linoleic acid, 10% palmitic acid and 5% stearic acid. Several pharmaceutical applications of sesame seeds or oil have been used. In folk medicine it has been used as a laxative, and in the treatment of dizziness and headache. The oil is known to reduce cholesterol due to its high content of polyunsaturated fatty acids and sesame seems to have a capacity to prevent cancer.

Ecology and agronomy

Sesame thrives best in warm areas with a long growing season, with daytime temperatures of about 26°C. Below 20°C, growth will decrease considerably and below 10°C there is hardly any growth at all. A minimum rainfall of 500–600 mm per growing season is required to obtain a good yield. Although sesame is rather drought-tolerant, the species requires moist soil for germination and seedling development. Sesame can be grown on many soil types; however, it grows best on well-drained fertile soils of neutral pH. The species has a low salt tolerance and does not tolerate waterlogging (Fig. 6.32).

Sesame is propagated from seed. Different sources suggest different seed rates (1–10 kg/ha), different row spacings (45–100 cm) and different plant densities (25–75 plants/m^2). However, Turkish research about optimizing plant densities showed that in Mediterranean types of environment the highest seed yield was obtained from about 500,000 plants/ha with row spacing of 70 cm. Seeds are usually planted 2–5 cm deep, although in loose soil it may be 10 cm, to ensure the availability of water.

In the cultivation of sesame early weed control is required, because of the slow early growth. Only pre-emergence herbicides are available, so weeding has to be done manually or mechanically. Two or three weedings are often necessary in the juvenile stage of the plants, followed by another two or three weedings in later stages. Shallow weeding is needed to avoid root damage. Intercropping with other crops can also be used as weed control, but this is usually done only by smallholders.

Application of fertilizer depends on the nutrient pool in the soil and the expected yield. The amount of nutrients removed by a crop per tonne of seed is about 30 kg N, 14 kg P and 5.5 kg K. The amount of nutrients can best be split into two equal applications, one just before planting and the other at the beginning of flowering.

Harvesting can start 80–150 days after planting, when the lower fruits start to dry out, the colour of the leaves and the stems change from green to yellow and red, and the leaves begin to fall from the plants. Shattering cultivars are mostly harvested manually. The harvested plants are often placed upright in sheaves to let them dry in the field. After about 2 weeks, the crop can be threshed. Non-shattering cultivars are often harvested with combine harvesters.

Seed yields vary considerably from 0.35 to 2.5 t/ha. In 2005, the total world production of sesame seed was about 3.3 million t. Today, India and China are the main producers of sesame.

Sunflower

Sunflower – *Helianthus annuus*; Daisy family – *Asteraceae*

Origin, history and spread

The name '*Helianthus*' is created from the Greek words 'helios' (sun) and 'anthos' (flower). Sunflower is native to the south-west USA–Mexico area. It is the only important crop of indigenous origin which has evolved within the USA. At present, the wild species still can be found as a weed in grasslands or along the roadsides. Wild sunflower is highly branched with small flowering heads. Native Americans used the seeds of the wild plants for food; the seeds were roasted, made into flour or used for producing oil. Some Indian tribes used the plant as a medicine and it played a part in their ceremonies.

When sunflower was brought into cultivation is not completely clear, but it is believed to be around 1000 BC in the east of the USA. When the first European colonists came to North America, sunflower was already cultivated. The plant reached Europe from Mexico into Spain in the 16th century. First it was grown as an ornamental plant; the Russians were the first to cultivate sunflower as an oil crop. The first commercial oil production was in 1830–1840.

At present, sunflower is cultivated around the world both in tropical and temperate regions.

Botany

The cultivated sunflower (Fig. 6.34) is an erect, often unbranched, coarse, annual herb, with a varying height up to 4 m. The stem is robust, circular in

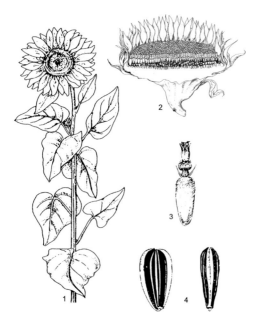

Fig. 6.34. Sunflower: 1, upper part of a flowering plant; 2, head cut lengthwise; 3, flower with inferior ovary; 4, two views of seed (line drawings 2–4: L. van de Burg).

section, 3–6 cm in diameter, curved below the head, and woody when mature. The stem is filled with white pith that often becomes hollow with age. Leaves are usually alternate (lower leaves opposite), ovate, cordate, with three main veins, 10–30 cm long, 5–20 cm wide, margin serrate, and carried on long petioles. The colour of the leaves is usually dark green. The disc-shaped flowering head is borne terminally on the main stem, 10–50 cm in diameter, sometimes drooping, and containing 800–8000 bisexual florets. Around the margin of the head there are individual ray flowers, which are sterile, brightly coloured, usually yellow but varying from deep yellow to red. The brown or purplish disc florets are spirally arranged, flowering from the outer to the centre. The ovary is inferior with a single basal ovule. The fruit is an achene, black, brown or white, and it can also be striped. Seed is compressed, flattened oblong, the top truncated and base pointed, 10–25 mm long, 7–15 mm wide. The 1000-kernel weight varies from 50 g to many times this. The flowering head is heliotropic (rotating to face the sun). The root is a taproot, which can penetrate the soil to a depth of about 3 m, with a large lateral spread of surface roots; however, most of the roots generally remain in the first 50 cm.

Cultivars, uses and constituents

Sunflower cultivars can be divided into three types: (i) giant types, 1.8–4.2 m tall; (ii) semi-dwarf types, 1.3–1.8 m tall; and (iii) dwarf types, 0.6–1.4 m tall. In connection with final uses, cultivars can also be divided into three groups: (i) for seed production; (ii) the whole plants as feed for livestock; or (iii) as ornamentals. Seed is used for oil production, as salted and roasted snacks, or as birdseed. Sunflower oil is considered a premium oil due to its light colour, low

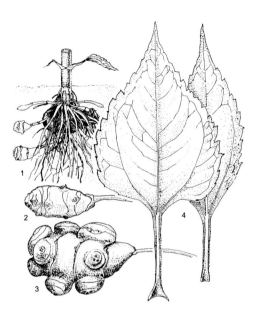

Fig. 6.35. Jerusalem artichoke: 1, subterranean part of the plant with seed-tuber, root system and stolons with young tubers; 2, tuber; 3, tuber with second growth; 4, leaves.

level of saturated fats, mild flavour, good taste and ability to be used at high cooking temperatures. Oil content of the seed varies from 25 to 65%. Oil composition depends on temperature and can be 20–60% linoleic acid and 25–65% oleic acid, protein content is 15–20%. Sunflower oil is mainly used for food purposes. Inferior grades of oil are used for the production of paint, varnish and soap. The remaining material after oil extraction has a protein content of 28–45% and is used as cattle feed. Sunflower is sometimes grown as a silage crop as feed for livestock, when the crop has to be harvested when half of the flowering head of the plant has mature seeds.

Belonging to the same genus is Jerusalem artichoke (*Helianthus tuberosus*). This plant resembles sunflower, but is a hardy perennial, has many branches, no or much smaller flowering heads, and forms tubers (Figs 6.35 and 6.36). The texture of the tuber flesh is more or less like young potatoes and the taste is slightly sweet. They can be eaten raw or boiled like potatoes, or can be used as feed for livestock. The tubers contain inulin, which can be used to produce fructose.

Ecology and agronomy

Sunflower prefers the warm temperate regions (Fig. 6.37), but is cultivated from 40°S to 55°N. Sunflower is unsuitable for humid climates. New cultivars are adapted to a wide range of environments. The plant grows well at temperatures of 20–30°C, although a range of 8–34°C is tolerated. A frost-free period of 120 days is usually necessary for commercial crops. Good yields can be obtained with 500 mm of rainfall or irrigation water. The plants are quite drought-resistant, except during the flowering period. Sunflowers grow well in any well-drained, neutral to slightly alkaline sandy soil. Seed should normally

Fig. 6.36. Lower part of Jerusalem artichoke plant with tubers.

be planted 3–8 cm deep, depending on cultivar. A seedbed temperature of 25–30°C is best but should not be below 15°C. Seed rate, spacing and number of plants per hectare vary widely, depending among other things on seed size: for example, the number of plants per hectare varies from 28,000 to 75,000. To obtain high seed yields, application of fertilizer is necessary. It was found that

Fig. 6.37. Field with flowering sunflowers in Pfalz, Germany.

1000 kg seed removed 60 kg N, 10 kg P and 180 kg K. Sunflower seedlings are poor competitors to weeds, and young plants are easily damaged by mechanical weed control. Moreover, the period of effective machinery use is limited. Several herbicides are registered for use with sunflower.

The growth cycle is usually about 4 months but ranges from 60 to 180 days depending on the environment and genotype. Sunflower can be harvested by hand or by combine harvester, and has to be done when the back portion of the heads turns brown. In 2005, the average seed yield was 1.2 t/ha and the total world seed production was about 31 million t. Much higher yields, of up to 3.4 t/ha, may also be achieved. Main production countries are Argentina, Ukraine and China.

7   Protein Crops

Chickpea and Pigeonpea

Chickpea – *Cicer arietinum*, Pigeonpea – *Cajanus cajan*; Pea family – *Leguminosae*

Chickpea

Origin, history and spread

Chickpea, also called 'Bengal gram' or 'garbanzo beans', may have originated from south-eastern Turkey where the wild subspecies *Cicer arietinum* subsp. *reticulatum* occurs. The wild subspecies and cultivated *Cicer* can easily be crossed, and have many similarities in appearance. Larger chickpea seeds than wild ones, which have been found in archaeological sites in Jericho dating back to 6500 BC, suggest that domestication happened some 8500 years ago. Chickpea has been cultivated in India, the Middle East and parts of Africa since ancient times. The crop was known to the ancient Egyptians, Greeks and Romans. In the Middle Ages, the species was described in Charlemagne's *Capitulare de Villis* as 'C. italicum'. The botanic Latin name of the species, 'arietinum', means 'ram-like', because the seed resembles a ram's head.

Botany

Chickpea (Fig. 7.1) is an erect or a spreading annual herb with a square stem and numerous branches, the height ranges from 25–90 cm. The whole plant is covered with glandular hairs. The leaves are pinnately compound, 7–10 cm in length, with nine to 17 small leaflets which are about 2 cm in length. The midrib terminates with a leaflet. The leaflets are ovate or

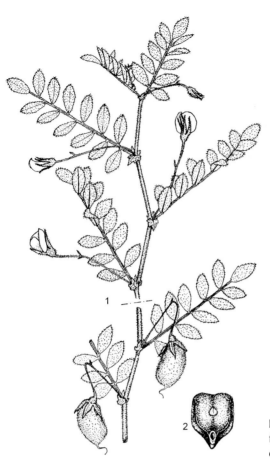

Fig. 7.1. Chickpea: 1, branch with flowers and fruits; 2, seed (line drawing: PROSEA volume 1).

elliptic with a serrated margin. The whole foliage has a feathery appearance. The flowers are borne axillary and usually solitary, on peduncles that are 2–4 cm in length. The flower is a characteristic pea-family flower (see Pea, this chapter), calyx united with five teeth, corolla about 1 cm in length, white, green, pink or blue in colour; standard broad and clawed, wings free, and keel incurved. The fruit is a swollen, pubescent pod, oblong, up to 3 cm × 2 cm, containing one or two seeds (Fig. 7.2). The seeds vary considerably in colour, shape and size. They may be white, yellow, green, brown or black in colour, smooth to very wrinkled, having a 1000-seed weight ranging from 110–380 g. Chickpea has a well-developed taproot and an extensive lateral root system, which usually bears numerous large indeterminate nodules caused by *Rhizobium* bacteria.

Varieties, uses and constituents

There are many chickpea cultivars, which can be classified as either 'desi' or 'kabuli' types. Desi types have smaller, angular and pigmented seeds; kabuli

Fig. 7.2. Close-up of chickpea flower (left) and unripe fruit (right).

types have larger seeds, which are more rounded and lack pigmentation. Desi types are grown mainly in India, Pakistan and Ethiopia, kabuli types predominantly in other countries.

Chickpea is mainly eaten as a pulse; whole dried seeds can be cooked or boiled. In India it is often eaten in the form of 'dhal', which is a thick soup made from split seeds without their seedcoats. Seeds ground to make flour called 'baisin' are used in confectionery, soup and bread-making. Green pods and young shoots are eaten as a vegetable. Green pods can also be shelled for the peas, which are eaten as a snack food or as a vegetable. Sprouted seeds can be consumed as a vegetable, or added to salads. Chickpea is usually marketed dried or canned; the latter may be either whole seeds or mashed.

After threshing and producing dhal, the remaining dried leaves, husks and stems can be used as animal feed. Dry seeds contain approximately 10% water, 17% protein, 60% carbohydrates, 5% fat, 4% fibre and 3% ash. Fresh seeds are a source of vitamin C. From the glandular hairs on the leaves and the stems, an acrid liquid is sometimes collected. The liquid contains about 94% malic acid and 6% oxalic acid and is used as vinegar or in medicines. It has been used in the treatment of dyspepsia and constipation. Oxalic acid may irritate the skin, although cooking reduces the oxalic acid content.

Ecology and agronomy

Chickpea can be grown in tropical, subtropical and temperate regions (Fig. 7.3). Optimum temperature for the day is 20–27°C, for the night 18–21°C. It prefers full sunlight. As a cool-weather crop, it is rather frost-tolerant. The required

Fig. 7.3. Habit of chickpea, Malawi (photo: J.D. Ferwerda).

annual rainfall is 600–1000 mm; the species does not withstand waterlogging. Because of the deep root system, chickpea is fairly drought-tolerant. Chickpea thrives best on well-drained, fertile sandy soils and loam soils, pH 5.5–6.5.

Chickpea is propagated from seed, which can be broadcast or drilled in rows, 25–60 cm apart, and a spacing in the row of 10 cm. Depending on soil, seed depth varies from 2 to 10 cm. Seeding rates vary from 25 to 120 kg/ha, depending on cropping system and cultivar. If the crop is planted for the first time, inoculating the soil with a specific strain of N_2-fixing rhizobium bacteria for chickpea may be required (rhizobium bacteria can fix atmospheric N_2 in symbiosis with chickpea). Chickpea is grown both as a sole crop and intercropped.

Weed control has to start before the crop emerges, and subsequently be carried out until 4 weeks after sowing. It is usually done by manual or mechanical weeding. After 4 weeks the crop will usually be able to repress weeds. Weeds can also be controlled chemically, a few herbicides are available.

If the roots are nodulated with the correct rhizobium bacteria, N fertilizer is not needed. Depending on soil fertility, P and K are applied at rates of 0–24 and 0–105 kg/ha, respectively.

Chickpeas mature in 3–7 months. For dried seeds the plants are usually harvested shortly before full maturity by cutting them close to the ground. The plants are left in the field for some days, then the crop is threshed, generally by trampling or beating with a flail. Tall cultivars can be harvested with combine harvesters. Dry seed yields vary from 0.4 to 2.0 t/ha, although experiments demonstrate that yield can be increased up to 5.0 t/ha. In 2004 the world total production of chickpea was 8.6 million t of dry seeds, making it the third most important pulse crop in the world. At present, chickpea is grown mainly in India

(80%). The remainder is grown in Pakistan, the Mediterranean region, the USA, Australia, Mexico and Ethiopia.

Pigeonpea

Origin, history and spread

Cajanus cajan, pigeonpea, is probably native to India. The spread of the species is unclear, but it is found in South-east Asia from ancient times and it reached Africa around 2000 BC. Seeds have been found in Egyptian tombs. Another theory is that pigeonpea originated in Africa and spread from there to India. Both India and Africa can be considered as centres of diversity for the genus *Cajanus*. The wild progenitor may be *Cajanus cajanifolius*, which occurs in India. As a result of the slave trade, it was brought to the Americas in the 16th century. In the West Indies, the seeds were used as bird feed, which led to the name 'pigeonpea'. At present, it is widely grown in the tropics.

Botany

Pigeonpea (Fig. 7.4) is a woody, short-lived perennial shrub, which is also grown as an annual. There are spreading and erect types, their height may vary from 0.5 to 4 m. Stems are angled and covered with fine hairs when they are young. Basal stem diameter is 1–4 cm. The plants may be branched, or almost un-branched. There is a wide variation in the point where branches arise on the

Fig. 7.4. Pigeonpea: 1, branch with inflorescence; 2, fruits; 3, seed (line drawing: PROSEA volume 1).

main stem. The compound leaves are trifoliolate and spirally arranged on the stem. The leaflets are entire, lanceolate to elliptical, pubescent, and up to 14 cm × 3 cm. The leaflets are greyish beneath with yellow resin glands. The terminal leaflet has a long petiole; the lateral leaflets have short petioles. Petioles are grooved above. The inflorescence is a terminal or axillary raceme, 4–12 cm long, carrying about seven flowers or more. Flowering extends over several months, up to nearly throughout the year. The flower is a characteristic pea-family flower (see Pea, this chapter), up to 2.5 cm long. Calyx is 10–12 mm long with five linear teeth, corolla is mostly yellow in colour, although the standard can be flecked or streaked with red or purple. The fruit is a straight or sickle-shaped pod with diagonal depressions between the seeds, flattened, up to 10 cm × 1.5 cm, and green, dark red or purple-coloured, or it may be blotched with dark red or purple (Fig. 7.5). The fruit often contains three or four globose to ellipsoid and more or less rectangular seeds, but number may vary. Seeds are smooth, up to 8 mm in diameter, and if mature, yellow, grey, red, purple or black in colour, entire or spotted. The 1000-seed weight varies considerably, from 40 to 260 g.

The species produces a root system with many laterals, which can be 2 m deep; spreading types have a more extensive root system than erect types. If compatible rhizobium bacteria occur in the soil, the roots will bear numerous effective nodules.

Cultivars, uses and constituents

Especially in India, extensive research and breeding with pigeonpea have been carried out, so numerous cultivars are available. Most of the cultivars can be

Fig. 7.5. Flowering and fruiting branch of pigeonpea (photo: K.E. Giller).

divided into two commercially important types: (i) 'arhar', a type that is late, tall, and has long pods containing many seeds; and (ii) the 'turn' type, a type that is early, small, and has small pods containing only a few seeds. Cultivars are also classified into (seed) maturity groups: extra early-, early-, medium- and late-maturing groups, under Indian conditions corresponding to 120, 145, 185 and 200 or more days after planting, respectively.

Pigeonpea is primarily grown as a crop for mature dry seeds for human consumption and stock feed. However, the seeds may sometimes be harvested when the seeds are still green and unripe, to use the peas as a vegetable or for canning. Unripe pods are also consumed as a vegetable. In India, ripe split seeds are made into the typical pulse dish 'dhal'. The seeds can be ground into flour to use in soups and other dishes. The foliage can be fed to livestock, both fresh and conserved, the feed value is excellent. Although grazing of the crop occurs, the plant is not suitable for grazing because the branches are brittle and break easily; grazing causes excessive damage. The wood of the older stems is hard and brittle and can be used as firewood. The species is highly valued as green manure.

There are many folk medicinal uses for different parts of the pigeonpea plant. For example, the flowers have been used to treat bronchitis and pneumonia, and the leaves have been used to treat genital and skin irritations, toothache, dysentery, and many more.

Dry seeds contain 7–10% water, 15–30% protein, 36–66% carbohydrates, 5–9% fibre and 3–4% ash, as well as provitamin A, vitamin B-complex and vitamin C.

Ecology and agronomy

Pigeonpea thrives best at day temperatures ranging from 18 to 30°C, and prefers full sunlight. Optimum rainfall is between 600 and 1000 mm/year with moist conditions for the first 2 growing months and drier conditions for flowering and harvest. The crop is remarkably tolerant of drought and high temperatures; it does not tolerate frost, extreme soil salinity, or waterlogging for a longer period. Pigeonpea is cultivated at altitudes ranging from near sea level up to 3000 m. Pigeonpea is a short-day plant and is grown on a wide range of well-drained soil types ranging from acid sandy soils to alkaline clay soils with a pH between 5 and 7.

Pigeonpea is propagated from seed, with the seeding rate varying from 10 to 25 kg/ha for rows. Seed is sometimes broadcast but often drilled. Seeding depth varies with soil condition and seed size, from 2.5 to 10 cm. If pigeonpea is grown for the first time, it may be useful to inoculate the seed with a compatible strain of rhizobium bacteria to ensure N_2 fixation. In pure stands the plants can be grown in rows 40–130 cm apart with 30–90 cm between plants within the rows. When mixed with other crops, often in alternate rows, a wider spacing is needed.

Seedlings emerge 2–3 weeks after sowing; growth is rather slow during the first 3 months. During this period weed control is needed, by hoeing or cultivating mechanically or by using herbicides. Although pigeonpea is an

N$_2$-fixing species, it may respond to application of N. P is usually the most limiting growth factor and most cultivars are susceptible to Zn deficiency, which can be applied as a foliar spray. Fertilizer rates are of the order of 20–25 kg N and 20–100 kg P per hectare. The time from seeding to harvesting depends on the type of end product (vegetable, seed, fodder) and cultivar. Late- and early-maturing cultivars may differ more than 100 days in time to reach full seed maturity. As flowering extends over several months, harvesting also takes a prolonged period. Unripe pods are always harvested by hand. In most parts of India and Africa whole plants are harvested with sickles and then dried and threshed. Only uniformly flowering cultivars can be combine-harvested. Pigeonpea can be grown as an annual crop but cropping can also be continued for 2 to 4 years. Pigeonpea can be intercropped with cereals because of the shallow root system of the latter. The crop is sometimes used as a windbreak and shade for other plants, for example young coffee trees.

Annual yields of dry seed vary from 0.5 to 5.0 t/ha; the world average yield is about 0.65 t/ha. In 2004, the world total production of dry seed was about 3.6 million t, produced mainly in India and East Africa.

Faba bean

Faba bean – *Vicia faba*; Pea family – *Leguminosae*

Origin, history and spread

Vicia faba is probably one of the world's oldest cultivated vegetable crops. It is thought to be native to the Mediterranean and south-western Asia. It was probably also domesticated in this region; it was well known to the ancient Egyptians, Greek and Romans. However, small-seeded faba beans have been found together with Iron and Bronze Age relicts, all over Europe. It is not clear how the species was spread within Europe. The ancient Greeks and Romans were of the opinion that eating faba beans could damage their vision. Therefore, Pythagoras even ordered his students not to eat faba beans. Recently, it was discovered that eating faba beans can cause the haemolytic disorder favism. This disorder occurs among people from the Mediterranean region, elsewhere it is rather rare. European settlers took faba beans to their colonies. Nowadays it is widely grown throughout the temperate regions and coastal regions in the tropics.

Botany

V. faba (Fig. 7.6) is an erect, stiff, glabrous, hardy, annual herb, up to 2 m high. The stem is four-ribbed, square, hollow, slightly winged, and with only a small number of branches at the base. The pinnately compound leaves alternate, and are composed of two to six large, ovate leaflets (3–10 cm × 1–4 cm), which bear large stipules at their bases. The top of the

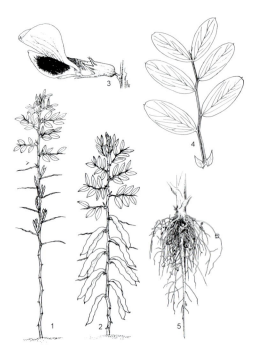

Fig. 7.6. Faba bean: 1, growth habit, variety *minor*; 2, growth habit, variety *major*; 3, flower; 4, leaf; 5, root system with nodules.

leaf is transferred to a short tendril, 1–2 cm long. The characteristic pea-family flowers (see Pea, this chapter) are white or white with black or purplish spots. The inflorescences are borne on short axillary racemes. An inflorescence (Fig. 7.7) may contain one to six flowers, each flower 3–4 cm long. The fruit is a pod

Fig. 7.7. Inflorescence of faba bean in side-view (photo: D.J. Schuiling).

Fig. 7.8. Full-grown fruits of faba bean, variety *minor* (photo: D.L. Schuiling).

(Fig. 7.8), sub-cylindrical to flattened, 5–25 cm long, and containing four to nine seeds. The interior of the pod is filled with a velvety covering. The pods turn dark brown to black when they mature. Seeds vary considerably in shape and size, from strongly compressed to almost spherical, 1–2.5 cm long. The colour of the seeds varies from almost white, beige, green, brown to almost black. The 1000 seed-weight is 300–1400 g. Characteristic for the seeds of *V. faba* is that the hilum is located on one of the short sides of the seed. The plant develops a strong taproot, and a branched lateral root system.

Species, varieties, uses and constituents

Species within the genus *Vicia* vary in appearance and other characteristics. Variation is found in seed size, number of leaflets of the compound leaves, and the length of the tendrils. Seed size ranges from 1000-seed weight of 33 g to 1000-seed weight of about 1400 g. Leaves range from seven to nine pairs of leaflets and distinct tendrils to only two pairs of leaflets and a tendril of 1–2 cm. The range begins with *Vicia villosa*, followed by *Vicia sativa* and *Vicia narbonensis*, which have many leaflets, and ends with *V. faba* var. *minor* and *V. faba* var. *major*, which have only two leaflets. The first three species are vetch species and are fodder crops or green-manure crops.

The small-seeded *V. faba* var. *minor* is mostly grown as a source of protein for livestock. It can be harvested green for silage. It can also be harvested when the seeds are ripe. After grinding the seeds, they can be used in concentrates for livestock. The feeding value of faba beans is very good. *V. faba* var. *major* is often grown as a vegetable; it is harvested when the seeds are still unripe and whitish-green. For the fresh market the seeds (mostly named 'broad beans') are sold when they are still in the green pods. Large amounts of shelled beans are canned or quick-frozen. For food, the beans can also be harvested when they are ripe. Ripe beans must be ground before they can be eaten. Faba bean is considered to be a meat extender or substitute. In India, the beans are sometimes roasted and eaten like peanuts. As well as containing 10% water, dried beans also contain 56% carbohydrates, 26% protein, 6% fibre, 2% ash and 1% fat. In folk medicine faba bean has been used as a diuretic and an expectorant.

Ecology and agronomy

V. faba is primarily a crop for temperate regions. It is a quantitative long-day plant. During growth, a temperature of 15–25°C is required. Most cultivars tolerate frost but they do not endure extreme heat. Germination is possible from 2°C and higher. Rainfall of about 700 mm/year will be sufficient, although water requirement is highest 9–12 weeks after establishment and after flowering. *V. faba* is grown as a winter crop or a spring crop. It can be cultivated on a wide range of soils, but prefers rich loam or clay soils with pH ranging from 6 to 7 (Fig. 7.9).

Faba bean is propagated by seed. Number of plants per square metre is eight to 25. The seed rate depends on cultivar, variety and cropping system, but varies widely: 90–340 kg/ha. Plant spacing also varies: 8–25 cm between plants in the row and 50–75 cm between rows. Seed is usually sown with precision

Fig. 7.9. Faba bean fields at about 2900 m altitude, Guatemala (photo: D.L. Schuiling).

sowing machines, 5–10 cm deep. Sometimes, the seed must be inoculated with proper *Rhizobium* bacteria before sowing.

As N_2 is fixed by the legume–*Rhizobium* symbiosis, application of N fertilizer is usually not necessary, although P and K are mostly applied, often at seeding. On light soils, about 30 kg N/ha may be applied as a 'starter dose' at seeding.

Both mechanical and chemical weed control are often used. Whether mechanical control is possible depends among other factors on spacing between the rows; if possible, cultivation throughout the growing period is recommended. Chemical weed control is usually done by pre-emergence and post-emergence herbicides. Flowering starts 4–6 weeks after crop establishment. Time of harvest depends on what product is desired: whole plants for silage as feed for livestock, green unripe beans as vegetable, or dry beans as feed or food. Harvesting dry beans has to be done when the lower pods are mature and dark brown or black and the upper pods are fully developed and start to change colour. Waiting with harvesting until the upper pods are fully ripe results in large losses from shattering. Maturing of faba beans varies from 100 to 220 days after planting. For silage, the crop can be harvested at various stages, either unripe and green or almost mature. Unripe beans as a vegetable for the fresh market, canning or quick-freezing are harvested when the pods are still green. Special machines for harvesting faba beans are available. When grown as a fodder crop, faba bean is sometimes intercropped with cereals, peas or vetch species.

There is a wide variation in the yield of dry beans between different regions: 0.8–6.5 t/ha. There is also a wide variation in yield within a region, from year to year. This yield instability has not been solved to date. In 2004, the world total production of dry beans was about 4.4 million t; China, Egypt and Ethiopia were the main producers. In 2004, the world total production of green beans was about 1.4 million t, China and Morocco being the main producers.

Pea and Lentil

Pea – *Pisum sativum*, Lentil – *Lens culinaris*; Pea family – *Leguminosae*

Pea

Origin, history and spread

The primary centre of origin of pea is probably the mountainous region of southwest Asia, particularly Afghanistan and India. The genus *Pisum* is represented by several species. For the development of cultivated peas, two species seem to be important: *Pisum fulvum* and *Pisum sativum*. *P. fulvum* is a wild species that originates from the eastern Mediterranean region. *P. sativum* is a cluster of wild and cultivated varieties that is native to the Mediterranean region and Near East. Hybrids between the two species occur, and when *P. fulvum* is the maternal parent, the seeds are wrinkled. Within the wild forms of *P. sativum*, two groups can be distinguished that were called *Pisum elatius* and *Pisum humile*, the latter probably the main ancestor of the cultivated pea.

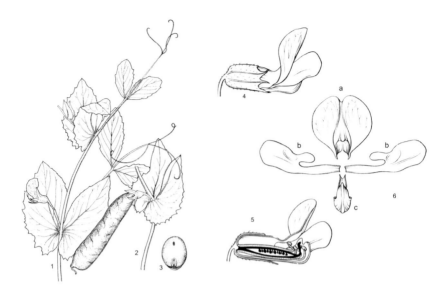

Fig. 7.10. Pea: 1, branch with flower; 2, branch with fruit; 3, seed (line drawing:
PROSEA volume 1); 4, flower; 5, flower in longitudinal section; 6, petals (a: standard,
b: wings, c: keel) (line drawing: CIBA GEIGY).

Dried peas have been a good source of nutritious food since Neolithic times.
Carbonized remains have been found in several archaeological sites in the Near
East, which suggests that humans were eating peas at least 9500 years ago. There
is some evidence that peas were cultivated about 8500 years ago. Wild populations of *P. humile* in Turkey and Syria are found to be chromosomally similar to
cultivated peas, which suggests that this area is most likely an origin of domestication. Pea cultivation was known in the Nile Valley in 5000 BC and in India in
4000 BC. The plant probably followed the spread of Neolithic agriculture into
Europe, first in Central Europe around 4000 BC and by 2000 BC throughout
Europe.

P. *sativum* is morphologically very variable and is mainly self-pollinating, which
makes it easy to develop different true breeding lines. These aspects formed the
basis of the success of the genetic experiments performed by Gregor Mendel.

At present peas are grown worldwide, but because of sensitivity to extremes
in weather, they are largely confined to temperate regions and to the higher altitudes or cooler seasons of warmer regions.

Botany

Pea (Fig. 7.10) is a climbing annual herb, up to 3 m tall; however, modern cultivars are often much shorter, about 30 cm. The weak stem is mostly hollow and
semi-vining; the tall cultivars usually cannot climb without support (Fig. 7.11).
The alternate, pinnate leaves are borne along the whole length of the stem and
have a length of 15–20 cm. They usually consist of two or three pairs (rarely

Fig. 7.11. Habit of a pea plant of a bushy variety (left) (photo: D.L. Schuiling); inflorescence of pea (right).

up to seven pairs) of broad, ovate leaflets and terminate in tendrils, at the base with two leaf-like stipules. However, some cultivars have a reduced number of leaflets, or no leaflets at all (semi-leafless and leafless types), but an enlarged number of tendrils and large stipules. The margins of the leaflets are entire to serrated. Inflorescences are in axillary racemes with one or two flowers (Fig. 7.11). The calyx consists of five green united sepals. The corolla consists of five white, purple or pink petals. The size of the petals differs considerably. The top of the flower is one large conspicuous petal called the 'standard'. Underneath the standard are four petals; the two small petals in the centre are joined together and are known as the 'keel', because of their boat-shaped appearance. On both sides of the keel there is a petal that is broad above and tapering towards the base, which is called the 'wing'. Within the keel lie ten stamens, nine forming a tube that surrounds the pistil, and one loose stamen. The ovary contains four to 15 ovules. The fruit is a pod, 3–11 cm long, often with a tough inner membrane. The ripe seeds are round and smooth or wrinkled, colour varies from green, yellow, beige, brown, blue-red, dark violet to almost black. Some types have spotted seeds. The 1000-seed weight varies from 150 to 450 g. The whole plant is glabrous and has a bluish-green waxy appearance.

The root system consists of a taproot and laterals. Nodules are formed on the roots (Fig. 7.12), caused by penetrating *Rhizobium* bacteria, which fix atmospheric N_2 in a symbiotic process with the pea plant (see Clover, Chapter 5).

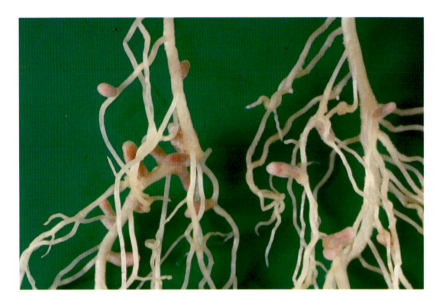

Fig. 7.12. Detail of the root system of pea showing nodules formed by infection with the N$_2$-fixing bacterium *Rhizobium leguminosarum* (photo: W. Lommen).

Varieties, uses and constituents

Many pea varieties exist: they vary in plant height; appearance of the leaves; colour of the flowers; size, tenderness, colour, smoothness or wrinkliness, and sweetness of the seeds; season of maturity; and other properties.

The type *P. sativum* var. *arvense* includes field pea varieties, which are harvested when the seeds are mature and dry. The seeds are mainly used for animal feed, often ground into flour and added to concentrates. Field pea can supplement the small amount of protein present in feed processed from cereal grains. Intercrops of peas and cereals may be grown and machine-harvested together for animal feed. A portion of the field peas, as whole, split or ground peas, is used as food, such as in soups and as a cooked vegetable.

Next to field peas, there are garden peas, classified as *P. sativum* var. *sativum*. Garden peas are a major vegetable crop in countries of the temperate region. Garden peas are harvested when the seeds are immature. A proportion of these peas is marketed fresh, although large amounts of garden peas are canned or frozen. Within this type, both wrinkled and smooth peas exist; wrinkled peas have higher sugar content than smooth peas. Sugars attract water by osmosis and, when the seeds dry out, they wrinkle due to the loss of water.

The type *P. sativum* var. *macrocarpon* is called the 'sugar pea' or 'snow pea' and includes edible-podded varieties. Sugar pea is used for eating both pod and seed as a fresh, green vegetable when the seeds are quite immature and before the seeds start to accumulate starch. The pods can also be eaten because they are fleshier and less fibrous than other peas and they have a tender inner membrane. Another edible-podded pea is 'snap pea', a type with thick edible pods and sweet full-size seeds.

The round and smooth, mature seeds contain 10–14% water, 50–60% carbohydrates (mainly starch, about 6% sugars), 15–35% protein and 4% fibre. Wrinkled mature seeds have approximately the same composition, although the starch content is lower and the sugar content is higher, about 10%. Furthermore peas contain high concentrations of the amino acids lysine and tryptophan, which are essential for the diets of humans and monogastric animals. Types with a dark testa contain tannin, which can lower their digestibility.

The green foliage of pea plants is used as a vegetable in parts of Asia and Africa. The straw of peas after threshing is a nutritious feed. Peas are also often grown together with cereals for silage and green fodder. Peas are sometimes grown as green manure.

Ecology and agronomy

A cool growing season (optimum 20°C) is necessary for obtaining good pea production. Hot weather during flowering may reduce seed set. Young pea plants can withstand some frost without damage; depending on the cultivar they may resist −10°C. When the plants are covered by snow it may even be up to −30°C. Although pea tolerates rainfall as low as 400 mm/year, for good development about 1000 mm is required. Peas tolerate a wide range of soils, from light sandy to heavy clay, but grow best on a well-drained and slightly acid soil, pH optimum 5.5–6.5. Peas do not tolerate waterlogged soils; plants may die after just 1 or 2 days in a waterlogged situation.

Peas are propagated from seed. Seed is sown at a depth of 4–7 cm at a rate of 65 to 280 kg/ha. Plant densities vary considerably depending on cultivar, type, soil, seed size, climate, and several biotic factors such as diseases. Peas can be seeded in row spacings from 15 to 30 cm. It takes 10–14 days for emergence of the seedlings. Peas have hypogeal emergence, which means that the cotyledons remain below the soil surface. Flowering starts about 45 days after planting and lasts 2–4 weeks. The pea-growing season varies from 80 to 150 days depending on temperature and rainfall.

Peas usually do not need much fertilizer, particularly N, because of their ability to obtain much of the N through fixation of atmospheric N_2. The amount of N fixed through symbiosis can be up to 125 kg/ha. Therefore the seed has to be inoculated with the appropriate strain of *Rhizobium* bacteria, if not present in the soil. N additions will be necessary when nodulation is poor or lacking; some N is also often given at the beginning of the development of the plants to ensure a prosperous start. Usually nodules will appear on the roots 2 to 4 weeks after the emergence of the seedling. Depending on soil fertility and removal of nutrients, an addition of P and K is usually necessary. Per tonne of harvested seeds, about 8 kg N, 0.6 kg P and 3 kg K are removed. When whole plants are harvested and removed from the field, 47 kg N, 5.5 kg P and 30 kg K per tonne will be removed.

Pea is a poor competitor with weeds. A rapid emergence of the seedlings and a correct plant density help to make pea more competitive. In addition, pre-emergence and early post-emergence tillage helps in reducing weed pressure. Furthermore, some herbicides are available.

Mature or immature seeds of field and garden peas are often harvested by cutting the plants and letting them pass through threshers which remove the peas from the pods and the vines. For dry peas the moisture content has to be less than 13%.

Garden peas are also often picked by hand before the seeds are fully matured and the seeds stay in the pod when they are sold on the fresh market.

Annual yields of dry peas vary greatly between countries; in 2004, the world average yield was about 2.6 t/ha, although it can be up to about 5.0 t/ha. The total world production of dry peas in 2004 was 12.2 million t. Canada, the Russian Federation and India are the largest producers.

Lentil

Origin, history and spread

Lentils probably originated in the Near East or the Mediterranean region, which is also the area where lentils were domesticated, probably as the oldest pulse crop of ancient cultivation. The species spread across Europe and Asia in early times; it was already being grown in Spain and Germany about 7000 years ago and in India 4500 years ago. It was probably first introduced into the USA around 1900. *Lens culinaris* is most likely derived from the wild species *Lens orientalis*. At present, lentil is widely cultivated in subtropical regions and some tropical regions at higher altitudes. Turkey, the USA and especially India are the major producers.

Botany

Lentil (Fig. 7.13) is a highly branched, erect or sub-erect annual herb with slender, square stems. The stems are covered with fine hairs. The plant may reach a height of about 50 cm. The leaves are pinnate with four to eight pairs of leaflets and usually ending in a tendril. The leaflets, which may be opposite or alternate, are obovate to lanceolate, and average 13 mm × 3.5 mm. The small (approximately 6 mm) typical pea-family flowers are white, pink to bluish, and borne in axillary racemes, up to four flowers together (Fig. 7.14). The fruit is a small, flattened pod, up to 2 cm long and containing one or two seeds. The seed is lens-shaped, 3–9 mm in diameter, and green, greenish-brown or reddish in colour, sometimes mottled. The 1000-seed weight is 20–90 g.

Types, uses and constituents

Two types of lentils can be discerned: (i) macrosperma, grown mainly in the Mediterranean region and America; and (ii) microsperma, which is grown mainly in Africa, the Near East and India. The seed of the macrosperma type has yellow cotyledons and is 6 to 9 mm in diameter. The seed of the microsperma type has yellow or orange cotyledons and is 2 to 6 mm in diameter.

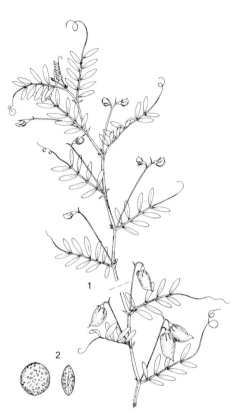

Fig. 7.13. Lentil: 1, flowering and fruiting branch; 2, seeds (line drawing: PROSEA volume 1).

Fig. 7.14. Flowering lentil plants (left) and inflorescence of lentil (right) (photos: J. van Zee).

Whole or split seeds are used in soups, stews and salads. Seeds and young pods are used as vegetables. Seeds can be ground into flour and mixed with flour from cereals; the flour mix can be used for producing cakes or infant food.

Seed consists of 12% water, 24% protein, 1–2% fat, 52% carbohydrates, 1.8% fibre and 2.2% ash. Lentils contain anti-nutritional factors that can cause flatulence, but heating of the seeds reduces the risk. The husk contains tannins, thus the husk is often removed. Crop residues are excellent feed for livestock, because of the high protein content.

Ecology and agronomy

Lentil tolerates a wide range of soil types, although it thrives best on loamy sand and clay soils. It tolerates a pH range of 4.4–8.2, but prefers pH 5.5–7. The optimum temperature for production is 24°C. At high yield level, moisture requirement is 750 mm per growing period.

Lentil is propagated by seed, which can be broadcast or sown in drills. For monoculture, 25–90 kg of seeds per hectare is required. The spacing for drilled rows can be 20 cm apart, although in the USA it may be 60–90 cm. Spacing of the plants in the rows ranges from 20 to 30 cm. When lentils have not been grown on the field recently, the seed may have to be inoculated with the proper strain of *Rhizobium* bacteria, prior to sowing. Lentil is often grown in intercropping, for example with mustard, rice or barley. Lentil mostly does not require much fertilizer, but this also depends on soil fertility. N requirement can be fulfilled through the symbiosis with the *Rhizobium* bacteria that fix N_2 from the air. However, if there is not yet enough N available in the soil, a small amount of fertilizer N (25 kg N/ha) as a starter application can be considered. Application of P and K for most soils can be 30–60 kg/ha and 170–225 kg/ha, respectively.

As lentil is a poor competitor due to the fragile and open plant growth habit, weed control is necessary. Weed control has to be done before sowing and early in the growing season. The treatments can be mechanical or chemical.

The duration of the growth cycle depends on the cultivar; it ranges from 80 to 125 days. At maturity the plants turn yellow. The plants are cut or pulled by hand and left in the field to dry for up to 10 days. After that the crop can be threshed or combine-harvested. Seed yields vary, depending on cultivar and growing factors, and range from 0.4 to 3.0 t/ha.

Phaseolus Beans

Common bean – *Phaseolus vulgaris*, Runner bean – *Phaseolus coccineus*, Lima bean – *Phaseolus lunatus*; Pea family – *Leguminosae*

Common bean and runner bean

Origin, history and spread

Phaseolus species all originate from the New World. Nowadays it is widely accepted that common bean was first cultivated about the same time,

Fig. 7.15. Common bean: 1, fruiting branch; 2, inflorescence; 3, two views of seed (line drawing: PROSEA volume 1).

independently, in two different regions both in Central America and South America. Cultivated beans some 7000–8000 years old have been found in Mexico and Peru. Maize, another important crop from the same regions, was probably domesticated at the same time.

The wild species *Phaseolus aborigineus* is probably the immediate progenitor of *Phaseolus vulgaris*. Runner bean was probably domesticated about 2000 years ago in Central America. Domestication involved a considerable number of changes, such as the shift from the perennial to annual habit, increases in yield, larger and softer seeds, a shift from a short-day response to a day-neutral response for a number of varieties, and a loss of dormancy. Additionally, a large variation in appearance developed.

The genus name '*Phaseolus*' is derived from the Greek name for bean, which is 'pháselos'.

Over the course of many centuries, common bean spread widely across North and South America. Between AD 1500 and 1600, explorers and slave traders brought it to Europe and Africa and it then spread throughout the world. Nowadays, beans are grown in temperate regions, the subtropics and tropics.

Botany

Common bean (Fig. 7.15) is an annual herb; runner bean is an annual or perennial herb, but nearly always grown as an annual one. Both species have highly variable plant height. Some cultivars have twining stems, up to 4 m long, called 'pole beans', while others develop into more erect, bushy plants, about 50 cm high (Fig. 7.16). Intermediate types occur. The stem is angular or nearly cylindrical. Seen from above, runner bean plants twine clockwise,

Fig. 7.16. Common bean: field of bushy type, Guatemala (left) (photo: D.L. Schuiling); mixed crop of a twining type and maize, Nariño, Colombia (right) (photo: K.E. Giller).

whereas most other beans twine anticlockwise. The plant is sometimes slightly pubescent.

The usually dark green leaves are alternate, trifoliolate compound, bilaterally symmetrical, and borne on petioles that can be up to 15 cm long. Each leaflet is broad oval, entire, with pointed tip, and about 12 cm long. Leaves have a marked pulvinus at the base.

Flowers are 1.5–2.5 cm wide and borne on axillary or terminal racemes (Fig. 7.17). The structure of the flower is typical of the pea family (see Pea, this chapter). There are five green sepals, more or less united. Next there are five petals: a large upper standard and two lateral wings, which are free-growing petals; the two lowermost petals are united in a boat-shaped keel. Flower colour can be red, pink, violet, cream or white. Some flowers are bicoloured. A flower has ten stamens.

The fruit is a pod, which can be linear, straight or slightly curved, with a prominent beak. The length of the pod depends on the cultivar and ranges from 10 to 30 cm. The pod is usually fleshy when immature and mostly green or yellow, sometimes purplish or reddish, spotted or striped (Fig. 7.18). Pods contain from two to 12 seeds. Seeds are highly variable in size, shape and colour. They may be round, ovoid, sub-spherical or kidney-shaped, plain or speckled or flecked, white, yellow, beige, brown, black, beige or white with a brown spot around the hilum, and many more colours and combinations. The 1000-seed weight varies between 200 and 2000 g.

Fig. 7.17. Inflorescences (left to right): common bean; runner bean; lima bean (photo, right: B.P. Schuiling).

Fig. 7.18. Common bean: infructescence (left) and fruits of several varieties (right).

The species have well-developed taproots with lateral and adventitious roots, or tuberous roots. Nodules, caused by symbiotic N_2-fixing bacteria, can be found on the roots.

Species, varieties, uses and constituents

There are several species of *Phaseolus* beans, including *P. vulgaris*, *Phaseolus coccineus*, *Phaseolus lunatus* and *Phaseolus angularis*. *P. vulgaris* and *P. coccineus* are important crops and closely related, hybrids between the two species occur.

A large number of bean cultivars exist. From *P. vulgaris* there are more than 2500 cultivars recorded. They can be grouped into determinate bush types, intermediate and indeterminate climbing types. The different types can be divided into snap beans, also called 'string beans', for harvesting immature pods with very small seeds; green shell beans for harvesting bigger green seeds; and dry beans for harvesting ripe beans. The green immature pods are eaten as a vegetable that may be fresh and cooked, or conserved, which means canned or frozen. Mature ripe beans, such as white beans, field beans and red kidney beans, are stored dry and then re-hydrated, cooked, and eaten as vegetables or used in soups and several dishes, for example 'chilli con carne'. White-seeded beans are also canned, sometimes with tomato sauce added. Mature or dry beans are important grain legumes to provide plant protein for human diets. In many parts of the world, such as Central and South America and many parts of Africa, beans are staple food. Commercially grown beans are usually bush types; pole types or climbers are usually grown in gardens. Runner bean is always a pole type. Pole beans have to be trailed on poles or trellis. Runner beans have mostly long pods, up to 30 cm. The pods are harvested when they are immature and eaten as a vegetable. Before cooking or conserving the pods, they are usually sliced. In Central America, not only pods are eaten but also green and/or dry seeds. The fleshy root tubers of runner beans can also be eaten. Runner bean is sometimes grown as an ornamental for the showy red and white flowers.

Ripe seeds contain 10% water, 55–60% carbohydrate, 20–25% protein, 4.3% fibre, 2% fat and 3.7% ash. Green pods contain about 90% water, 6.6% carbohydrate, 1.8% protein, 0.2% fat and 0.7% ash. Fresh green parts of the plant contain the precursors of vitamins C and A.

Beans are better not eaten fresh because they contain intestinal protease inhibitors. These interfere with digestion; however, they are destroyed by cooking. Beans also contain tannins, black beans and dark red beans usually being richer in tannins than lighter coloured ones.

Ecology and agronomy

The species prefer air temperatures from 18 to 25°C, when temperature is too high, especially with low relative humidity, pod set will be reduced; they are sensitive to frost. Well-distributed rainfall of 300–400 mm is required per ripe-seed-crop cycle. There are both short-day and day-neutral cultivars of *P. vulgaris*. Beans grow on many different well-drained soils with a pH between 6.0 and 7.5 and rich in organic matter. However, as beans thrive best when soil temperature is about 16°C, loose sandy loam soils are most suitable because they warm rapidly; beans are not well adapted to heavy clay soils.

Beans are propagated by seed. After establishment of the plants, symbiotic rhizobium bacteria, which are usually present in the soil, induce the function of nodules on the roots. These nodules are colonized by the rhizobia which fix N_2 from the atmosphere.

Commercially grown beans are often bushy types, the climbing pole types are often grown in home gardens; however, in South America climbing types are also grown commercially. The seeds are drilled or planted in rows at a depth of 3–6 cm. When soil temperature is about 16°C, seedlings will emerge within 1 week.

The number of plants per hectare of the bush-type beans varies depending on the type of soil, cultivar and time of sowing, and ranges from 14 to 40 plants/m². Spacing can be 30–75 cm between the rows and 15–30 cm between plants in the row. In tropical countries, common beans are often interplanted with other crops such as maize, cotton and sweet potato.

In spite of the ability to fix N_2 from the atmosphere, some N fertilizer is required, especially at the start of the cultivation and prior to flowering. An excellent crop of snap beans removes about 45 kg N, 8 kg P and 50 kg K per hectare.

Wide rows allow cultivation for weed control but result in poor canopy closure and uncovered soil stimulates weed development. Cultivation should be carried out shallow and carefully to avoid root damage, because beans are shallow-rooting. As well as mechanical weed control, it can also be done by herbicides.

Harvesting immature pods of snap beans can start about 50–60 days after planting; after that, pods can be picked every 3 or 4 days for a number of weeks. However, the pods of commercially cultivated snap beans are often harvested all at once by machines. It requires a great uniformity of maturity, which can be achieved by choosing determinate cultivars and a cropping system with high plant density.

The maturing of dried beans is about 100–150 days. When the pods have turned yellow, the plants are pulled and left on the field for drying in windrows for some days before threshing. In general, pole beans take longer to mature than bush beans. Harvesting of commercially cultivated dry crops can be completely mechanized.

Yields of dry common bush beans vary widely with cropping system, cultivar and region, ranging from 0.3 to 2.8 t/ha. The average yield of immature snap beans is about 4.5 t/ha.

In 2004, the world total production of dry beans was about 18.3 million t. Brazil, China, Myanmar and Mexico were the main producers. The world total production of green beans was about 6.4 million t. The main producers were China, Indonesia and Turkey.

Lima bean

Another bean species that has some importance for a number of tropical or subtropical areas is lima bean, sometimes called 'butter bean'. The species is also native to Central and South America. Lima bean (Fig. 7.19) is perennial in nature but usually grown as an annual. There are short, bushy cultivars and climbing ones. The cultivars are very variable in the sizes of the leaves (leaflets 5–19 cm long), pods (5–12 cm long), flowers (corolla 0.6–1.0 cm wide) and seeds (1000-seed weight 450–3000 g).

The colour of the flower may be white, pale green, violet, or a combination of two of these colours (Fig. 7.17). Pods contain two to four seeds. Seeds are

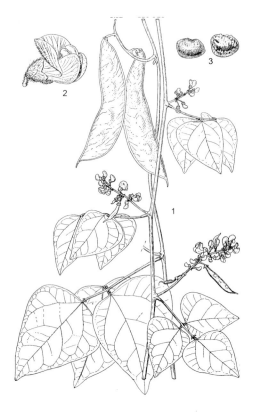

Fig. 7.19. Lima bean: 1, flowering and fruiting branches; 2, flower; 3, seeds (line drawing: PROSEA volume 1).

also variable in shape and colour. Shape varies from kidney-shaped to round; colour may be uniform or mottled, green, white, cream, yellow, brown, red, purple or black.

Lima bean is grown mainly for its green or mature seeds. In Asia, young sprouts, leaves and pods are also consumed. A part of the bean production is canned or frozen. Lima beans are used in soups, salads, stews and as a vegetable. In China, sprouted lima beans are used in many dishes.

Soybean

Soybean – *Glycine max*; Pea family – *Leguminosae*

Origin, history and spread

Soybean originated probably in the eastern half of northern China, and was domesticated at least 3000 years ago. However, soybean may have been domesticated much earlier because there are indications that the oldest description of the plant can be found in Chinese literature that is 4800 years old. *Glycine max* is probably derived from the wild species *Glycine soja*, which is a recumbent vining plant and can be found in northern China, Korea and Japan.

Cultivated soybeans and the wild species can still be hybridized easily. From China it spread to other parts of Asia through trade routes such as the 'Silk Road'. In many Asian countries, for example Indonesia, Japan, Vietnam, Thailand and India, hundreds of landraces developed. These countries can be considered as secondary gene centres. The plant was highly valued by the Chinese as shown by the names they gave different varieties, such as 'great treasure', 'heaven's bird' and 'yellow jewel'.

Soybean was not introduced into countries outside Asia before the 1700s and 1800s. The plant must have reached The Netherlands a while before 1737, because in that year Linnaeus stayed in The Netherlands and described soybean in the *Hortus Cliffortianus*. Although the introduction and the cultivation of soybean outside Asia happened rather late, soybean products were known much earlier. In the 1600s, many visitors to China and Japan described several products made out of soybeans. One of these products was soy sauce, which became a valuable trade product for Europe. A merchant named Samuel Bowen brought soybeans from China to Georgia, USA in 1765. The name '*Glycine*' is derived from the Greek word 'glykerós', which means 'sweet' in the meaning 'pleasant'.

At present, soybean is the most cultivated species from the pea family. It is also the world's major oilseed and the source of the majority of the protein used to feed pigs and poultry in the Western world and Japan.

Botany

Soybean (Fig. 7.20) is an erect annual herb. The entire plant is usually covered with grey or brown hairs. The stem is 20–50 cm tall, sometimes vine-like, more or less

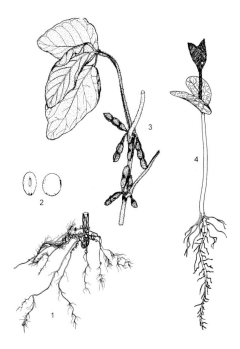

Fig. 7.20. Soybean: 1, root system with nodules; 2, seeds; 3, detail of fruiting branch with one leaf; 4, seedling.

Fig. 7.21. Small soybean flowers growing in the leaf axils (photo: J. van Zee).

angled and grooved to sub-quadrangular above. The stem becomes woody in the course of the growing season. Axillary buds on the main stem may develop into branches, but it depends on cultivar and plant density. The leaves are alternate and pinnately trifoliolate with petioles that can be 2–20 cm long. Lower leaves have longer petioles. The stipules are 3–6 mm long and pointed. The three leaflets are broadly ovate, oval or elliptic-lanceolate, 3–10 cm long and 2–6 cm broad, entire, with a rounded base. The leaves drop when the fruits are ripening. Inflorescences are borne on axillary racemes, which contain three to 15 flowers (Fig. 7.21). Some cultivars bear terminal racemes, which have more flowers than the laterals. Flowers are small, about 5 mm, and white or lilac in colour. The flowers are typically papilionaceous with a rather broad standard and small wings, the keel is shorter than the wings. The fruit is a small hairy pod, slightly curved and compressed, 5–7 cm long and 1–1.5 cm wide, containing one to three globose to spherical seeds (Fig. 7.22). The colour of the seeds varies enormously; many different colours or mottled combinations of colours can be possible, such as green, cream-yellow, yellow-brown, green-brown, brown, black, etc. However, a considerable part of the soybean is cream-yellow in colour, often with a black hilum. The size of the seeds also varies considerably, 1000-seed weight varies from 50 to 400 g. Soybean has a substantial taproot up to 2 m long with wide-spreading laterals. Effective root nodules can be formed when the soil contains the proper N_2-fixing rhizobia. Soybean is nodulated largely by slow-growing *Bradyrhizobium* strains, but also by some fast-growing rhizobia named *Sinorhizobium fredii*.

Cultivars, uses and constituents

In the Orient numerous cultivars are grown which have a wide genetic diversity. They may vary in many characteristics; for example in time to maturity, size,

Fig. 7.22. Detail of a fruiting soybean branch; black and yellow seeds of soybean (inset) (photos: J. van Zee).

plant type, number of pods per plant, colour of the seed, and content of oil and protein in the seed. In North and South America and in Europe, many fewer cultivars are grown. In these regions classification of cultivars is mainly based on adaptation to different climatic factors, the time needed to reach maturity, and the use of the crop. For oil production and many industrial products, yellow seeds are preferred. Soybeans used as a vegetable often have yellow or green seeds, whereas soybeans grown for animal feed often have brown or black seeds. Vegetable soybeans are harvested when the pods are still green.

In the Orient most soybeans are eaten directly as food, whereas in the Americas and Europe most beans are used as animal feed and in a variety of industrial and food products.

The seeds contain edible oil, which has to be refined before it can be used for cooking, as a salad oil, and processing margarine. The oil can also be used to produce soap, paints and plastics. The oil is used for numerous industrial purposes. Lecithin, a by-product obtained from oil processing, is an important emulsifier and antioxidant for the food, cosmetics and pharmaceutical industries.

In the Orient, unripe or dried beans are eaten as vegetables, and the sprouts, which are about 6-day-old seedlings that grew in the dark, are much used in cookery. The seeds can be ground into flour and mixed with wheat flour for use in bakery products. This mixture can be useful in special diets because, compared with wheat flour, soybean flour has high protein and relatively low carbohydrate content.

Soybean 'milk', which is derived after the seeds are successively soaked, ground, cooked and filtered, is used in Asian cooking and also as a milk substitute. The 'milk' is a protein supplement in infant feeding. Soybean 'milk' can be processed into a curd by coagulating and pressing the 'milk'. The curd is called 'tofu'. The process of tofu-making is almost the same as cheese-making. Tofu is an important source of protein in Oriental diets. It may be fashioned into cakes and loaves, or made into candy; it may be shredded or sliced before use; it may be smoked, marinated or fermented. Tofu is an ingredient of many different dishes.

As well as tofu, numerous Oriental dishes can be prepared with soybeans such as 'sufu', 'red sufu', 'tempeh', 'natto', 'red miso', 'white miso', 'hamanatto' and 'shoyu'. These products are mostly the result of fermentation processes, from 1 day to 9 months.

Spun fibres of soybean protein can be given meat colour and meat flavour, for use as a substitute for meat or to use in vegetarian diets. Vegetarian diets are common in the Orient, encouraged by Buddhist monks. The monks have strict vegetarian diets, so soybean products became important foods in monasteries. In the Western world, soybean products are also used in breakfast cereals and desserts.

Soy sauce is a dark brown, salty liquid made from a mixture of fermented soybeans and grain. Two fungi play a major part in the fermentation process: *Aspergillus oryzae* and *Aspergillus sojae*; yeasts and bacteria are also involved. Soy sauce produced in the USA is usually a non-fermented, synthetically manufactured product that is produced from defatted soybean meal, grain and chemicals. However, the synthetically manufactured soy sauce lacks the savoury flavour of the fermented soy sauce, which is much used to flavour Oriental foods. Furthermore, soybean is used extensively in health foods. The whole plant, but also residues, such as residual oilseed after oil extraction, can be used as animal feed.

Soybeans, especially black ones, are used in the preparation of food supplements, which are thought to improve the action of the heart, liver, kidneys and stomach.

Dry soybeans contain approximately 10% water, 35–54% protein, 14–27% oil, 30% carbohydrates, 4% fibre and 5% ash. Soy oil is rich in polyunsaturated fatty acids, especially linoleic acid, and contains no cholesterol. Soybeans are rich in Ca and vitamin E, and contain vitamins B_1 and C.

Ecology and agronomy

Soybean is chiefly a crop for the tropics and the subtropics, although it is also grown in warm temperate regions (Fig. 7.23). High light intensities and day temperature of about 30°C are required to obtain good yields. Temperatures below 21°C and above 32°C can reduce floral initiation and pod set. For good growth, soybean requires warm moist weather and 500–750 mm of water in the growing period. Drought stress during periods of flowering and pod filling may reduce yield considerably. Soybean thrives best on fertile, well-drained clay soils with pH 6–6.5. The plant responds strongly to photoperiod and is

Fig. 7.23. Field with soybean, Tanzania (photo: J. Ferwerda).

fundamentally a short-day plant. However, many cultivars have been developed with different day-length requirements. This enables soybeans to be grown in a wide range of day-length zones. Some cultivars are almost completely insensitive to photoperiod.

Soybeans are propagated by seed. A deep, loose seedbed is required, so before sowing the soil has to be cultivated thoroughly. The seeds need inoculation with a culture of N_2-fixing bacteria, unless the bacteria are already in the soil. Soybeans are often planted with a drill or a row planter, sometimes by broadcasting. Spacing between the rows varies from 40 to 90 cm, and in rows from 3 to 10 cm. Smallholders often use spacing of 25 cm × 25 cm or 20 cm × 20 cm. High plant densities tend to increase plant height and lodging. Planting depth has to be 4–5 cm. When the soil temperature is 18–20°C seedlings can be expected to emerge 5–7 days after planting. Minimum temperature for germination is about 10°C.

The production of 1 t of beans removes approximately 70 kg N/ha, half or more of which is provided by symbiotic fixation by the rhizobium bacteria. The rest has to come from the soil or from fertilizer. The removal of P and K per tonne of beans is approximately 14 and 75 kg, respectively. Recommended application of fertilizer is 0–40 kg N/ha, 13–26 kg P/ha and 42–63 kg K/ha.

Weed control is essential because competition with weeds may reduce yield by up to 50%. Weed control begins with pre-emergence treatment, which can be mechanical or by herbicides. Close spacing can be used to control weeds, but that may include the risk of increased lodging. During emergence seedlings can be damaged easily by cultivating equipment. Later on there is less danger of damaging the plants. For chemical weed control, several herbicides are available. Smallholders in Asian countries often control weeds by hand-weeding. If precipitation during the growing season is not sufficient, irrigation may be required for obtaining high yields and good seed quality.

At present, genetically modified varieties are often used for growing a soybean crop that is resistant to some herbicides. This is rather controversial and not accepted in all countries, because the effect in the long term is not clear.

Depending on cultivar, climate and cropping system, the time from planting to harvesting may vary from 60 to 150 days. Almost all seeds ripen simultaneously; the same applies to shedding of leaves and drying of the stems. Most soybeans are combine-harvested under large-scale agriculture. Asian smallholders usually harvest by hand, cutting the plants near the ground and leaving them in the field to dry. When the pods are brown or black, the plants can be threshed. For long storage, the moisture content of the beans should not be higher than 10%. Higher moisture contents may cause moulding and heating.

Annual yields of dry beans range from 0.7 to 5.0 t/ha; in 2004 the world average yield was about 1.7 t/ha. In 2004 the total world production of dry beans was about 204.5 million t. The USA is responsible for about 75% of the world's soybean production, with Brazil, Argentina and China, respectively, ranking next.

Vigna Beans

Cowpea – *Vigna unguiculata*, Mung bean – *Vigna radiata*; Pea family – *Leguminosae*

Cowpea

Origin, history and spread

Cowpea, also called 'black-eye pea', originated in Africa, although where the plant was first domesticated is uncertain. It is suggested that it was first cultivated in African savannahs at least 6000 years ago. Early cultivation was probably closely associated with those of sorghum and pearl millet. Two centres of diversity occur, where both wild and cultivated forms can be found: Africa and India. Wild and cultivated forms cross easily. Arabic traders spread it into the Mediterranean region and Asia. Cowpea was already known in India in Sanskritic times and the ancient Greeks and Romans grew it. Spaniards brought it to the West Indies in the 16th century and it reached North America around 1700. Cowpea is now widely cultivated throughout the tropics and the subtropics, although mainly in Africa.

Botany

Cowpea (Fig. 7.24) is an erect to sub-erect, or prostrate, climbing, glabrous, to 80 cm or taller, annual herb. The stems are somewhat square and ribbed, often with violet nodes. The leaves are alternate, trifoliolate and have a grooved, 5–25 cm long petiole. The leaflets are usually entire, ovate-rhomboid, apex acute; the lateral leaflets are opposite and asymmetrical, top leaflet is symmetrical,

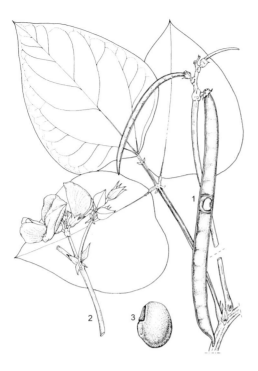

Fig. 7.24. Cowpea: 1, fruiting branch; 2, inflorescence; 3, seed (line drawing: PROSEA volume 1).

6.5–16 cm × 4–11 cm. The inflorescences are axillary, usually with two to four flowers at the top of the peduncle, in alternating pairs on thickened nodes. The peduncles are stout, grooved and 2.5–15 cm long; pedicels are very short. Flowers are characteristic for the pea family (see Pea, this chapter) (Fig. 7.25). The calyx is campanulate; the corolla is dirty white, yellow or violet; the standard 2–3 cm in diameter. The pods are variable: erect, ascending or pendent, more or less inflated, hard and firm or flabby when young, linear, and almost circular in transverse section; length varies, depending on cultivar group, from 8 to 100 cm with a diameter of 0.8–1 cm; number of seeds per pod ranges from eight to 20. The seeds (Fig. 7.26) are very variable in size, shape and colour, 2–12 mm long, globular to kidney-shaped, smooth or wrinkled, white, yellow, green, red, brown or black, or variously speckled or mottled; the hilum is white, surrounded by a dark ring. The 1000-seed weight is 50–300 g. Cowpea has a well-developed root system, consisting of a stout taproot with numerous spreading lateral roots in the top layer of the soil. Roots have large nodules due to the action of rhizobia (see Pea, this chapter).

Cultivars, uses and constituents

Within the species *Vigna unguiculata* three cultivar groups can be distinguished.

1. Cultivar group *Unguiculata* or common cowpea: 15–80 cm high, pods 10–30 cm long, grown in many tropical or subtropical countries although mainly in Africa, used as described below.

Fig. 7.25. Flowers (left to right): mung bean; cowpea; yard-long bean (photos: D.L. Schuiling).

Fig. 7.26. Seeds (left to right): mung bean; cowpea; yard-long bean.

2. Cultivar group *Biflora* or catjan cowpea: 15–80 cm high, pods 7.5–12 cm long, grown mainly in India and Sri Lanka, grown for fresh or dried seed, as a vegetable and a forage crop.

3. Cultivar group *Sesquipedalis*, asparagus bean or yard-long bean: climbing plant, 2–4 m high, pods 30–100 cm long, grown in tropical and subtropical countries, mainly for the long, succulent young pods as vegetable, sometimes for dry seeds (Figs 7.25–7.28).

Some authors consider the three groups to be three different species. Cowpea is primarily grown as a pulse, but also as a vegetable, both for the greens and the green seeds; it is also grown as a livestock feed. As a vegetable, cowpea can be used in all stages of development. Young green shoots and leaves are eaten in Africa as a pot herb; immature pods and green seeds are boiled as fresh vegetables, and also frozen or canned. However, most cowpeas are marketed as dry seeds or as a boiled and canned product. Dry seeds may be split, or ground into meal and used in soups and other dishes. Cowpea provides a high-quality feed for livestock, both fresh and as hay. Yield and digestibility of several cultivars are comparable to lucerne. Cowpea is also used as a green-manure crop, highly valued for its N_2 fixation, and as a cover crop.

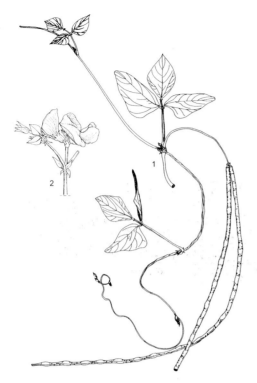

Fig. 7.27. Yard-long bean: 1, fruiting branch; 2, inflorescence.

Dried seeds contain approximately 9% water, 22% protein, 60% carbohydrates, 1.5% fat and 4% fibre. The protein is rich in the amino acids lysine and tryptophan. The seeds also contain Ca, thiamine, riboflavin, niacin and ascorbic acid.

In folk medicine ground seeds mixed with oil have been used to treat stubborn boils.

The bambara groundnut, *Vigna subterranea*, is a pulse crop, native in West Africa and still under-utilized as a protein-rich food crop.

Ecology and agronomy

Cowpea (Fig. 7.28) is grown over a temperature range of 21–36°C at daytime and 16–31°C at night; it is sensitive to cold and cannot stand frost. It is a short-day or day-neutral species. The species is considerably drought-resistant and can withstand some shade. The crop can be grown at annual rainfall ranging from 280 to 4100 mm; it is not tolerant of waterlogging. Cowpea can be grown on many kinds of soils, from highly acid to neutral, but it performs best on well-drained, deep, sandy loam or sandy soils with pH 5.5–6.5.

Cowpea is propagated from seeds. Before sowing, the soil has to be cultivated deeply, to ensure that the taproot can penetrate the soil, unhindered by impenetrable layers, and also as weed control. At sowing time, soil temperature has to be at least 18°C, and soil moisture should be adequate for germination

Fig. 7.28. Crop details: mung bean (left) (photo: K.E. Giller); cowpea (middle); yard-long bean (right).

and growth. Inoculating the soil with the correct strain of rhizobium bacteria may be required (see Pea, this chapter). When sown mechanically, it is usually established in rows spaced 30–90 cm apart and 7–10 cm within the rows; for yard-long beans, 20 cm apart within the rows. For forage, seeds are often broadcasted. Seeding rates depend on the aim of the cultivation and may range from 20 to 100 kg/ha. Cowpea is often grown intercropped with cereals such as maize, millet and sorghum. For weed control, usually three weedings are required in the first 2 months after germination. Weeding can be carried out manually or mechanically; using herbicides is an alternative. When the crop is grown to produce yard-long beans, they have to be staked when they are about 30 days old. Much N fertilizer is usually not required because of the plant's ability to fix N_2. Cowpea–rhizobium symbiosis may fix up to 200 kg N/ha, which is about 90% of the N the crop requires; about 30 kg N/ha can be applied for early plant development. Some P and K may be beneficial. On soils of medium fertility about 30 kg P/ha and 50 kg K/ha is recommended. However, in practice, no fertilizers are generally applied.

Depending on climate and cultivar, the time from sowing to maturity (dry seed) ranges from 90 to 240 days. Green pods have to be harvested earlier, when they are still immature and tender. As development of new pods continues over several weeks, there are several stages of pod maturity at the same time; smallholders often hand-pick ripe pods when available. When the whole crop has to be combine-harvested, the right moment is when most of the pods are mature, but just before the ripest ones shatter.

Annual yields of dry seeds vary widely; in Africa it can be as little as 0.3 t/ha, although it can be up to 4.0 t/ha. Annual yields of green pods of yard-long bean vary from 6 to 8 t/ha. In 2004, the total world production of dry seeds was almost 4 million t. The main cowpea-growing countries are Nigeria, Niger and Burkina Faso.

Mung bean

The mung bean is native to the north-eastern India–Burma region. It has been grown in India since ancient times, and India is still the main mung bean-growing country.

Mung bean (Fig. 7.29) is an erect to sub-erect, highly branching and hairy annual herb, up to 75 cm tall. The leaves are trifoliolate, with ovate leaflets, 12 cm × 10 cm, and long petioles. The inflorescence is an axillary raceme on a long pedicel, with five to 20 flowers. The flower is a characteristic pea-family flower (see Pea, this chapter) (Fig. 7.25), but small (8 mm), and usually yellow in colour. The pods are straight, covered with hair, 10 cm × 0.6 cm, yellow, grey or dark brown, and may contain up to 15 seeds. The shape of the seeds varies from globular to square in section, and the colour may be green, yellow or black (Fig. 7.26). The 1000-seed weight ranges from 30 to 90 g. The species is deep-rooted.

Within the species *Vigna radiata*, two varieties can be distinguished: variety *aureus*, called 'golden gram' or 'green gram', has longer plants, longer pods containing more seeds than variety *mungo*, the latter called 'black gram'. Mung

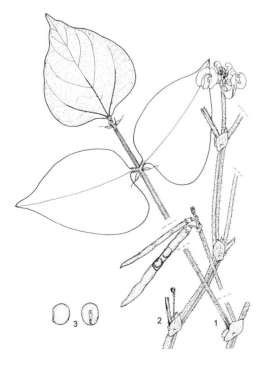

Fig. 7.29. Mung bean: 1, fruiting branch; 2, flowering branch; 3, seeds (line drawing: PROSEA volume 1).

beans are grown mainly for the dry seeds. They can be boiled and eaten as a vegetable, added to soups and other dishes; they are also ground into flour and added to noodles, biscuits and bread; or, especially in China and the USA, used for the production of sprouts. Sprouts can be obtained by germinating the seeds in the dark, until the sprouts have the right length. One gram of dry seeds produces 6–8 g of fresh sprouts. The sprouts are popular in certain dishes, including Chinese dishes. Sprouts are an important source of vitamins A and C. Green pods are sometimes used as a cooked vegetable. The species can also be grown as green manure or as forage for livestock.

As mung beans are closely related to cowpea, the ecology and agronomy of the species are more or less comparable (Fig. 7.28). Meticulous data on yield and production of mung bean are scarce. It is estimated that the annual yield of dry seed varies from 350 to 850 kg/ha and the world total production of mung bean (green gram and black gram) is about 0.5 million t. India accounts for about 70% of the total world production.

8 Spices and Flavourings

Bay laurel

Bay laurel – *Laurus nobilis*; Laurel family – *Lauraceae*

Origin, history and spread

Bay laurel, also called 'sweet laurel', 'sweet bay', 'Greek bay' or just 'laurel', is native to the Mediterranean region and Asia Minor. The ancient Greeks and Romans used it as condiment and medicine. In Greek mythology the nymph Daphne caught the eye of the god Apollo but she turned herself into a bay laurel shrub to escape Apollo's affection. After that, Apollo considered bay laurel to be sacred. That is why in ancient Greece, receiving a wreath made of bay laurel leaves was considered to be an honour. Olympic winners, poets, victors and heroes were given the wreath to wear on their heads. This habit was also accepted by the Romans. The name '*Laurus*' is derived from the Latin word 'laureola', which means a 'crown of laurel', and the Latin word 'laureatus' means 'crowned with laurel'. Romans spread the species throughout parts of Europe; early settlers introduced it into the New World. Nowadays, the species is both cultivated and collected from the wild in the Mediterranean region; it is also cultivated in Russia, Central America and the south of the USA.

Botany

Bay laurel is an evergreen shrub or tree up to 12 m high, however, in cultivation usually pruned to 2–3 m high (Fig. 8.1). The species is naturally multiple-trunked. The bark of the stem and the branches is dark brown to almost black. The plant has alternating, thick, leathery, simple, dull green, oblong-elliptical to oblong-lanceolate, aromatic leaves. Leaves are 5–12 cm long and 3–4 cm

Fig. 8.1. Habit of a flowering bay laurel bush, Tuscany, Italy.

broad, with serrated and wrinkled margins, petioles 0.5–1.5 cm long. Bay laurel is usually dioecious, so male and female flowers are found on different plants. The inflorescence is an axillary umbel, composed of four or five flowers; umbels are sometimes solitary although mostly two to five are arranged in a short raceme (Fig. 8.2). Each umbel has four bracts. Flowers have four greenish-yellow, ovate-oblong to rounded petals, which are 4–6 mm long. Male flowers have eight to 12 fertile stamens. Female flowers have a one-celled ovary and a short style with an enlarged stigma. The fruit is a globular to oval drupe, 1–2 cm long and dark violet to black at maturity. The fruit contains a single seed.

Uses and constituents

Bay laurel leaves, and sometimes the fruits, are used mainly for culinary purposes. They are used to flavour several dishes, stews, soups, sauces, fish, meat and beverages. Bay laurel is one of the components of the well-known 'bouquet garni'. The leaves can be one of the flavouring agents in marinades for meats. Bay laurel has a fruity, camphorous, heavenly fragrance. As fresh leaves are bitter and the bitterness decreases by drying, the leaves are usually dried before use. Dried and powdered leaves are used industrially in various foodstuffs. Bay laurel contains an essential oil that can be obtained from the leaves by steam

Fig. 8.2. Leaves (left) and inflorescences (right) of bay laurel.

distillation; the oil is used in industry to scent candles, perfumes, creams and soaps, it is also used to flavour foodstuffs. Both bay laurel leaves and the oil can be used to repel insects. The leaves are also used for smoking; however, smoking may negatively influence the ability to concentrate and performance capacity. Since ancient times, bay laurel has been used medicinally and it is still used in processing several pharmaceutical products. It is considered a stimulant, astringent and stomachic. Bay laurel oil is used in medicines to treat colds, flu and angina.

The leaves contain 5–10% water, 65% carbohydrates, 8–11% protein, 5–9% fat and 4% ash. The oil comprises over 140 different components. The species is very slow-growing, which makes it suitable for growing in containers, tubs or large pots. *Laurus nobilis* is often grown as an ornamental; it responds well to trimming, it can be used for making nice hedges and the plant can be sheared into distinctive shapes.

Ecology and agronomy

Ideally, day temperatures should be between 10 and 27°C, and night temperatures not below 10°C. It can withstand several degrees of frost. Severe frost may kill the above-ground parts, although regeneration from subterranean parts of the plant may be possible. Bay laurel thrives best in well-drained, rich, deep and moist soils; however, the species can grow more or less adequately on a wide range of soils with a pH of 4.5–8.3, in areas with an annual rainfall of 300–2200 mm.

The species is propagated from seed, cuttings or by layering. For successful germination the seeds should be soaked in warm water for about 24 h before they are sown in pots or seedbeds. As germination of the seed and juvenile growth of the seedling happen best in partial shade, it can be carried out in a greenhouse. Germination takes quite a long time, it may be up to 4 months. The seedling grows slowly and is mostly transplanted when 2 years old. Bay laurel likes deep watering to get started, but when the species is established it becomes somewhat drought-hardy. Spacing between the plants depends on several factors such as cropping system and availability of water. It ranges from 0.5 to 6 m between plants in the row, and from 2 to 6 m between rows. Bay laurel can be planted as solitary shrubs or trees or as hedges.

Weeds can be controlled manually or mechanically or with herbicides. Alternatively, natural vegetation can be allowed to grow under the trees and shrubs, which after slashing can serve as mulch. For good growth a balanced NPK fertilizer is required, but in practice little fertilizer is used. After fruit set, the plants will usually make flushes of new growth. The new soft and bright green leaves are rather vulnerable and may easily be burnt by harsh sun.

The frequency of harvesting depends among other things on fertilizer use and moisture supply; it may be two or three times per year. Harvesting is usually done manually, but can be done mechanically, especially when bay laurel is planted in hedges. Yield varies but can be up to 5 t of dried leaves per hectare. Turkey is the major producer of bay laurel leaves. World production statistics are not available.

Cinnamon and Cassia

Cinnamon – *Cinnamomum verum*, Cassia – *Cinnamomum* spp.; Laurel family – *Lauraceae*

Origin, history and spread

Cinnamomum verum originated in Ceylon (now Sri Lanka), Burma (now Myanmar) and southern India, where it still occurs in the wild. It is assumed that cinnamon reached Europe and Egypt in the 5th century BC, by means of Arabic trade. The spice was known to Herodotus and other classical authors. It is mentioned in the Bible and was used on funeral pyres in ancient Rome. In antiquity, cinnamon bark was used to flavour food, to perfume, or to burn it for its pleasant scent. Arabic traders sold cinnamon to Venetian traders, who held a monopoly on the spice trade in Europe for many years. Until the Middle Ages, the source of cinnamon was unknown to Europeans because it was concealed by the traders. This means that all the reported cinnamon until the Middle Ages may have been from different origins. Cinnamon probably included the cinnamon-related species called 'cassia', which was found in Indonesia, Vietnam, Burma and China.

The trade in cinnamon was disrupted by the rise of other Mediterranean powers such as the Ottoman Empire, which forced other countries to search for

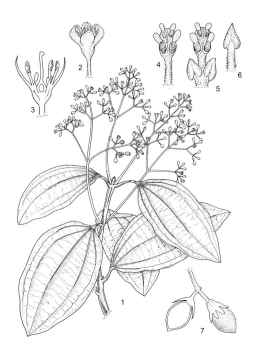

Fig. 8.3. Cinnamon: 1, flowering branch; 2, flower; 3, schematic longitudinal section through flower; 4, stamen of the first and second whorl; 5, stamen of the third whorl with glands; 6, staminode of the fourth whorl; 7, fruit and schematic longitudinal section through fruit (line drawing: PROSEA volume 13).

sea routes to reach the cinnamon-growing regions. The Portuguese occupied Ceylon in 1536 and took over the monopoly on cinnamon trade. In 1656 it was taken over by the Dutch. The bark was collected in the wild until 1770, when commercial cultivation began in Ceylon. The Dutch in turn were ousted by the British in 1796, who carried on the trade monopoly until 1833. Cinnamon was introduced into Java in 1825. Since then, it has been cultivated in southern India, the Seychelles and Brazil. The best-quality cinnamon is still produced in Sri Lanka (formerly Ceylon).

Botany

C. verum (Fig. 8.3) is a much-branched evergreen tree of 8–17 m height in the wild state. In cultivation, it is usually a dense bushy plant about 2–3 m high (Fig. 8.4). The trunk is low-branching and the wild form can have a diameter of 60 cm. Bark and leaves are strongly aromatic. The bark on young shoots is smooth and pale brown; on mature stems and branches it is rough and dark brown or greyish-brown. The leaves are opposite and have a grooved petiole 1–2 cm in length. The leaf-blade is stiff, ovate or elliptical, 5–25 cm × 3–10 cm, with three outstanding, whitish-green veins from the base to the tip, and two or more lateral veins to about three-quarters the length of the leaf; the base is rounded and the tip acuminate. Growth occurs in flushes; new leaves are reddish and limp, turning bright green above and dull greyish-green beneath later (Fig. 8.5).

The inflorescence is an axillary or terminal panicle, drooping, about 5–10 cm long, on a soft hairy, white, peduncle (Fig. 8.5). The flowers are small, only about 3 mm in diameter, with a foetid smell. Each flower is subtended by a

Fig. 8.4. Cinnamon stool or bush (photo: M. Flach).

Fig. 8.5. Cinnamon (left to right): inflorescence; flush of new leaves; mature leaves (photos, left: J. van Zee, middle: D.L. Schuiling, right: J. van Zee).

small, ovate, hairy bract. The flower consists of a six-pointed, campanulate, pubescent, yellow-white calyx; corona is absent; there are nine stamens with glands at the base; ovary is superior, single-celled, with a single ovule and a short style. The fruit is a fleshy, ovoid and pointed drupe, about 15 mm long, and with a persistent calyx around the base.

Cinnamon has a deep and extensive root system, with a well-developed taproot and numerous spreading laterals; after pruning new shoots arise from the roots.

Species, uses and constituents

As well as *C. verum*, there are more *Cinnamomum* species that are also grown for the aromatic bark. These species are jointly called 'cassia', but often, mistakenly, also called 'cinnamon'. In many statistics, cinnamon and cassia species are joined together due to the slight differences between the species, concerning morphology (Fig. 8.6), cultivation and utilization. More or less important cassia species, yielding spice and essential oil, are the following.

1. *Cinnamomum cassia* – Chinese cassia: a native of Myanmar (Burma) and grown in south China and Myanmar for the bark and dried unripe fruits; cassia oil is extracted from the leaves.
2. *Cinnamomum burmanni* – Indonesian cassia or Padang cassia: grown in Indonesia for the bark, the species has a smooth bark and no cork.
3. *Cinnamomum tamala* – Indian cassia; *Cinnamomum loureirii* – Saigon

Fig. 8.6. Cassia: 1, flowering branch; 2, flower; 3, stamen of the first and second whorl; 4, stamen of the third whorl with glands; 5, staminode of the fourth whorl; 6, fruits (line drawing: PROSEA volume 13).

cassia; *Cinnamomum oliveri* – Olivers bark: these species are all minor crops grown for the aromatic bark, sometimes for the leaves, but they are all an inferior substitute for cinnamon.

C. cassia and *C. burmanni* are by far the most important cassia species. Their bark is mainly exported to the USA, South-east Asia and Central Asia. Often the whole bark of cassia is used (not only the inner bark as in cinnamon). Cassia is an important spice in Chinese cookery. Americans prefer cassia to cinnamon; however, cassia is sold labelled as 'cinnamon' in the USA. Most of the real cinnamon is used in Europe.

The thin inner bark of real cinnamon is, both whole and in ground form, used as a spice. It is applied as a condiment in cookery; it is used for flavouring desserts, pies, candies, stewed fruits, liqueurs and all kinds of confectionery. In general, real cinnamon is mainly used in sweet foods or foods that require a mild flavour, whereas cassia is more suitable for spicy dishes, pastries, curries and meat dishes. Compared with cinnamon, cassia tastes more bitter and astringent and is more robustly flavoured. However, the species are often used interchangeably. Dried cassia buds are also added to food for flavouring.

The aromatic substances in the bark are mainly cinnamic aldehyde and eugenol. The bark is also used for the extraction of oil and oleoresin. Both oil and oleoresin are used in industry for flavouring processed foods, beverages and pharmaceuticals. The oil content of the bark varies from 0.5 to 2%; the oil contains 60% cinnamic aldehyde and 10% euganol. Furthermore, the bark contains tannins, proteins, cellulose, pentosans, mucilage, starch and minerals.

Leaves are also used for the extraction of oil, which is used in flavourings, perfumery, and extensively as a fragrance component in soaps, detergents and cosmetics. The dried leaves contain 0.7–2% oil; leaf oil contains 65–95% eugenol and about 3% cinnamic aldehyde.

In folk medicine, cinnamon and cassia, including the oil, have been used as a stimulant and to treat colds, nausea, flatulence and digestive problems. The volatile oil is used in some inhalants. Cinnamic aldehyde has good antifungal properties. Cinnamon was also considered to be an aphrodisiac.

Cinnamomum camphora is a native of China, Japan and Taiwan. This species is grown for the production of camphor. Camphor is an important essential oil used in the production of celluloid, in disinfectants, in chemical preparations, in medicines, in perfumes and soaps. The oil is obtained by distilling the wood.

Ecology and agronomy

Cinnamomum is adapted to a wide range of climatic conditions. It thrives at a well-distributed annual rainfall of 2000–2500 mm. Wild cinnamon trees are adapted to tropical evergreen rainforests. The plants grow best at an average temperature of 27°C. These ecological aspects are more or less the same for both cinnamon and cassia, although cassia is somewhat less specific than cinnamon. It grows well on different soils in the tropics. However, soil conditions have a great effect on bark quality. The species occurs up to 2000 m altitude, on well-drained hillside soils of low fertility and pH 4–6.

Cinnamon is often propagated by seed. Seeds are usually sown in a nursery; fresh seeds germinate in about 25 days. After 4 months, the seedlings are transplanted into containers. After another 4–5 months the young plants are transplanted into the field. Seeds can also be sown directly in the field. Cinnamon is also propagated by cuttings of young shoots, or by layering shoots. The spacing between the plants varies from 0.9 to 4 m. Spacing of 3 m × 3 m is often used, which results in about 1100 plants/ha. Weeding has to be done two to four times per year, until the weeds are shaded out. Farmyard manure, green manure or plant residues are usually applied as fertilizer; in practice, artificial fertilizers are little used. However, applying phosphate at planting is usually profitable, and annual application of an NPK 2:1.5:1.5 mixture is recommended at a rate of about 50 kg/ha to young trees and 100 kg/ha to mature trees.

Depending on species, first harvest is 3–7 years after germination. The species coppice well, so the stumps will regrow into a new stand. The shoots have to be kept straight by regular pruning. Subsequent harvests take place every 3 or 4 years. Shoots can be harvested when they are about 2–3 m high with a diameter of about 1.5–5 cm. Cutting has to be done in a wet season, when the stool is actively growing, and peeling of the bark is easier. The bark is removed by cutting it into two vertical, long strips. The two strips are taken off the stem; next they are heaped and covered with wet fabric for 24 h, to facilitate fermentation of the bark. Subsequently the epidermis, cork and green cortex are scraped from the inner bark, and can be used for oil extraction (cassia bark is often processed unscraped). Residues after processing may be used as mulch and fertilizer. Next, the inner bark is dried slowly. During the drying process, the inner bark becomes rolled into so-called 'quills' (Fig. 8.7). Before

Fig. 8.7. Bark of cassia laid out on a floor to dry, Indonesia (photo: D.L. Schuiling).

marketing, the quills are often inserted one inside another to form compound quills, about 115 cm long (later, the quills are cut into 10 cm pieces). The quills are graded into five qualities based on thickness of the bark, aroma, colour and appearance. The bark may not extend a thickness of 0.5 cm. The quills can be marketed in whole pieces or ground into powder.

Commercial cinnamon plantations can be productive for 20 to 40 years. Yields of quills depend on crop cycles, for example a 3-year or a 10-year cycle. Yields range from 70 to 875 kg/ha, with the world average being 762 kg/ha. In 2004, the total world production of dry cinnamon, including cassia, was 133,000 t. Nowadays, the major cinnamon (including cassia)-growing countries are Indonesia and China.

Clove

Clove – *Syzygium aromaticum*; Myrtle family – *Myrtaceae*

Origin, history and spread

The clove tree is indigenous to the islands of Ternate, Tidore, Bacan and Makian and their smaller adjacent islands in the Moluccas (Indonesia). These were the original 'Spice Islands' that enticed Europeans into the region. The crop and its trade have a long and fascinating history going back to the Han Dynasty in the 3rd century BC. By means of transit trade by Arabs, the clove was known already to the Romans in the 1st century. The story of the clove trade and the spread of the crop are full of intrigue and brutality. Early in the 17th century, when the Dutch ousted the Portuguese from the Moluccas, clove cultivation had spread to many islands. In order to establish a monopoly in the clove trade, the crop was forcibly eradicated everywhere and concentrated on Ambon and three nearby small islands. The decline of the Dutch monopoly started in the latter half of the 18th century. At that time the French smuggled some offspring from trees that must have escaped the Dutch axe, from the North Moluccas to Mauritius. In the 19th century the British took plants from the Moluccas to Malaysia, Sumatra (Indonesia), India and Sri Lanka. In the 20th century much material spread throughout Indonesia. The plants from Mauritius gave rise to the clove populations outside Asia, in Zanzibar (Tanzania), Madagascar and Grenada.

Botany

The clove tree is a slender evergreen tree, up to 20 m tall, conical when young, later becoming cylindrical; in cultivation it is usually smaller and branched from the base. Shoot growth, appearing in flushes, forms a dense canopy of fine twigs. The leaves are opposite, simple and glabrous; petiole 1–3 cm long; blade obovate-oblong to elliptical, coriaceous, shining, gland-dotted, 6–13 cm × 3–6 cm, base very acute, apex acuminate. The inflorescence is terminal, paniculate, about 5 cm long, with three to 20 (40) bisexual flowers, usually borne in cymose

Fig. 8.8. Clove: 1, branches with flower buds and flowers; 2, a flower bud or clove (line drawing: PROSEA volume 13).

groups of three; flower buds 1–2 cm long, constituting the cloves just before opening (Figs 8.8 and 8.9). The numerous stamens are up to 7 mm long. The four unopened petals, forming a small globule, are enclosed in four tooth-like lobes of the calyx. The flower buds are at first of a pale colour and gradually

Fig. 8.9. Clove: inflorescence with flower buds (left) (photo: M. Flach); close-up of inflorescence with almost harvestable flower buds (middle) (photo: D.L. Schuiling); dried flower buds, the cloves of commerce (right).

become green, after which they develop into a bright red, when they are ready for collecting. The cloves which are traded are the dried, still unopened flower buds. The Greek word 'syzygium' means 'closed together', referring to the closed petals of the flower when it is harvested. The word 'clove' comes from the Latin word 'clavus', for nail.

Types, uses and constituents

The three major commercial clove types – 'Zanzibar', 'Siputih' and 'Sikotok' – differ in tree habit, leaf size and colour but few trees are true to type; transitional forms are common. The genetic base, however, appears to be narrow. 'Siputih' produces large cloves, valued for the spice trade. In Indonesia, young trees are mainly of the reintroduced 'Zanzibar' type.

Dried cloves are marketed either whole or ground and used as a spice for culinary purposes, particularly for meats and bakery products. It is one of the ingredients of curry powder and Worcestershire sauce. In Indonesia clove is also used as a stimulant in so-called 'kretek' cigarettes. Kretek cigarettes, mainly manufactured in Java, are made by mixing cloves with tobacco in the proportion of one-third of shredded cloves and two-thirds of tobacco.

Cloves contain 14–20% of the strongly pungent, essential oil eugenol, which is obtained by steam distillation of cloves as well as flower stalks and leaves. An increasing portion of the clove production is being used for essential oil and oleoresin. Clove oil is used as a sweetener or intensifier in synthetic vanilla and in other flavourings. It is also used in perfumes, soaps and mouthwashes. Eugenol is a powerful natural antiseptic and is also frequently used as a local anaesthetic for toothache.

Ingestion of clove in too large amounts, especially the oil, can cause minor skin irritations and upset stomach. Smoking kretek cigarettes is more harmful than smoking regular tobacco cigarettes.

Ecology and agronomy

The clove tree grows best in the humid tropics at low altitudes, preferably on well-drained deep soils (Fig. 8.10). Although a climate with a marked dry season promotes flowering, clove does not withstand drought stress at all. In Indonesia, cloves for 'kretek' cigarettes are said to require 3 months with less than 60 mm rainfall, whereas for cloves for spice, rainfall should not drop below 80 mm in any month. Strong wind and waterlogging are not tolerated. Cloves are almost exclusively grown on islands, but proximity of the sea may not be as necessary as once thought.

The clove tree is propagated by seed. Shade is necessary for seedlings and after transplanting for young trees until they are firmly established. The standard spacing is 8 m × 8 m, but smallholders often plant much closer. The rectangular pattern facilitates intercropping in the early years. Banana and cassava are common intercrops. Intercrops may also provide shade but, near the young clove tree, shade trees such as *Erythrina*, *Gliricidia* and *Leucaena* species are preferred, since these trees can be pruned to even out irradiance through the

Fig. 8.10. Clove plantation, Bali, Indonesia (photo: D.L. Schuiling).

year. The first crop can be harvested when the trees start flowering at 5–8 years. The right stage for harvesting lasts only a few days and the flower buds are picked three to eight times a season (Fig. 8.11). Yields vary considerably but they are generally low, about 4 kg of dried cloves per tree. Top yields of 50 kg have

Fig. 8.11. Harvesting cloves, Java, Indonesia (photo: M. Flach).

been reported for individual trees in different parts of the world. Probably the oldest clove tree in the world grows on Ternate. This old giant ('afo', in Indonesian), 36 m tall and 4 m in diameter, produces 150 kg of dried cloves each year. At harvest, the inflorescences with unopened buds are removed from the trees by hand. After harvest the inforescences are separated into flower stalks and buds. The buds are then sun-dried.

In 2005, the world total clove production (including flower stalks) was about 0.15 million t. The main producers were Indonesia (61%), Madagascar (18%) and the Tanzanian islands of Zanzibar and Pemba (12%). Although today Indonesia is the world's largest producer, consumer and importer, world clove trade is centred on Zanzibar.

Ginger

Ginger – *Zingiber officinale*; Ginger family – *Zingiberaceae*

Origin, history and spread

Ginger originated in tropical Asia, probably in southern China and India. However, the exact origin is unclear because it has been cultivated for millennia in both China and India, and has never been found in the wild. Ginger is mentioned in the earliest Chinese herbals. The plant has a long history of culinary and medicinal use in the cultures of these countries. The refreshing smell and the pungent, warm taste were highly valued. Ginger was known by the ancient Greeks and Romans. The Greeks ate ginger wrapped in bread to prevent nausea after an orgy. The plant is described in writings from Dioscorides and Pliny. The spice was known in Germany and France in the 9th century. Marco Polo, who visited China and Sumatra in the 13th century, saw ginger in these regions and took some with him to Europe. In addition, in the 13th century, the Arabs took the plant from India to East Africa. In the 16th century, the Portuguese took it to West Africa. The Spaniard Francesco de Mendoza introduced the cultivation of ginger into Mexico. In Europe, during the Middle Ages, ginger was one of the spices used to flavour beer. The English botanist William Roscoe gave the plant the name '*Zingiber officinale*' in 1807. The name '*Zingiber*' is via the Greek word 'zingiberis' derived from the Sanskrit word 'shringavera', which means 'shaped like a deer's antlers'; '*officinale*' indicates medicinal properties of the plant.

Ginger is now grown as a commercial crop in Africa, Latin America and South-east Asia.

Botany

Ginger (Fig. 8.12) is an upright, slender, perennial herb, about 1 m tall. Pale green leafy shoots, which are mainly formed by the leaf-sheaths, arise from subterranean rhizomes. Each shoot has eight to 12 leaves. Scales cover the basal part of the shoots. The sheath of the leaf is distinctly veined, blade linear to

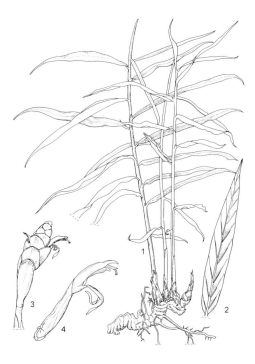

Fig. 8.12. Ginger: 1, habit; 2, leaf; 3, inflorescence; 4, flower (line drawing: PROSEA volume 13).

lanceolate, closely parallel-pinnate veined, up to 30 cm × 2 cm, and acuminating at the apex. Ligule is up to 5 mm long. The leaves are arranged in two ranks on the stem (or 'pseudo-stem').

Inflorescences do not arise from the leafy, vertical growing shoots or stems, but direct from the nodes of the rhizomes. Flowers grow in dense spikes, which are 4–8 cm long. Spikes are cylindrical and cone like. The bracts in the spikes are green with translucent margins. The flowers are surrounded by a bracteole; calyx tubular-spathaceous, 10–12 mm long and white in colour; corolla tubular, pale yellow, widening at the top into three lobes, tube 18–25 mm long with purple tips. Below the spike, the inflorescence is covered with scales. Most ginger cultivars are sterile and flowers are rarely seen; reproduction is usually vegetative (Fig. 8.13).

The rhizomes are up to 3 cm thick, fleshy, and often irregularly branched. They are formed by many short internodes with thin scales, which leave behind ringing scars. The epidermis is corky and yellow; the flesh is pale yellow and aromatic (Fig. 8.14). Rhizomes grow underground at shallow depth, sometimes partly above the ground. The plant develops relatively few roots, which arise from the nodes of the rhizomes.

Species, uses and constituents

As well as *Z. officinale*, the genus *Zingiber* contains another 100 species, all aromatic herbs. Only three of them have any commercial importance, usually grown in home gardens. The three species are *Zingiber montanum*, *Zingiber*

Fig. 8.13. Ginger rhizome with three flowering shoots between the leafy shoots (photo: J. van Zee).

spectabile and *Zingiber zerumbet*. They are usually cultivated for their medicinal properties, for essential oils, as a spice or as an ornamental.

As *Z. officinalis* is always propagated vegetatively, the number of different clones is limited; however, each production centre developed a distinctive, typical type or cultivar.

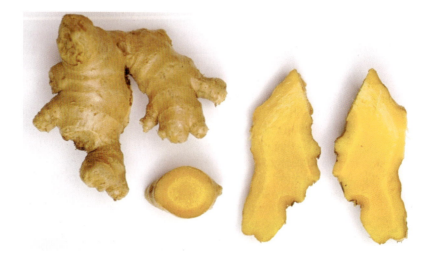

Fig. 8.14. The spice ginger: the rhizome, whole (left) and in longitudinal and cross-section (right) (photo: D.L. Schuiling).

Ginger rhizomes are widely used around the world as a spice or food additive, especially in Chinese, Indian and South-east Asian cookery. Ginger is used in three different forms: (i) fresh ginger; (ii) dried and ground ginger; and (iii) preserved ginger. Fresh ginger can be immature or mature rhizomes. It is usually grated or finely chopped and added to many different dishes to achieve a fresh, spicy and pungent taste. It is used fresh, boiled or fried, each of the methods giving a different taste to the dish. Fresh ginger is used in producing ginger ale and other beverages. Fresh ginger is sometimes eaten raw, as a vegetable. It is used in curries, spice pastes, sauces, to rub on meat and to make ginger tea. Dried and ground ginger is rather different in taste to fresh ginger; it is more aromatic and pungent than fresh ginger. It is also used to flavour various dishes, gravies and soups. Dried ginger is a component of curry powders and several spice mixtures; it is used in gingerbread, biscuits, cakes and desserts. Mature rhizomes are used for dried ginger. Preserved ginger is prepared by boiling peeled, immature rhizomes; the rhizomes are cut into pieces and preserved in sugar syrup. It can be used as snack food; it is also used in the manufacture of jams, marmalades and confectioneries. Candied ginger is made from young rhizomes, which are cleaned and peeled first, next boiled, soaked in syrup, and dusted with sugar. The rhizome contains an essential oil, which is used in producing flavouring essences, in pharmaceuticals and in perfumery. In Japan, very young rhizomes are sometimes pickled and served with sushi.

Dried rhizomes contain approximately 10% water, 10–20% protein, 40–60% carbohydrates, 10% fat, 2–10% fibre and 6% ash. The pungent taste is caused by non-volatile phenols; the essential oil is responsible for the odour and flavour.

Ginger is still widely used in folk medicine to treat various ailments; it is also considered to be an aphrodisiac. Chinese sailors ate ginger to avoid sea sickness. Rhizomes contain oleoresin, which is nowadays used as ingredient in digestive, laxative, antitussive, antiemetic and carminative preparations.

Ecology and agronomy

Ginger is cultivated mainly in the humid tropics, from sea level up to an altitude of 1500 m. It requires an annual rainfall of at least 2000 mm; optimum is 2500–3000 mm, well distributed over the year and preferably with a short dry period before harvest. Ginger cannot stand waterlogging. Although high temperatures are required, at least during a part of the season the plant thrives best in partial shade. Full sun may cause damage to the leaves and subsequently poor plant growth. Ginger grows best on well-drained medium loams with a good supply of organic matter, pH 6–7.

Ginger is usually propagated from rhizome cuttings; the rhizomes that are used for cuttings are mostly stored in covered pits. The cuttings have to be disease-free, 3–4 cm long, weighing 20–30 g each, and should have at least one bud ('eye'), preferably two. The rhizome pieces, usually called 'seed rhizomes', have to be cut a few days ahead of planting to let the cut surfaces dry. This prevents rotting and other forms of decay. Planting rates of seed

Fig. 8.15. Ginger field with mulch layer, Bali, Indonesia (photo: D.L. Schuiling).

rhizomes range from 850 to 2500 kg/ha. The rhizome pieces have to be planted in well-prepared soil at a depth of about 5–10 cm. The soil has to be moist and about 25°C. Spacing can be 40 cm in the row and 40 cm between the rows, but narrower spacings are also used. Seed rhizomes can be planted on beds or ridges. Well-prepared soil means that the soil has to be ploughed several times or dug thoroughly; good tilth is required to avoid malformed rhizomes. Cattle manure or compost has to be applied before planting. During the growing season, fertilizers should be applied in split doses, 100 kg N, 22 kg P and 42 kg K per hectare are recommended. During establishment of the plants, two or three weedings are usually required, or herbicides can be used. Manual or mechanical weeding should be shallow to avoid damage to the rhizomes. Earthing up the rows or ridges, which is done to prevent exposure of the rhizomes, is also a method to control weeds. Ginger fields are often mulched with green leaves or organic wastes to prevent erosion and repress weeds (Fig. 8.15). Ginger is sometimes grown as an intercrop of, for example, coconut and coffee. About 9–10 months after planting when the leaves turn yellow and dry up, the rhizomes can be harvested. Harvesting of immature rhizomes begins 5 months after planting. The rhizomes are usually lifted with a spade or other tool, and separated from leaves and roots. Next, the rhizomes are washed and sundried.

Yields vary considerably, from 6 to 30 t of green ginger per hectare, depending among other things on differences in cultural practices. Ginger is grown mainly by smallholders. In 2004, the total world production was about 984 million t of green rhizomes. The main producers were China, India, Indonesia and Nigeria.

Hop

Hop – *Humulus lupulus*; Hemp family – *Cannabinaceae*

Origin, history and spread

Hop most likely originates from China because the genus *Humulus* consists of three species, *Humulus lupulus*, *Humulus japonicus* and *Humulus yunna-nensis*, all native to China. Migration westwards resulted in the appearance of hop in Europe. A number of authors indicate Eurasia as the centre of origin.

The cultivated hop is *H. lupulus*. German brewers started using the cone-like mature female inflorescences of hop, called 'hops', to flavour their beer in the Middle Ages. Before that, various combinations of bitter herbs were used. From Germany this procedure spread throughout Europe. Hop was introduced into the British Isles during the 16th century, from where it was introduced into the USA in the 17th century. Today hop is grown at various places in the temperate regions where climatic and soil conditions are suitable.

Botany

The hop is a perennial vine without tendrils. The flexible rough shoots twine in a clockwise direction with the aid of strong hooked hairs located on the angles of the stem (Fig. 8.16). The shoots grow rapidly to a length of up to 8 m. At the end of the growing season, the shoots die to near ground level, where the perennial crown, which consists of stem tissue, is situated on top of a rootstock system. The crown produces new shoots in spring. The leaves with toothed

Fig. 8.16. Hop plantation, Hereford and Worcester, UK (photo: D.L. Schuiling).

Fig. 8.17. Hop: detail of twining stems (left) and female inflorescences (right) (photos: D.L. Schuiling).

margins are heart-shaped with three to seven lobes, 7–16 cm × 6–11 cm, with petioles of 2–7 cm. They grow mostly in pairs at each node. The hop is dioecious, which means that male and female flowers are found on separate plants. Male flowers are in loose panicles. The flowers have five green sepals and five anthers, which produce large quantities of wind-borne pollen. Female flowers grow in cone-like inflorescences (Fig. 8.17). On each node of the central axis of the cone there is a pair of bracts. Each bract encloses a pair of bracteoles and each bracteole contains a small flower. Ripe cones are 2.5–5.0 cm long, with papery yellowish scales. The pendulous cones are born in clusters. Particularly on the bracteoles, lupulin glands develop, which contain resins. The glands resemble pollen. The perennial crown, becoming woody with age, produces an extensive, fibrous root system to a depth of 1.5 m and spreading up to 2.5 m.

Cultivars, uses and constituents

The cone-like female inflorescences are economically the most important part of the hop, especially the glands, which are of great value to the brewer. Glands contain alpha acid and essential oils, which give the bitterness and flavour of beer, respectively.

   Potential bitterness of the various hop cultivars is expressed in alpha acid units (AAU). According to bitterness, three groups of cultivars can be distinguished.

1. Low bitterness, which means 2–4% alpha acid content in the hops.
2. Medium bitterness, 5–7% alpha acid content.
3. High bitterness, 8–12% alpha acid content.

Most of the commercially grown hops are seedless hybrids. At present, efforts are also being made in breeding dwarf-type cultivars. The selected hop cultivar and the time of adding the hops to the brewing process influence the taste of the beer. In some countries, the young shoots of hop are eaten as a vegetable. Hop is often used in folk medicine; it is reported to be antiseptic, diuretic, hypnotic, nervine and stomachic.

Ecology and agronomy

The best results for growing hop are achieved in areas having an abundant rainfall during the growing period and abundant sunlight during the fruiting period. The length of the growing period and the temperature affect the alpha acid content. Hop is grown as a hardy perennial species, preferably planted in tilled, rich, loamy soil. Poorly drained, strongly alkaline or saline soils should be avoided.

Hop is usually grown from cuttings taken from female plants. The cuttings are planted in rows about 2 m apart. When the shoots have reached a length of about 0.5 m, they are trained up a framework of long poles and wires.

Although only female cones are harvested, many hop plantations have a small number of male plants included. The reason is that the presence of a few male plants is profitable because fertilized female inflorescences grow larger and more rapidly. In 2005, the world average yield of hops was 1.6 t/ha/year. In the same year, the world total production of hops was around 1 million t. The USA and China were the main producers.

Labiate Herbs

Basil – *Ocimum basilicum*, Rosemary – *Rosmarinus officinalis*, Thyme – *Thymus vulgaris*, Spearmint – *Mentha spicata*; Mint family – *Lamiaceae*

Introduction

A number of species from the mint family are popular culinary herbs. The ancient Greek and Romans used some spices from the herb family to produce 'garum', which is a sauce made of a mixture of fermented herbs and fish. The sauce was highly valued for flavouring dishes. Fresh or dried sprouts or leaves are also used to flavour dishes, salads and soups. Some species of the mint family are among the bundled herbs of the famous French 'bouquet garni'. In contrast with the umbelliferous herbs, most of the *Lamiaceae* are only grown for the sprouts or the leaves and not for the seeds. The leaves are covered with glands, which contain an aromatic essential oil. The cultivated species from the mint family have a number of botanical characteristics in common: the stems are quadrangular, with opposite, simple, paired leaves, which are arranged successively at right angles to each other. Flowers are arranged in clusters in the axils of the upper leaves or in a terminal spike. The colour of the flowers varies from white, pinkish-white, pink, lilac to purple. Flowers are symmetrical in one plane; they have two lips, the upper ones two-lobed, and the lower ones three-lobed, stamens mostly four. The fruit contains four one-seeded nutlets.

Four of the more important, cultivated species are described here: basil, rosemary, thyme and mints. Yield and production statistics of each of these labiate spices are unknown.

Basil

The origin of basil, also called 'sweet basil', is Central Africa, western Asia and India. It is a very old spice. The ancient Egyptians knew the plant; they used it to flavour dishes, and basil wreaths have been found in several burial chambers. The ancient Greeks and Romans used it also as a medicinal plant. In India it is a sacred herb, used in Hindu temples. '*Basilicum*' is probably derived from the Greek word 'basileus', which means 'king' or 'royal'. In some languages basil is called 'king's herb'. *Ocimum basilicum* reached northern and western Europe in the 16th century and North America in the 17th century. It is now cultivated throughout the world, in temperate and tropic regions.

Basil is an erect, pilose, annual herb, up to 1 m high (Fig. 8.18). There are tall and dwarf types. The stem is about 6 mm in diameter, much-branched, light green to dark purple. The decussately opposite, simple, leaves have ovate to elliptical blades (1–8 cm × 0.5–4 cm). The acuminated apex is densely glandular punctuate, and light green to purplish-green in colour. The inflorescence is terminal and about 30 cm long, consisting of three flowered cymes, arranged in

Fig. 8.18. Labiate herbs: 1, mint; 2, basil; 3, thyme; 4, rosemary (photo: The Greenery, The Netherlands).

whirls, 1–3 cm apart. The colour of the flowers is whitish-yellow to reddish. The plant has a rather thick taproot with many secondary roots.

Fresh or dried leaves of basil are used to flavour salads, soups and vegetable dishes, and combined with lamb and chicken. It is an important herb in Italian cookery, in pizzas, pastas and pesto. There is some basil oil production, mainly for industrial use in baked products, pickles, sauces, liqueurs, etc. The oil can be used to repel insects. The oil is also used in cosmetics. The extract of the leaves is used to produce a number of medicines such as a carminative, and to treat cough, dysentery, diarrhoea, etc. Basil with purple leaves is sometimes also grown as an ornamental. There are many other *Ocimum* species apart from *O. basilicum*, and within *O. basilicum* there are many varieties which can vary considerably in growth habit, size, colour and aroma.

Basil is adapted to a wide range of climatic conditions, in the tropics, sub-tropics and temperate regions; however, it is susceptible to frost. It grows best on fertile, light, well-drained soils, pH 5.5–6.5. Water requirement is rather high.

Basil is propagated by seed, seldom by cuttings. The seed can be sown in rows, at the most 1 cm deep and ultimately a plant spacing of 20 cm. In the tropics it takes 4–6 days to germinate, in temperate regions 8–14 days. Weed control is important, especially in the juvenile phase of the crop; it has to be done mechanically or by manual weeding because of the lack of suitable herbicides. A mulch layer can also give sufficient control. Application of fertilizer depends on soil fertility, but will be necessary because the crop has a high N requirement.

Where basil is grown for the leaves, it is harvested just before the appearance of the flowers. In some regions it is cut earlier to obtain more cuts; up to five is sometimes possible, depending on climatic conditions. For essential oil, basil is harvested when the plants are flowering.

Rosemary

Rosemary is native to the Mediterranean Sea coast, where it grows on dry scrub, rocky and sunny places. It has been used as a spice and medical herb since ancient times. The Greeks thought that rosemary had a positive effect on the functioning of the brain and strengthening of the memory. That is why they stuck a twig in their hair during exams. The name 'Rosmarinus' is probably derived from the Latin words 'ros', which means 'dew', and 'marinus' (mare), which means 'belonging to the sea'. Rosemary was grown in parts of medieval Europe – it was one of the plants in Charlemagne's *Capitulare de Villis*.

Rosemary is a hardy evergreen shrub, up to 2 m tall and wide (Fig. 8.18). The stem is grey pubescent. Leaves are opposite, sessile to short petiolate, growing in whorls, leaf-blade linear, 1–5 cm × 1–2 mm, similar to pine-needles. The top side of the leaves is dark green; beneath they are almost white. The leaves are very fragrant, especially when they are crushed. The inflorescence is racemose, axillary and up to ten-flowered, terminating short branches. The colour of the flowers is white, pale blue or blue. Flowering usually begins when the plants are 2 years old (Fig. 8.19).

Fig. 8.19. Detail of a flowering rosemary plant.

Rosemary is a strongly aromatic, resinous and bitter herb. When it is dried it is even more powerful than fresh. Next to the leaves, rosemary is also a source of an essential oil. The oil can be distilled from flowering tops and the leaves. Rosemary's leaves and oil have many uses, such as culinary, medicinal and cosmetic uses. Fresh or dried leaves can be used to flavour food such as soups, salads, stews, meat, poultry, fish, vinegar, etc. Leaves and twigs are components of the French spice mixture 'herbes de Provence' and the French herb bundle 'bouquet garni', respectively. The herb is very popular in Mediterranean cooking. Different from many other spices, the leaves do not lose the flavour and odour by long cooking. The oil is used in large amounts in the food-processing industry.

The medicinal uses of leaves and oil include among many others the following functions: antispasmodic, cardiac, carminative, diaphoretic, stomachic, diuretic and nervine. It is also used to treat headaches, bruises, colds and nervous diseases. The whole plant is antiseptic. The oil is also used as an ingredient of soaps and shampoos, in perfumery and to scent the air. Oil can also be an ingredient of products to repel insects. Rosemary is often grown as an ornamental.

The ecological amplitude of rosemary is rather wide but it prefers well-drained, calcareous soils on dry sunny slopes near the sea, pH 6–7.5. When the plant is frequently exposed to fog and salt spray, little water needed. Rosemary is mostly propagated by cuttings or by air-layering, because the species seldom produces seed. The plants are often raised in containers. Fertilizer applications must be moderate because too much can cause weak growth; however, for good oil production, adequate K is needed.

When the plants are large enough, 25-cm-long terminal shoots can be cut once or twice each growing season. When the plants grow older, it has to be done carefully because cutting into old wood may cause damage to the plant. Nowadays, rosemary is grown mainly in the Mediterranean area, especially in Spain, Tunisia and Morocco, and the USA.

Thyme

Thyme is native to the Mediterranean, mainly the European part, and Asia Minor. The herb has been well known since classical antiquity, because the name of the plant, '*Thymus*', was first given by the ancient Greeks and derived from the word 'thymon', meaning 'to fumigate', probably because they used it as a perfume, to scent the air. For Greeks, thyme was a symbol of elegance. When a Greek said: 'You smell of thyme', it was considered a compliment. The herb was probably first cultivated by the Assyrians. The ancient Egyptians used thyme as one of the herbs for embalming corpses. The Romans used it as one of the herbs in their famous 'garum' sauce. Pliny mentioned thyme in his writings as a medicine. In northern and north-western Europe it was probably first cultivated in the Middle Ages. In spite of the popularity of the herb, wild harvesting has been an important source of thyme supply for a long time.

Thyme is a small, perennial, evergreen, dwarf shrub (Fig. 8.18). The stems are much-branched, quadrangular, stiff, 10–30 cm high, and woody at the base. Branchlets are densely coated with short grey hairs. The opposite leaves are very small, linear to elliptic, 3–8 cm long and 1–2 mm wide. Leaves have short or no petioles, the margins are curved inwards, the colour is greenish-grey, and the surfaces are covered with short hairs and oil glands. The inflorescence is composed of axillary many-flowered whorls, forming a terminal spike. The flowers are pale purple. Thyme has a woody, fibrous root.

Within the cultivated thyme, several varieties can be distinguished which vary in hardiness, shape of the leaves and fragrancy; for example, lemon thyme, winter thyme, orange thyme and caraway thyme.

Thyme is one of the most important culinary herbs. It has a delightful aromatic smell and a warm pungent taste. The used parts of the plant are the fresh or dried leaves or flowering tops, and the essential oil. It is used in soups, salads, sauces, cheeses, various stuffings, Cajun dishes, etc. It is often combined with roasted meat, especially lamb, poultry and fish.

The essential oil has a wide range of uses in the manufacture of liqueurs such as 'Benedictine', perfumes, pharmaceutical products and cosmetic products. The active ingredient in the oil is thymol, which has antiseptic and expectorant properties. Thyme is used to treat cough, bronchitis, catarrh-colic, sore throat, spleen disorders and uterine disorders, among others; it is considered antispasmodic and carminative.

Lately, research has shown the value of the oil in reducing the severity of degenerative diseases of old age. The oil and the smoke of burning leaves repel insects. The oil is also used in fungicides and insecticides. Thyme is attractive to bees and butterflies, the flowers deliver famous honey. The plant is often grown as an ornamental, especially in rock gardens.

Thyme can grow between 4 and 28°C, but 16°C is ideal. The species is rather drought-tolerant. It can be cultivated on a wide range of soils, but it is best to grow it on calcareous, light and dry soils, because then the herb becomes most aromatic. Soil pH requirements are 6.6–8.5.

Thyme is propagated by seed, by division or from cuttings; it can be grown as an annual or as a perennial. Germination takes 2–3 weeks at 21°C. Spacing between plants in the rows and between the rows can be 15 and 25 cm, respectively. Later, thinning may be necessary, depending on cropping system. Flowering starts about 3 months after sowing. Weeds can be controlled by chemical and mechanical systems, but a mulch layer covering the soil is often used. Thyme can be grown without applying fertilizer, but for sufficient yield some manure or fertilizer is required. The crop is harvested prior to or during flowering, once in the first growing season and up to four times in the following years. To avoid damage to the plants they should not be cut too low, about 10 cm above the soil is ideal.

At present, thyme is cultivated primarily in the Mediterranean area, Germany, Russia, the USA, Canada, Japan and China.

Spearmint

Spearmint (or green mint) is native to central and southern Europe; it was well known to the ancient peoples of the Mediterranean and appreciated by them. There are many references to mint in old writings. According to Pliny, the smell of this herb did stir up the mind. In Athens, it was used to scent bath water and to perfume the body. The ancients also valued mint highly for its medicinal properties. The Romans brought it to northern and north-western Europe. The name 'Mentha' is probably derived from a Greek mythological character: the nymph Minthe. Hades, the god of the underworld, seduced Minthe, which made Hades' wife Persephone jealous and that was for her reason enough to change Mirthe into a (mint) plant. In the Middle Ages it was probably grown in the gardens of monasteries and royal domains, because mint was one of the cultivated plants on Charlemagne's list *Capitulare de Villis*. The Pilgrim Fathers probably brought mint to North America.

Spearmint is a perennial herb, up to 50 cm high, with glabrous, quadrangular stems (Fig. 8.18). The opposite leaves are sessile or very short-stalked, lanceolate or oblong-lanceolate, toothed, 7 cm × 3 cm, glabrous on top, beneath as well glabrous or sparsely pubescent. The leaves have a delightful and strong fragrance when they are bruised. The conical inflorescence is composed of spikes of lilac flowers in whorls.

Like all rhizomatous mints, spearmint is a vigorous plant which is very invasive. There are several species, hybrids and varieties of mints: as well as spearmint, field mint (*Mentha arvensis*, especially in Asia) and peppermint (*Mentha × piperita*) have significance, the latter is a hybrid between *Mentha aquatica* and *Mentha spicata*.

Spearmint is very much connected with English cookery. Lamb and mint sauce is a famous combination. Mint is used to flavour meat, soups, salads, teas and candies. Mint tea is very popular in North African and Arab countries. It is a component of 'bouquet garni', which is served as a garnish to dishes and desserts. Spearmint is also cultivated for the distillation of the oil.

Peppermint and field mint are grown mainly for the essential oil, obtained by distillation from the fresh plants, although a part of the field mint production is also grown for direct consumption. The oil has the typical, pure and refreshing peppermint scent, and a pungent and burning taste. Menthol is the main component of the oil. The oil or menthol is used to produce chewing gum, liquors, peppermint and other confectioneries. Menthol is used in many pharmaceutical preparations such as itch-relieving creams and cough syrups. It is also used in cosmetics, for example to perfume soaps or in toothpaste. The plant is attractive to bees and butterflies.

Spearmint is primarily a crop for temperate regions. It prefers moist to wet soil and pH of 5.6–7.5. Mint plants produced from seeds are not uniform, and hybrids such as peppermint are sterile and do not produce seeds at all. Thus mint is propagated by dividing rhizomes into cuttings. They can be planted 5 cm deep. A good spacing of the plants is 15 cm between the plants in the rows and 50 cm between the rows. A top dressing of manure in spring and after the first cut is essential. Water supply may be needed. As mint grows vigorously, weeds are not a big problem; only shortly after planting may some weeding have to be done. Mint is often harvested when the shoots are young and tender, or later, when the plants are starting to flower. To obtain a sufficient mint production, the plantation should be re-made every 3 or 4 years. Nowadays, mint species are cultivated in Europe, North America, Asia and North Africa, especially in Morocco.

Mustards and Horseradish

White or yellow mustard – *Sinapis alba*, Black mustard – *Brassica nigra*, Brown or Indian mustard – *Brassica juncea*, Horseradish – *Armoracia rusticana*; Wallflower family – *Brassicaceae*

White mustard

White mustard is native to the eastern Mediterranean region and the Middle East. It was domesticated in the same area in ancient times, about 2000 BC, and used as a salad oil, spice and medicine. The Greek mathematician Pythagoras wrote about the medicinal application of the seed of white mustard. The name 'mustard' is derived from two Latin words: 'mostum' and 'ardere'. 'Mostum' means 'fresh grape-juice' and 'ardere' means burning or heating. It refers to the process of mustard-making with mustard seed and wine. White mustard was introduced into northern Europe in the Middle Ages, and subsequently brought to the New World by the colonists. Nowadays it is grown mainly in temperate and subtropical regions.

White mustard (Fig. 8.20) is an erect, annual, hairy herb, up to 125 cm tall. The stem is ribbed and branching in the upper part. The leaves alternate

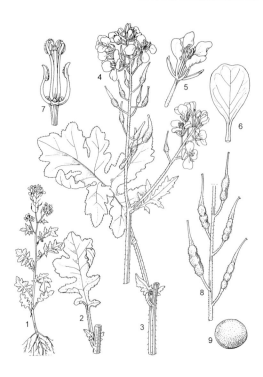

Fig. 8.20. White or yellow mustard: 1, habit of flowering and fruiting plant; 2, detail of stem with lower leaf; 3, detail of stem with central leaf; 4, flowering and fruiting branch; 5, flower; 6, petal; 7, stamens and pistil; 8, part of infructescence; 9, seed (line drawing: PROSEA volume 13).

and are deeply divided. The leaf-blade is elliptical, ovate or obovate in outline, pale green in colour and up to 15 cm long. The inflorescence is an axillary or terminal raceme (Fig. 8.21). The flowers are bisexual and about 1 cm long. They exist of four green, narrowly elliptical sepals; four yellow, obovate petals; six stamens; and one elongated pistil. The fruit is a ribbed and hairy, two-valved silique containing two or three globose seeds. The fruit is about 4 cm long, and nearly half the length consists of a flat and slightly curved, more or less sabre-like (seedless) beak. The angle formed by the position of the pedicel of the fruits, compared with the inflorescence axis, is between 45 and 90°. The seed is yellow; 1000-seed weight is 4–8 g. The species has a thin taproot with laterals.

White mustard can be a component of table mustard, which is mostly a mixture of crushed seeds of one or several different mustard species and vinegar or wine. Sugar, salt and herbs can be added; there are numerous recipes. Table mustard is one of the cheapest spices; for the poor it was often the only spice they could afford. In cooking, white mustard powder is used for flavouring meat and sauces. Whole seeds are used in the preparation of several pickles such as gherkins. The oil that can be pressed from the seed is used in the manufacture of mayonnaise; it is also used as a lubricant and an illuminant. Seedlings can be eaten raw, in salads, with sandwiches or used as garnish.

In temperate regions, white mustard is often grown as green manure. Some cultivars can be used in rotation with sugarbeet because they reduce the population of beet cyst nematodes in the soil. As with all cruciferous plants, the seed

Fig. 8.21. Inflorescences of mustards (left to right): white or yellow mustard; black mustard; brown or Indian mustard (right photo: J. van Zee).

of white mustard contains glucosinolates; furthermore it contains 5% water, 26% protein, 23% carbohydrates, 35% fatty oil, 5% fibre, 4% ash, the minerals Ca, P and Fe, and (pro-) vitamins A and B. White mustard is reported to have many medicinal properties; for example, diaphoretic, emetic and diuretic.

White mustard is primarily a cool-season crop for temperate climates with some humidity (Fig. 8.22). It is moderately tolerant to mild frost. The species thrives best on well-drained loamy soils, pH 6–7.5. White mustard is

Fig. 8.22. Field of white (yellow) mustard, The Netherlands.

propagated by seed. Seed has to be sown about 1–2 cm deep in rows, spacing between the rows may be 25–30 cm, or broadcast. Seed rate varies from 3 to 14 kg/ha. The higher rate is often used on heavy soils where emergence can be difficult. White mustard is a rapidly growing plant that requires rich soils with good N availability. Ample fertilization with N fertilizer can be applied at sowing; soil P and K are mostly adequate. Young plants do not compete well with weeds. It is therefore important to start with a clean field. Weeding will be necessary until the crop has established. Most weeds can also be controlled by herbicides. The crop can be harvested when the seeds are yellow and hard, which may be 80–120 days after sowing. White mustard for salad greens is usually grown in greenhouses and harvested when the plants are a few centimetres tall. Seed yield may be up to 2.4 t/ha. Statistics of mustard seed production are generally not species-specific. In 2004, the world total seed production of all mustard species was about 0.7 million t. Western Europe, Sweden, Canada and the Near East are important production areas.

Black mustard

Black mustard is native to the Mediterranean region and Asia Minor. It was probably already domesticated in the Neolithic period. The use of mustard seeds as spice or medicine was known in ancient Babylonia and India, mustard is often mentioned in old Greek and Roman writings, and in the Bible. The spread of black mustard is often supposed to be the result of contamination of cereal grain with mustard seeds. This is because the species very often goes wild because of the easy-shattering seeds, and then acts as a problematic weed. This makes black mustard less suitable for large-scale cultivation than white mustard. Nowadays, black mustard is grown in central and south Europe and other areas with a temperate climate.

Black mustard (Fig. 8.23) is a heavily branched, annual herb, about 1.5 m high. The stem is erect, terete, glabrous or covered with hair. The leaves are glaucous, alternating and becoming smaller towards the top of the plant. The lowest leaves are compound, pinnate, mostly with two lower lobes, about 15 cm × 6 cm; the upper leaves are oblong to obovate, entire or with a few shallow teeth, about 6 cm × 2 cm. Inflorescences are axillary or terminal racemes (Fig. 8.21). Flowers are bisexual, yellow and up to 0.8 cm long; in all other characteristics, they are typical crucifer flowers, like the ones of white mustard. The fruit is a silique, more or less four-sided, with a short beak, 1.5–2 cm long, and containing six to eight, light to dark brown, globose seeds. The fruits grow almost parallel to the inflorescence axis. The 1000-seed weight is 3–4 g. The species has a firm taproot, which may grow 1.5–2 m into the soil under dry conditions.

Black mustard is cultivated almost solely for the seeds. The leaves are seldom used as a vegetable. The seeds can be processed into table mustard, which can be obtained by mixing ground seeds with water or vinegar, eventually also with sugar, salt and herbs. White, brown and black mustard are often blended to obtain the desired aroma and taste. In the blends, white mustard is used for

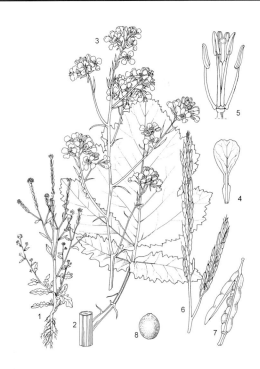

Fig. 8.23. Black mustard: 1, habit of flowering and fruiting plant; 2, detail of stem with lower leaf; 3, flowering branch; 4, petal; 5, stamens and pistil; 6, fruiting branches; 7, dehiscing fruit; 8, seed (line drawing: PROSEA volume 13).

achieving the right taste, whereas brown and black mustard are also used for aroma. The seeds of all mustard species, or seed products such as oil, are widely used as a condiment in the food industry. On a smaller scale the oil is also used as an illuminant and a lubricant. In folk medicine, mustard species are considered to be diuretic, emetic and stimulant. The well-known mustard poultice is applied in medicating rheumatism, for example. Mature seed contains about 7% water, 28% protein, 28% fatty oil, 29% carbohydrates and 9% fibre.

Black mustard has adapted to a range of temperate and subtropical climates. It is chiefly grown as a rainfed crop in areas with an annual rainfall of 300–1700 mm; it is moderately drought-tolerant because of the deep-growing root system. Waterlogged soils are not tolerated. The crop grows best on well-drained, fertile, sandy loam soils, pH 5–8.

Black mustard is propagated by seed. Seeds have to be sown about 1–1.5 cm deep. They may be drilled in rows 20–30 cm apart at a rate of 3–4 kg/ha. Seeds can also be broadcasted at a rate of 8–10 kg/ha. Application of fertilizer depends on soil fertility but some N fertilizer at sowing will usually be required. Seed will germinate when the soil temperature is at least 4.5°C. Pre-emergence weed control, as well as control during crop establishment, will mostly be necessary. The crop will flower 4–7 weeks after sowing and can be harvested in another 4–8 weeks.

Harvesting can be mechanized. The crop is cut when it is ripening but not completely ripe and left in the field for further drying and ripening. After some time it can be threshed. The crop has to be harvested when the fruits are still closed, to avoid as much seed loss as possible. However, the right moment for harvesting is difficult to determine, because the fruits do not mature at the same time: maturing of the seeds starts at the base of the plants and proceeds upwards.

Because of the problematic tendency of black mustard to shatter, the importance of this species has decreased in favour of brown mustard. Seed yield of black mustard can be up to 1.1 t/ha.

Brown mustard or Indian mustard

Brown mustard is thought to be native to central and south-western Asia, where the species originated from the hybridization of *Brassica nigra* and *Brassica campestris*. In this area, the natural distribution of the latter two species overlaps. Brown mustard has been cultivated in India and China for a long time. Chinese traders and Indian contract labourers probably played a part in spreading brown mustard over the world; nowadays the species is grown mainly in Asia, in North America and on a small scale in Europe.

Brown mustard (Fig. 8.24) is an erect, perennial, somewhat glaucous herb that may grow up to 160 cm tall. The species is usually grown as an annual. The stem is branched in the upper part, some cultivars have swollen stems. Leaves are very variable in shape and size, depending on cultivar. Lower leaves are petioled and often elongated. They may be lyrate-pinnatifid or entire and oval, with frilled or crumpled margins, or the whole leaf may be crumpled, smooth or pubescent, heading or non-heading. The colour of the leaves can be pale to dark green, red or purple, sometimes with white veins. The leaves may be 10–60 cm long. Upper leaves are reduced, sessile, linear-lanceolate, usually with entire margins, and 3–6 cm long. The inflorescence is a loose raceme with numerous bright yellow flowers (Fig. 8.21). The flowers are more or less similar to those of white mustard, though somewhat smaller. The fruit is a linear siliqua, 3–8 cm long, terminating in a conical beak and containing ten to 20 seeds.

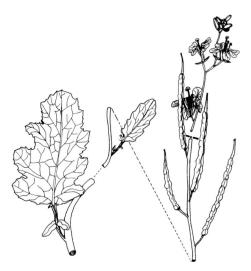

Fig. 8.24. Brown or Indian mustard: flowering and fruiting branch.

Seeds are globose, 1.5 mm in diameter, yellow and brown to dark grey; 1000-seed weight is about 2 g. The species has a taproot, which is in some cultivars swollen like a turnip.

Brown mustard is valued for its sharp and peppery flavour, which is caused by the volatile mustard oil that can be found in all parts of the plant. There are many cultivars of *Brassica juncea*, although they are grown for different uses. This species can be grown for greens, or as a spice, or for the seeds.

Leaves are used fresh in salads; the leaves can also be cooked with ham or cooked like spinach, or used in soups and stews. Leaves are often sold fresh; they can also be canned or frozen. From some cultivars, the swollen stem or turnip-like roots are used as vegetables. In Asia, mustard leaves and stems are sometimes pickled. Other cultivars are grown for the oil, which can be extracted from the seeds. The oil can be used as salad oil, all the more because the relatively high level of polyunsaturated fatty acids makes it much valued as being good for the health. The oil is also used for hair-oil and in lubricants.

The pungent seeds are used to flavour meat and several dishes. Ground seeds can be processed into table mustard, often blended with other mustard species. The constituents in the seed and uses in folk medicine are similar to those of *B. nigra*. Fresh leaves and stems contain approximately 90% water, 5% carbohydrates, 2.5% protein and 0.5% fat and also Ca, Fe and vitamins A and C.

Ecology and agronomy of brown mustard for the production of seeds are comparable with those of the other mustard species; however, brown mustard is more tolerant to drought and heat. Cultivation practices of vegetable mustard cultivars (greens) are more or less similar to those of the vegetables of *Brassica oleracea*. The vegetable mustard cultivars are grown mainly in most Asian countries, especially China. The oilseed cultivars are grown mainly in India, Bangladesh, Japan and China. Important countries for the production of seeds for processing table mustard are Canada, the USA, Great Britain and Denmark.

Seed yield can be up to 1.25 t/ha; the average yield of mustard greens is about 12 t/ha. Because black mustard seeds shatter so easily, *B. juncea* has partially replaced *B. nigra* for seed production.

Horseradish

Horseradish is native to south-eastern Europe, where this perennial species grows naturally on river banks. Nowadays it is cultivated in Europe and the USA on a small scale, although usually as an annual. It is a winter-hardy herb that grows 40–50 cm high in cultivation, and up to 1 m as a perennial. It is a tall vigorous plant with large (30–50 cm), basal, oval, dark green, toothed, shiny leaves. The leaves resemble those of the weed curly dock. The flowers are cruciferous, whitish, and produce hardly any seeds (Fig. 8.25). The species has deep, large, fleshy and pungent roots.

Horseradish is grown for the roots, which contain a volatile oil with a strong pungent odour and a hot spicy taste. After the roots have been harvested, they should be processed immediately because of the high volatility of the oil. The

Fig. 8.25. Horseradish: inflorescence (left); leaves (right).

roots are often grated and mixed with vinegar, salt and herbs, into a mustard-like condiment. The roots are also often dried and ground into a powder.

It is a plant for temperate regions. It prefers deep, rich, loam soils, high in organic matter, pH 6–7.5. The species grows well in moist and semi-shaded environments. Horseradish is propagated by root cuttings. Spacing between plants has to be about 40 cm. The planted root cuttings do not grow in length, only the diameter increases.

Harvesting should be done carefully, because small pieces of roots that are left behind may develop adventitious buds and generate new plants, thus causing a weed problem.

Horseradish has been used in many medical treatments. The presence of the volatile oil makes horseradish poisonous for livestock.

Nutmeg

Nutmeg – *Myristica fragrans*; Nutmeg family – *Myristicaceae*

Origin, history and spread

The true nutmeg tree, producing the two spices nutmeg and mace, is a native of the Indonesian archipelago of the southern Moluccan islands, especially Banda and Ambon. It is seldom, if ever, found truly wild and is mainly known from cultivation. From there it spread throughout South-east Asia. The first recording of nutmeg in Europe dates back to the 6th century and by the end of the 12th century, nutmeg and mace were generally known in Europe although they were very costly. The further history of the trade and cultivation of the valuable nutmeg is closely related to an aggressive way of colonization. The Arabs were the exclusive importers of the spice to Europe until 1512, when the Portuguese explorer Vasco da Gama reached the Moluccas and claimed the islands. The Portuguese subsequently obtained a monopoly on nutmeg. In the 17th century they were ousted by the Dutch who took over the monopoly, and held to it rigorously, even by uprooting trees grown elsewhere, a policy similar to that of the clove trade. In 1772, the French broke the monopoly and succeeded in introducing the tree into Mauritius and French Guiana. In 1802, the monopoly was further broken by the British who sailed, with Captain Bligh on the *Providence*, to the Moluccas and collected 70,000 plants which were sent to Penang (Malaysia) and a few to Kew, Calcutta and Madras. However, the plantations in Penang and peninsular Malaysia disappeared, probably due to pests and diseases. In 1843, nutmeg was successfully introduced into Grenada (West Indies), which led to large-scale production.

Botany

Nutmeg (Fig. 8.26) is an evergreen, mostly dioecious tree of 5–13 m high, sometimes reaching as high as 20 m. The bark is dark grey-green to olive-green, and when the bark is wounded it exudes a sticky yellow sap which oxidizes to red. The crown of the tree is heavily branched with dense foliage and slender

Fig. 8.26. Nutmeg: 1, flowering and fruiting branch; 2, opened female flower; 3, staminal column; 4, seed with aril; 5, seed without aril; 6, section through seed without aril (line drawing: PROSEA volume 13).

Fig. 8.27. Fruit-bearing branches of the nutmeg tree (photo: M. Flach).

twigs towards the top. The leaves are alternately arranged, elliptic to oblong-lanceolate, pinnately nerved, dark green, 5 cm × 15 cm and aromatic when bruised; petiole about 1 cm. The flowers are pale yellow, waxy, glabrous; the calyx bell-shaped with three reflexed triangular lobes; petals are absent. The male flowers are 5–7 mm long with eight to 12 stamens; female flowers up to 1 cm long with a one-celled ovary. The male and female inflorescences are similar, axillary in umbellate cymes, with one to ten flowers in males and one to three flowers in females. The fruit is a peach-shaped, fleshy berry or drupe-like (Fig. 8.27), 5–9 cm long, turning yellow and splitting open into two valves when ripe, containing one ovoid seed, 2–3 cm long. At opening of the fruit, the exposed seed is purplish-brown, shiny and partly surrounded by an incised, bright red aril which is attached to its base. When dried, mostly in the sun, the aril becomes yellowish-orange to give the mace of commerce. The dried seed constitutes the nutmeg of commerce (Fig. 8.28). The nutmeg tree has a very shallow root system. The taproot, however, may penetrate the soil to depths of 10 m. Vegetatively propagated trees do not develop a taproot.

Species, uses and constituents

The genus *Myristica* consists of about 100 species extending from India to Polynesia and north-eastern Australia; however, only few of them are cultivated. Next to *Myristica fragrans*, being the most important species, several other species can be found as adulteration or substitute of the true nutmeg. *Myristica argentea*, a good second species, originates in New Guinea, where it occurs both wild and cultivated. It is also cultivated in the Moluccas. The shelled seeds, called 'Papua nutmeg' or 'long nutmeg', and the dried arils, called 'Macassar

Fig. 8.28. Nutmeg (above): branches with leaves and mature fruits (photo: J. Sipasulta); (below): 1 and 2, halved fruit showing aril-enveloped seed inside; 3 and 4, seed with separate aril (photo: D.L. Schuiling).

mace', are used as spices. The seeds are oblong-cylindrical, broadening at the base and up to 4 cm long. *Myristica succedanea*, the Halmahera nutmeg, is found in the wild and cultivated in the northern Moluccas, on the islands of Ternate, Tidore and Bacan. The small, shelled seeds, broadly oblongoid, 3 cm– 2.5 cm, are used like those of *M. fragrans*.

The two spices, nutmeg and mace, are strongly aromatic, resinous and warm in taste. Mace is generally said to have a finer aroma than nutmeg. Nutmeg is grated in small quantities to flavour confectionery, milk, punches, possets and various dishes. In Europe it is used mainly in meat dishes, soups and for flavouring vegetables. Mace is preferably used in savoury dishes, pickles and ketchups. Nutmeg oil is a mixture of volatile essential oils and is obtained by distillation of the fat from broken seeds and mace which are not good enough for the spice trade and sometimes also from bark, leaf and flower. Nutmeg oil is used in the canning industry, in soft drinks, cosmetics, medicinally (externally) and as a flavour component in food products. The maximum permitted level in food is about 0.08%. The seeds yield also nutmeg butter, which is used in ointments and perfumery. Nutmeg butter, a fixed oil at room temperature, is obtained by pressing the seeds between hot plates. Both nutmeg oil and butter contain myristicin, which is a narcotic with hallucinogenic effects and poisonous, so the use of nutmeg and mace should be moderate. A consumption of 3–4 g of nutmeg and mace can produce symptoms of poisoning in humans. It is said that the consumption of two grated nutmegs (about 8 g) can cause death. On Zanzibar nutmegs are chewed as an alternative to smoking marijuana. Nutmeg is used in folk medicine to induce hypnosis, to suppress fever and coughs, to treat diarrhoea, and it is said to have aphrodisiac, purgative, carminative and stimulant properties. The young husks are made into marmalades, jellies, sweets and preserved foods and are very popular in Malaysia and West Java. In Indonesia, the old husks are sometimes used as a substrate to grow the popular edible mushroom 'kulat pala' (*Volvariella volvacea*). The red sap, present in the bark, can be used as a dye that gives a permanent brown stain.

The edible portion of nutmeg contains approximately 10% water, 7% protein, 33% fat (nutmeg butter), 5% essential oil, 30% carbohydrates, 11% fibre and 2% ash. Mace is generally higher in water content (16%), carbohydrates (48%) and essential oil (10%) and lower in fat (22%). The red pigment in mace is lycopene and is identical to the red colorant in tomato.

Ecology and agronomy

The natural habitat of nutmeg is found in the lowlands of the tropical, evergreen rainforest up to 800 m altitude, where it usually forms a part of the second storey. Nutmeg in cultivation is grown in a warm and tropical, non-seasonal climate with average temperature of 25–30°C and average annual rainfall of 2000–3500 mm (Fig. 8.29). Temperatures above 35°C and hot, dry monsoon winds adversely affect flowering. It cannot tolerate waterlogging or excessive drying out of the soil. Shade is favourable only in early growth. Frost is always detrimental and may kill the tree. Therefore, in the tropics, the crop can be grown only at altitudes of up to 500 m. Nutmeg is preferably grown on soils of volcanic origin and soils with a high content of organic matter, with pH of 6.5–7.5.

Propagation is usually by seeds in the shell, sown in shaded nurseries at a depth of 5–7 cm and in rows 30–50 cm apart. The seeds should be sown soon after collecting, because their viability decreases quickly. Seeds which rattle in the shell are too dry and will not germinate. Germination takes 4–6 weeks; when the seedlings are about 15 cm high, after 5–6 months, they are transplanted to the field. Determining the sex of the trees is possible only after first flowering, which usually occurs some 6 years after planting. Therefore two or three seedlings are usually planted at the same spot and the surplus males are cut later on, leaving one male to ten female trees. Young nutmeg plants are usually grown under 50% shade; with increasing age the shade can be reduced to zero after 6–7 years. Planting distance is usually about 6 m × 6 m and thinned or replaced later when needed. The plant spacing for full-grown trees is around 10 m × 10 m. Well-spaced trees start full production after 15–20 years, which may

Fig. 8.29. Approximately 25-year-old nutmeg trees, Sumatra, Indonesia.

continue for 30 to 50 years or more. Usually no fertilizer is used. On the island of Banda, plantations on volcanic soils have remained productive for hundreds of years. In non-seasonal climates nutmeg can be harvested during the whole year. On Banda, the fruits are harvested when they are open, using a small basket on a long pole with a bifurcated top end, to which a sharpened piece of iron is attached. This careful method of harvesting favours particularly the quality of mace. In Grenada, the seeds with attached aril are collected from the ground after they are released from the split fruit. After harvest the seeds are removed from the fruits and the arils are separated from the seeds.

The seeds in shell are often dried in barns which are specifically built for this purpose. The seeds are spread on a floor of loosely layered bamboo slats, while underneath is a slow-burning and smoking fire. The drying temperature should not exceed 40°C. The arils are mostly sun-dried on mats, to give the mace of commerce. After the seeds are dried, the shells are cracked to free the dry kernels (the nutmegs), which are graded according to quality, size and weight.

For centuries past and to the present time, the centre of nutmeg cultivation has been found in Indonesia, on the island of Banda and the nearby surrounding islands. Recent yield and production statistics are hardly available. The present, annual world consumption of nutmeg and mace, estimated at some 20,000 t, could be produced on a well-managed 20,000 ha and possibly even less. Indonesia and Grenada dominate the production and export of both products with a world market share of 75% and 20%, respectively.

Pepper

Pepper – *Piper nigrum*; Pepper family – *Piperaceae*

Origin, history and spread

Piper nigrum is native to the Malabar region in India, which is today a part of the Indian state of Kerala. Pepper has been cultivated for millennia. The wild form and closely related species can still be found in the hills of Assam in India, and in Myanmar. After Alexander the Great reached India in the 4th century BC, new trading routes were established. Arabic traders soon monopolized pepper trade; they brought the spice to the Arabian peninsula and the Mediterranean countries. Within a short time, pepper became very popular but also extremely expensive. Using pepper became a status symbol. The ancient Romans used pepper intensively in their cookery. In a war with the Romans, Atilla the Hun demanded a ransom of about 1500 kg of pepper from the Romans, to secure the release of Roman hostages. About 2000 years ago, the plant was brought to Indonesia and Malaysia by Hindu colonists, migrating from India. Marco Polo reported pepper in Malaysia in 1280. In the Middle Ages traders from Venice bought pepper from Arabic traders and distributed it in Europe. From the end of the 15th century onwards, European explorers sailed to India and South-east Asia to search for pepper, bypassing Arabic and Venetian traders. Soon, in the 16th century, Venice lost its position as the European pepper metropolis to Lisbon. About 100 years later Lisbon had to give up this position to the advantage of Amsterdam. In the following years, pepper cultivation and pepper trade were dominated by the Dutch and the English. For a long time, an extreme amount of money was involved in pepper commerce. Yale University in the USA was founded with the money of Elihu Yale, an American who earned his money in pepper commerce. The name 'pepper' is via the Greek 'peperi' and the Latin 'piper' is derived from the Sanskrit name of long pepper ('pippali'; long pepper is *Piper longum*). Nowadays, pepper is probably the most widely used spice in the world.

Botany

Pepper is a dioecious, perennial, evergreen, woody vine; vines have swollen nodes and are up to 15 m in length. In cultivation, plants are usually grown on poles, 4 m high, which often appear as bushy columns (Fig. 8.30). Two types of branches can be distinguished: (i) orthotropic, vegetative, climbing branches with roots on the nodes, forming the framework of the plant; and (ii) plagiotropic generative branches that bear axillary inflorescences and have no roots. Leaves alternate; they are simple, glabrous and leathery; the petiole is 2–5 cm long. The leaf-blade is ovate 8–20 cm × 4–12 cm, entire, rounded at the base, with an acuminate tip, shiny dark green above, pale beneath, with five to seven pronounced veins. The inflorescence is a spike, 3–25 cm long, with 50–150 flowers. Flowers are unisexual or hermaphrodite, although most cultivars have hermaphrodite flowers. The flowers are borne in the axils of ovate fleshy bracts;

Fig. 8.30. Habit of pepper plants when supported by wooden stakes.

calyx and corona are absent; a hermaphrodite flower has two to four stamens, and a globose ovary with one cell and one ovule. The fruit is a sessile globose drupe, 4–6 mm in diameter, with a fleshy, thin, mesocarp and a red exocarp when ripe (Fig. 8.31). The seed is about 3–4 mm in diameter; 1000-seed weight

Fig. 8.31. Infructescences with ripe and unripe pepper fruits.

varies from 30 to 80 g. The root system consists of a number of main roots, which grow up to 4 m deep, and a dense mat of feeder roots in the upper 60 cm of soil.

Species, cultivars, uses and constituents

As well as *P. nigrum* there are more *Piper* species that have some importance, often only local. *P. longum* is grown and used as a spice in India and Sri Lanka, and it was known to the ancient Greeks and Romans. *Piper betle* is grown in Zanzibar, India, Malaysia, Indonesia and Oceania. The leaves are chewed together with betel nut (*Areca catechu*), acting as a stimulant. This chewing habit was already known to Herodotus in 340 BC. *Piper methysticum* is grown in Oceania, the peeled roots are made into the national beverage of the Polynesians; it acts as a sedative, soporific and hypnotic. Furthermore, there are several wild *Piper* species which are collected and used as substitutes of *P. nigrum*.

From *P. nigrum* several cultivars have been selected, which are mainly hermaphrodite, and these are maintained by cuttings. Important cultivars are 'Ballancotta' and 'Kalluvalli', grown mainly in India; 'Kuching', grown in Sarawak; and 'Belantung', grown in Sumatra.

Pepper is used to flavour all kinds of food, all over the world. It is widely valued for its highly seasoned flavour and distinctive aroma. It is also used in pickles, ketchups, confectionery and sauces. Pepper can be obtained in different forms. When unripe pepper fruits are harvested, fermented and dried, the outer skin of the fruit turns black and wrinkled. This product is called 'black pepper'. When the pepper fruits are allowed to ripen, they turn red. After harvesting almost ripe pepper, the fruits can be soaked for about a week, and subsequently the outer skin and the mesocarp can be washed away, which leaves the creamy-white endocarp with seed inside, called 'white pepper'. Both black pepper and white pepper are marketed as whole peppercorns or ground to powder. Flavour and pungency differ for black and white pepper; black pepper is more pungent, although pungency is also influenced by region and cultivar. Next to the important black and white pepper, there are two minor commodities: green pepper and red pepper. Freshly harvested unripe green peppers and ripe red peppers can be pickled in salt or vinegar or dried quickly at high temperature, so that the colours are retained. Green pepper has little pungency; red pepper is considerably more pungent. Red pepper (*P. nigrum*) must not be confused with the spicy chilli (*Capsicum frutescens*), which is also called 'red pepper' but belongs to another family.

Dried black pepper contains approximately 10% water, 11–13% protein, 26–45% carbohydrates, 10–17% fibre and 3–6% ash. Dried white pepper contains 10–14% water, 11–12% protein, 54–60% carbohydrates, 4–5% fibre and 1–3% ash. The pungent bite of pepper comes mainly from the alkaloid piperine. Its content varies from about 5 to 8% and is usually highest in black pepper. From the fruits, an essential oil can be extracted; it is used in perfumery. The oil is also responsible for the characteristic pepper odour. Pepper has been used in folk medicine as a carminative, antipyretic, anthelmintic and as a stimulant. Pepper is also made into an effective insecticide against flies.

Ecology and agronomy

Pepper requires a hot, wet, tropical climate; it grows best at low altitudes in a monsoon climate with a well-distributed annual rainfall of 1750–2500 mm. Mean day temperature has to be 25–30°C, with a relative atmospheric humidity of 65–95%. It grows on a wide range of soils, from heavy clay to light sandy clay. However, it prefers deep, well-drained alluvial soils, rich in organic matter, with ample water-holding capacity and a pH above 5.5. The species cannot stand waterlogging.

Pepper is usually propagated by cuttings taken from terminal orthotropic shoots, with seven nodes and about 60–70 cm long. The cuttings can be placed in a shaded nursery to let them root. After 2 months, the young plants have enough roots and can be transplanted into the field. Cuttings can also be planted directly in the field. Micro-propagation by using short shoot-tips also occurs. Transplanting or planting has to be done in a rainy period; plants have to have three or four nodes below the surface. Before planting, hardwood poles are placed at 2–4 m × 2–4 m; the poles have to stick out about 3.5 m above the soil. Around the base of the poles, the soil is often mounded; mounds are usually enriched by organic matter. A cutting is planted in each mound. In subsequent years, mounds have to be replenished to provide room for continuous dense rooting (Fig. 8.32). As well as poles, trees and trellis can be used to support pepper plants. Pepper is sometimes planted as an intercrop, for example in coconut and coffee plantations. A few months after planting, usually three climbing stems are tied to the poles and pruned regularly, to stimulate the development of fruiting branches. When the stems reach the top of the pole, which is after about 3 years, the terminal buds are removed periodically to prevent further

Fig. 8.32. Pepper plantation. Wooden stakes for expansion of the plantation are being put up, Sarawak, Malaysia (photo: D.L. Schuiling).

vertical growth. During the first 2 years, inflorescences are regularly removed and sometimes selective leaf pluck is utilized, both to stimulate productive side branching. The crop is often kept clean-weeded, but as pepper gardens are often on slopes, erosion may occur. It may be prevented by a mulch layer, which also contributes to maintaining soil moisture and soil fertility and it suppresses weeds. Next to weeding, manually or mechanically, herbicides can be used.

Soil fertility is maintained in many ways, by mulching or by application of farmyard manure, organic manures (including guano), burnt earth, wood ashes, blood-and-bone meal or soybean cake. Artificial fertilizer can also be used, NPK mixtures in the proportion of 12:5:14, with additional Mg and trace elements, are recommended. In the first year each plant receives 0.5 kg of the NPK mixture, in the second year 1 kg, and in the following years 1.5–2 kg, always split into four applications. The first crop is taken in the third year and the economic production life of the vines is 12–15 years. It takes about 5–6 months from the emergence of the inflorescence, and 4 months after flowering, to ripen the fruit. The whole fruit spikes are hand-picked when the fruits are mature but still green for black pepper; and when a few fruits have turned red and the rest are yellow and green, for white pepper. Harvesting can be repeated six to eight times per season.

Yields of dried fruits per hectare vary widely, from 300 up to 1900 kg. In 2004, the total world production of dried fruits of *P. nigrum* was about 408,500 t. Main producers were Vietnam, Indonesia, Brazil and India.

Umbellifer Herbs

Celery – *Apium graveolens*, Fennel – *Foeniculum vulgare*, Caraway – *Carum carvi*, Coriander – *Coriandrum sativum*, Cumin – *Cuminum cyminum*, Dill – *Anethum graveolens*, Parsley – *Petroselinum crispum*; Parsley family – *Apiaceae*

Introduction

Many umbelliferous plants are important herbs and have been used for a long time to flavour dishes. The ancient Greeks and Romans used umbelliferous herbs as a component of a sauce called 'garum'. In garum a special herb blend is fermented together with fish. Usually small quantities of this sauce were used for flavouring dishes. The odour of these herbs has also been used for more than 2000 years in perfumes and suchlike. For instance, the three kings visiting the infant Jesus offered myrrh, a perfume derived from *Myrrhis odorata* (sweet cicely).

The stems of *Umbelliferae* are seldom tough and ligneous but nearly always mellow, solid or hollow. The leaves are alternate, compound, often finely divided, and mostly sheathing at the base. The flowers are small, formed in simple or compound umbels and often white or yellow. The sepals are small or absent, the number of petals and stamens is five. The fruit is a schizocarp and consists of two, ribbed, one-seeded carpels, which are attached only at the top and often split at maturity. Carpels are often inaccurately called 'seeds'.

Especially the fruit, but also the root, stem and foliage contain essential oil. Recent yield and production statistics of labiate herbs are unknown.

Celery

Cultivated celery, also called 'celeriac', is derived from the wild celery, which is spread over a large part of the temperate regions as a species of wet, brackish soils. It was probably domesticated in the south of Europe. The cultivation is now spread all over the temperate regions and at higher altitudes in the tropics.

In the first year the species forms a rosette of divided leaves with long petioles. The leaves have mostly three pairs of leaflets and a terminal one. In the second year celery forms a flowering stem with terminal and axillary umbels. The flowers are small and greenish-white in colour.

Three different types of celery can be distinguished: (i) a type grown for its enlarged, succulent, blanched, leaf-stalks or petioles, about 30 cm long; (ii) a type called 'celeriac', grown for its turnip-like swollen base of the stem, which is mostly sub-globose in shape and the size of a tennis ball; and (iii) a type grown for the divided, fragrant leaves; this type does not have a swollen stem or enlarged leaf-stalks (Figs 8.33 and 8.34). The flavour of the three types is more or less similar to each other. The three types can be eaten raw in salads or boiled as vegetables or to flavour stews and soups.

Fig. 8.33. Celeriac: 1, view from above of a crop; 2, plant with tuber consisting of enlarged stem and root. Parsley: 3, view from above of a crop; 4, turnip-rooted parsley type (photo: D.L. Schuiling).

Fig. 8.34. Young leaves of umbellifer herbs: 1, parsley (curly-leaved type); 2, celery; 3, coriander; 4, dill.

Celery is in principle a cool-season crop. It thrives best, on deep, fertile moist soils, pH 6.0–6.8. It is a biennial herb but mostly grown as a long-season annual, because the purpose is seldom producing seed.

Celery is propagated from seed. It can be field-sown but seedlings are often raised in greenhouses and planted in the open when they are strong enough. Planting rate is about 120,000 plants/ha. The crop is mostly grown in rows to make it possible to earth up the plants in order to blanch the leaf-stalks (only the first type). Celery requires ample fertilizer for a good production.

In cases where celery is grown for the seed, it needs vernalization and the harvest is at the end of the second growing season. Then, the produced seed is intended for sowing, or flavouring dishes and pickles, or ground and mixed with salt in celery salt.

Fennel

Fennel is native to southern Europe and the Mediterranean. It was probably cultivated by the ancient Greeks and Romans, because fennel was one of the so-called 'garum' herbs. Nowadays it is grown widely in Europe, North America, North Africa and Asia.

Fennel (Fig. 8.35) is a biennial or perennial herb with erect stems, when older, with hollow internodes. The leaves are much divided and have a hair-like, almost feathery appearance. The foliage may be green or bronze-green in colour. In the vegetative stage, the plant is about 50 cm tall; in the generative stage the plant may reach a height of 180 cm. In the generative stage the plant

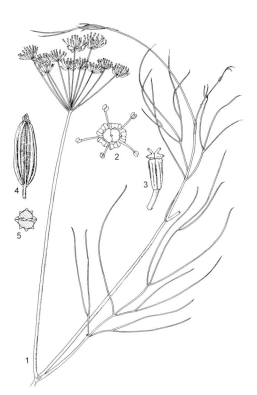

Fig. 8.35. Fennel: 1, flowering branch; 2, flower at male flowering stage; 3, flower at female flowering stage; 4, fruit; 5, cross-section of fruit (line drawing: PROSEA volume 13).

forms small yellow flowers, which are borne on flat-topped umbels. The plant has an elongated and enlarged taproot. All parts of fennel contain essential oil, which causes a fragrant scent.

Within *Foeniculum vulgare*, three different groups can be distinguished: (i) Florence fennel, which is grown for the enlarged basal parts of the petioles, which is also the consumed part of the plant; (ii) sweet fennel, grown for the fresh leaves and the sweet-tasting fruits; and (iii) bitter fennel, grown for the bitter-tasting fruits, although mainly for cosmetics.

Raw or cooked leaves and ground or whole fruits are used for flavouring dishes, especially fish dishes and pickles, and for garnishing. The enlarged basal part of the petioles of Florence fennel, called the 'bulb', is mainly used as a cooked vegetable. Fennel oil is utilized in cosmetic products, beverages, desserts, candies and suchlike. Since ancient times, fennel has been used medicinally to treat a great number of ailments such as gastroenteritis, indigestion and dyspeptic disorders. It is also thought to stimulate milk production. This species is, just like dill, often grown as an ornamental.

Fennel thrives best on sunny locations in mild climates; it is primarily a cool-season crop. It prefers well-drained moist loams or loamy clay soils with pH between 6.3 and 8.3. Moisture stress should be avoided.

Fennel is propagated from seed. Seed is often sown in rows, 40–120 cm apart. The spacing between the plants in the row may be 15–50 cm. Weed control is necessary, especially in the juvenile growth period. Nutrient supply is required, depending on soil fertility, but may be 80–100 kg N/ha and 40 kg P/ha. As the basal parts of the petioles of Florence fennel enlarge, they can be earthed up to stimulate blanching of the 'bulb'. It takes about 100 days from germination until harvest. India, China, Egypt and Italy are the main producers of fennel.

Caraway

Caraway has its origin in Europe, Asia and the eastern part of the Mediterranean region. It is believed to be one of the oldest spices in Europe. It has probably been utilized for more than 5000 years, as proved by archaeological findings in Mesolithic sites. Ancient writers mentioned caraway many times. Caraway is mostly a biennial, sometimes an annual, glabrous, much-branched herb, up to 1.5 m tall. The stem is hollow and may have a diameter of 1.5–2 cm. The grass-green leaves alternate and are bipinnate, with deep, linear-lanceolate lobes. Petioles are up to 15 cm long in the basal part of the plant, becoming shorter or absent towards the top of the plant. The inflorescence is a compound umbel, 10 cm in diameter, containing small, white or pinkish flowers. The fruit is about 4 mm long, with two brown carpels, each with five ridges. The taproot is fleshy and long.

Two types of caraway can be distinguished: annual spring caraway and biennial winter caraway.

Caraway is grown for its seeds, in the testa there are several small channels containing essential oil. Caraway seeds are used for flavouring stews, soups, cakes, candies, bread, cheese, pickles, meat, and alcoholic beverages such as 'Kümmel' and 'Aquavit'. The oil is used in several cosmetic products such as soap and toothpaste; it is also used in products with medical purposes, among others as a carminative, stomachic, antispasmodic, and for gargles and mouthwashes. The positive effect can be attributed to the main component of the oil, which is carvone. Lately, carvone has been found to be effective as an anti-sprouting agent in stored potatoes.

Caraway prefers a cold-season climate. It thrives best in well-cultivated loam or loamy clay soils, which have average moisture and are slightly acidic to neutral, pH 6.5–7.

Caraway is propagated by seed. Sowing can be done in autumn or in spring, but the annual spring type only in spring. The biennial winter caraway is sometimes sown in autumn, but often in spring together with an annual crop, for instance seed spinach. In that case, caraway will grow slowly beneath the annual crop until that crop is harvested. After that, caraway will start growing faster and develop a firm plant before winter. In this system, seed spinach is called the 'nurse crop'.

The seed should be sown 1.5–2.5 cm deep, in rows 20–30 cm apart. Germination is slow, as is the growth of the plants in the juvenile stage, which makes weed control necessary during this stage. Application of nutrients depends on soil fertility; 75–100 kg N/ha and other nutrients in proportion is often applied, also to encourage root growth. Caraway can be harvested when the umbels and the other parts of the plants have turned brown. To store the seeds without damage, the moisture content should not be more than 12%. Caraway is cultivated in eastern European countries and The Netherlands.

Coriander

Coriander is probably a native of Asia Minor and the eastern Mediterranean. The name '*Coriandrum*' is derived from the Greek name of the plant, which is 'koriannon'. An element of this name is 'kori(s)', which means 'bug' and refers to the foetid, bug-like smell of the bruised plant. It has been known in Asia and parts of the Mediterranean for millennia; Pliny wrote about it, and the Romans used it and brought it to northern Europe. It has been cultivated for over 3000 years.

Coriander (Fig. 8.36) is an erect, glabrous, annual herb, about 70 cm high. The species has solid, finely grooved stems, and is richly branched. The leaves alternate, are compound, and most of them are pinnate or bipinnate (Fig. 8.34). The plant develops leaves of two different shapes: the basal leaves, which start more or less as rosette leaves, are usually simple and broader than the leaves attached to the stem. The white or pink flowers are borne in umbels, the middle flowers are small and infertile, and the outer flowers are larger and fertile. The fruit is globose, consists of two carpels and is 3–4 mm long. Each carpel is clearly ribbed and brown when ripe. Coriander has a well-developed, fleshy taproot.

Within the species *Coriandrum sativum* there are several types, varying in oil content and the size of the seed. Both leaves and seeds, sometimes even the roots, are used for flavouring dishes. However, leaves and seed are totally different in flavour, so they cannot substitute each other. Coriander is used in stews and soups, cakes, breads, pickles, sausages, and combined with meat, etc. The leaves are always used fresh; the seed is often dried and usually used ground or powdered. Coriander seed is much used in Asia; it is an important ingredient of curry powder. It is also an important spice in South America and Africa. Both

Fig. 8.36. Coriander: 1, inflorescence (umbel); 2, part of infructescence and fruits; 3, various leaves (photos: D.L. Schuiling).

leaves and seeds are characteristic of Arab cookery, for instance for spicing lamb. However, many inhabitants of North America and Europe do not value the flavour of coriander, especially the leaves. Coriander is still one of the bitter herbs which are eaten by Jewish people at the Passover, when they remember the journey out of slavery from Egypt to their homeland. Coriander oil is used in cosmetics such as soap and perfumes. Coriander also has, like many other umbelliferous herbs, medicinal uses. For example, it enhances the digestive system and it may reduce flatulence.

Coriander thrives well in temperate regions, tropical highlands and the sub-tropics. It prefers sandy loam but can also be grown successfully on well-drained loam and clay soils. For germination and growth, temperatures between 17 and 27°C are required. During the seedling and juvenile stage, the plants need adequate water supply. After stem elongation coriander is resistant to drought.

Coriander is propagated by seed. Number of plants varies from 15 to 75/m². This depends on cultivar and whether leaves or fruits are the aim of the cultivation. Weed control is often carried out twice, especially in the juvenile stage because the young plants are not very competitive. P–K fertilizer will mostly be required; the demand for N is low. For seed production it takes about 90–140 days from sowing to harvesting. For the production of leaves, several cuts are possible when the right cultivars are used. Coriander is now grown mainly in India, South America and North Africa.

Cumin

The origin of cumin is thought to be the area from the east of the Mediterranean to Central Asia. It has been known since antiquity. Cumin seeds have been found in Egyptian pyramids and there are Biblical references to the species. Cumin was probably first cultivated in western Asia. It was a common spice in ancient Greek and Roman cookery, and they also used it medicinally. Nowadays, it is a popular spice all over the world, especially in Asia, North Africa and South America, less so in Europe and North America.

Cumin is an erect or sub-erect, small, slender, annual herb, 15–50 cm high. The plant tends to droop. The stem is finely grooved and branching. The blue-green compound leaves alternate and are divided in thread-like segments. The whole plant is glabrous, although mostly covered with a bloom. The small white or pink flowers are borne in few-flowered compound umbels, with thread-like bracts; the umbel is about 3.5 cm in diameter. The fruit is ovoid-oblong, 3–6 mm in length, and contains two separate carpels. The testa of each carpel has four distinct ridges and five, less pronounced, oil canals. The seeds are hairy and yellow-brown in colour. Cumin has a thin taproot.

Cumin has a strong flavour and odour, which resemble caraway, although cumin is more powerful. It is often put to similar uses as caraway. Cumin is traded in several forms: as dried fruits, in powdered form and as essential oil. It is used as a culinary spice for flavouring a large number of different dishes; it is used in soups, pickles, sausages, beverages, liquors, cosmetics and perfumery. It is an important ingredient of curry powder and 'chilli con carne'. In The

Netherlands and France it is used to flavour cheese. It is considered to be an appetite stimulant and is also used in herbal medicines such as digestives, stimulants, sedatives and many others.

Cumin prefers a mild and rather dry climate; its temperature range is 9–26°C. In a warmer climate, such as in the Middle East, it is grown as a winter crop or at higher altitudes. It is a short-day plant. Cumin thrives best on rich, well-drained, sandy loam to heavy loam soils, pH 6.8–8.3.

Cumin is propagated by seed. Sowing rate is about 20 kg/ha, and it takes 2–4 weeks to germinate. Spacing between plants can be 15 cm. Since cumin is a bad competitor because of its open growth habit, regular weed control is required. As fertilizer, an application of 130 kg N/ha is often used, P and K proportionally. It takes 3–4 months for the crop to mature. The crop can be harvested when the plants wither and the fruits turn yellow. Yields vary greatly, up to 1.2 t/ha. The main production countries or areas are India, Iran, China and the Mediterranean.

Dill

Dill is probably native to the Mediterranean and south-eastern Asia. The leaves and the fruits of the plant have been used since ancient times; the Egyptians, Jews, Greeks and Romans used it as a medicine. The ancient Greeks and Romans also used it to flavour dishes; it was one of the so-called 'garum' herbs. It is assumed that monks brought dill to the rest of Europe. The name 'dill' is probably derived from the Saxon word 'dilarm', which translates roughly into 'rock to sleep', referring to the supposed carminative properties of the plant.

Dill is an upright, glabrous, blue-green, annual or biennial herb, up to 1.5 m high. It has gleaming, hollow, much branched, subterete stems, 10–15 mm in diameter. The leaves are greatly compounded, the leaflets almost thread-like (Fig. 8.34). Lower leaves have mostly long petioles, higher leaves without or with short petioles. The inflorescence is a compound umbel, up to 15 cm in diameter, containing greenish-yellow flowers. The fruit is lens-shaped, 3–6 mm long, brown in colour, with a pale brown margin. Each carpel usually has three prominent ridges.

The used parts of dill are the leaves, whole or ground fruits, and the essential oil. The oil can be distilled from the whole plant or just from the fruits. Leaves have a very delicate flavour, although they have to be used fresh because they lose flavour and odour during cooking. Dill is used to flavour dishes, especially fish dishes; it is used in sauces, salads, pickles, sauerkraut and bread; it is often combined with young potatoes and tomatoes; and it is used in producing dill-vinegar. The herb is also valued as an ornamental; inflorescences are often used in flower arrangements. The plant is very attractive to bees and butterflies.

As a medicinal herb, dill has been used to treat many different illnesses, for example as an antispasmodic, carminative, stomachic and diuretic. The oil is also used in cosmetics. In the Middle Ages, dill was assumed to have magical properties, and therefore it was used in or against witchcraft.

Dill is primarily a long-day and cool-weather plant; it grows best in temperate regions. For sufficient development it requires an average temperature of 16–18°C. The species thrives well on organic, deep fertile, sandy loam soils, mildly acidic to neutral, pH 6.1–7.8. It is sensitive to low moisture, high temperatures, hard rains and strong winds during the generative phase.

Dill is propagated from seed; the sowing rate is 5–7 kg of seed per hectare. It is usually grown as an annual crop. Plant spacing depends on what part of the plant will be harvested; it can be 15–45 cm between rows and 10 cm within rows. Germination occurs in 10–20 days; flowering starts 2–3 months after sowing. To control weeds, frequent weeding or spraying with herbicides is indispensable, and irrigation can be necessary. An application of fertilizers, depending on soil fertility, of 80 kg N, 30 kg P and 30 kg K per hectare is recommended.

Dill leaves are harvested before the plants flower; the fruits can be harvested when most of the fruits are mature and turn brown, which can be 5–6 months after sowing. Dill is now cultivated in India, Germany, The Netherlands, Poland, England and the Americas.

Parsley

Parsley is native to the Mediterranean and western Asia. Like most of the cultivated, umbelliferous species, it is an old crop, and was already known to the ancient Greeks and Romans. The species was probably first cultivated by the Romans. In the Middle Ages it became popular in the rest of Europe, first grown in the gardens of monasteries. Parsley was also on Charlemagne's list of cultivated plants, *Capitulare de Villis*. The name is probably derived from the Greek word 'petros', which means 'stone' or 'rock'. It refers to the place where wild parsley grows, which is between rocks and stones. Nowadays, parsley is grown in the temperate regions all over the world.

Parsley (Fig. 8.37) is a glabrous biennial or perennial herb, in vegetative stage forming a rosette, up to 35 cm tall. The leaves alternate and are multicompound, curly or flat, bright green, and with long petioles. The leaflets are finely divided. In the second growing season it produces an erect stem, up to 75 cm high, with flat-topped, terminal or axillary, compound umbels consisting of small yellow flowers. The generative stage is not that important because parsley is almost exclusively grown as an annual crop. The plant has a thin, fibrous or a tender, fleshy, swollen taproot.

Three different types of cultivated parsley can be distinguished: (i) parsley with curled and crisped leaves; (ii) a type with flat, non-crisped leaves; and (iii) a type with flat non-crisped leaves and an edible, swollen taproot, also called 'turnip-rooted parsley' (Figs 8.33 and 8.34). The fresh and chopped leaves of the cultivars with the curled and crisped leaves are used mainly for garnishing dishes and window dressings in, for example, butcher's shops. The fresh or dried leaves of the flat-leaved parsley have a more pronounced flavour and are used in salads, sauces and soups, to flavour several different dishes and processed foods. Parsley with the swollen taproot is used as a vegetable; the firm

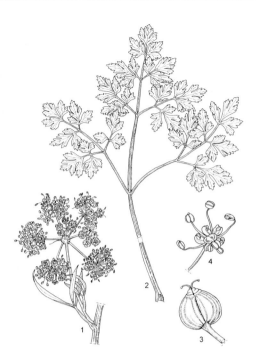

Fig. 8.37. Parsley, flat-leaved type: 1, umbel; 2, leaf; 3, fruit; 4, flower (line drawing: PROSEA volume 13).

white taproot is the eaten part, and the taproot can also be cooked in soups. The plant contains an essential oil, which can be used similarly to the oil of other umbelliferous crops, although the production of parsley oil has little significance. The plant is a rich source of vitamin C and Fe. Parsley is one of the herbs of the famous French 'bouquet garni', a bundle of different herbs that is used to flavour a variety of dishes. Like other umbelliferous crops, parsley is also used as a medicinal plant, to heal the same illnesses; it has been used as a stomachic, carminative, diuretic, etc.

Parsley is a cool-weather plant; it grows best between 7 and 16°C. It is sometimes grown in the tropics at higher altitudes. The crop thrives in well-drained fertile loams, pH ranging from 4.9 to 8.2. Especially during germination and in the juvenile stage, ample moisture is required.

Parsley is propagated by seed. Sowing rate depends strongly on the aim of the culture and varies widely from 10 to 60 kg/ha. The seed may be sown in rows, 30–60 cm apart. Germination takes 10–25 days. Weed control, especially in the early growth, is necessary. It can be done mechanically or chemically. Pre-emergence and post-emergence herbicides are available. Fertilizer can be applied before sowing and repeated after germination and the first leaf harvest. The total application depends upon soil type and soil fertility, 130 kg N/ha with P and K in proportion, will usually be sufficient. Harvesting of the leaves may start about 75 days after sowing; in temperate regions it is possible to achieve three cuts in a year. The cultivation of turnip-rooted parsley is similar to that for carrot. Germany and France are the main producers of parsley.

9 Starch Crops

CEREALS

Barley

Barley – *Hordeum vulgare*; Grass family – *Gramineae*

Origin, history and spread

Hordeum vulgare is probably native to the Middle East, from Afghanistan to northern India, the area in which barley was domesticated. Barley is derived from the wild species *H. vulgare* subsp. *spontaneum*, which is still widely distributed in the area of cultivation today. The cultivated plant is probably as old as agriculture itself: cultivation of barley dates back at least 9000 years. In the Near and Middle East, archaeologists have found grains or fragments of grains which were carbon-dated to between 7000 and 5000 BC. Barley reached Spain around 5000 BC and spread over the Old World between 5000 and 2000 BC; it reached China around 2000 BC. In Europe, many remains suggest that barley was probably the main cereal during the Neolithic and Bronze Age. Ancient Greek and Roman coins from about 500 BC have been found that show impressions of four- or six-rowed barley. Gladiators in Rome were fed on barley. Already in ancient times, the use of barley for brewing was known. This is recorded in Egyptian paintings dating to 3000 BC and on Babylonian clay tablets dating to 2000 BC. Columbus probably introduced the first barley varieties into North America in 1492, but it did not take hold in the New World. In the 16th century, various types of barley were brought to America: the English probably introduced two-rowed types, while the Dutch brought six-rowed types from continental Europe. Probably the first commercial brewery in North America was built by the Dutch

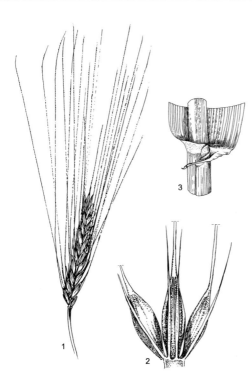

Fig. 9.1. Barley: 1, spike of two-rowed barley (one fertile and two sterile spikelets at each node); 2, detail of a spike of multiple-rowed barley (three fertile spikelets at each node); 3, auricles at the base of the leaf-blade of barley.

in 1625 on Manhattan Island in what was then New Amsterdam (now New York).

At present barley is grown all over the world and over a broader environmental range than any other cereal, from 70°N to 44°S. In many countries, barley is cultivated higher on the mountain slopes than other cereals, and in many areas it is the only possible rainfed cereal.

Botany

Barley (Fig. 9.1) is an annual grass. The stem is erect, with solid nodes, five to seven hollow internodes, and up to 120 cm tall. The basal internode is the shortest; they increase in length and are progressively smaller in diameter towards the top. The leaves are borne alternately, are linear-lanceolate, with short ligules and big, overlapping auricles. The leaves mostly have 15–20 veins. The inflorescence is a terminal spike, up to 12 cm long, excluding the awns. At each node of the spike there are three spikelets, not singly as in wheat and rye. If all three spikelets contain one fertile floret each and finally form three seeds, it is called 'six-rowed' or 'four-rowed'. If only the middle of the three spikelets is fertile, and the two sterile florets are represented only by their glumes and reduced lemma and palea, it is called 'two-rowed' (Fig. 9.2). If one looks on top of the spikes of the three groups, it shows two, four or six rows of kernels. The two glumes are narrow and terminate in an awn-like point. The lemma is lanceolate with five ribs and often terminating in a long, serrated awn, up to 15

Fig. 9.2. Detail of a ripe crop of two-rowed barley (photo: D.L. Schuiling).

cm long, but there are awnless forms. The palea is slightly smaller than the lemma; the margins are inflexed. Lemma and palea are mostly tightly fused to the pericarp; however, there are naked or hull-less forms, which have seeds loosely attached to the lemma and palea. The floret has two lodicules, three stamens and one ovary with two stigmas. The fruit is a caryopsis, elliptical in frontal view, convex on the embryo side. The colour of the husks can be yellow or black. The 1000-kernel weight is 25–50 g.

Barley has two sets of roots. The first, produced as the seed germinates, are the primary set. The primary roots grow outwards and form a fibrous, branching mass. The second set of roots is adventitious and grows from the nodes at or just below the soil surface. The adventitious roots are thicker and less branched than the primary roots.

Cultivars, uses and constituents

Hundreds of barley cultivars exist, and based on vernalization requirements, they can be classified as winter or spring barley. Plant breeders have succeeded in producing short-strawed cultivars, which are capable of very high yields as they resist lodging. In the tropics and the subtropics, barley is cultivated for food. The straw is often used as feed for livestock. In temperate regions, barley is grown for feed; it may be fed as a sole cereal grain to swine or poultry, or blended with other cereals. When the grains are fed ground, it is a good source of energy for both cattle and sheep. While still green and vegetative, barley can be used as forage for animals, either by grazing or as hay. At a later stage, when the grain is hardening, the crop can be cut and used for whole-crop silage. Two-rowed barley is often used for malting. Malt is a product for the brewer, distiller or food processor.

Malt is produced by germinating barley grains and growing the small seedlings for 4–6 days. After drying, the grains are brushed to remove the rootlets. The short coleoptiles have then attained a length of 75–100% of that of the kernel and are protected by the hulls. The remaining product is a source of α- and β-amylase, which are needed for hydrolysing starch. Most of the malt is used for brewing beers, including lager, stout and ale. At the beginning of the brewing process, crushed malt is combined with warm water to the consistency of a thin porridge called 'mash'. The temperature of the mash is increased to 80°C to inactivate enzymes. The hot mash is transferred to a filtration vessel, in which husks (lemmas and paleas) serve as a filter bed. The filtrate is called 'wort'. Wort is then transferred to boiling kettles where hops are added. Hop is used in the form of dried cones of the female hop plant or as extracts. Hop has a mild antiseptic action towards a wide range of bacteria and it flavours the beer. After cooling and filtering, the wort is inoculated with yeast and fermented at relatively low temperatures. After that, the remains of the yeast are removed and the so-called 'green beer' is transferred to the lagering cellar. The barley residue after the brewing process is used as feed for livestock. In food processing and distillery, malt is used for the manufacture of malt beverages like a cooling drink in India called 'sattu', gin and whisky. Malt extracts are also used for flavouring biscuits and sweets.

To use barley grains for human food, the husks have to be removed (Fig. 9.3). If the seedcoat and the aleurone layers have also been removed, the final product is called 'pearl barley'. The whole pearled kernels can be cooked and eaten; the kernels can also be milled into flour and used in breakfast cereals, bread and infant food. Barley is sometimes used as a coffee substitute after roasting and grinding the kernels. Barley grain contains approximately 72–79% carbohydrates, 8–11% protein, 1–2% fat, 1–7% fibre and 1.5% ash. The grain is a good source of vitamins of the B-complex and vitamins D and E. In folk medicine, barley had many applications; it is reported to be demulcent, diuretic, emollient and stomachic.

Fig. 9.3. Barley grains within their husks (left) and with the husks removed by polishing to give barley groats (right).

Ecology and agronomy

Barley has a wider ecological range than any other cereal. It grows in tropical countries as well as in northern countries like Norway. However, barley does not thrive in the humid tropics. It is adapted to areas with annual rainfall from 200 mm to more than 1000 mm. Barley does not tolerate acid soils, but is more salt-tolerant than other cereals. Barley has a shorter growing season than wheat or oats; some forms survive under extreme conditions and mature in 60–70 days. Barley is not very winter-hardy, so it is often grown as a spring crop. In countries with relatively mild winters, it can be grown as a winter crop. It thrives best in areas with a temperate climate where seasons are cool and moderately dry, on a well-drained fertile loam or light clay soil, and a soil pH of 6–8.5.

Barley is propagated by direct seeding. The seed is drilled 2–6 cm deep in rows 15–35 cm apart. Seeding rates vary from 70 to 150 kg/ha. Average plant density is 200–250 plants/m^2 (Fig. 9.2). Application of NPK fertilizers increases yield of straw and grain. Barley removes about 30 kg N, 22 kg P and 28 kg K per tonne of seeds. N increases the protein content of the kernels, which affects the brewing quality; for brewing the protein content should be about 11%. Consequently, too large applications of N should be avoided.

Because of the wide range of environments and because of its spring and winter habit, a broad spectrum of weeds may cause serious economic losses. So weed control is needed, either mechanical or chemical. Post-emergence spraying of 2,4-D is widely used to control broad-leaved weeds. Care must be taken to select herbicides because barley is sensitive to many herbicides at certain growth stages.

Barley is ready for harvest in about 4 months after sowing. Moisture content has to be 14% or less. However, delaying the time of harvest until the safe storage moisture content is reached increases the chance of losses due to shattering. Most of the barley is harvested by combine harvester. Barley yields vary from 0.3 t/ha under marginal conditions in dry years, to 10 t/ha in high-input agriculture. In 2004, the world total production of barley seeds was about 154 million t. The average yield is 3 t/ha in North America and 4 t/ha in Europe. The Russian Federation, Canada, Germany, Ukraine and France were the main producers, each of these countries producing more than 10 million t.

Maize

Maize – *Zea mays*; Grass family – *Gramineae*

Origin, history and spread

The history of the evolution of modern maize is difficult to describe because its direct ancestor does not occur in the wild. Several theories about the origin of maize exist; the most widely accepted one is that maize was derived by human selection from the wild grass teosinte (*Zea mexicana*). Teosinte is a wild grass that grows in Mexico and northern Central America. Teosinte and maize are still interfertile. Maize pollen has been identified in excavations in Mexico City, dating back some 8000 years. Some researchers suggest that pod corn, a primitive form of maize in which each kernel is enclosed by floral bracts, and which was grown in Mexico in prehistoric times, played a part in the development of modern maize. However, others think that modern maize was speciated from teosinte into a separate gene pool many thousands of years ago from which a number of different species arose. Archaeological finds from the Tehuacan caves in Puebla, Mexico, suggest that Indians were using maize rather than teosinte from about 5000 BC. At that time, maize resembled teosinte quite closely. Maize was originally a much smaller plant than the present varieties. In the course of thousands of years of cultivation numerous forms of maize were developed. Many of these forms still exist in Central America, called 'Indian maize'. Columbus brought maize kernels back to Europe, where it was grown for the first time in 1493. Maize-growing spread rapidly in southern Europe. In the 16th century maize was introduced into Africa. In this continent maize displaced the traditional grain crop sorghum in all but the drier regions. The Portuguese introduced maize into Asia where it became widely grown. Breeders developed cultivars that are adapted to a wide range of climatic conditions. In the cooler regions, maize may not ripen to dead ripe stage, but if grown as forage for livestock, fresh or ensiled, the grains should not be too hard and it can be harvested in waxy ripe stage.

At present, maize is grown throughout the world between 58°N (Canada, Russia and Denmark) and 42°S (New Zealand and Argentina), and from altitudes below sea level up to 4000 m.

Botany

Maize (Fig. 9.4) is a large annual grass, up to 4 m tall. It varies greatly in size, depending on cultivar and growing conditions. The stem is usually simple, solid,

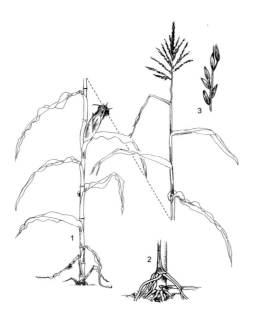

Fig. 9.4. Maize: 1, plant with female inflorescence (cob) in a leaf axil and terminating in a male inflorescence; 2, detail of the stem with adventitious roots growing from the nodes; 3, detail of a male inflorescence.

Fig. 9.5. Maize: 1, male inflorescence; 2, male flowers at anthesis; 3, detail of the female inflorescence with many styles (photo: D. van der Linde); 4, a ripe cob (photo: D.L. Schuiling).

with clearly apparent nodes (eight to 21). Incidental side shoots (suckers) may arise at the base of the stem. The eight to 21 leaves are smooth with linear-lanceolate blades of 70 cm × 7 cm in average size. They have clearly visible ribs with a pronounced midrib and colourless ligules, about 5 mm long. The leaves are borne alternately on either side of the stem, usually bending downwards. The colour of the leaves is usually green but may be variegated white, yellow or purple-red.

A unique characteristic of maize among the grasses is its monoecious flowering, which is expressed by forming female ears (cobs) and male tassels (Fig. 9.5). The stem is terminated by the tassel, which is an inflorescence of male spikelets. The tassel consists of erect or spreading racemes forming a panicle, which can be 30 cm long. The spikelets are borne in pairs, one sessile, the other pedicelled, and both 8–12 mm long. Spikelets consist of two male florets enclosed by papery glumes. A floret consists of a lemma and palea, two lodicules and three green, yellow or purple anthers. The female inflorescence, called the 'cob' or 'ear', occurs in the axil of a leaf in the mid-region of the stem, usually one to three per plant. Each cob consists of a number of compressed internodes called the 'shank', and attached to the nodes a number of modified leaves, together forming a tightly packed structure called a 'husk'. Actually a husk is a leaf-sheath, sometimes bearing a reduced leaf-blade. The husks envelop the female inflorescence. The upper part of the cob comprises an enlarged axis bearing sessile spikelets in longitudinal double rows. Each spikelet contains two florets, one sterile and the other fertile; lemma, palea and glumes are highly reduced in size. Thus after removing the husks from a mature cob, one can see only the rows of naked caryopses. Long thread-like styles and adhered stigmas arise from

Fig. 9.6. Variability of kernel colour in free-crossing of an on-farm maize population, Guatemala (photo: D.L. Schuiling).

the ovaries and emerge at the top of the cob. This bundle of styles and stigmas can grow up to 45 cm in length and is known as 'silk'. Mature cobs can reach a length of 45 cm and contain up to 1000 kernels. Colours of the kernels (Fig. 9.6) and 1000-seed weight vary greatly, but the latter is 250–300 g for most of the dent and flint maize cultivars.

The root system consists of a primary or taproot and associated lateral roots, and adventitious roots, some of them growing from nodes just above the soil surface called 'prop roots'. Roots can penetrate the soil to a depth of about 2 m.

Varieties, uses and constituents

Maize is, after wheat and rice, the third most important starch crop in the world. Dent maize, flint maize, popcorn and sugar maize are the most widely grown varieties. The dent and flint types are very old. They can still be found among the so-called 'Indian maize', which is a mixture of different types and a great variety of colours, all harvested from one field, and even from one plant. Indian maize is now also grown as an ornamental. Indians selected popcorn from flint maize already in early civilizations. The endosperm of maize consists mainly of starch, except in sugar maize. The different types (Fig. 9.7) vary in starch composition as follows.

- Dent maize: when this variety is ripe, it has a depression at the top of the caryopsis, caused by a hard form of starch at the sides of the kernel and a soft form in the centre. The soft starch shrinks when the kernel is drying, which

Fig. 9.7. Several types of kernels (left to right): flint type; dent type; sugar type; popcorn; flour type.

results in a dent. The shape of dent maize kernels can vary from long and narrow to wide and shallow.

- Flint maize: the outer layer of the kernel consists entirely of hard starch, which means uniform shrinking of the entire layer in drying, thus without a dent.
- Popcorn: this variety has usually small, pointed grains with a very high content of hard starch. When the kernels are heated, the moisture in the starch expands and causes the kernel to burst explosively. Kernels increase in volume 20- to 40-fold after 'popping'. Popcorn is mainly used as snack food. Like flint maize, popcorn kernels have a smooth skin without dents.
- Sugar maize: sugar maize or sweet maize arose in the 19th century as a mutant in a maize field. It can be distinguished from other maize varieties by the high sugar content of the endosperm at the early 'dough' stage. The pericarp of the seed is thinner than that of other corns, which makes it tender. When the seeds dry, they become wrinkled and more or less translucent. For human consumption, sweet maize must be harvested when the seeds are fully developed but still in an immature 'dough' stage. Mature seeds lose sweetness and become tough. In past decades, breeders have introduced new cultivars with increased sugar content, which are known as 'supersweets'. Sugar maize is mainly used as a vegetable, cooked and eaten on the cob. Kernels are often canned or frozen.
- Flour corn and waxy corn: these are two more maize varieties from the USA, which are small in acreage. The kernels of flour corn contain almost entirely soft starch, surrounded by a very thin layer of hard starch. Flour corn was the cereal grown originally by American Indians. Waxy corn has a wax-like appearance on cross-section, caused by the very high amylopectin content of the endosperm. Waxy corn is a main raw material for use in industry.

Dent maize and flint maize are grown as food for both people and animals. In the USA and Europe, maize is grown mainly as feed or fodder crop; in Central and South America it is mainly used for human consumption; in Africa it is used both as feed or fodder crop (yellow grains) and for human consumption (white grains); in Asia it is used for both people and animals.

Maize contains approximately 10% water, 70% carbohydrates, 10% protein, 4.5% fat, 2% fibre and 2% ash. The starch exists of about 75% amylopectin and 25% amylose. Yellow maize contains carotene (provitamin A).

Maize has a low content of essential amino acids such as lysine. At present, plant-breeding technology has led to the development of high-lysine varieties.

For human consumption, maize has to be ground or pounded first. The meal can be boiled or roasted. As maize flour lacks gluten, it is not possible to bake leavened bread from it. In Latin America, the meal is often used to cook flat cakes called 'tortillas'. In Africa it is often boiled into porridge, called 'ugali' in Swahili. Maize is used all over the world as breakfast cereal such as 'corn flakes', and also processed in snack food. Other industrial products are starch, meal, oil, corn-oil meal, glucose, gluten feed, and alcoholic beverages. In the preparation of starch and glucose, the germ is separated from the rest of the seed. The germ contains about 50% oil, which can be gained as an excellent salad or cooking oil. In cooler temperate regions where maize is grown as feed for animals, the whole plants, sometimes only the cobs, are cut off and chopped mechanically, for ensilage. Maize silage forms palatable feed, which is high in nutritive value.

At present, most maize is sown from single-crossed or double-crossed seed. The crop that arises from this seed is very uniform and highly productive. About 90% of the maize production for feed in the European Union and the USA is from single-crossed seed. The disadvantage of single-crossed seed is that it is more expensive than double-crossed seed.

The production of single-crossed seed is as follows:

1. Two inbred lines are created by self-pollination through several generations, resulting in two inbreds with reduced vigour and production, but increased uniformity.

2. The last two inbreds, one from each line, are crossed after one inbred is detasselled, resulting in a single-crossed hybrid with restored vigour.

The successive steps in the making of a single hybrid are presented in Fig. 9.8.

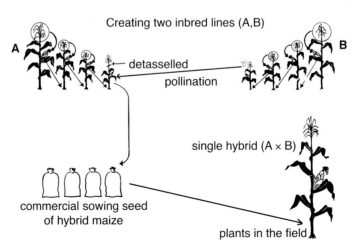

Fig. 9.8. Maize: producing commercial sowing seed by creating two inbred lines to obtain high-yielding plants.

The production of double-crossed seed can be as follows:

1. Two inbred lines are created by self-pollination through several generations, resulting in two inbreds with reduced vigour and production, but increased uniformity.
2. The two inbreds are crossed after one inbred is detasselled, resulting in a single-crossed hybrid with restored vigour.
3. The single-crossed hybrid is crossed with another, similar developed single-crossed hybrid, after one of the plants is detasselled, resulting in a double-crossed hybrid.

The successive steps in the making of a double hybrid are presented in Fig. 9.9.

Ecology and agronomy

Maize is essentially a crop of subtropical regions, but after selecting and breeding, a large number of cultivars are now also adapted to temperate conditions. However, maize has the characteristic C4 cycle photosynthetic development, which means vigorous growth and high yields under warm and bright conditions. Maize does not thrive well in semi-arid or equatorial climates. For germination the minimum soil temperature has to be 10°C. For growth and development the

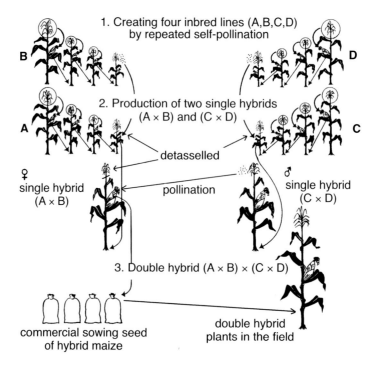

Fig. 9.9. Maize: producing commercial sowing seed by creating four inbred lines to obtain high-yielding plants.

average day temperature should be at least 15°C. It is easily killed by frost. Maize grows on a wide range of soil types, but prefers deep, naturally rich, easily tilled soil, pH 5.5–6; it is sensitive to salinity. Maize is rather drought-tolerant; however, annual rainfall of at least 750 mm is required for adequate moisture: especially around the time of tasseling and fertilization, maize plants are very sensitive to water deficiency.

Maize is nearly always planted through direct seeding. Soil temperature has to be about 10°C, below that temperature the seeds do not germinate until the desired soil temperature is reached. If that takes too much time, soil pathogens can cause damage to the seeds. The result may be that the seeds do not germinate at all or produce abnormal seedlings. Plant density depends on cultivar, cropping system and cropping purpose (for silage, higher density than for grain), soil condition and climate; it ranges from 20,000 to 100,000 plants/ha (Fig. 9.10). Sowing has to be done very carefully because maize has poor tillering capacity; it is mostly not able to fill open spaces in the field. The seeds have to be placed 2.5–6 cm deep, depending on how deep the moist soil is located below the dry surface layer.

Maize usually responds well to fertilizers. Each produced tonne of grain removes about 30 kg N, 5 kg P and 35 kg K. How much fertilizer has to be applied depends on inherent soil fertility. Modern hybrids only reach their high yield potential when sufficient nutrients are applied. Weed control will be necessary, especially in the juvenile stage. In this stage the crop is most sensitive to competition with weeds for nutrients, light and moisture, which can decrease yield dramatically. Weeds can be killed mechanically before the maize plants are up or while they are small enough (up to 10 cm) to stand harrowing without injury. When the plants grow up, herbicides can be used. Ridging or earthing-up is sometimes used as a form of weed control.

Fig. 9.10. Mixed cropping of maize and common bean.

When the crop can be harvested differs for the various forms. Sweet maize is harvested about 75 days after emergence of the plants. In this stage the kernels produce a milky liquid when punctured, the husks should be tight and the silks somewhat dried. When maize is harvested for grain, the moisture content has to be 15–20%. Yellowing of the leaves, yellow and dry husks, and hard kernels can show the stage of ripeness. Harvesting can be done by hand but most maize is harvested mechanically. Popcorn is harvested when the moisture content is 16–17%. Harvesting the whole plant for silage is done in many ways, from young plants, 8 weeks old and a dry matter content of 16%, up to plants in dough-ripe stage and a dry matter content of 50%.

In the USA, the average yield of sweet maize for the fresh market is about 9 t/ha and the grain yield of popcorn is about 3 t/ha. Grain yields of maize vary greatly, from 1–2 t/ha in Africa up to the average grain yield in the USA of about 7 t/ha. Yields of up to 14 t/ha have also been achieved. The yield of silage maize ranges from 11 to 16 t of dry matter per hectare. In 2004, the world total grain production was about 725 million t. By far the most important producer is the USA, accounting for about 32% of the total production. China is the second largest producer of grain maize.

Millets

Pearl millet – *Pennisetum glaucum*, Finger millet – *Eleusine coracana*, Foxtail millet – *Setaria italica*, Common millet – *Panicum miliaceum*, Little millet – *Panicum sumatrense*, Barnyard millets – *Echinochloa* spp., Kodo millet – *Paspalum scrobiculatum*; Grass family – *Gramineae*

Introduction

'Millet' is a collective term referring to a number of small-seeded annual grain crops from different genera in the grass family. Although millets represent only approximately 1.5% of world cereal production, they are important staple foods in a large number of countries in the semi-arid tropics of Africa and Asia, where low precipitation and poor soils limit the cultivation of other major food crops. Millets (and sorghum) are still the principal sources of energy, protein, vitamins and minerals for millions of the poorest people in these regions.

All millet species provide an appropriate food for patients intolerant to gluten causing coeliac disease.

Among the millets of the world, pearl millet (*Pennisetum glaucum*) is the most important species. Minor millets, also referred to as small millets, have received far less attention in terms of cultivation and utilization but are certainly of regional importance. They include finger millet (*Eleusine coracana*), foxtail millet (*Setaria italica*), common or proso millet (*Panicum miliaceum*), little millet (*Panicum sumatrense*), barnyard millet (*Echinochloa colona* and *Echinochloa crus-galli*) and kodo millet (*Paspalum scrobiculatum*).

In 2004, the world total seed production of millets was about 28 million t. Africa accounted for around 55% with major producing countries being Nigeria,

Niger, Burkina Faso and Mali. Asia accounts for around 40%, with major producing countries being India and China.

Pearl millet

Origin, history and spread

Pearl millet, also known as 'bullrush millet', almost certainly originated in tropical western Africa, where the greatest number of both wild ancestors and cultivated forms occurs. It was domesticated along the southern margins of the Saharan central highlands at the onset of the present dry phase some 4000–5000 years ago. About 2000 years ago the crop was carried to India, where it became established in the drier environments because of its excellent tolerance to drought. In India it is the fourth most important food crop after rice, wheat and sorghum. It has recently been introduced as a grain crop in the south-eastern coastal plain of the USA, where it has been used as a green fodder crop. In Asia, millet is restricted almost exclusively to India and China, although Myanmar, Nepal and Pakistan also produce small quantities. In Africa millet production is distributed among a much larger number of countries of which Niger, Nigeria and Sudan are the most important.

Botany

Pearl millet is an annual, erect, robust bunch grass, of 2–2.5 m tall or even higher. Dwarf varieties are only to about 1 m tall. The stem is slender, 1–3 cm diameter at the base, solid and with prominent nodes which may be glabrous or hairy. The total number of tillers can be very large and up to 40 in late cultivars. Leaf-sheaths are open above, overlapping at the base and usually glabrous. The leaf-blade is linear to linear-lanceolate up to 1.5 m × 5–8 cm. The ligule is short, membranous and fringed with hairs. The inflorescence is a terminal, cylindrical, compact and stiff spike-like panicle, 5–150 cm long (Fig. 9.11). The spikelets are usually in pairs but vary from one to five per cluster. The spikelet groups are subtended by and deciduous with clusters of scabrid bristles. Seeds are white, yellow, blue-white, pale grey or brown, occasionally purple in colour; hilum marked by a distinct black dot at maturity. The ovoid grains are about 3–4 mm long and much larger than those of other millets. The size of the pearl millet kernel is about one-third that of sorghum. The 1000-seed weight ranges from 2.5 to 14 g with an average of 8 g. The root system of pearl millet is extremely profuse.

Cultivars, uses and constituents

Pearl millet may be considered as a single species but it includes a number of cultivated races and a complex of natural hybrids. Different cultivar groups are based on grain shape and growth duration. Dwarf types are also distinguished. In addition to genotype differences, an adverse environment or photoperiod can profoundly alter the plant habit. Cultivars vary in time to maturity from 55 to 280 days, but mostly from 75 to 180 days. The 'Iniadi' cultivar from northern

Fig. 9.11. Inflorescences of seven millet species: 1, pearl millet; 2, finger millet; 3, foxtail millet; 4, common millet; 5, little millet; 6, barnyard millet; 7, kodo millet.

Togo and Ghana has been frequently used as a parent in pearl millet breeding. Selections from it have been successfully introduced as cultivars in India, Namibia and Botswana.

Of the 28 million t of all millets produced in the world in 2004, about 90% is utilized in developing countries. Exact production and consumption data are not available for most countries, but it is estimated that a total of 20 million t are consumed as food, the rest being equally divided between feed and other uses. Pearl millet is mostly grown as a subsistence crop and is the staple food for about 90 million people in parts of tropical Africa and India. After threshing it is usually ground for flour, porridge or gruel in Africa, and for flat unleavened bread in India. Flour is usually produced as it is needed because it tends to turn rancid. There are various other regional products, such as couscous, ricelike products, fermented and non-fermented beverages.

Average composition of the edible portion of the seed is 12% water, 10–12% protein, 3–5% fat, 60–70% carbohydrates, 1.5–3% fibre, and 1.5–2% ash. Its nutritional value is somewhat superior to maize, rice, sorghum and wheat. The relative proportion of germ to endosperm is higher than in sorghum. Livestock are an important component of most millet production systems and millet crop residues contribute significantly to fodder supplies. Stalks are also used for thatching, as fuel, and made into mats for winnowing. Outside India and Africa, in particular Australia and the USA, it is mostly cultivated for grazing, green fodder and silage, and seed is used as feed for poultry and as a component of birdseed.

Ecology and agronomy

In Africa and Asia pearl millet is grown as a rainfed crop in areas with an annual rainfall of 430–500 mm. Where also dry matter is used for fodder, more rain is required. It is adapted to a wide range of soils. Its large and dense root system allows it to grow on soils with a low nutrient status. The crop is normally grown in areas where maize and sorghum fail because of low rainfall or adverse soil factors.

Short-duration cultivars (55–65 days) require less water than long-duration cultivars (up to 280 days). Land preparation can improve the soil moisture content with noticeable effects on yield. Hand tillage with a hoe to a depth 5–10 cm and ploughing with a traditional plough to a depth of 15–20 cm are usual methods. Optimum temperature for germination is 33–35°C, for tiller production 21–24°C and for spikelet development about 25°C.

Propagation is by seed, usually sown directly in the field into open furrows, broadcasting and drill sowing. In higher rainfall areas transplanting may be practised. Depth of sowing is critical for the establishment of pearl millet and shows an optimum depth of 3.5–4 cm. Row spacing varies between 35 and 70 cm. Seed rates vary with techniques and requirements, ranging between 3 and 11 kg/ha. Seeds are ready for harvest 3–4 weeks after anthesis. Pearl millet is usually harvested by hand. The spikes can be cut or the whole plant cut and the spikes removed later. In strongly tillering cultivars several pickings are required due to unevenly ripening of the spikes. In Africa, after being dried in the sun for a few days, the spikes are commonly stored in elevated store houses, built of mud or plant materials and covered with thatch. The spikes are threshed only as required (Fig. 9.12). In India the grain is usually threshed soon after harvesting and drying. The grain may then be stored in granaries or sometimes in pits in the ground. Seed may then be covered with sand and leaves of the neem tree (*Azadirachta indica*) to reduce insect attack. Average grain yields in West Africa and India are about 600 kg/ha. Under favourable conditions, yields of improved cultivars range from 3 to 3.5 t/ha, those of hybrids can reach 5–8 t/ha.

Fig. 9.12. Stack of bundles of harvested pearl millet spikes, Cameroon (photo: E. Westphal).

Finger millet

The term '*Eleusine*' is derived from Eleusis, an old epic city sacred to Demeter, the Greek goddess of agriculture, marriage and fertility. The term '*coracana*' is derived from 'kurukkan', the Singhalese name for this grain. Finger millet, also called 'African millet' or 'ragi millet', is grown throughout eastern and southern Africa, but especially in the sub-humid uplands of Uganda, Kenya, Tanzania, Malawi, Zaire, Zambia and Zimbabwe. The crop originated about 5000 years ago somewhere in the area that today is Uganda. It was taken to southern Africa about 800 years ago and to India at a very early date, probably over 3000 years ago. In India finger millet is grown mainly in the southern peninsular region of the country. The major producing states are Karnataka, Tamil Nadu, Andhra Pradesh and Maharashtra.

Finger millet (Fig. 9.13) is an annual grass with ascending to erect, compressed stems, 60–120 cm tall. Due to the profuse tillering and extensive branching the plants are often lodged or prostrate; the root system is fibrous and remarkably strong, permeating soil thoroughly. The leaf-sheath is flattened; blade linear, 30–75 cm × 1.0–1.7 cm, often folded, with a strong midrib, glabrous or with few hairs. The inflorescence is a whorl of two to eight (normally four to six), digitate, straight, or slightly curved spikes 12.5–15 cm long, about 1 cm in diameter (Fig. 9.11); spikelets about 70 per spike, arranged alternately on the rachis, each containing four to seven seeds, varying from 1 to 2 mm in diameter; caryopsis is nearly globose to somewhat flattened, smooth

Fig. 9.13. Field of finger millet, Malawi (photo: J. Ferwerda).

or rugose, reddish-brown to nearly white or black. The pericarp remains distinct during development and at maturity appears as a papery structure surrounding the seed. The 1000-seed weight ranges from 2 to 2.5 g.

The composition of the grain is approximately 13% water, 8% protein, 1.3% fat, 72% carbohydrates, 3% fibre and 2.7% ash. The grain is rich in Ca (up to 0.3%). Finger millet grains may be consumed as roasted green in dough-ripe stage; the ripe dry grain is ground into flour and made into porridge, cakes or puddings. The whole grain, particularly the white-seeded types, is germinated to make malt for direct consumption or for brewing. The straw is a valuable fodder for both working and milking animals or used as mulch. The multiple use of the grain and the storability of the ripe grain are the most important assets of finger millet. The grain will retain its viability and quality longer than any other cereal crop in areas with marginal storage facilities.

Finger millet takes 3.5–6 months to mature, depending on cultivar, photoperiod and climatic conditions. The climatic conditions range from cool temperate moist to tropical very dry. Finger millet thrives at higher altitudes than most other tropical cereals. Annual precipitation tolerance ranges from 300 to 4000 mm and a temperature range of 11–28°C is tolerated. Many cultivars and landraces exist, with different ecological requirements and morphological characteristics which are based primarily on type of inflorescence. Well-known cultivars are 'Engenyi' from Uganda, 'EC593' from India, '25C' from Zimbabwe and 'Lima' from Zambia. Grain yield ranges from 0.4 to 2 t/ha. Under optimal conditions grain yield can reach 3 t/ha.

Foxtail millet

It is generally considered that foxtail millet was domesticated in the highlands of central China and, together with proso millet (*Panicum miliaceum*), was probably among the first crops to be domesticated in central and eastern Asia. It is stated that foxtail millet has been known as a cultivated cereal since 5000 BC in China. The world production has declined substantially since the 1950s. At present the main areas of production are China, parts of India, Afghanistan, Central Asia, Manchuria, Korea and Georgia.

Foxtail millet is a tufted, annual grass with erect culms, 60–120 cm tall. The inflorescence is a spike-like panicle, 8–18 cm × 1–2 cm (Figs 9.11 and 9.14). The spikelets fall singly, without subtending bristles. The caryopsis is tightly enclosed by the lemma and palea.

Foxtail millet is mainly a dry-land crop and can be grown in semi-arid regions with rainfall of less than 125 mm. Growth duration depends on cultivar and growth conditions and may vary from 2 to 4 months. Dry conditions are required for harvesting. A rainfed crop usually yields 800–900 kg of grain per hectare and 2500 kg of straw. Under optimal conditions much higher yields can be obtained.

Fig. 9.14. Close-up of the panicles of a stand of foxtail millet (left); detail of a fruiting panicle of common millet (right) (photos: J. van Zee).

Common millet

Common millet, also known as 'proso millet', was domesticated in central and eastern China and has been cultivated in this area since 3000 BC. The true wild form of subspecies *ruderale* is native to this area and considered to be the ancestor of the cultivated forms. In the Bronze Age it spread widely in Europe and remains of the seed have been found in Celtic fields in north-western Europe. It was the 'milium' of the Romans and one of the millets in the Old Testament. As a cereal it is widely grown in northern China, Mongolia, Korea, south-eastern Russia, Afghanistan, Pakistan, India and southern Europe. It was also introduced into North America.

Common millet is an annual grass up to 1 m tall, with a rather shallow root system and weakly tillering. The inflorescence is variable: an open or compact, drooping or erect panicle up to 45 cm long (Figs 9.11 and 9.14). Spikelets are two-flowered; lower floret sterile, upper floret fertile. The grains are ovoid, smooth and glossy; colour varies from white, yellow, brown, red to almost black. The caryopsis is broadly ovoid, up to 3 mm × 2 mm and tightly enclosed by palea and lemma.

The grain of common millet is very nutritious, containing, as well as carbohydrate, 10–18% protein and 1–4% fat. The 1000-grain weight is 5.5 g for husked grain and 7.1 g for unhusked grain. The husked grain can be eaten whole after roasting or cooked and eaten like rice. The flour of milled grains is used for making porridge, unleavened bread, beer or wine. The grain is also feed for animals, poultry and cage birds.

Common millet is adapted to hot and dry conditions and shallow soils which are not suitable to most other cereals. Average yield of common millet is

400–600 kg/ha. Under optimal growing conditions a grain yield of 1–2 t/ha can be reached.

Little millet

Little millet is grown throughout India, in particular in the states of Madhya Pradesh and Uttar Pradesh. Elsewhere it is of minor importance. It is an annual tufted grass with slender stems up to 1.5 m tall. Depending on cultivar the plant habit may be decumbent or erect. The inflorescence is an open to compact panicle, 14–40 cm long, composed of five to 50 branches each 3–9 cm long (Fig. 9.11). Spikelets are flattened, 3–4.5 mm long. The caryopsis is glabrous, mostly brown, sometimes shiny white to almost black.

The husked grain may be cooked and eaten like rice and it can also be made into flour to make bread. The straw and green plants are used as forage. Little millet will thrive under such adverse conditions that it can be grown on soils which otherwise produce little or nothing.

The growth period varies from 2.5 to 5 months. Yield ranges from 200 to 600 kg/ha, increasing to 900 kg/ha in a good season.

Barnyard millets

For barnyard millets numerous botanical names as well as common names are mentioned in the literature. However, the names *Echinochloa frumentacea* for Japanese barnyard millet and *Echinochloa crus-galli* for barnyard millet are approved by most authorities.

Japanese barnyard millet is a robust tufted annual, cultivated in Japan, Korea and China, and grown for its forage as well as its grains in places where rice does not grow well. In the USA it is grown for forage and wildlife feed. The grains are also used for feeding cage birds. The inflorescence is up to 15 cm long, densely branched and usually purple-tinged, with awnless scabrous spikelets (Fig. 9.11). It is the quickest growing cereal of all millets and will produce a crop in about 6 weeks, yielding 700–800 kg of grain and 1000–1500 kg of straw per hectare.

Barnyard millet is an annual tufted grass which grows in swampy places. It provides a useful fodder and is sometimes cultivated for this purpose. Especially in times of food scarcity it is used as a grain crop. Young shoots are eaten as a vegetable in Java. It is one of the worst weeds in paddy rice in Asia and California.

Kodo millet

Kodo millet occurs throughout the tropics in Asia and Africa, but is cultivated for grain in only India and confined to Gujarat, Karnataka and parts of Tamil Nadu. Elsewhere it is a common grass and provides useful forage but is also regarded as a weed of annual and plantation crops. Compared with other small millets it is a long-duration crop: 105–120 days of growth. It is one of the hardiest among small millets and grows well on shallow as well as on deep soils. It is well adapted to waterlogged soils. The seed can remain dormant for several months after maturity and can be stored for many years. Kodo millet is an annual tufted grass that grows to 90 cm high (Fig. 9.11). Some forms have been reported to be poisonous to humans and animals, possibly because of a fungus infecting the grain. The grain is enclosed in hard, corneous, persistent husks that are difficult to remove. The grain may vary in colour from light red to dark grey. Average grain yield can vary considerably. At Bangalore, India, a grain yield of 850 kg/ha was obtained without fertilizer compared with 1600 kg/ha with moderate N and P fertilizer.

Oat

Oat – *Avena sativa*; Grass family – *Gramineae*

Origin, history and spread

The origin of oat is not clear, but it is believed to originate mainly from Asia Minor. Different species probably came from different regions. Oat is most likely a secondary crop, which means that it probably reached Europe as a weed with wheat and barley, and at a certain time, when growing conditions became suitable for oat, it was selected as a crop in its own right. Cultivation was initiated by the Celts and the Germans in the Bronze Age, about 2000 BC. So, as a cultivated crop oat appears to be younger than wheat and barley. The oat cultivars arose by crossing and selection from wild oats, believed to be *Avena fatua* and *Avena sterilis*. Oat was well known to the lake dwellers of Switzerland and has also been described in ancient Greek and Roman writings. In medieval times, oat was an important crop. The earliest colonists brought oat to North and South America and Australia. Now, oat is grown throughout the temperate regions, especially in cool, humid coastal regions and in Mediterranean-type climates.

Botany

Oat (Fig. 9.15) is an annual grass, adapted either to autumn planting or spring planting. The stem is smooth, sometimes scabrous beneath the inflorescence, from 65 to 130 cm high. The leaf-blade is linear-lanceolate, up to 35 cm long and about 1 cm wide, and rolled in an anticlockwise direction. The leaf-sheath is loose, auricles absent, ligule large and membranous. The inflorescence is a multiple-branched panicle, which may be spreading or one-sided (Fig. 9.16). The spikelets are two- or three-flowered, up to 3 cm long, mostly pendulous but sometimes erect. When the spikelets are three-flowered, the upper floret may be rudimentary. The spikelets are subtended by two, seven- to ten-nerved glumes which are longer than the spikelet. A fertile floret consists of the palea and

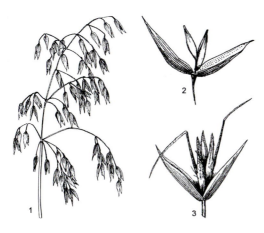

Fig. 9.15. Oat: 1, panicle; 2, spikelet of cultivated oat; 3, spikelet of common wild oat with twisted awns.

lemma, which enclose the three stamens and the single ovary. The palea is sometimes awned. The lemma is glabrous and short. The caryopsis is 6–12 mm long, elongated, hairy, grooved and tightly enclosed by the hard lemma and palea (hull), which cannot be removed by threshing. In so-called hull-less oats (*Avena nuda*) the caryopsis is loose within the palea, when ripe deciduous without the hull. The colour of the hull may be white, yellow, grey, brown or black (Fig. 9.17). The 1000-kernel weight (including the hull) is 30–42 g, the hull being about 35% of the weight.

Species, uses and constituents

As well as oat or common oat (*Avena sativa*), there are nine other *Avena* species recognized.

Fig. 9.16. Close-up of a stand of unripe oat with one-sided spreading panicles (left); detail of a field of ripe oat with panicles spreading in equilateral fashion (right).

Fig. 9.17. Several 'seeds' of black and white coloured oat varieties; caryopsis is covered by the glumes.

1. Wild red oat – *A. sterilis*: characterized by large hairy lemmas with strong twisted awns.
2. Red oat – *Avena byzantina*: a number of cultivars are included in red oat; they have a slender stem and are reddish in colour, the flowers adhere tightly to each other and separate by fracture of the rachilla, the lemmas have weak, non-twisted awns, panicles small, narrow and erect. Red oat probably originated from wild red oat.
3. Large naked oat – *A. nuda*: characterized by the caryopses which are loose within the hulls, three to seven flowers may be produced per spikelet, cultivated to a very limited extent.
4. Small naked oat – *Avena nudibrevis*: like large naked oat but with smaller stems, leaves, panicles and caryopses, probably no longer cultivated.
5. Desert oat – *Avena wiestii*: stem slender, medium high and stiff, slender lemmas that terminate in two bristle-like glume points, not in cultivation.
6. Slender oat – *Avena barbata*: small weak stems resulting in decumbent growth habit, panicles equilateral, large and drooping, seed falls to the ground at maturity, not in cultivation.
7. Sand oat – *Avena strigosa*: small, erect stems, panicles near equilateral, lemmas lance-like, extending to two distinct points, grown in hilly regions where conditions are unsuitable for common oat.
8. Abyssinian oat – *Avena abyssinica*: similar in habit to desert oat, but differs in chromosome number, cultivated to some extent in North Africa.
9. Wild oat – *A. fatua*: resembles common oat but often taller, panicles large and drooping, lemmas carry long twisted awns, spikelet separate from their pedicels by abscission, drooping when ripe, troublesome weed in grains throughout temperate and subtropical regions.

Oat is used mainly for feeding livestock, not only as ripened grain but also in the unripe, green stage as silage or hay. Oat is highly rated as a feed for horses. The plant is a good source of protein, carbohydrates, fibre and minerals. A relatively small proportion of the grain production is used as human food, in breakfast cereals, baked products and oat flour. For human food, the hulls first have to be removed. The caryopsis contains 11–14% protein, 65% carbohydrate and about 7% fat. Oat hulls are used in the production of the aldehyde furfural, which is widely used as a solvent and reagent in the dyestuffs, plastics and other industries. Oat hulls are, on a small scale, also used as a filter in breweries. Oat straw is more nutritious than wheat straw and is used as a supplementary feed for horses and cattle. The seed contains β-sitosterol, which epidemiological studies suggest has a cancer-protective effect.

Ecology and agronomy

Oat species are long-day plants, grown in cool and moist climates of the temperate region. They thrive on a wide range of soils of ample, but not excessive, fertility. The crop does not make heavy demands on fertilizer except N. Well-drained neutral soils in regions where annual rainfall is 700 mm or more are best. They prefer silt and clay loams and a pH of below 6.5. Oats are not as frost-hardy as rye, wheat and barley. Since seed weight varies, precise planting rates should be given in seeds per square metre rather than in weight. For the USA, 240–360 seeds/m^2 are recommended, which means about 72–108 kg/ha. The seeding rates in Europe are considerably higher: 110–220 kg/ha. Under dry-land conditions in the subtropics, a seed rate of 30–70 kg/ha is recommended. However, it is difficult to identify the optimum seeding rate because oat plants adjust their tillering in dependence of plant density.

Planting depth should be 2–4 cm. For maximum grain yield it is important to use good-quality seed, which means that lightweight seeds have to be removed because larger seeds germinate better and produce more vigorous seedlings. For optimal fertilization, a soil test is needed because fertilizer requirement depends on the soil condition. Per tonne of seeds about 42 kg N, 22 kg P and 26 kg K are removed. N applied at mid-boot stage or earlier increases yield, whereas later applications usually do not affect yield but will raise grain protein percentage. Most N is applied before or at seeding. Lodging is a serious problem in a number of oat cultivars. Application of herbicides has to be done carefully because oats are more sensitive to herbicides than the other cereals.

Harvesting oats is not easy because the straw is often fully ripe later than the kernels. Waiting with harvesting until the whole plant is fully ripe can cause seed losses. Direct combine-harvesting of the standing oat has to be done when kernels are in the hard dough stage, when the moisture content of the kernels is 14% or less and the straw is still slightly green.

World oats production has declined since the 1970s. The average production during 1970–1974 was 51.9 million t but declined to about 26 million t in 2004. The Russian Federation, Canada, the USA, Poland and Germany were the main producers.

Pseudo Cereals

Buckwheat – *Fagopyrum esculentum* (Dock family – *Polygonaceae*); Amaranth – *Amaranthus* spp. (Cockscomb family – *Amaranthaceae*); Quinoa – *Chenopodium quinoa* (Goosefoot family – *Chenopodiaceae*)

Buckwheat

The genus name '*Fagopyrum*' is derived from the Latin word 'fagus' meaning 'beech' and the Greek word 'pyron' meaning 'wheat'. The triangular fruit resembles the beechnut. Buckwheat was described for the first time in China in the 5th century. It probably evolved from the wild buckwheat species *Fallopia convolvulus*. It became an important crop in the Himalayan region and Japan. In the early Middle Ages it was introduced into Europe. European colonists took it to the New World.

Buckwheat (Fig. 9.18) is an erect, annual herb growing to a height of 40–120 cm. The stem is branched, angular, smooth, hollow and often partially red in colour. On each node there is a membranous stipule called the 'ocrea'. The leaves alternate, upper ones sessile, lower ones with petioles, stipules tubular, blade triangular. The inflorescence is compound and consists of terminal clusters of white or rose flowers (Fig. 9.19). The flowers are small with five petals and eight stamens, alternating at the base with eight honey glands. The fruit is a greyish-brown, triangular, three-sided nutlet. The 1000-seed weight is 22 g. In addition to normal buck-

Fig. 9.18. Buckwheat: 1, upper part of a flowering plant; 2, flower; 3, fruit; 4, inflorescence with flowers and fruits; 5, membranous ocrea.

Fig. 9.19. Detail of a buckwheat plant with leaves and inflorescences including flowers and flower buds.

wheat there is a related species called 'tartary buckwheat' or 'green buckwheat' (*Fagopyrum tataricum*). It is a native of Siberia and grown in China and India under cooler conditions and on the poorest soils. It differs from normal buckwheat by having more biomass, green flowers and stems and dented seed margins.

The seed of normal buckwheat has a cereal-like composition. The composition of hulled seed is approximately 75% carbohydrates, 13% protein, 2% fat, 1.5% ash and 1.5% fibre. The protein is very valuable because of its high lysine content. Leaves and stems can be used to extract rutin, which is a drug for vascular disorders. Most of the buckwheat is used for livestock or poultry, some for green manure. About 25% of the grain production is milled into buckwheat flour. The flour is used for the preparation of pancakes, porridge, biscuits, breakfast cereals and infant food. In Japan it is used for making noodles, called 'soba'. Buckwheat is an excellent nectar source; bees like its sweet-smelling flowers and produce a dark, good-tasting honey.

Buckwheat grows best when the climate is moist and cool. It can be grown on a wide range of soil types and fertilities. Compared with cereals, it produces a better crop on poor and wet sandy soils. Buckwheat has a high tolerance to soil acidity. The growing period is short, 10–12 weeks. The plant does not withstand frost. When about 75% of the seeds are brown or almost black, and the plants have lost most of their leaves, it can be harvested. In 2005, the world average dry seed yield was about 1 t/ha. Higher yields were obtained in Croatia (3.1 t/ha) and France (3.5 t/ha). In the same year, the world total seed production was 2.5 million t. It is grown in many countries but especially in the Russian Federation and China.

Amaranth

Grain amaranths (*Amaranthus* spp.) originate from Mexico, South and Central America. For the Aztecs, amaranth was an important staple. Domestication appeared probably about 4000 BC. It is an erect, much-branched plant, up to 2.5 m tall. The leaves are simple and entire, with long petioles. The male or female flowers are in axillary clusters (Fig. 9.20). The fruit is a laterally compressed utricle. The seed is lenticular and pale, ivory, brown or black in colour. The composition of the dry seed is approximately 13–18% protein with a high lysine level, 7–8% fat and 62–65% starch, mainly consisting of amylopectin. Milled and popped seeds are used in many food products such as sweet snacks, infant food and breakfast cereals. A traditional Mexican drink called 'atole' is made from milled and roasted seed. Grain amaranth is also grown as a forage crop. Although this text is focused only on grain amaranth, one should not overlook the fact that many species of *Amaranthus* are cultivated as leafy vegetables or ornamentals throughout the world. At present there is limited cultivation of grain amaranth in Mexico, the Andes, India, Nepal, China, the USA and Argentina. However, there has been increasing interest in this crop. A number of countries have started amaranth research projects, for example China, Venezuela, the USA, Canada and Nigeria.

Fig. 9.20. Inflorescences of two types of amaranth, Chosica, Peru (photos: C. Almekinders).

Quinoa

Quinoa is also called 'Peruvian rice' or 'Inca rice'. To the ancient Incas of South America it was a sacred food. It was probably domesticated between 3000 and 1000 BC, at the high altitudes of the Andes. It is an annual herb, growing up to a height of 3 m. The lower leaves are rhomboidal, the upper ones lanceolate. Young leaves are green, but when they mature the colour turns yellow, red or purple. The inflorescence is a panicle, consisting of flower clusters. The fruit is a utricle; the seed is translucent, white, brown or black (Fig. 9.21). The dry seed has an average starch content of 48% and a protein content of about 18%. The protein is high in lysine and other essential amino acids. The seeds can be toasted or ground into flour. They can be boiled and added to soups, breakfast foods or made into pastas. The leaves are sometimes eaten as a vegetable. The crop is grown predominantly in the Andes region, yielding approximately 0.8 t of dry seed per hectare. In 2005, the total quinoa seed production was 51,152 t, which was produced entirely in Peru, Bolivia and Ecuador.

Fig. 9.21. Field with quinoa (left) and collected infructescences waiting to be threshed (right), Puno, Peru (photos: K.E. Giller).

Rice

Rice – *Oryza sativa*; Grass family – *Gramineae*

Origin, history and spread

The genus *Oryza* is among the most ancient grasses and was able to spread before the continents drifted too far apart. The result is that different *Oryza* species are spread over the tropical regions of the world, including South America and Australia. Only one species in Asia (*Oryza sativa*) and one in Africa (*Oryza glaberrima*) were domesticated.

Botanical evidence concerning the distribution of cultivated species is based chiefly on the range and habitat of wild species that are believed to have contributed to the cultivated forms. The greatest variety of such species is found in the zone of monsoonal rainfall extending from eastern India through Myanmar, Thailand, Laos, northern Vietnam, and into southern China. This diversity of species, including those considered by many to have been involved in the original domestication process, lends support to the argument for South-east Asia as the heartland of rice cultivation.

Rice has been cultivated for about 9000 years. In Indonesia, Malaysia and the Philippines, rice cultivation began some time after 1500 BC. Rice is planted throughout the tropics and subtropics, as well as in areas with a Mediterranean climate.

Botany

Rice (Fig. 9.22) is generally grown as an annual crop. However, if the environment and cultivars are favourable, ratoon growth (regrowth after the first harvest)

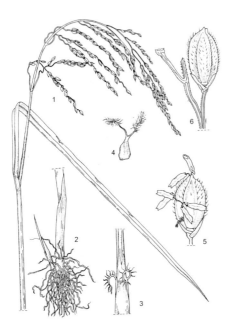

Fig. 9.22. Rice: 1, panicle with leaf; 2, plant base with roots; 3, ligule with auricles; 4, pistil; 5, flowering spikelet; 6, inflorescence branch with mature spikelet (line drawing: PROSEA volume 10).

is also made use of. Depending on the cultivar, the stems are 50–150 cm tall and up to 5 m long in deep-water rice. The shallow root system consists of fibrous roots with aerenchyma tissue. The leaf-blade is linear, 30–60 cm × 1.2– 2.5 cm, glabrous to somewhat pubescent; ligule triangular to linear-lanceolate, 1–1.5 cm long, often split; auricles often present, sickle-shaped and hairy.

The inflorescence is a terminal panicle, 9–40 cm long, with 50–500, usually about 100 spikelets. A spikelet contains a single floret with two small lower bracts (glumes) and two larger upper bracts (lemma and palea, together forming the husk of the rice grain to-be). Lemma often is with an awn. Unlike most grasses there are six stamens. The fruit (caryopsis) is varying in size, shape and colour; ovoid, ellipsoid or cylindrical, 5–7.5 mm × 2–3.5 mm, often whitish-yellow or brown to greyish-brown.

Cultivars, uses and constituents

The widespread dispersal of the Asian cultivated forms led to the formation of two major eco-geographical cultivar groups (Fig. 9.23).

1. The *Indica* cultivar group: cultivars are mostly from the humid tropics; traditional cultivars are tall, leafy, high tillering and prone to lodging; they respond poorly to fertilization, particularly to N. *Indica* cultivars are generally photo-period-sensitive and flower when day length is short (critical day length of 12.5– 14 h). The panicle is open and the grains are long.
2. The *Japonica* cultivar group: cultivars are mostly from areas with a subtropical or Mediterranean climate with a frost-free period of at least 130 days; traditional cultivars have relatively short stems with leaf-blades wide and erect, moderately tillering; they respond well to fertilization. *Japonica* cultivars are

Fig. 9.23. Rice plant habit and panicle: 1, *Indica* type; 2, *Japonica* type.

generally photoperiod-insensitive. The panicle is compact and the grains are short.

Modern improved cultivars originate from crossing between cultivars from each group. They are often cultivars with short stems (favourable yield index, resistance to lodging); erect leaves (high light interception); short growing period and photoperiod-insensitive (requirements for double cropping). In 1966, the International Rice Research Institute released its first N-responding, high-yielding variety 'IR8', which was later followed by varieties with specific resistance for pests and diseases and a higher yield potential. China successfully developed hybrid rice from 1976 onwards. 'Shanyou 63' is one of the most widely used hybrids in the rice production in China.

The African rice species *O. glaberrima* is largely being replaced by the introduced *O. sativa*. In West Africa, a hybrid between *O. sativa* and *O. glaberrima* has been developed. This new hybrid, labelled 'Nerica' which is derived from 'New rice for Africa', has been successfully introduced in West and Central African countries.

Specific terms are used in the further processing of the rough rice grains. 'Paddy' or 'rough rice' is defined as rice in the husk, whether gathered or growing. Husks are removed by pounding or milling and then winnowing. What remains is 'brown rice', even though there are varieties with red, black or white outer layers. By polishing the brown rice, the outer layers (pericarp and the aleurone layer) and the embryo are removed and what remains is 'white rice' (endosperm). 'Rice bran' is a by-product from the milling, consisting of the outer layers of the grain, the embryo and part of the endosperm. Grains may break during the process; these grains are sold as 'broken rice'. Very small broken rice is called 'brewers' rice', generally used for industrial purposes. On milling paddy gives approximately 20% husk, 50% brown rice, 16% broken rice, 14% bran and meal. 'Parboiled paddy' is rice in the husk that has been specially processed by soaking, steaming or boiling and drying. Parboiled paddy can be milled to various degrees in the same way as ordinary paddy and will then be called 'parboiled rice'.

The grain is the important economic yield component of the rice plant and its endosperm is the final product consumed. The endosperm consists mainly of starch granules embedded in a proteinaceous matrix. The endosperm may be glutinous or non-glutinous depending on the content of amylose and amylopectin. The higher the amylopectin content, the more glutinous is the product. Glutinous rice has a peculiar stickiness regardless of how it is cooked. Analyses of brown (and white) rice give the following composition 12% water, (6.7–)7.5% protein, (0.4–)1.9% fat, 77.4(–80.4)% carbohydrates, (0.3–)0.9% fibre, and (0.5–)1.2% ash. Milling and polishing result in a substantial loss of protein, fat, minerals (P and K) and vitamins (B_1, B_2), but storability increases. Parboiling results in the retention of more minerals and vitamins. Aromatic rice is mainly produced in India (Basmati types) and Thailand (Jasmine types). Scented rice or aromatic rice is popular in Asia. Because of their aroma, flavour and texture, aromatic varieties command a higher price in the rice market although traditional aromatic rice varieties are low-yielding.

Beers, wines and liquors like the Japanese 'sake' are also manufactured from rice. Rice bran contains 14–17% oil and is, besides a source of vitamin B, a valuable feed for livestock and poultry. Rice straw is used as roofing and packing material, feed, fertilizer, fuel and as raw material in the paper industry.

Ecology and agronomy

Rice is grown from 53°N in northern China to 35°S in Australia. Low temperature limits the growth of rice and the average temperature during the growing season should be between 20 and 38°C. It grows best on fertile heavy soils. Because rice is grown primarily in submerged soil, the physical properties of the soil are relatively unimportant as long as sufficient water is available. The optimum pH for flooded soils is 6.5–7. According to the growing conditions rice can be classified as follows.

1. Upland rice, grown as a rainfed crop, requires an assured rainfall of at least 750 mm in 3–4 months. Only 30% of the total area under rice is rainfed (dry rice).

2. Lowland rice is grown on flooded or submerged fields surrounded by dikes in flat lowlands, in river basins and also on terraced hillsides. In South-east Asia one irrigated rice crop requires about 1200 mm of water. When the main crop is grown in the rainy season, only supplementary irrigation is needed.

3. Deep-water rice, also known as 'floating rice', grown in areas of deep flooding, up to 5 m or more, in which the rapid growth of the internodes keeps pace with the rising water.

Propagation is by seed only, directly in the field or in the nursery followed by transplanting in flooded fields at a spacing of 15–25 cm. Prior to transplanting soils are ploughed, flooded and puddled. Puddling is needed to form an impermeable layer, at a depth of about 20 cm, to reduce water losses by percolation. Land preparation for dry seeded rice is restricted to ploughing, levelling and harrowing (Figs 9.24–9.26).

Weeds are one of the major causes of low yields. They compete with rice for nutrients and light and serve as alternative hosts for pests and diseases. In lowland rice the thorough land preparation and flooding initially reduces the weed population, but later on several rounds of hand weeding are needed. Specific herbicides are effective against broad-leaved weeds, sedges and grasses, which can be used pre-plant, pre-emergence and post-emergence.

In the continuous rice-cropping system pests and diseases are major problems. In the past brown and green plant hopper attacks have destroyed entire crops. The advent of resistant varieties of rice has largely solved this problem, but chemical control for other pests is still needed.

N is the first-limiting nutrient under most conditions. An average yield of 4 t/ha requires 80–90 kg N per crop. Usually about half of this amount can be provided by soil organic matter and biologically fixed N_2. Some blue-green algae, living in submerged fields, are able to fix N_2 from the air. The same counts for the water fern *Azolla*, common in Eastern Asia, when interacting with a specific

Fig. 9.24. Levelling of a flooded field to prepare it for planting rice, Indonesia.

blue-green alga. The remaining half of the N requirement has to be applied preferably in a basal dressing at planting and a top dressing just before flowering. Application of soluble phosphate fertilizers may also be needed.

The degree of mechanization varies. In lowland rice animal traction and small tractors are commonly used for land preparation. Harvesting is usually manual; in some countries it is fully mechanized. Prior to harvest a dry period

Fig. 9.25. Transplanting rice plants in a flooded field, Bali, Indonesia (photo: T. Huiberts).

Fig. 9.26. Terraces with flooded rice fields, Indonesia (photo: E. Westphal).

is required while submerged fields should be drained. Depending on variety and environmental conditions, the time from seeding or transplanting to harvest varies from 3.5 to 7 months. Because of a higher solar radiation, irrigated dry-season crops give the highest yields.

From the beginning of the 1960s to the middle of the 1990s, the world mean yield per hectare increased from 2000 to 4000 kg and the total world production from 240 million t to 600 million t. This spectacular increase has mainly resulted from the introduction of high-yielding varieties (and the accompanying improved technology) in South-east Asia and hybrids in China. This phenomenon has rightly been named a 'green revolution'.

China, India and Indonesia are the largest producers. In 2005 world rice production amounted to 615 million t, Asia accounting for almost 90% of it. Most of the world's rice production is consumed in the areas where it is grown, only about 5% enters into international trade.

Rye and Triticale

Rye – *Secale cereale*, Triticale – × *Triticosecale* Wittmack; Grass family – *Gramineae*

Rye

Origin, history and spread

Most authors believe that *Secale cereale* originated from *Secale montanum*, a wild rye that grows in southern Europe and Morocco, although *Secale anatolicum*, found in Syria, Turkey, Iran and the Kirghis steppe, has also been proposed as the ancestor. Rye was and still is found as a weed in wheat and barley fields in Central Asia. Rye probably developed as a secondary crop, first occurring as a weed and at a certain time being accepted as a crop in its own right. Because of its tough constitution, rye may have performed better than other grains under pressure of hard winters or on poor soils and therefore attracted human attention. Thus rye was domesticated later than wheat and other grains. The earliest finds of cultivated rye in mid-Europe came from Hallstatt (1000–500 BC). The English and the Dutch introduced rye into the New World.

At present rye is grown primarily as a spring or winter cereal in northern and continental Europe, and to a lesser degree in North America.

Botany

The culm is tall (1.5 m), erect, slender, glabrous except for pubescence near the spike, and has a hollow structure. The colour is bluish-green with a coat of waxy bloom. The number of nodes is six or seven for spring rye and ten to 12 for winter rye; one leaf is formed at each node. The leaf-sheath is long and loose; ligule short; auricles small and short; blade linear-lanceolate; colour is bluish-green. The coleoptile of the germinating plant may be violet. The root system is fibrous and penetrates the soil rather deep, up to 2 m. The inflorescence of rye is a spike, which may contain up to 50 spikelets. The spikelets usually have two florets; the lemmas are narrow, with stiff hairs on the keel and long-awned (Figs 9.27 and 9.28). The stamens are three in number; the pistil is pubescent with two feathery stigmas. Rye flowers are cross-pollinating. The fruit is a caryopsis, in which the seedcoat is fused with the ovary wall. The caryopsis is oblongoid, 5–9 mm long and light brown or greenish-brown. The 1000-seed weight ranges from 30 to 40 g.

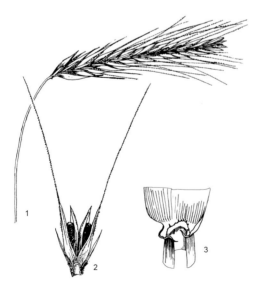

Fig. 9.27. Rye: 1, inflorescence (spike); 2, spikelet with two fertile florets; 3, base of a leaf-blade with ligule and auricles.

Fig. 9.28. Detail of a field with fruiting rye.

Cultivars, uses and constituents

There are many landraces, usually with longer culms and smaller grains, and because of the long culms they often lodge. Today there is a great variety of rye cultivars, which are bred for particular geographical areas. Well-known cultivars include 'Petkus', 'Pearl', 'King II' and 'Hancock'. New developed hybrids ('Tetra-rye') yield on average up to 20% more (Fig. 9.29).

Fig. 9.29. Inflorescences and 'seeds': 1, diploid rye; 2, tetraploid rye; 3, triticale.

Rye is often used as feed for livestock, for hay and pasturage; immature rye is sometimes harvested as a whole crop (whole-crop silage) or used for green manure. When rye is grown over the winter it will prevent soil erosion (cover crop) and the leaching of nutrients from the soil (catch crop). The deeply penetrating and extensive root system makes rye useful for stabilization of sandy soils.

The grain contains 8–10% protein (containing lysine and threonine), 70% starch and 2% fat and, as well as a feed for livestock, is used for making blackbread, cake and in Scandinavia the well-known crispbread. Because rye does not have gluten like wheat, bread made from rye has a compact structure. However, certain substances in rye flour, pentosans, give its bread consistency and elasticity.

Rye is also used as a grain for distilling, whisky in Canada and the USA and 'geneva' in The Netherlands. In Russia, rye is used for brewing beer.

The tough straw of rye cannot be used as fodder but is valued for many other purposes; for example, thatching for roofs, fuel, making paper, bedding, packing material, weaving mats, mushroom compost, and making straw hats.

Ergot (*Claviceps purpurea*) is the most conspicuous fungal disease of rye and poisonous to man and livestock. Eating ergot may cause gangrene, abortion and hallucinations. Ergot is characterized by large purplish-black sclerotinia that replace the caryopsis in the spikelet.

Ecology and agronomy

Rye can be grown in a wider range of environmental conditions than any other small grain. Winter rye is the most winter-hardy of all cereals (tolerating temperatures as low as −30°C). It is the most productive cereal on infertile, sandy or acid soils, as well as on poorly prepared land. However, for the best results rye should be grown on well-prepared, fertile, well-drained soils with a pH of 5.6 or higher. Rye grows better on light loams, sandy and peaty soils than on heavy clay soils. It is also able to germinate in relatively dry soils and at low temperatures (2–3°C). Rye is fairly tolerant to drought.

Sowing the seed can best be done by drilling it mechanically at a depth of 2–4 cm in rows 10–25 cm apart. Depending on whether winter or spring rye is grown, the seed rate ranges from 100 to 200 kg/ha to obtain a density of 200–300 plants/m^2 (Fig. 9.28). The application of fertilizer depends on the expected yield; each tonne of dry matter yielded removes 20 kg N, 4 kg P and 13 kg K from the soil. Winter rye is rather effective in controlling weeds. Midsummer is harvest time, when the leaves are yellowish-brown and the spikes are bowing. For combine-harvesting, the moisture content of the kernels has to be below 16%. Kernel yields vary widely, from less than 1 t/ha in South Africa up to 8 t/ha in Western Europe. In 2004, the total world production of rye grain was about 18 million t. Poland, Germany, the Russian Federation, Ukraine and Belarus were the main producers.

Triticale

Triticale is a cereal derived from a genetic cross (not a hybrid) of rye and wheat; it shows characteristics in between wheat and rye (Fig. 9.29). Crossing was aimed at combining the hardiness and high lysine content of rye and the high yield and protein content of wheat. The first cultivar was released in 1970 in Canada. Triticale is one of the few examples of successful artificial polyploidization. Triticale has not become an important crop for food, but it is becoming increasingly important as a forage crop. In 2004, the world total production of triticale seed was about 14 million t. Poland, France, the Russian Federation and Germany were the main producers.

Sorghum

Sorghum – *Sorghum bicolor*; Grass family – *Gramineae*

Origin, history and spread

The centre of origin of sorghum lies in Africa. Domestication took place in Sudan, Chad and Ethiopia between 5000 and 7000 years ago. It probably spread with trade through the Middle East to India at least 3000 years ago. When sorghum reached China is not known, but as there was movement of rice between India and China by 200 BC, it is likely that grain sorghums moved at the same time. Grain sorghum was first brought to the USA with the slave trade from West Africa. During the late 19th and early 20th centuries it was reintroduced from North and South Africa and India. At present sorghums are found throughout the drier areas in the tropics, the subtropics and warm temperate areas.

Botany

Depending on variety and environmental conditions, the morphological characteristics may vary considerably. Generally, sorghum (Fig. 9.30) is an annual

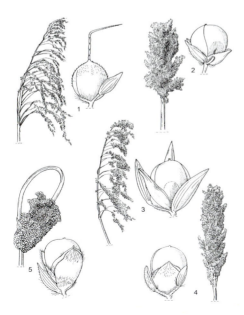

Fig. 9.30. Habit of infructescences with spikelets of the five basic sorghum cultivar groups: 1, Bicolor; 2, Caudatum; 3, Guinea; 4, Kafir; 5, Durra (line drawing: PROSEA volume 10).

grass 0.5–5.0 m tall, with usually erect, solid, insipid or sweet stems. Leaf-blades are up to 30–60 cm long, 4–8 cm wide, the prominent midrib may be white or yellowish-white; ligule 1.5–5.5 mm long with rounded apex, initially transparent later becoming dry and torn. Sheaths and stem are glabrous, with a waxy bloom. In cultivation sorghum is often grown as a single-stem type. Some varieties tiller early, while others do not tiller until after flowering. Sorghum has an extensive and deep root system.

The inflorescence is a panicle; the central rachis may be long or short, with primary, secondary and sometimes tertiary branches bearing racemes of spikelets. The spikelets are paired, one sessile and hermaphrodite, the other pedicelled and male or sterile, the terminal spikelets of a raceme borne in threes (one sessile and two pedicelled).

There are numerous varieties and many classification systems have been used. At present the International Board of Plant Genetic Resources recognizes five main varieties and ten hybrid combinations on the basis of sessile spikelet, panicle and grain characteristics (Fig. 9.31).

1. Bicolor sorghums have loose panicles and small grains which are completely covered by large, closed glumes. These are grown throughout Africa but are widespread in Asia.
2. Guinea sorghums have loose panicles and the spikelets have generally open glumes which enclose an elliptical grain. These are grown in West Africa.
3. Caudatum sorghums have panicles variable in shape. The grain is asymmetrical, flat on one side and convex on the other. The glumes are usually much shorter than the grain. These are widely grown in Chad, Sudan, north-eastern Nigeria and Uganda.
4. Kafir sorghums are small and have often cylindrical and relatively compact panicles. The grains are elliptical and tightly enclosed by the usually much shorter glumes. They are cultivated mostly in southern Africa.

Fig. 9.31. Sorghum fruit types after threshing: without glumes (left) and covered with glumes (right).

5. Durra sorghums have very compact panicles with an erect or sometimes recurved peduncle, a so-called 'goose neck'. The grains are globular and tightly enclosed by small glumes. The lower glume is often wrinkled. These are grown mainly in East Africa, the Middle East and India.

The most commonly grown intermediate varieties have combinations of characteristics derived from the main varieties.

Uses and constituents

Sorghum is the fifth most important cereal in the world and an important staple food in the semi-arid tropical areas in Africa and Asia (Figs 9.32 and 9.33). The simplest use of sorghum is as boiled whole grain. Small, corneous grains may be pearled by removing the pericarp, and then boiled to give a product resembling rice. Normally the grain is ground into flour. Flour can be made into a thin or thick porridge and unleavened bread. Sorghum is widely used for brewing beer, particularly in Africa. The sorghum wort may be drunk before fermentation or may be fermented for 4–5 days when it contains 2–10% alcohol. Sorghum beer is a valuable dietary supplement because of its high vitamin B content. An alcoholic sorghum wine is made in China. The average composition of air-dried sorghum grain is 8–16% water, 8–15% protein, 2–6% fat, 70–80% carbohydrates, 1–3% fibre and 1–2% ash. The protein contains no gluten and the flour will not make good leavened bread unless mixed with other cereals.

Fig. 9.32. Field with flowering sorghum of cultivar group Bicolor, Ethiopia (photo: C. Almekinders).

Fig. 9.33. Collected sorghum infructescences waiting to be threshed, Ethiopia (photo: J. Ferwerda).

Sorghum grain is a significant component in animal feeds in the Americas, Australia and China. The grains have to be processed before being fed to cattle, else a large proportion of the grains will be swallowed whole and the waxy bran covering the grain will make digestion difficult. Grinding is the simplest, least expensive method of preparing sorghum grain for cattle. Sorghum is also grown for forage, either for direct feeding for ruminants or for preservation as hay or silage. Sweet-stemmed varieties are used for animal feed and for extraction of sugar and syrup. Dried sorghum stems are used for thatching and for fences. Harvest residues are used for animal feed and fuel. Cultivars known as 'broomcorn' have inflorescences with a much shortened rachis but with long, up to 90 cm, fibrous branches and are used for making brooms. The 1000-seed weight is 13–14 g.

Ecology and agronomy

Sorghum is grown at altitudes from sea level up to 1000 m and between latitudes of 40°N and 40°S. However, it is primarily a plant of hot, semi-arid tropical environments with an annual rainfall of 400–600 mm. The great advantage of sorghum is its drought resistance, and it can therefore be grown in areas which are too dry for maize. The waxy layer on the sheaths and stem contributes to reducing evaporation. After a period of drought it can resume growth when stress is relieved. Sorghum is adapted to a wide range of soils, temperatures and soil moisture conditions. It tolerates a soil pH from 5 to 8.5 and will survive temporary waterlogging; it does not grow well in shade. Sorghum is a short-day plant, although temperatures below 20°C can delay head differentiation.

Sorghum is normally propagated by seed, which is usually sown directly into the plough furrow. Sometimes small hand-drills are used. Sophisticated grain and fertilizer drills are used in advanced agriculture. The seed rate depends on seed size, soil type and availability of soil moisture, and may vary from 3 kg/ha in very dry areas to 10–15 kg/ha under irrigation. Usual sowing depth on heavy soil is 2–3 cm and 3–5 cm on sandy soil. Plant spacing under favourable conditions is 45–60 cm between rows and 12–20 cm within rows, resulting in plant densities of about 120,000 plants/ha. Forage sorghums are planted at closer spacing.

Where grown as a subsistence crop, it is usually weeded by hand. The period from sowing to harvest ranges from 3 to 6.5 months depending on cultivar and growing conditions; the amount of rainfall and its distribution in particular.

Too frequent growing of sorghum on the same land, in Africa in particular, but also in India and Myanmar, often leads to a build-up of the parasitic weed *Striga*, which can cause land to be abandoned.

In India and Africa under rainfed conditions, the annual grain yield of sorghum ranges from 0.3 to 2.0 t/ha; in the USA and Australia, for hybrid types, under irrigation the annual grain yield ranges from 4.5 to 6.5 t/ha. The world production of grain sorghum was 60.2 million t in 2004. The major producing countries are the USA (20%), Nigeria (13%), India (11%), Mexico (10%), Sudan (9%) and China (5%).

Wheat

Wheat – *Triticum* spp.; Grass family – *Gramineae*

Origin, history and spread

The evolution of modern wheat species is not yet completely resolved. However, based on the number of chromosomes, three major groups of *Triticum* species are distinguished.

1. Diploids ($n=14$):
   Triticum monococcum – einkorn
This species arose through mutation from the wild species *Triticum boeoticum*.
2. Tetraploids ($n=28$):
   Triticum dicoccum – emmer
   Triticum turgidum – rivet or English wheat
   Triticum durum – durum or macaroni wheat
   Triticum polonicum – Polish wheat
The tetraploid group is probably derived from the hybridization of *T. monococcum* and another diploid wild species, but it could have resulted from hybridization with *Aegilops speltoides*. One of the wheats belonging to this group is *T. dicoccum*. Findings of this crop in archaeological sites of the Near East were dated to be 9500 years old. Another tetraploid species which evolved within this group, *T. durum*, was dated to be about 6000 years old.

Fig. 9.34. Inflorescences of *Aegilops squarrosa* and nine *Triticum* species: 1, *A. squarrosa*; 2, *T. boeoticum*; 3, *T. monococcum*; 4, *T. dicoccum*; 5, *T. durum*; 6, *T. turgidum*; 7, *T. polonicum*; 8, *T. aestivum*; 9, *T. spelta*; 10, *T. compactum*.

3. Hexaploids (*n*=42):

Triticum aestivum – common or bread wheat

Triticum spelta – spelt or dinkel

Triticum compactum – club wheat

The hexaploid group, including *T. aestivum*, is derived from hybridization between tetraploid wheats and the diploid wild grass *Aegilops squarrosa*. This grass is commonly found in the wheat fields of Asia Minor and could easily have crossed with cultivated tetraploids (Fig. 9.34).

The probable origin of domesticated wheats is summarized in Fig. 9.35. Domestication of wheat began at least 9500 years ago in the Fertile Crescent region of the Middle East, which means that wheat was one of the first cultivated plants. Wheat was cultivated around the Mediterranean and throughout Europe in prehistoric times. It reached China around 3000 BC. In 1529 the Spanish introduced wheat into the New World. Wheat is now grown successfully throughout the world.

Botany

Wheat species are annual herbs with cylindrical stems or culms with four to seven solid nodes and mostly hollow internodes (the internodes in the upper part of the stem of English wheat are often solid, i.e. filled with pith). Plants are 40–150 cm high, often strongly tillering. Leaves are situated in

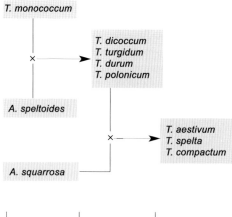

T. monococcum

× ⟶ T. dicoccum
T. turgidum
T. durum
T. polonicum

A. speltoides

× ⟶ T. aestivum
T. spelta
T. compactum

A. squarrosa

diploids (n = 14)	tetraploids (n = 28)	hexaploids (n = 42)

Fig. 9.35. Summary of probable origin of domesticated wheats (*A. = Aegilops, T. = Triticum*).

two vertical rows at the nodes. At the top of the leaf-sheath there are two small auricles sparsely fringed with hairs and a short membranous ligule. The leaf-blade is linear, flat, parallel-veined, and 15–40 cm × 1–3 cm. The uppermost leaf is called the 'flag leaf', which plays an important part in grain production.

The inflorescence of wheat is a distichous spike (Figs 9.36 and 9.37), which has a zigzag jointed central axis, the rachis, which can be brittle (einkorn, emmer and spelt) or tough (remaining cultivated wheats). The rachis has internodes with spikelets attached at each node. The internodes can be long, which results

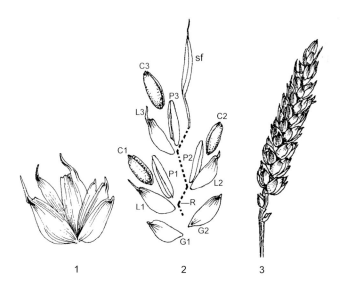

Fig. 9.36. Wheat (*Triticum aestivum*): 1, spikelet with three fertile florets; 2, parts of spikelet schematically: G1–G2, glumes; L1–L3, lemmas; P1–P3, paleas; C1–C3, caryopses; R, rachilla; sf, sterile floret; 3, inflorescence (spike).

Fig. 9.37. Field of maturing wheat of a bearded cultivar at 2300 m altitude, Guatemala (photo: D.L. Schuiling).

in a lax spike (spelt), or shorter, resulting in a dense spike (common wheat). When the internodes are extremely short, the spike will be very compact (club wheat).

A spikelet has a small axis called the 'rachilla'. Situated on opposite sides at the base of the rachilla there are two bracts, the glumes. Above the glumes, three to nine flowers or florets are arranged alternately on opposite sites of the rachilla. Each floret consists of two bracts called the 'lemma' and the 'palea', the lemma is attached slightly below the palea and can be awned (durum wheat) or awnless (common wheat); pistil with two plumose stigmas; three stamens. The fruit is an ellipsoid caryopsis with a tuft of hair at the apex. The endosperm forms the major part of the fruit. The characteristic fruit of the grass family is called a 'caryopsis', in which the seedcoat is fused with the ovary wall. The 1000-kernel weight is 30–50 g.

Einkorn develops only one caryopsis per spikelet, emmer two, and the other wheats mostly three or more. The glumes of einkorn, emmer and spelt fit tightly around the caryopsis, making it difficult to remove them through threshing. These wheats are therefore called 'hulled' wheats; the remaining cultivated species produce 'naked' grains on threshing.

Wheats form seminal roots and many adventitious roots.

Cultivars, uses and constituents

There are many different cultivars of wheat, which vary especially in the length of the culms. Compared with old cultivars, newer cultivars have much shorter culms. Breeders have succeeded in developing semi-dwarf and dwarf types, which are only about 30 cm tall. Short culms are important to avoid lodging of the wheat plants by wind and rain. Severe losses may result from lodging, because when the spikes touch the soil, the caryopses in the spikes may germinate and become worthless.

Wheat is the world's most widely consumed source of food. It is processed into a great number of products such as bread, pasta, crackers, biscuits, breakfast cereals, noodles, cakes, puddings, alcoholic beverages, and a great number of minor products. The various forms of wheat represent about 30% of the world's grain production. About 40% of the world population uses wheat as a staple food. By far the most important wheat is *T. aestivum* or common wheat. This includes many varieties, which show great diversity in agroecological adaptation and utilization. Groups are characterized on the basis of milling properties (hard or soft), dough rheology (strong or weak), bran colour (red or white) and vernalization requirement (spring or winter). Almost 90% of the world's wheat production is common wheat; the remaining 10% is durum wheat. Durum wheat is mainly used for pasta production. All remaining wheat species are minor crops.

Common wheat differs from other cereals because its grain contains proteins such as gliadin and glutenin with unique chemical and physical properties. When wheat flour is mixed with water the proteins bind in such a way that a coherent mass called 'gluten' is formed. Gluten creates strength and elasticity, which means that it is capable of deformation. Dough made from wheat flour can therefore expand to accommodate the gas which is produced by the added yeast. This makes it possible for the dough to swell up or to rise and to bake leavened bread with it. Loaf volume is an important quality criterion.

In a number of Asian countries wheat is utilized to produce rolls, which are steamed. This process results in a product with a thin white skin and a dense crumb. In India, Pakistan and Arabian countries wheat is consumed in the form of unleavened bread, as 'chapatis'. This bread is made by placing thin sheets of dough on a hot plate. In this process no yeast is added to the dough.

Wheat produced in warm, dry, continental climates, so-called 'hard wheat', is highly suitable for bread-making. In cool temperate regions yields are mostly considerably higher but the grain is softer and the protein content is smaller, making it less suitable for bread-making. The constituents of the grain are 11–15% water, 8–16% protein, 1–1.2% fat, 74–76% carbohydrates, 0.4% fibre and 0.4% ash.

Wheat is also an important feed grain for livestock worldwide. By-products of flour milling, particularly the bran, are used as feed. Wheat germ is often used as a human food supplement because of its relatively high content of protein and vitamins B and E. The straw of wheat is used as a feed for ruminants or as bedding material, laid on soil to protect the soil against erosion, and used for thatching, manufacture of cardboard and various other uses. Wheat has also been used in folk medicine and it is reported to be, for example, demulcent, intoxicant, laxative and sedative.

Ecology and agronomy

Wheat, although essentially a temperate and subtropical climate crop, tolerates a very wide range of growing conditions, from sea level up to altitudes of more than 4500 m and from the equator up to the Arctic Circle (Fig. 9.37). It grows best when temperatures are 15–25°C. Temperatures above 35°C will reduce photosynthesis and growth. Spring and winter cultivars occur. Winter cultivars need vernalization for generative development. In tropical countries it is possible to grow wheat at higher altitudes. For a good production a temperate climate is needed for at least 90 days; 1475–1600 growing degree-days are required for a complete crop cycle. Wheats differ in their water requirements depending on species, cultivar, climate and soil conditions, although high production is achieved at 400–600 mm. The ideal soil for growing wheat has good humus content (0.5% or more), the soil particles must be aggregated into small porous grains, it must be well-aerated and well-drained, and pH should range between 5.5 and 7.5. Wheat is sensitive to soil salinity.

   Wheat cultivars have different requirements in day length (photoperiod) to bring about optimum flowering and reproduction. Three groups can be distinguished.

1. Long-day cultivars: optimum flowering at day lengths longer than 14 h.
2. Short-day cultivars: optimum flowering at day lengths shorter than 14 h.
3. Day-neutral cultivars: flowering will happen over a range of day lengths.

Wheat is propagated by seed. Seeding rates vary depending upon local conditions and methods of production, but are commonly 100–150 kg/ha. Sowing depth varies from 2 to 12 cm and depends on the depth of the moist soil layer. Good soil moisture content is necessary for good germination and establishment of the juvenile plant. However, a sowing depth of 12 cm or more is undesirable because after germination the seedling has to develop a long, elongated internode before the growing point reaches the surface. This process consumes a large part of the food reserves of the endosperm at the cost of the development of roots and leaves. Sowing is mostly done by machine drilling in rows, spaced at 10–35 cm. After germination the following six stages of growth can be recognized.

1. Emergence: the plant appears above the surface 7–14 days after the seed has been sown.
2. Tillering: when the first few leaves have appeared, branches known as 'tillers' grow from the buds just below or at the soil surface.
3. Stem extension: in this stage the internodes and the leaf-sheaths elongate rapidly (shooting).
4. Heading: the spike or head is distending the sheath of the flag leaf.
5. Anthesis: a few days after the head emerges from the sheath, flowering – the release of pollen – begins.
6. Ripening: six stages of increasing hardness of the caryopsis can be distinguished – watery ripe, milky ripe, mealy ripe, waxy ripe, fully ripe, and dead ripe.

Supply of fertilizers depends on soil fertility. The average removal per tonne of seed per hectare is 40–43 kg N, 5–8 kg P, 25–35 kg K, 2–4 kg S, 3–4 kg Ca, 3–4 kg Mg, and a small amount of micronutrients. N is often applied as 'split-dressing', which means that the total amount is split into portions and applied to the crop at several stages of development. Early N supply increases tiller number and spikelet number. Application at the onset of the heading stage increases grain size and N content of the grain.

Weeds may cause yield losses by competition in the first 4 or 5 weeks of the development of the crop, so weed control should be done in this period. Weeds can be controlled by hand-weeding, proper crop rotation, machine cultivation, or application of chemical herbicides.

The crop can be harvested when the grain is fully ripe. The right moment is important because harvesting too early means high moisture content and harvesting too late (in the dead ripe stage) can cause seed losses. The ideal moisture content at harvesting is about 16%. Harvesting is mostly done by combine harvester, although in Asia and Africa sickles are still in use.

The world average seed yield of wheat is 2.5 t/ha, varying from 0.3–2.5 t/ha in South-east Asia up to 12 t/ha in North-west Europe. In 2004, the world production of wheat was about 630 million t. The major wheat-producing countries, in successive order, were China, India, the USA, the Russian Federation and France.

ROOT AND TUBER CROPS

Cassava

Cassava – *Manihot esculenta*; Spurge family – *Euphorbiaceae*

Origin, history and spread

Cassava is a domesticated species of unknown origin, with no wild forms of this species being known. It is thought to have first been cultivated in South America in the area reaching from Paraguay to north-eastern Brazil. The area, including western and southern Mexico and parts of Guatemala, is considered to be also a centre of domestication. In Mexico, remains of cassava leaves have been found that are 2500 years old, and cassava starch has been identified in archaeological finds that are 2100–2800 years old. Some scientists suggest that bitter cassava may have been domesticated in the northern part of South America and that sweet cassava was independently domesticated in Central America. After its domestication, cassava spread throughout the tropical areas of the Americas. By the time the Europeans arrived, cassava was already grown across the current area of cultivation.

The Portuguese introduced and spread cassava in Africa through trade with the isles of São Tomé and Fernando Po (Bioko). Around 1660 cassava was already an important crop in Angola. Its spread to other parts of Africa increased during the 20th century. Although cassava has now spread to all tropical regions of the world, Africa today grows more cassava than all the rest of the world.

Botany

Cassava is a perennial woody shrub up to 4 m tall (Fig. 9.38). The shoots show strong apical dominance, which suppresses the development of secondary shoots. When the main shoot becomes reproductive, the apical dominance is broken and two to four of the axillary buds immediately below the apex begin to develop, resulting in the typical branching habit of the cassava plant. However, some clones never branch. The colour of the woody stems varies with cultivar but is usually greyish or brownish. The leaf of cassava is deciduous and exists for only a few months. The prominent leaf scars indicate the positions that were once occupied by leaves.

The petiolate leaves are spirally arranged, petiole 5–30 cm long and in general longer than the lamina. The lamina is simple but deeply palmate with usually five to seven lobes, each lobe 4–20 cm × 1–6 cm and widest at about one-third in from the tip. Like most other members of the spurge family, all parts of the plant contain laticifers and produce white latex.

Usually five to ten storage roots per plant develop as swellings on adventitious roots a short distance from the stem by a process of secondary thickening. The increase in girth occurs after the relatively thin and fibrous root has penetrated the soil. However, most of the fibrous roots remain thin and continue to function in nutrient absorption. Habit and colour of the storage roots vary greatly and may be rough or smooth, white, light to dark brown, or reddish. The storage roots are usually tapering but also long and slender, up to 100 cm × 15 cm (Fig. 9.39).

Fig. 9.38. Habit of vegetatively propagated cassava plants, cultivated on ridges.

Fig. 9.39. Base of cassava plant, cultivar 'Aspro', with 8-month-old tubers, freshly uprooted (photo: D.L. Schuiling).

The peel of the storage root is composed of the cortex with the outer periderm layer adhering to it. The central portion of the storage root consists mostly of parenchymatous cells containing large amounts of stored starch. The pith may be white, yellowish or reddish in colour. Due to the absence of adventitious buds on the storage roots, unlike sweet potato, it is impossible to propagate cassava by storage roots.

Flowering in cassava depends on cultivar and varies from frequent to rare or non-existent. The inflorescence is a terminal raceme, 3–10 cm long, with female flowers near the tip and male flowers closer to the base. In each inflorescence, the full bloom of female flowers is 1 week prior to that of male flowers. Cross-pollination, mainly by insects, is therefore the rule. The fruit matures 3–5 months after fertilization. The fruit is a three-seeded, almost round capsule, 1–1.5 cm in diameter, with six narrow longitudinal wings (Fig. 9.40). Seeds are ellipsoid, about 12 mm long, light grey, brownish or dark grey, with darker blotches. A pronounced caruncle is located near the hilum of the seed.

Cultivars, uses and constituents

Several attempts have been made to classify the enormous number of cultivars. Due to the continuous variation in every plant characteristic, a definitive classification has proved to be elusive.

The usual distinction into 'sweet' varieties with low glucoside content and 'bitter' varieties with high glucoside content in the tubers is of only historical and local practical value, since the glucoside content changes within a variety

Fig. 9.40. Inflorescence of cassava with flower buds and open male flowers (left); part of infructescence of cassava with close-up of full-grown immature fruits (right) (photo: D.L. Schuiling).

depending on soil and climate. Several breeding programmes have resulted in the release of improved cassava varieties which are disease- and pest-resistant, low in cyanide content, drought-resistant, early-maturing and high-yielding. Varieties improved by the International Institute of Tropical Agriculture (IITA), Ibadan, Nigeria, are now used in most cassava-growing countries in Africa. All cassava in Thailand comes from varieties introduced by the International Center for Tropical Agriculture (CIAT).

The major processed forms of the cassava tuber are meal, flour, chips, pellets and starch. About 65% of the world production is used for human consumption. There are innumerable methods to prepare a tasty dish of cassava flour and meal. Cassava starch is used in industry for the production of glucose, confectionery, crackers, cookies and snacks. The chips and pelleted forms are used mainly for animal feed. Cassava starch is also used as a binding agent, in the production of paper and textiles, and to make monosodium glutamate, an important flavouring agent in Asian cooking. Cassava is also used for the production of bio-ethanol.

Before the cassava root is used it is almost invariably peeled. The peel comprises 10–20% of the storage root. The average composition per 100 g of the remaining edible portion is water 62 g, protein 1.0 g, fat 0.3 g, carbohydrates 35 g and minerals 1.0 g.

All parts of the cassava plant contain the cyanogenic glucoside linamarin. This glucoside is highly soluble in water and tends to decompose when heated. Under the influence of the enzyme linamarase, which is also present in the cassava plant, the linamarin is hydrolysed to produce the highly poisonous prussic acid (HCN). Only when the storage cells are crushed is the glucoside released and makes contact with the enzyme linamarase. This is the key to the methods of getting rid of HCN. The volatile HCN should be allowed to escape. Boiling is not always a guarantee that the product is safe, as the HCN can be trapped in the starchy paste. Grating and slowly drying the resulting product is a widely used and effective method. Although traditional consumers are aware of the risks, accidents still occur, especially with children. Consumption of more than 1 mg HCN per kilogram of body weight per day is poisonous for humans. The glucoside content (as HCN) in the central part of fresh storage roots varies from 10–490 mg/kg. The toxicity is considerably influenced by soil and climatic conditions and moving cultivars from one country to another can affect the toxicity. In Africa and South-east Asia the leaves and tender shoots are sometimes boiled and eaten as a vegetable, which provides protein (up to 7%) and vitamins A and B.

Ecology and agronomy

Cassava is a crop of the tropical lowlands. Worldwide, it is distributed in regions between 30°N and 30°S. Near the equator cassava can be grown up to 1500 m altitude. Cassava is unable to survive periods of frost. Optimum temperature ranges from 20 to 30°C. Cassava can be grown in areas with 500–6000 mm of annual rainfall with an optimum range of 1000–1500 mm. Except at planting, it can withstand prolonged periods of drought. Almost all soil types can be used as long as they are permeable, not too shallow and not stony. Growth performance is best on sandy soils with reasonable soil fertility. Cassava tolerates a soil pH ranging from 4 to 8. It performs best in full sun. Soil preparation varies from deep thorough cultivation, which favours tuber development, to practically zero under shifting cultivation.

Cassava is propagated exclusively from stem cuttings, 20–30 cm long, well-lignified and preferably taken from the middle of the stems of plants 8–14 months old. The time between cutting stems and planting should not exceed a couple of days. Whole stems can be stored in shady places for 3 months. To overcome cassava's low multiplication rate, the IITA has developed a technique to make a two-node cutting that can make 50 plants from each parent cassava instead of ten as before. These mini-stakes are easily moved and protected in plastic sacks until they can be grown on and hardened off in individual plastic bags or nursery beds before being planted in the field. Propagation from seeds is confined to breeding work.

For large-scale production cassava may be planted in pure stands at densities of 10,000–15,000 plants/ha. Cassava for home production is often planted in mixtures with other crops such as maize, groundnuts, grain legumes, bananas or coconuts. Cassava is seldom manured although it responds well to farmyard manure. Regular weeding is necessary until 2–3 months after planting. In traditional fallow systems, cassava is one of the last crops in the cultivation cycle because it can survive (grow) longer than other crops against weed and shrub competition on declining soil fertility.

Because cassava is a perennial crop there is no distinct period of harvesting. The time of tuber harvest depends mainly on cultivar and use. For short-season cultivars and if the tubers are used for human consumption, harvest is usually 6–12 months after planting. In times of food shortage it is harvested even earlier. Long-season cultivars and tubers to be used for starch production may be harvested after 18–24 months or even later.

When cassava leaves are harvested for human or animal consumption it will reduce the root yield. Leaves can be harvested 50–70 days after planting;

yields of 20 t of fresh leaves per hectare per year have been reported. In 2005, the world average annual yield of fresh roots was almost 11 t/ha. Under optimal conditions an annual yield of 35 t of fresh roots per hectare can be achieved. CIAT now has varieties that produce up to 80 t of fresh roots per hectare. In practice the yields on the farm are far below the potential yields, because cassava is often grown on poorly fertile soils with low inputs. In 2005, the world total production of fresh roots was about 203 million t. Nigeria, Brazil, Thailand and Indonesia are the main producers of cassava. Thailand is the most important exporting country of both dried cassava, mainly for animal feed in European countries, and cassava starch.

Potato

Potato – *Solanum tuberosum*; Nightshade family – *Solanaceae*

Origin, history and spread

The cultivated potato originated at high altitudes in the South American Andes. The original area of domestication was probably in the high plateau of Bolivia–Peru. The oldest evidence of potato use dates back to 11,000 BC.

The tuber-bearing species are only a small part of the very large genus *Solanum*. The tetraploid potato has a confusing ancestry, based to a large extent on the high degree of interspecific hybridization occurring in the genus. Its basic chromosome number is $n=12$, although ploidy levels range from diploid to hexaploid. Tetraploid *Solanum tuberosum* ssp. *tuberosum* is by far the most grown taxon. The related tetraploid *S. tuberosum* ssp. *andigenum* was first brought into Europe. In 1570–1572 still unknown missionaries brought the potato from Peru/Colombia to Seville (Spain). There, the potato was first grown in the garden of the monastery 'Los Remedios' and the first well-documented use at the 'Hospital de la Sangre' in Seville dates back to 1573. In the 17th century it was exceptionally eaten and for years it was a rare food which was served only at monasteries, hospitals and palaces. From the 18th century onwards the use and spread of the potato increased and it became an important crop until the fungus disease late blight (caused by *Phytophtora infestans*) almost eliminated it in the 1840s. The famine of 1845–1847 in Ireland, which was caused by this plant pathogenic organism, was responsible for over a million deaths from starvation and initiated large-scale emigration from Ireland to the USA. A clone of resistant *S. tuberosum* ssp. *tuberosum* was then introduced in the mid-1800s from Chile and filled the void left by subspecies *andigena*. This one clone played an important role in the subsequent development of most European and North American potatoes. Later on, supplementary introductions from South America were made to both these regions.

Currently the crop is grown on a significant scale in about 130 countries, and covers around 18 million ha worldwide. About 80% of the world tuber production is located in the temperate regions in Asia, Europe and North America.

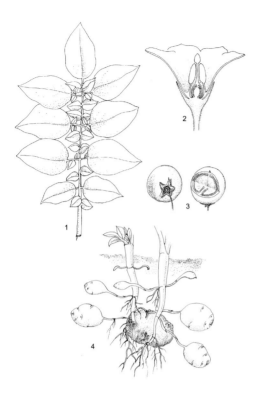

Fig. 9.41. Potato: 1, odd-pinnately compound leaf with pairs of unequally sized leaflets; 2, longitudinal section of flower; 3, whole fruit (left) and cut to show inside (right); 4, tuberization approximately 12 weeks after emergence.

Botany

The potato (Fig. 9.41) is a perennial herb, although in cultivation used as an annual crop. The species can readily be distinguished by its large, swollen tubers that are borne at the ends of slender subterranean rhizomes. The tubers are globose to ellipsoid, and very variable in size and colour. The tuber skin has scars of scale leaves ('eyebrows') with axillary buds ('eyes'). In practice, the rhizomes are often called 'stolons', although this is morphologically incorrect.

Potato has various, more or less erect, angular, branched, and rather weak and juicy stems (30–80 cm). The alternate leaves are pinnate compound with three or four pairs of ovate leaflets with smaller ones in between, in outline 10–30 cm × 5–15 cm. The leaflets are opposite or alternate, the largest ones stalked, ovate to ovate elliptical, 2–10 cm × 1–6 cm. The smallest leaflets are sub-sessile, ovate to suborbicular, 1–2 cm wide. The terminal leaflet is usually the largest. The inflorescence is a many-flowered cymose panicle (Fig. 9.42), which occurs in the leaf axil. Some cultivars never flower. The flower consists of a greenish campanulate calyx composed of five sepals, five white, yellow or pale violet petals, and five stamens; the yellow anthers of the stamens are joined laterally to form a cone-shaped structure, which conceals the ovary. The flower is about 2.5 cm across. The fruit is a globular, green or yellowish berry, about 2 cm across, two-carpellate, many-seeded, and similar to unripe tomato. The fruit is poisonous.

Potato has a fibrous, rather weak root system.

Fig. 9.42. Inflorescence of potato.

Cultivars, uses and constituents

Considerable morphological and qualitative differences can be distinguished between cultivars. A selection of such characteristics is: colour of skin; shape, size, number and uniformity of tubers; colour of flesh; cooking type (ranging from solid to very mealy); precocity; relative yield and dry matter content; resistance to various diseases and harvest damage. The Recommended List of Varieties of The Netherlands (2004), for example, comprises 191 different ware (table-use) and starch potato cultivars.

The potato is the most important non-cereal food crop in the world. It is also grown for animal feed (mainly Eastern Europe), industrial use in production of starch and alcohol, and for seed potato production. The tubers are consumed in many forms, which vary from simple cooking to baking, roasting, frying (French fries) or as thinly sliced crisps and other processed forms. Potato is an ingredient of many different dishes. Starch potatoes are processed into pure starch, which can be sold pure or be the basis of various starch derivatives. The freshly harvested, raw tubers consist of 20–30% dry matter of which 65–80% is starch. In general cultivars for consumption are lower in starch content than cultivars used for the industrial production of pure starch. The raw tuber contains 2–6% protein, 19% carbohydrates and 2% fibre; it is a source of Ca, P, K and Fe, and contains niacin, riboflavin and ascorbic acid. Solanine, a toxic glycoalkaloid, is mainly concentrated in the skin (0.01– 0.1% of the dry matter) and increases when exposed to light.

Ecology and agronomy

Optimum day temperature for dry matter production ranges from 20 to 25°C; however, for optimal tuberization cooler nights are required. A well-distributed rainfall of 500–700 mm per growing period is required. Potato is tolerant to a wide variety of soils except heavy, waterlogged clays. Deep soils with good water retention and aeration, and a soil pH ranging from 4.4 to 6.7, give the best growth and yields.

Worldwide there is a considerable variation in production systems, varying from small-scale and comparable to vegetable growing, to intercropping, to larger-scale (Figs 9.43 and 9.44). However, the systems of the temperate regions are more or less the same: on a larger scale, in monocultures and almost completely mechanized.

The crop is mostly propagated vegetatively by small tubers. To obtain an acceptable yield of good quality, it is important to start with healthy 'seed' potatoes. Vegetative propagation implies the risk of gradual degeneration due to infection by viruses. A recent development is producing mini-tubers from disease-free *in vitro* plants. Propagation of these mini-tubers has subsequently to be done in isolated fields, in a healthy environment and in areas protected from insects.

Storage conditions for seed tubers influence the physiological age of the tubers and can affect future production considerably. During storage the tuber may go through four physiological stages: (i) dormancy; (ii) apical dominance; (iii) normal sprouting; and (iv) senility. The duration of each of the stages depends among other factors on storage conditions. The higher the storage temperature the sooner senility is reached, meaning that the tuber is partly exhausted and will be too weak to produce a well-developing plant. For a

Fig. 9.43. A flowering potato crop.

Fig. 9.44. Harvest of a local potato crop, Ollantaytambo, Peru. Inset shows genetic variability (photos: C. Almekinders).

well-developing crop and therefore sufficient yield, the 'seed' potatoes must be planted in the stage of normal sprouting. Generally, low-temperature storage is best. To increase the precocity and the yield of the crop, pre-sprouting of the 'seed' potatoes in diffuse light prior to planting is an option.

In several regions in Asia and Africa there is a lack of storage capacity; moreover, purchasing healthy planting material is often not an option, due to economic circumstances. These factors cause an increasing interest in propagating potatoes from true potato seed. The advantages of using true seed are that it does not transmit most potato diseases, and it is very light and therefore is easy to transport (less than 1 kg of true seed per hectare). As the multiplication rate of 'seed' potatoes is low (10–20), 5–15% of the yield is required for propagation. In regions with food shortage, that is hardly an option. Disadvantages of using true potato seed are lower yield and the necessity of a longer growing season. To prevent the disadvantages, the seedlings can be raised in nurseries and then transplanted to the field. A different approach is growing seedlings in nurseries at high densities to produce small seed potatoes.

Potatoes are generally planted in rows 75–100 cm apart with a spacing of 20–40 cm within the row. The exact quantity of the required planting material depends on the aim of the production (ware potatoes, seed potatoes, starch production); the sizes of the 'seed' potatoes; cultivar and type of soil. The number of required tubers varies also (30,000–80,000/ha). At planting, the tubers must be covered by a soil layer of 5–15 cm depth. Planting depth is greater under warm, dry conditions than under cool, moist conditions. Tubers that grow too close to the surface and intercept light develop chlorophyll and become green. Such tubers are not edible because of the high solanine concentration. Earthing-up is often carried out to avoid the tubers developing chlorophyll. After planting, the seed tuber develops stems and roots. For sufficient root formation and a well-developing plant, potato needs a moist soil. Dry soil and low soil temperature delay emergence. After emergence, the stem and root growth develop more or less simultaneously. Tuber growth may start about 3 weeks after emergence and continues at a constant rate over a rather long period (Fig. 9.45). Weed control has to be started pre-emergence; it can be done by manual or mechanical hoeing or by herbicides. After that, during the growing season, weed control is often carried out by earthing-up or making ridges. Doing so, the earth will cover and subsequently kill the weed plants. Besides these options, several herbicides are also available.

Radiation and the uptake of nutrients mainly determine the final yield. Potato demands a large proportion of applied nutrients, because the uptake is not optimal, due to a relatively poor root system. How much fertilizer has to be applied is very variable. It depends on many factors such as climatic zone, duration of the growing season, cultivar, purpose of the crop and the amount of nutrients in the soil. In The Netherlands, 140–300 kg N/ha and 50–60 kg P/ha are often applied. Application of K has to be done very carefully because too much K will decrease the dry matter content of the tubers. K application has therefore to be based on soil tests. A yield of 45 t of tubers removes about 200 kg K.

The length of the growing period depends among other things highly on climatic zone. Comparing the different climatic zones, there are differences in growing periods and yields. For example, the duration of the growing season in Washington (USA) is about 7 months with an average yield of about 65 t/ha; in Pakistan, the duration of the growing season is about 3.5 months with an average yield of about 15 t/ha. In 2005, the total potato tuber production was about 321 million t. China, the Russian Federation, India, Ukraine and the USA were the main producers, together accounting for about half of the total production.

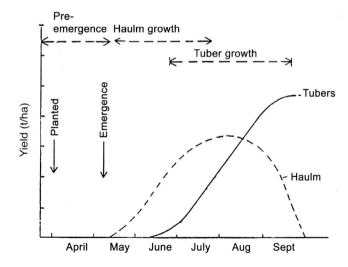

Fig. 9.45. Growth of a potato crop in The Netherlands (after Beukema and Van der Zaag, 1990).

Sweet potato

Sweet potato – *Ipomoea batatas*; Morning glory family – *Convolvulaceae*

Origin, history and spread

Sweet potato originates from Central America and the north-eastern part of South America where it was cultivated by 2500 BC. In pre-Columbian times it had already reached eastern and southern Polynesia as well as New Zealand. Based on archaeological findings it is generally accepted that the species did not evolve convergently in America and Polynesia. Currently there is no sound theory about the prehistoric dispersal of the sweet potato to Polynesia and adjacent regions, although most sources do mention a natural spread as most likely. The introduction into Africa and Asia was through Europe from the first voyage of Columbus. Sweet potato was transferred by the Spaniards from Mexico via Hawaii and Guam to the Philippines in the 16th century. At present sweet potato is widely cultivated in tropical, subtropical and warm temperate regions, between latitudes 48°N and 40°S.

Botany

Sweet potato (Fig. 9.46) is a perennial trailing herb, usually annual in cultivation, with milky juice in all of its parts. The stems are mostly vine-like trailing, occasionally twining, 1–8 m long, branching, rooting at the nodes if in contact with the soil, and producing a dense mass of foliage. Some bunch varieties are more erect and have shorter stems. The leaves are very variable, even on the same plant, ovate-cordate with 5–30 cm long petioles and spirally arranged on the stem. Depending on cultivar the lamina can either be entire (American type) or palmately lobed (Chinese type) and green or purplish in colour (Fig. 9.47). Per plant, three to 12 storage roots develop by secondary thickening of adventitious roots. On these storage roots adventitious buds can be found arranged in rows. The outer skin may be either white or a shade of red or purple. The flesh may be white, yellow, orange or purple (Fig. 9.48). The flowers are funnel-shaped, white or purplish, deeper in colour in the throat and paler at the margin, borne singly or in cymes on 3–15 cm long peduncles (Fig. 9.49). Flowering is more abundant in the tropics. The fruit is a 5–8 mm long capsule with one to four black seeds. Seedless fruits are common.

Cultivars, uses and constituents

There are numerous landraces, mostly farmers' selections in populations, resulting from natural hybridization and spontaneous mutation. In the USA two forms are common; those with dry mealy flesh on cooking and those with soft gelatinous flesh. A breeding programme in Japan is focusing, among other things, on cultivars for table use and cultivars for processing.

Fig. 9.46. Sweet potato: 1, flowering branch; 2, storage roots (line drawing: PROSEA volume 9).

In most tropical countries the storage roots of sweet potato are mainly used for human consumption (70–100%). They are usually eaten boiled or baked, sometimes candied with syrup or used as purée. Lesser quantities are used as feed for livestock (10–30%). The vines are widely used as feed. In temperate Asia 30–35% is produced for industrial purposes, mainly for starch and alcohol. The average composition varies with cultivars, environmental and cultural conditions. Freshly harvested storage roots consist of 60–84% water. The dry matter consists of 60–80% starch, 4–30% sugar and 1.3–10% protein. Sweet potato storage roots have on average a 40% higher energy value than potato. As well as being an important starch provider, sweet potato storage roots are rich in vitamin C, especially the varieties with yellow or orange flesh, which are rich in provitamin A. The young shoots are often consumed as a green vegetable.

Fig. 9.47. Leaves of sweet potato types: 1, Chinese type with palmately lobed leaves; 2, American type with entire leaves.

Fig. 9.48. Tubers of the sweet potato cultivars 'Tox Paars' (left) and 'Egeida' (right), entire and cut to show flesh colour (photo: J. van Zee).

Fig. 9.49. Flowers of sweet potato (photo: D.L. Schuiling).

Ecology and agronomy

Sweet potato is grown between latitudes of 48°N and 40°S. At the equator it is grown at altitudes ranging from sea level to 3000 m. Temperature for growth ranges from 12 to 35°C with an optimum temperature above 25°C. A frost-free growing period of 4–6 months is essential. Sweet potato grows well in full sunlight, although it can withstand a 30–50% reduction of full solar radiation. It requires a well-distributed annual rainfall of 600–1600 mm. Dry weather favours the formation and development of storage roots. However, it cannot withstand long periods of drought without irrigation nor is it tolerant of waterlogging. To avoid losses due to waterlogging sweet potato is usually grown on mounds or ridges (Fig. 9.50). The crop prefers well-drained sandy loam soils. The optimum soil pH is 5.6–6.6, but it grows well even in soils with a pH as low as 4.2. It is sensitive to alkaline and saline soils.

Sweet potato is usually propagated vegetatively from short stem cuttings, preferably the ends of sturdy vines about 20 cm long, which root easily and give rise to an extensive, fibrous, adventitious root system. The stem cuttings are inserted into the soil horizontally or at an angle, with three or four nodes covered by soil. Normally stem cuttings are planted 25–30 cm apart in rows and 60–100 cm between rows, with an average result of 45,000 plants per hectare at which the highest total yield may be expected. In areas where the plant cannot grow all year round, slips or sprouts obtained as cuttings from storage roots are used. Propagation from seed is used only for breeding purposes. Weeding during the first 2 months of growth is essential. Soon thereafter vigorous growth of the vines results in rapid and effective coverage of the soil and smothering of weeds. Although sweet potato responds well to chemical fertilizers, they are seldom applied in the tropics. In smallholdings and traditional agriculture, organic manure is used instead.

Fig. 9.50. Sweet potato grown on ridges, North Cameroon (photo: E. Westphal).

In the tropics the crop receives little cultivation after it is established and the storage roots are usually harvested as required and consumed within a few days. For long-term storage the roots are sliced and sun-dried. In temperate regions harvested storage roots are normally cured for 4–7 days at temperatures of 29–35°C and a relative humidity of 85–90%. After curing, storage roots are stored at 13°C and a relative humidity of 80–90%. Under these conditions they can be kept for 12 months or longer, depending on cultivar.

Farmers' yields, especially in the tropics, vary greatly, from 2 to 25 t/ha. In 2005, the total world production of storage roots was almost 130 million t and the mean yield was 14.9 t/ha. China is by far the greatest producer of sweet potato. Substantial quantities are also produced in Indonesia, Vietnam, Uganda and Nigeria. The USA is the largest exporter of sweet potato.

Yams

Winged yam, greater yam, water yam, 10-months yam – *Dioscorea alata*, Aerial yam, potato yam, bulbil-bearing yam – *Dioscorea bulbifera*, Yellow (Guinea) yam, 12-months yam – *Dioscorea cayenensis*, White (Guinea) yam, 8-months yam – *Dioscorea rotundata*, Cush-cush yam – *Dioscorea trifida*; Yam family – *Dioscoreaceae*

Origin, history and spread

The yams are among the oldest recorded food crops and their domestication is thought to have taken place separately in Africa, Asia and America, with different species involved in each region. The break-up of the world-continent Pangea seems to have separated the Old and New World species of yams. Separation of the Asiatic from the African *Dioscorea* species probably occurred in the late Miocene or early Pliocene. In West Africa man began to gather yams for domestic use as early as 50,000 BC. It is estimated that yam-based agriculture started in West Africa around 3000 BC. Major species among the earliest domesticated yams in West Africa and Central Africa are *Dioscorea cayenensis* and *Dioscorea rotundata*. About 90% of the world yam output is produced in the so-called 'West African yam belt', stretching from west of the Cameroon mountains to central Ivory Coast.

Dioscorea alata originates in South-east Asia where domestication took place around 3000 BC and spread to India and the Pacific more than 2000 years ago. *Dioscorea trifida* originates in the northern part of South America and spread to the Caribbean in pre-Columbian times. *Dioscorea bulbifera* is the only *Dioscorea* species that is common in the wild in both Africa and Asia, and is now widely distributed through tropical Africa and Asia.

Intercontinental distribution of the yam species did not begin until the last 500 years with the advent of long-distance ocean travel and the establishment of trade routes. The most relevant features of the various yam species are summarized in Table 9.1.

Table 9.1. The most important cultivated yam species accounting for over 90% of the food yams produced in the tropics.

Dioscorea spp.	English name	Area of origin	Important areas of cultivation
D. bulbifera	Aerial yam, potato yam	West Africa and South-east Asia	Tropical Asia and Africa
D. cayenensis	Yellow yam, yellow Guinea yam	West Africa	Nigeria, Ivory Coast
D. rotundata	White yam, white Guinea yam	West Africa	Nigeria, Ivory Coast
D. alata	Greater yam, water yam, winged yam	South-east Asia	South-east Asia, Africa, America
D. trifida	Cush-cush yam	Northern South America	Northern South America and the Caribbean

Botany

The genus *Dioscorea* consists of about 600 species, mainly occurring in the tropics and subtropics, and about 60 species are gathered or cultivated for their edible tubers. However, the most important edible yams belong to only a few species. The genus is further divided into a number of sections. The important edible yam species, *D. rotundata*, *D. alata* and *D. cayenensis*, belong to the section *Enantiophyllum*. They are characterized by vines which twine in a clockwise direction. Two other important yam species, *D. bulbifera* and *D. trifida*, belong to different sections but are both characterized by vines which twine in an anticlockwise direction.

The term 'yam' is often confused with several other species. In the USA the name 'yam' is commonly used for the sweet potato (*Ipomoea batatas*). In fact, almost any edible starchy root, tuber or rhizome that is grown in the tropics has at one time or another been described as a yam. Coursy (1967) suggested a formal definition of yam: 'Any of the economically useful plants of the botanical genus *Dioscorea*, or the tuber or rhizome of these plants'.

All of the yams that are of economic importance as food crops are dioecious, perennial and tuberous plants. The branching stems, 2–15 m or more long, climb entirely by twining; there are no tendrils or other specialized organs (Fig. 9.51). Leaves are usually alternate, simple, cordate, acuminate (apex pointing downwards), palmately veined or palmately compound with three or five lobes or leaflets and always petiolate. The inflorescence is unisexual, racemose or spike-like, occasionally both male and female flowers are borne on the same plant. The fruit is a dehiscent capsule, three-winged or strongly angled, 1–3 cm long. The seed is flattened, partly or completely winged at its margin. Axillary buds sometimes develop into a bulbil. The tuber is an annual organ which eventually decays when regrowth is commenced. New tubers are formed when regrowth starts or later in the season,

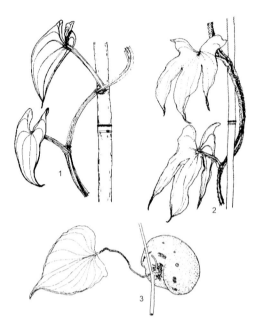

Fig. 9.51. Yams: 1, leaves and part of twining stem of winged yam; 2, cush-cush yam; 3, bulbil and leaves of aerial yam.

and vary greatly in number, size and form. During the dry part of the year the plant remains in a dormant state. Also, the stems and roots are renewed annually.

Characteristics of the most important edible yams are as follows:

1. *D. alata*: the tubers are usually single. Under normal conditions of cultivation tubers weigh 5–10 kg, usually cylindrical but also globular or long and serpentine. The flesh is white, cream or purplish. The stem is square in cross-section and usually conspicuously four-winged (alation), from which the specific name is derived (Fig. 9.52). Axillary bulbils are abundant but the number varies with cultivar. The leaves are variable in size and shape but essentially ovate with a deep basal sinus, 10–30 cm × 5–20 cm, and sharply pointed.

2. *D. bulbifera*: most varieties produce mainly aerial tubers (bulbils) and few or no underground tubers. Bulbils are produced in leaf axils of the vine, grey to brown, often kidney-shaped and usually weighing 0.5 kg but up to 2 kg; flesh usually pale yellow, tinted with violet and very mucilaginous. The flesh oxidizes to orange when cut. Stems are cylindrical, twining and up to 6 m long. Leaves are simple, broad or long heart-shaped and pointed, 20–32 cm × 20–32 cm.

3. *D. cayenensis*: tubers usually have pale yellow flesh; very variable in size and shape (Fig. 9.53). The head of the tuber is a hard, woody, corm-like structure. Stems are cylindrical, twining and 10–12 m long. Leaves are long heart-shaped and pointed, 8–10 cm × 3–6 cm, rather leathery and usually light green.

4. *D. rotundata*: the tubers are usually cylindrical with rounded or pointed ends, with smooth brown skin and white flesh. Occasionally the tubers may

Fig. 9.52. Field of unstaked winged yam in Ivory Coast (photo: E. Westphal).

develop into quite fantastic shapes. Tuber weight varies from 2 to 5 kg, although under good conditions 10 kg is not uncommon and even 20–25 kg has been recorded (Fig. 9.54). Stems are cylindrical, twining and 10–12 m long, usually spiny, especially at the base. Leaves are broadly heart-shaped and pointed, 10–12 cm × 6–8 cm, usually dark glossy green.

5. *D. trifida*: the tubers are usually positioned in a group of many small individuals with lengths of 15–20 cm. The flesh may be white, yellow, pink, or purple (Fig. 9.53). The stem is spineless and square in cross-section and

Fig. 9.53. Tubers of cush-cush yam as sold on the market, some have been freshly cut to show the purplish flesh (left); tuber of yellow Guinea yam with one finger cut to show the yellow flesh (photos: D.L. Schuiling).

Fig. 9.54. Tubers of white Guinea yam on a market in Ivory Coast.

frequently winged. The leaves are three- to five-palmately lobed, 15–25 cm long and broad, petioles 10–15 cm long and winged.

Uses and constituents

It is the tuber, containing the food reserves mainly in the form of starch, which provides a staple carbohydrate food for humans. In a few cases, notably *D. bulbifera*, the axillary bulbils are used for food. The tubers are peeled and mostly eaten boiled; whole small tubers may be roasted or pieces may be fried in oil, sometimes after parboiling. Tubers can also be processed into yam flakes or yam flour. The top ends of the tubers, from which new sprouts arise, are usually hard and unpalatable and therefore rejected. Yams are most important in West Africa, where it is usually eaten as 'fufu'. Fufu is prepared by boiling peeled and cut-up yams and pounding them in a large wooden mortar, with a wooden pestle, to produce a glutinous dough. Fufu is often eaten with meat or fish (or both together), green vegetables, spices and a considerable quantity of plant oil. The term 'fufu' is also used for similar doughs prepared from cassava, plantain (cooking banana), cocoyams (taro and tannia) or from mixtures of either of these with yam.

The edible portion of the tubers is very variable and depends on the thickness of the corky bark and underlying cortex. The approximate composition of the edible portion of the tuber is 58–80% water, 1–3% protein, 0.03–0.40% fat, 15–29% carbohydrates, 0.2–1.4% fibre and 0.5–2.1% ash. The vitamin C content in yams is substantial. The ascorbic acid content ranges from 5 to 28 mg/100 g edible portion, while the recommended daily dose is 60 mg. Most of the ascorbic acid is retained during cooking. Even early mariners appreciated the antiscorbutic properties of yams, although they did not understand the nature of the scurvy disease. The use of yams by mariners at that time most likely furthered the spread of yams.

Tubers of some wild yam species contain sapogenins and toxic alkaloids. Diosgenin, one of the most important sapogenins, is used in the preparation of oral contraceptives. Toxic alkaloids can cause a general paralysis of the central nervous system.

Yams are normally too expensive to be used as animal feed, although residues, such as peelings, are commonly fed to the livestock.

Ecology and agronomy

The most important edible yam species are essentially tropical plants and cannot withstand frost. Within the range of 25–30°C the rate of growth increases with temperature. The minimum yearly rainfall requirement is 1000 mm; optimum rainfall is 1500 mm/year or more, well distributed over the growing season and with a sharply demarcated dry season of 2–5 months. Yams do not tolerate waterlogging. During the dry season when the shoots die away, the tubers become dormant. The soil should be well-drained, preferably a deep sandy loam and rich in organic matter. The rather weak root system of most yam species demands a good soil fertility to obtain sufficient nutrients.

Yam species are usually propagated vegetatively by whole seed tubers, cuttings (so-called 'heads', 'tails' and 'middles') or bulbils, and planted towards the end of the dry season while they are still dormant. Dormancy can be broken with a 2–8% watery solution of ethylene chlorohydrin. The propagules are usually planted on mounds, hills, banks or ridges. The size of the mounds varies greatly from one district to another, and may be 50–100 cm high and 100–200 cm in diameter. Yams are often intercropped with maize, okra, cucurbits or other crops, but where sole cropping is practised spacing between rows is 50–100 cm and within rows 50–100 cm. The number of propagules required is 10,000–15,000/ha.

Essential for good yield is staking of the plants soon after emergence. Stakes, which may be bamboo, sticks or growing poles, should be at least 2 m high. Unstaked cultivation also occurs in areas where timber is scarce. Although weeds are better suppressed in unstaked cultivation, the tuber yield may be reduced by up to 60%.

Depending on species and cultivar, the period from planting to harvest ranges from 7 to 12 months. The vernacular names of some yam species refer to the length of the growing period.

In some areas the crop is left in the ground and harvested only when required for consumption or sale. The harvest of the yam crop is predominantly carried out manually. The harvest is simple, although very laborious. Wooden digging tools are often preferred to iron tools in order to avoid cutting or bruising the tubers. During the handling after harvest any physical damage of the tubers should be avoided. Where yam is a staple crop the tubers must be stored for several months in cool shady conditions. In West Africa this is usually done by tying the tubers to a shaded vertical framework in order to permit adequate ventilation. Cold storage below 12°C results in chilling injury of the tubers.

Yields vary extremely depending on species, cultivar and environment. The world average yield of yam tubers is 9 t/ha, which is about the same as the average yield in Africa. Much higher yields, 40–60 t/ha, are also recorded. Most yams are consumed in the country of production and very small amounts enter international trade. In 2004, the world total production was about 40 million t. Nigeria, accounting for 27 million t, was by far the most important producer.

MINOR TROPICAL STARCH CROPS

Plantain and Enset

Plantain – *Musa* spp., Enset – *Ensete ventricosum*; Banana family – *Musaceae*

Plantain

The origin of plantain is thought to be along the peripheral areas of the primary centre of origin of the dessert banana, which is Malaysia. From there the plantain spread eastwards to the South Pacific and westwards from Indonesia to Africa and tropical America. The botany of plantain is almost identical to that of the dessert banana, described in Chapter 2.

The primary use of *Musa* spp. is the edible dessert banana, which is eaten fresh because in the ripe state they are sweet and easily digested. The conversion process of starch to sugar during fruit ripening is much slower or almost absent in the plantain varieties. Therefore the plantains generally have starchy flesh and at maturity they are usually unpalatable unless boiled. However, there is a considerable overlap between bananas and plantains in the way they are consumed. Dessert bananas are often popular as a starchy food when cooked in an unripe state.

All modern banana and plantain cultivars are derived from either one or both of the two wild, inedible, seeded *Musa* species: *Musa accuminata* (with genome A, involved in the dessert type) and *Musa balbisiana* (with genome B, involved in the plantain type). The current, generally accepted classification of cultivars is based on the relative contribution of the two wild species to the constitution of the cultivar. The most important, predominantly triploid, cultivars of plantain may have AAB, ABB or BBB genome.

Within the AAB plantain subgroup there are two main types: (i) the 'French' plantain type, which is grown in Central and West Africa, and parts of India and Central America; and (ii) the 'Horn' plantain type, grown – as well as in the same areas – in the Philippines and the Pacific.

Cultivars of the ABB group are very vigorous and drought-resistant. The cultivar 'Bluggoe' is a starchy cooking banana with large fruits, and is an important source of food in Samoa, the Philippines, southern India, the West Indies and Tanzania. The cultivar 'Pisang Awak' is the most important banana cultivar in Thailand where it is eaten fresh or cooked.

Within the BBB group, the cultivar 'Saba' is the most important one, notably in the Philippines. It is a cooking plantain with medium to large fruits with creamy white pulp, and although the flesh becomes sweet on ripening, the fruits are always cooked before consumption. In some areas in South-east Asia there has been increased use of plantain cultivars belonging to the ABB or BBB group due to the lower susceptibility of these cultivars to diseases, pests and periods of drought in comparison with AAB cultivars.

Unlike the previously mentioned plantain genome groups, in which the B genome is always involved, the so-called 'East African highland cooking bananas' are considered to be a subgroup within the AAA group. They were taken to Africa in prehistoric times and for centuries these highland bananas have been a major staple food and a source of income for millions of people in the highlands and mid-altitude regions of Burundi, Kenya, Malawi, Rwanda, Tanzania and Uganda. The majority of the different varieties found in the highlands of East Africa grow only in this part of the world, as a result of which the area became an important centre of diversity of cooking bananas. The main method for preparing the bananas in this area is to steam the peeled green fruits, wrapped in banana leaves, into a mass called 'matoke', which is often combined with beans, meat or fish.

The morphology of plantain is similar to that of the dessert banana to a large extent and is described in more detail in Chapter 2. The fruits of many plantain cultivars are more angular, with a more elongated tip, than the fruits of many dessert banana cultivars (Fig. 9.55).

The fruits of plantain and cooking banana are used ripe or unripe in a wide range of dishes. Boiling whole bananas, and roasting or frying peeled and sliced bananas are the simplest and most popular ways of preparing the fruit. In Africa, beer is brewed from plantains. In the Philippines banana ketchup, made from a certain plantain cultivar, substitutes tomato ketchup.

The edible portion of plantains and cooking bananas contains approximately 55–58% water, 1.2–1.6% protein, 0.25–0.30% fat, 34–35% carbohydrates, 6–7% fibre and 0.8% ash. They are good sources of vitamins A and C and of K, Mg, Ca, P and Fe.

The most common production systems are home gardens, mixed intercropping and commercial smallholder plantations (Fig. 9.56). Average yields in home gardens may vary from 8 to 30 t/ha/year. In mixed intercropping systems, they are grown as a nurse crop for shade-loving trees such as coffee and cacao, or grown under coconut trees. The yields in commercial plantations can be very high, sometimes reaching 50 t/ha/year.

Today plantain is found wherever dessert bananas are grown. In 2003, the total world production of *Musa* was 103.9 million t of which 69% was classified as bananas and 31% as plantains. Of the 32 million t of plantains produced, only 0.4 million t was exported. These data indicate the significance of plantain as a staple food for domestic use. This is especially so for the equatorial zone of West and Central Africa, where around 70% of the world crop is grown and consumed.

Fig. 9.55. Young fruit bunch of plantain (photo: D.L. Schuiling).

Fig. 9.56. Plantains planted in a double-row pattern, Costa Rica (photo: H. Waaijenberg).

Enset

Enset is a valuable starch crop in the highlands of south and south-western Ethiopia where it occurs in the wild at altitudes ranging from 1500 to 3100 m. It is a regionally dominant staple food covering approximately 168,000 ha and is the main source of food for about 7–10 million people.

For optimum growth the plant requires an average annual rainfall of 1100–1500 mm and a mean temperature of 16–20°C. Enset thrives on well-drained and fertile soils with a pH of 5.6–7.3 and 2–3% organic matter.

Enset can be distinguished from banana by, among other things, the single pseudostem with an enlarged basal part of an underground structure known as a 'corm', the smooth and nearly globose seeds and the almost erect leaves (Fig. 9.57). Enset accumulates starch in the leaf-sheaths of the pseudostem and the corm. Depending on altitude (temperature), enset reaches full maturity in 5–10 years. Harvesting takes place before flowering and fructification, which use part of the stored carbohydrates.

The freshly cooked corm (Fig. 9.58) is locally called 'amicho' and can be consumed in a similar way to potato. However, the main food product is obtained by fermenting a mixture of pulverized corm and pulp from the

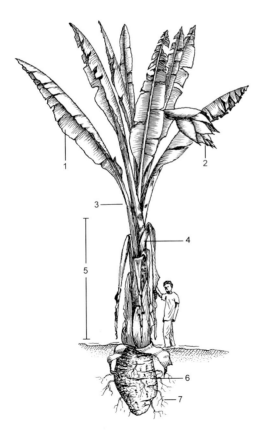

Fig. 9.57. Habit of enset: 1, leaf-blade; 2, inflorescence; 3, petiole; 4, leaf-sheath; 5, stem ('pseudostem' when the plant is in the vegetative stage); 6, corm; 7, root.

Fig. 9.58. Pulverizing the corm of an enset plant, Ethiopia (photo: A. Tsegaye).

pseudostem. The average yield of this fermented product, locally known as 'kocho', is about 30 kg per plant at 50% moisture. The leaves are used for thatching and the petioles to make fibres for cordage and sacking. The seeded fruits are small and not edible, although in time of famine the seeds may be used also as food. With an average annual dry matter yield of 10 t/ha, enset ranks among the more productive starch crops.

Sago palm

Sago palm – *Metroxylon sagu*; Palm family – *Palmae*

Origin, history and spread

The sago palm is thought to have originated from New Guinea and the Moluccas, where the greatest genetic diversity and the world's largest sago palm swamps and forests are still found. The total area in New Guinea of the vegetation type dominated by sago palm may well be over 6 million ha. Approximately 2.2 million ha of this area consists of largely wild, sago palm stands.

Beyond its area of origin the palm is now found in Indonesia: in parts of Sulawesi, Kalimantan, Sumatra and West Java, and on many smaller islands with a non-seasonal climate. In Malaysia, mainly in Sabah and Serawak, it is grown in (semi-) cultivated stands. It is also grown in Brunei, the Philippines (Mindanao) and a small area in southern Thailand.

Botany

Sago palm is a medium to tall palm with a trunk of 30–60 cm in diameter and 7–20 m tall. Suckers are produced from the leader palm, forming a cluster of palms in different stages of development (Fig. 9.59). The leaves are simply pinnate and usually 5–7 m long; petiole very robust and a stem-clasping sheath. Cultivars may differ in spininess of the petiole and sheath to various degrees. The needle-like spines are arranged in transverse combs and may be up to 20 cm long; spines are most prominent in the sucker stage of the palm. The root system consists of many long, thin and multiply-branching roots.

The inflorescence is a vast terminal panicle of 3–5 m or more high and wide, branched to the third order with spirally arranged pairs of hermaphrodite and staminate flowers on the last order branches. Flowering and fruiting occurs only once in its life: depending on variety and growing conditions, this is at the age of 4–15 years, after which the palm dies. The fruits are globose, 3–5 cm in diameter and covered with greenish-yellow scales, turning straw-coloured towards ripening. The seeds are sub-globose, about 3 cm in diameter, with a dark brown testa and a hard stony endosperm.

Fig. 9.59. Sago palm in vegetative trunk stage with trunkless suckers at the base, Saparua, Indonesia (photo: D.L. Schuiling).

Uses and constituents

The starch stored in the trunk of the sago palm provides a staple food, notably in New Guinea. Traditionally the most common preparation for consumption is by pouring, and meanwhile stirring, hot water over a slightly sour wet starch. The resulting glue-like mass is often eaten with fish and vegetables. As sago is very low in protein, fat and minerals, these supplements are essential in a well-balanced sago diet. Various kinds of home-made biscuits are made of sago starch, occasionally combined with ground groundnuts, other pulses or ground seeds of the kenari tree (*Canarium commune*). The wet starch may also be fried, roasted or dried with moderate heat to produce flour. Sago pearls are made from slightly wet starch by passing it through a sieve into a heated pan with a round bottom, while rolling the starch particles around until the outside has been gelatinized. In Indonesia and Malaysia, the starch is used industrially in the manufacture of cakes and biscuits, noodles and 'kerupuk' (crisps). The larvae of specific weevils, feeding on the pith of the trunk, are eaten raw, boiled or roasted in places where sago palm is a staple, thus providing a good source of protein.

The palm has many non-food uses. The leaves are used for thatching roofs; the petioles and adhered rachides for constructing walls, ceilings and fences; the bark as timber and fuel; young leaflets for making baskets; and the palm heart as a vegetable. Industrial uses include sizing pastes for glazing paper and finishing fabrics and the manufacturing of various starch derivates. Dry sago starch contains 10–17% water and 81–88% carbohydrates. Purified sago starch consists of 27% amylose and 73% amylopectin.

Ecology and agronomy

Sago palm grows best in humid, tropical, lowland areas with an evenly distributed rainfall of 2000–5000 mm/year and an average temperature of at least 26°C. The best yields come from sago palms below 400 m altitude, in full sunshine, a soil pH of 4 or higher, and a groundwater table not below 50 cm. Adult sago palms withstand temporary inundation, even with seawater. However, stagnant water inhibits growth. Daily flooding is harmful to seedling growth, as is salinity. Sago palms grow best on mineral soils with high organic matter content (up to 30%). Neither fertilizer nor manure is normally used, especially when only the starch of the sago palm is harvested and other plant parts are left in the field.

Sago palm has the advantage that it rejuvenates itself through suckers. In plantations sago palm is mostly propagated from rooted suckers of about 1 year old. The choice and preparation of the plant material demands special care, as follows.

- A large part of the runner should still be attached to the sucker.
- Sufficient well-developed, undamaged roots should remain.
- The cut surface should be clean and rubbed with wood ash to prevent it from rotting
- Leaves, except for the two youngest leaves, should be pruned off.

- The time between cutting the sucker and planting should be as short as possible.

Even when good care is taken of the propagated suckers, only about half of them may survive.

Recommended planting densities are 6 m × 6 m to 7 m × 7 m in a square pattern. Planting depth is very critical: the runner and the cut end of the sucker should be buried to prevent them from weevil attack and drying out, but the shoot should be above the water table or above flood-water level to prevent rot. During the first 3 months after planting plenty of shade should be provided after which the shade can gradually be removed. Weeding is necessary until the leaf canopy has closed. In order to achieve around 140 trunks/ha only one sucker is allowed to develop into a trunk every 1.5–2 years while all other suckers are pruned.

Just before flowering there is a maximum build-up of starch reserves in the pith of the trunk and usually at this stage the trees are felled. To maximize the starch production per unit time, however, trees should be felled before the inflorescence emerges. After felling the trunk, the crown and old leaf-sheaths are removed. Traditionally most of the handlings after felling are done at site (Fig. 9.60). The trunk is split lengthwise and the outer layer of fibre bundles is removed from the exposed pith, followed by pounding and pulverizing the loose pith. The pulverized pith is washed out with water over a sieve, after which the starch is recovered

Fig. 9.60. Traditional extraction of sago palm starch in the Moluccas, Indonesia: 1, pulverizing the pith manually (pounding), Ambon (photo: D.L. Schuiling); 2, pulverizing the pith mechanically (rasping); 3, washing out the starch from the pulverized pith; 4, after baking them in pre-heated earthenware moulds, small sago loaves are sun-dried, turning them into a long-keeping product, Saparua.

from the slurry by letting it settle. The wet starch, consisting of about 60% starch, is removed from the field in baskets made of sago palm leaflets.

In planted stands the trunks are usually cut into sections of about 1 m, weighing 80–120 kg, and transported to a central mill for further processing through a network of waterways by means of rafts made of these sago palm logs (Fig. 9.61).

The annual yield of air-dried starch depends largely on the used variety and type of sago palm stand. In cultivated stands of sago palm in Malaysia, the first crop of palms of short life cycle yields about 25 t/ha and subsequent ratoon crops yield about 15 t/ha. The production capacity of semi-wild stands is estimated at 10 t of air-dried starch per hectare. Most of the sago starch is consumed locally. Recent information on production and international trade is scarce. In 1992, Indonesia, the major producing country, exported 10,000 t of sago flour and meal to Japan, Hong Kong and Singapore. Malaysia (Sarawak) exports nearly all of its production; in 1992 the export of dry sago starch from Sarawak was 45,700 t.

Fig. 9.61. Transport of sago palm logs to central processing plant by floating the logs roped together down a river, Halmahera, Indonesia (photo: D.L. Schuiling).

Taro and Tannia

Taro – *Colocasia esculenta*, Tannia – *Xanthosoma sagittifolium*; Arum family – *Araceae*

Taro

Origin, history and spread

It is believed that taro originated in the wet tropical areas of southern central Asia, probably in India and Malaysia, where it still occurs in the wild state. Estimates are that taro was already cultivated in these areas before 5000 BC. From there it was spread eastwards into Myanmar (Burma), China and Japan and southwards to Indonesia, and thence to Melanesia and Polynesia. Eventually it reached ancient Egypt and spread further westwards across the Mediterranean and across Africa to the Guinea coast. From West Africa it spread, with the early slave trade, to the West Indies and to tropical parts of America. In these regions taro is known, depending on type, as 'dasheen' or 'eddoe'.

It was introduced to the southern USA in the early 20th century. Except in Hawaii and countries such as Nigeria, Indonesia, the Philippines and Egypt, where it has developed into a commercial crop, taro is generally produced as a local source of staple food. Today taro is grown in nearly all parts of the tropics, as well as in some subtropical regions.

Botany

It is generally accepted that *Colocasia esculenta* is the most important cultivated taro species. Various botanical varieties of taro exist and are generally divided into two main groups: (i) the eddoe type with small corms and large cormels; and (ii) the dasheen type with large corms and small cormels.

Taro is a perennial herb of 1–2 m tall (Fig. 9.62). In cultivation the plant is mostly grown as an annual crop. The underground starchy corm is a cylindrical or spherical, bulb-like fleshy structure with short internodes at the base of the stem (Fig. 9.63). The dasheen types are up to 30 cm long and 15 cm in diameter. Eddoe types are usually smaller. Few of the axillary buds may develop new cormels, suckers or stolons. Depending on cultivar the tuber flesh may be white, pink or yellow. The root system is shallow and fibrous. The large heart-shaped leaf-blades are 20–50 cm long and arise in a whorl from the apex of the corm. The long (sometimes over 1 m) petiole is attached to the lower surface of the leaf-blade instead of by its margin. This is in contrast with tannia. The inflorescence is a 6–14 cm long spadix. Many cultivars do not flower and seeds have seldom been reported. Due to the presence of aerenchyma tissue in all parts of the plant, flooded conditions are well tolerated.

Wild taro (*Alocasia macrorrhiza*) is a related species from tropical Asia. The above-ground corms are used for pig feed and for human consumption in times of food scarcity.

Fig. 9.62. Plant habit of tannia (left) and taro (right). Note the difference in leaf shape (photos: D.L. Schuiling).

Fig. 9.63. Corms of tannia (left) and taro (right) (photos: G. de Bruin).

The giant swamp taro (*Cyrtosperma chamissonis*) can be found throughout Micronesia and is the largest of the taros, with corms weighing as much as 45–90 kg after 10–15 years. It is basically a prestige crop and offered at feasts; the bigger the corm the greater the honour.

Uses and constituents

The large, central, starchy corm is the portion usually eaten and sold. It provides an important source of starch throughout the humid tropics, and particularly in the Pacific it is one of the most important common foods.

Some cultivars contain unpleasant amounts of needle-like crystals of calcium oxalate (raphides) in the plant cells, which cause oral and intestinal irritation. However, these raphides are destroyed by cooking. After peeling the corms are boiled, roasted or baked. In Hawaii and parts of Polynesia, the peeled and cooked corms are pounded and then fermented anaerobically in water to produce a sticky paste called 'poi'.

The starch grains of taro are extremely small, fairly rich in amylose and therefore easily digested. Taro flour is used in baby food and is good for people with lactose intolerance. The composition of the edible portion of fresh corms is approximately 70% water, 1.1% protein, 26% carbohydrates, 1.5% fibre and 15 mg of vitamin C per 100 g. Young taro leaves contain 4.2% protein and are used as a main vegetable throughout Melanesia and Polynesia.

Ecology and agronomy

Taro is essentially a tropical lowland crop and grows well at temperatures of 25–30°C and high humidity. The required rainfall is above 2000 mm/year and the best yields are obtained when rainfall is evenly distributed. Taro prefers a soil pH of 5.5–6.5 and can tolerate saline soil better than most other crops.

There are two main systems of taro cultivation:

1. Lowland, flooded or wetland cultivation: flooded cultivation can be found in situations where water is abundant and in swampy areas that are not suitable for most other crops. In general the corm yields are about double compared with upland cultivation despite the longer time to mature. Flooding is also an effective method for controlling weeds. The dasheen types of taro in particular do best under flooded conditions. Planting can be done throughout the year.

2. Upland, unflooded or dry-land cultivation: this type of cultivation is essentially rainfed. Planting is usually done at the onset of the rainy season and where rainfall is irregular, furrow or sprinkler irrigation may be practised. Weed control is essential in the first 3 or 4 months after planting. The eddoe types can withstand drier upland conditions but prefer well-drained clayey soils with a high water table, although temporal flooding and waterlogging are tolerated well. The dry-land cultivation of taro is by far the most important worldwide and more particularly in the Asia/Pacific region.

Taro is propagated only vegetatively and preferably by so-called 'head-sets'. Head-sets consist of about 1 cm of the corm apex plus attached petiole bases 15–30 cm long. Pieces of the corm, whole small corms, cormels, suckers and stolons can also be planted. Planting density ranges from 4000 to 49,000 plants/ha. Higher plant densities increase the total yield although the average yield per plant decreases significantly. At low plant densities weeds can cause a reduction in total yield of up to 85%. For lowland taro, planting may be done by hand insertion of sets into 2–5 cm of standing water in the mud to a depth of 5–7.5 cm. For upland taro, mechanical planters can be utilized if the field is large enough and well-prepared. Fertilizer recommendations are based on soil conditions. In cases where taro has been cropped several times, 50–100 kg N/ha (split into three applications at 5, 10 and 15 weeks after planting), and at planting 50 kg P/ha and 70 kg K/ha, is recommended.

The time from planting to harvesting varies with cultivar and cultivation method: 5–12 months for dry-land cultivation and 12–15 months for flooded cultivation. Harvesting is most commonly done by hand or by hand tools. The corms do not store well and are therefore usually harvested, particularly by small-holders, when needed for consumption.

Total yields of taro on farms range from 3 to 40 t/ha depending on soil fer-tility and availability of water. From research plots, yields of 40–123 t/ha have been reported. In 2005 the world average yield of taro was 5.8 t/ha and the world production was estimated at approximately 10 million t. Today taro is grown throughout the West Indies and in West and North Africa. It is widely grown in south and central China. It is the most important staple food in many islands in the Pacific, including Papua New Guinea.

Tannia

Origin, history and spread

Tannia, also called 'yautia' or '(new) cocoyam', originates from the northern part of South America and had already spread in pre-Columbian times to the Antilles and Central America. During the slave trade tannia was taken to West Africa, which is now the major producer, and then to Oceania and Asia in the 19th and early 20th centuries.

Botany

Tannia is a herbaceous perennial, resembling taro very much in its botanical characteristics (Fig. 9.62). Tannia is a more robust plant of 1–2 m or more in height. The long, thick and ribbed petiole is attached to the margin of the sagit-tate or hastate leaf-blade. The main corm is short and stout, or globose to cylin-drical, from which secondary shoots or cormels sprout (Fig. 9.63). The inflorescence consists of a cylindrical, about 15 cm long, spadix surrounded by a usually pale green spathe of about 20 cm long, which closes at its base and opens at the top into a concave lamina. Some cultivars seldom flower and the spadices are rarely fertile.

Uses and constituents

The corms and cormels can be consumed after being washed and peeled, boiled, baked, steamed, creamed, mashed or fried in oil. It can be used in soups, stews and salads. The pre-cooked corms may be peeled, dried and ground into flour. In general tannia is more nutritious than taro and potato. The edible portion of tannia corms contains approximately 70–77% water, 1.3–3.7% protein, 0.2–

0.4% fat, 17–26% carbohydrates, 2 mg carotene/100 g and 96 mg vitamin C/100 g. Like taro, some cultivars of tannia contain unpleasant amounts of needle-like crystals of calcium oxalate (raphides) in the plant cells, causing oral and intestinal irritation. Coloured cultivars may also contain saponins. Raphides become ineffective after cooking and saponins can be released into the cooking water. The starch grains of tannia are much larger than those of taro and therefore regarded as less digestible. The young leaves of tannia are also used as a vegetable after the large veins have been removed. The cooked corms can be fed to animals, particularly to pigs.

Ecology and agronomy

Optimum temperature for good growth of tannia ranges between 25 and 29°C. It requires high rainfall, preferably 2000 mm well-distributed over the year. Unlike taro it does not withstand waterlogging. Tannia can tolerate light shade and, to some extent, saline soils. It grows best on deep, well-drained, fertile soils with a pH range of 5.5–6.5.

Tannia is propagated vegetatively most commonly by planting the top of the central corm with at least four buds plus 15–30 cm attached petiole bases. Tissue culture has been developed to obtain virus-free planting material. Subsequent yields with this material showed a significant increase.

Land preparation varies from clearing, in shifting cultivation, to ploughing, raking and forming mounds or ridges in permanent cropping systems. The planting depth is 6–7 cm; shallower planting will result in numerous side shoots and a reduction in yield. Planting distances vary from 0.6 to 1.8 m between rows and from 0.4 to 1.2 between plants. In smallholdings and home gardens the propagules are often planted in mounds spaced at 1 m × 1 m or 1.3 m × 1.3 m. Tannia is often used in farmers' fields as an intercrop of tree crops such as cacao, oil palm and other food crops. Its use as a nurse crop to provide shade for shade-loving seedlings (e.g. cacao) is widely practised. Weed control is critical during the first 6 months after planting. As the plant needs to be earthed-up several times, this contributes also to the weed control.

Fertilizer is commonly applied in commercial plantations. Recommended quantities are organic manure 20–40 t/ha or 110 kg N, 45 kg P and 110 kg K per hectare, split into two applications and given at 2 and 6 months after planting.

Partial harvesting is practised in small farms and may start from 4 to 6 months after planting, mostly by hand or by use of hand tools. In commercial plantations harvesting is carried out from 9 months after planting onwards. The crop is harvested by hand or by a semi-mechanized method. However, the latter method can cause too much damage to the tubers. Average annual yields of tannia are about 12–20 t/ha. By means of the use of selected clones, yields of up to 37 t/ha/year have been reported. In 2005 the world production of tannia was about 0.5 million t. Unlike other aroid root crops, the use of tannia is increasing in West Africa and some Pacific Islands.

10 Sugar Crops

Sugarbeet

Sugarbeet – *Beta vulgaris*; Goosefoot family – *Chenopodiaceae*

Origin, history and spread

Sugarbeet is one of just two crops (the other being sugarcane) which constitute the only important resource of sucrose. The ancestral form of all beets is the wild sea beet (*Beta maritima*), which is distributed on the seashores of southern Britain, the Mediterranean and Near Asian areas. The wild form is very variable, with branched taproots and varying sugar content. The cultivation of beets goes back to about 800 BC in the gardens of the Babylonian kings. At that time, the leaves were eaten as a vegetable. Leaf beets were described in a number of Greek and later Roman writings. Romans used leaf beets rather extensively as a vegetable and animal feed. In the *Capitulare de Villis* issued by Charlemagne in about AD 800, *Beta* was mentioned as one of the species which should be cultivated in the gardens of imperial estates. Probably from about that time beet varieties with both edible leaves and enlarged roots were selected and grown in Western Europe; by the end of the 15th century probably all over Europe.

In 1600, the French agronomist Olivier de Serres described the juice of beets as 'sugar syrup'. At that time, the sweet-tasting juice of the plants was highly valued. There was only little cane sugar available, besides which it was very expensive. At that time, fruit juice and honey were the most commonly used sweeteners. In 1747, Marggraf discovered that sugar crystals obtained from beet juice were of exactly the same nature as those from cane sugar. His student Achard received a grant from the King of Prussia to develop a ommercial industry. The first sugar factory was built in 1801, at Cunern in Selesia.

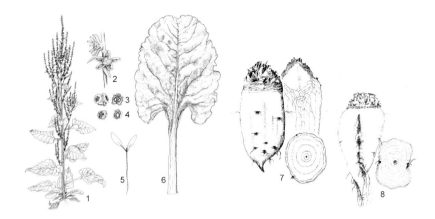

Fig. 10.1. Beet: 1, flowering plant (second growing season); 2, detail of stalk with two flowers; 3, multigerm 'seed' (cluster of one-seeded fruits); 4, monogerm 'seed' (one-seeded fruit); 5, seedling of sugarbeet with seed leaves; 6, full-grown leaf of sugarbeet; 7, mangel: entire, in longitudinal section and in cross-section; 8, sugarbeet: entire and in cross-section.

In 1806, Napoleon prohibited all imports of British goods into the European continent. That caused an enormous shortage of cane sugar on the continent and a replacement was vitally necessary. In 1811, Napoleon published his decree for the introduction of beet-sugar production in France and the occupied countries. That made the sugarbeet known all over Europe.

Sugarbeets were introduced into the USA in 1800. At present, sugarbeets are grown in most of the countries in the temperate climate zone.

Botany

Beet (Fig. 10.1) is a biennial herb. The leaves are glabrous, ovate to cordate, light or dark green or reddish, with prominent midribs and strong petioles; they form a rosette in the first year.

Sugarbeet develops more leaves than fodder beet. The taproot is diarch, swollen together with the hypocotyl, forming a tuber. The tuber or storage organ of the sugarbeet is usually called the 'root', although only about 90% is actually root-derived, the other 10% being derived from the hypocotyl. In fodder beet types, the contribution of the hypocotyl is varied but is larger. The root is mostly conical, sometimes globular or cylindrical (Fig. 10.2).

For flowering in the second year, vernalization is necessary. The flowering stalk is therefore produced in the second year, up to 2 m tall; the leaves of the stalk are small, flowers small and green. Flowers can occur in clusters and cohere so that a hard particle is produced. These clusters are said to be 'multigerm'. At present, most of the cultivars are 'monogerm', which means that the plant has only a single flower at each node of the inflorescence.

Fig. 10.2. Various varieties of beet: 1, sugarbeet; 2–4, three types of mangel.

Cultivars, uses and constituents

The enlarged beet root, including the hypocotyl, evolved gradually during the hundreds of years of selective breeding. Initially, most varieties were used as fodder crops. Around 1800, further selection and breeding resulted in the first sugarbeet varieties. Now there are five types of domesticated beets:

1. Sugarbeet: mainly enlarged root and relatively small part enlarged hypocotyl, an important sugar source.
2. Leaf beet: no enlarged hypocotyl and slightly swollen taproot, the leaves have thickened midribs and are used as a vegetable.
3. Beetroot: enlarged hypocotyl and enlarged part of the root are eaten as a salad vegetable, colour mostly red, although sometimes white or yellow (Fig. 10.3).
4. Mangel: enlarged hypocotyl and enlarged part of the root, the contribution of the hypocotyl is variable, used primarily as a fodder crop.
5. Fodder beet: used to feed livestock, hybrid of mangel and sugarbeet.

The early sugarbeet types contained approximately 6% sugar, while the modern cultivars have up to 20% on the basis of fresh weight (75–80% on basis of dry weight).

By far the most important type is sugarbeet. The National Recommended List of Varieties of The Netherlands (2002) describes 30 different cultivars, some having developed resistance to major diseases.

When harvested and processed for the production of sugar, the sugarbeet crop yields a number of by-products, as follows, which can be used as animal feed:

Fig. 10.3. Cross-sections of four beetroot colour varieties (vegetables).

- Crown and leaves: may be fed fresh or may be ensiled, can also be used as green manure, yield is about 5–6 t of dry matter per hectare.
- Sugarbeet pulp: in the factory beets are sliced in strips, next the sugar is extracted by diffusion, the spent strips may be fed wet or dried.
- Molasses: the residual syrup from the processing of sugarbeet from which no more sugar can be crystallized.

These by-products can be fed to ruminant livestock and pigs, where the major attributes are the energy, crude protein and digestible fibre content.

Ecology and agronomy

Sugarbeet is usually grown as a spring-sown crop in temperate regions (Fig. 10.4). It requires a cool climate and, except for seedlings, is able to withstand mild frost. The sugar content is highest in a temperate climate with an annual precipitation of 700–1000 mm and full sunlight. Although it tolerates temperatures between 5 and 30°C, an average day temperature of around 20°C is best for good growth. It grows well in a variety of soils, growing best in a deep, friable, well-drained soil abundant with organic matter. In northern Europe sugarbeet is often grown on clay soils, preferably at pH 6.5–8. The fields are ploughed in the autumn prior to sowing the crop. On clay soils, in regions with annual freezing, seedbed preparations are facilitated by the fine crumb structure which results from the action of the winter frosts on the surface of the plough layer. In a Mediterranean climate sugarbeet is sown in autumn and so the growing period is in winter.

At present most sugarbeet crops are grown from monogerm seeds, drilled to a stand using precision drills. It is important to place the seeds at the right

Fig. 10.4. Field with sugarbeets in July, The Netherlands. Inset shows detail of the plants.

depth, 2–3 cm. The aim is to achieve about 80,000 plants/ha. The seeds have to be spaced far enough apart for efficient harvesting of fully grown plants. The distance between the plants needs to be at least 15 cm to allow space for the topping mechanism. After germination and developing a seedling, there is a period of 6 weeks of leaf initiation and very little root growth. From the eight- to ten-leaf stage onwards, leaf and root growth occur simultaneously. Leaf production continues throughout the whole first growing season while the root swells and accumulates sucrose. After having developed 20 leaves, this number will be more or less fixed because new leaves appear but old leaves die.

The true root and the hypocotyl both contribute to the storage organ of the sugarbeet. The increase in diameter of this structure results from the activity of the cambia. The innermost cambium is produced between the primary xylem and phloem. Additional cambia are initiated centrifugally. Sugarbeet develops more cambium rings than fodder beet. The maximum number of cambium rings at harvest is 12–15 for sugarbeet.

The nutrient application varies, but since the 1970s the rate of N applied is decreasing. Nutrient application depends on soil fertility and the removal of nutrients at harvest. A yield of 55 t of sugarbeets per hectare removes 130 kg N, 30 kg P and 105 kg K.

For large yields it is crucial that the crop can intercept much of the sunlight in the period with long day lengths and high light intensities. Before this, the expansion of the leaf canopy has to be as fast as possible and so the plants need to have access to sufficient sources of N in the soil. An excessive N application, however, influences both sugar content and sugar extractability negatively. For sugarbeet it is not so easy to determine an exact harvest date. The plant is biennial and does not die at the end of the first growing season. Green leaves will still be present in autumn. Sugar amount will increase until the carbohydrate respired overnight exceeds the amount fixed during the day. That often occurs in the period when the days are shortening and the nights are long and still warm. On the basis of dry matter, sugar content is relatively constant. Fluctuation in sugar content is mainly due to fluctuation in the amount of water.

In 2005, the world average yield of sugarbeet was 44 t/ha. Much higher yields, of up to 90 t/ha, were also obtained. In the same year, the world total sugarbeet production was 243 million t. About one-third of the world total production of sugar is from sugarbeet; the remaining part is almost entirely from sugarcane. France, Germany and the Russian Federation are the main producers.

Sugarcane

Sugarcane – *Saccharum officinarum*; Grass family – *Gramineae*

Origin, history and spread

Sugarcane originated in New Guinea, where it has been grown since ancient times. It was probably grown as a native domestic garden crop for chewing. The dispersal of cultivated forms of sugarcane from New Guinea is closely related to migration in ancient times, which includes three main movements. The first brought sugarcane to the Solomon Islands, the New Hebrides and New Caledonia around 8000 BC. The second introduced the plant to Indonesia, the Philippines and India in about 6000 BC. The third main movement, occurring from AD 600 to 1100, brought sugarcane to various island groups including Fiji, Tonga, Samoa, the Cook Islands, Hawaii and other parts of Oceania. Some Hawaiian legends refer to the introduction of sugarcane around AD 750–1000. Alexander the Great brought sugarcane from India to Europe in 325 BC; however, at that time the species was already known to the Persians, Arabs and Egyptians. Introduction of sugarcane into tropical countries of the New World occurred in the 16th century by European settlers. Sugarcane plantations and the extraction of sugar as the basis of the modern sugarcane industry were developed soon after that. However, the preparation of sugar from the cane originated much earlier; it was developed in India around 400 BC, when the juice was removed from the stem by pressing. Next, the juice was boiled and allowed to thicken, and ultimately the sugar crystallized. A negative aspect of the plantation culture is its connection with slavery; labour in many plantations in the New World was often provided by slaves from Africa.

The first shipment of sugar from Brazil to Lisbon was in 1526; the first export of sugar from the Philippines occurred in 1565. Around 1750 European explorers found *Saccharum officinarum* in the South Pacific, which they referred to as 'noble cane' due to the high sugar content. At present cane is grown in more than 100 tropical and subtropical countries.

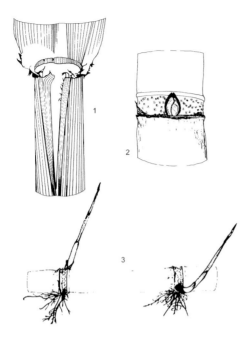

Fig. 10.5. Sugarcane: 1, junction of leaf-sheath and leaf-blade; 2, node with bud, band of root primordia and growth ring; 3, close-up of sprouting nodes.

Botany

Sugarcane (Fig. 10.5) is a giant, perennial, rhizomatous member of the grass family. The stems, 3–5 m tall and 1.3–5 cm thick, are divided into a number of joints, each consisting of a characteristic ring, the node, and an internode. At the base and the top, the internodes are short, and at the base they are also swollen. There is a wide variation in length and diameter of the other internodes. Internodes are usually coated with a waxy layer. Their colour may be green, yellow, brownish-red or red. The alternating leaves arise from the nodes, in two rows on either side of the stem. They consist of a leaf-sheath, and an elongated leaf-blade with a thick midrib, up to 2 m long and 4–10 cm wide, with sharp margins; the strong midrib is white above and green below. The sheaths enclose the stem, the lower ones greatly overlapping each other. The sheath may be smooth or covered with spiny hairs. At the junction of leaf-sheath and leaf-blade, a ligule and auricles can be seen. Auricles and the ligule are used to identify clones. One axillary bud is found on each node, but on alternate sides of the stem; they are usually covered by the leaf-sheath. The stem may terminate in an inflorescence, which is 25–50 cm long. However, flowering is considered as an undesirable characteristic in commercial cane crops because it stops vegetative growth and the development of sugar-storage capacity (Fig. 10.6). The inflorescence is an open-branched panicle in which there are some hundreds of tiny spikelets. These appear in pairs, one sessile and the other attached to a stalk. Flowers are bisexual. The fruit is a small caryopsis, about 1 mm long.

After planting a stem cutting, roots arise from the root primordia on the node at about the same time as a shoot arises from the axillary bud. These roots are thin and called 'sett roots'. They live only until the shoot forms roots of its

Fig. 10.6. Flowering sugarcane, Queensland, Australia (photo: D.L. Schuiling).

own, originating from the basal node of the shoot. This second root system develops into an extensive system with thin and thick roots. The thin roots occur in the top layer of the soil, up to 50 cm deep, spreading about 2 m from the plant. Thicker roots may grow downwards up to 6 m deep, if conditions are favourable.

Species, uses and constituents

The sugar industry around 1900 was based on a number of varieties, often called 'native canes', such as 'The Bourbon', 'Christalina', 'Tanna', 'Creole' and 'Cheribon'. These varieties were cultivated for many years until they were replaced by modern, newly developed hybrids with special characteristics including high sugar content and resistance to diseases. As well as *S. officinarum*, other species have been cultivated for sugar production: among them are *Saccharum barberi*, a species that originated in India, and *Saccharum sinense*, which originated in China. Since the 1920s, the cultivars have often been artificial interspecific hybrids between *S. officinarum* and the wild species *Saccharum spontaneum*. There are also interspecific hybrids developed between *S. officinarum* and *S. barberi*, to obtain disease-resistant cultivars with higher sugar content.

The main product of sugarcane is sucrose, a disaccharide consisting of one molecule of glucose and one molecule of fructose. The stem contains about 85% juice with a sugar content of about 15%; extractable sugar is 11–13%. At maturity the stems are cut and pressed to remove the sugar-rich juice. After that, the juice is heated, clarified and filtered. The filter cake, left over after filtering, consists of juice impurities and the lime that is used in the process is usually applied as fertilizer. Another by-product of sugar processing is wax, which occurs initially on the stem; it can be extracted from the filter cake.

After filtering follows evaporation and concentration to produce syrup; most of the syrup is poured into vacuum pans for crystallization. A small portion of the syrup is marketed as a sweetener (some cultivars are cultivated chiefly for syrup production). Subsequently, the mixture of sugar crystals and non-crystalline liquid, called 'molasses', is centrifuged to separate sugar crystals and molasses. At the end of this process the sugar may be refined, and is dried and packed. Sugar is highly valued as a sweetener for numerous food products and beverages. Sugar is also used as a preservative; it is an essential component of jams, marmalades and fruit sauces.

Molasses consists mostly of glucose and is often used as animal feed. It can also be used for producing alcohol, or fermented and distilled into rum. Yeasts can also be prepared from molasses. The remains of the stem after sugar processing is called 'bagasse'. It is fibrous material which can be used as a fuel, mulch, bedding for livestock, or in the manufacture of paper and cardboard. Sugarcane tops, which are often the highest fully formed nodes, are removed during harvesting and left in the field. The tops are valued as cattle feed, and can be fed fresh or as ensiled product. At present, a large portion of the sugarcane is processed into alcohol, especially in Brazil, for use as a vehicle fuel. In the regions where sugarcane is grown, people often chew fresh cane stem pieces. Fresh, millable cane contains 70% water, 15% fibre, 15% sugar and 1% minerals.

In folk medicine sugarcane is reported to be antiseptic, bactericidal, cardiotonic, diuretic, laxative and stomachic; it is said to be a remedy for arthritis, colds, dysentery, inflammation, sore throat, wounds, and more.

Ecology and agronomy

Sugarcane is grown in tropical and subtropical areas throughout the world, especially in humid lowlands. To obtain a satisfactory yield, temperatures must be above 21°C. The optimum average day temperature is about 30°C. Standing sugarcane plants freeze at –4 to –5.5°C; however, detrimental effect depends on cultivar and length of exposure. Annual rainfall has to be preferably 1800–2500 mm. The total water requirement of sugarcane is high, but its utilization efficiency is also high. If rainfall is not sufficient, irrigation is required. However, to obtain high sugar content in the cane, a 2-month dry period before harvest, together with a period with relatively low night temperatures below 18°C, is required. The species tolerates occasional flooding.

Sugarcane can be grown on a wide range of soils, but it prefers well-drained sandy loams to clay loams with adequate organic matter, a good water-holding capacity, and with pH ranging from 5 to 8.5. Liming is required if pH < 5; saline soils may cause yield reduction.

Sugarcane is propagated vegetatively; mature stems are cut into sections with two or three nodes with buds, known as 'setts', and laid horizontally in furrows. Next they are covered with a thin layer of soil. In most cane fields, the rows are earthed-up once or twice a year (Fig. 10.7).

Fig. 10.7. Planting stem cuttings of sugarcane, Ethiopia (photo: J. Ferwerda).

The spacing in the row varies with country, rainfall, soil type and cultivar, but the number of setts per hectare may be 21,000. The multiplication rate of sugarcane is about 8–10, which means that 1 ha of plants for propagation is needed to plant 8–10 ha of cane. Planting is in rows at about 1–2 m apart. Sugarcane is usually planted as a sole crop. Weed control is carried out manually, mechanically or by using herbicides. Weeding or cultivating has to be done about every 4 weeks and approximately four times per season. Herbicides can be sprayed twice, 1–2 weeks after planting and again 4–6 weeks later.

As sugarcane is a heavy feeder, the drain on soil nutrients is considerable. Fertile soils with heavy dressings of manure and fertilizers are required. Fertilizers are usually applied twice, at planting or some days later and a second application 4–6 weeks later. Actual application rates for each fertilizer depend on physical soil condition and fertility. A crop of 80 t of cane per hectare removes approximately 115 kg N, 40 kg P and 270 kg K. Foliar analysis can be used as guide to fertilizer applications.

Sugarcane is usually grown for many years in the same plantation without rotation. Regrowth from rhizomes, also called 'ratooning', is practised for two or more consecutive cane crops. In general, yields decrease with successive ratooning, and after two or three ratoon crops, replanting is usually required. When the crop can be harvested depends on cultivar and climate, it ranges from 11 to 14 months from time of planting; at higher altitudes growth cycles are longer. At maturity, the canes become tough and turn pale-yellow (Fig. 10.8).

Harvesting of cane can be highly mechanized, however much is still cut by hand. Cutting has to be done as close as possible to the ground because the basal part of the stem is richest in sugar. The leaves have to be removed; they can be burned from the stem by using a special flame thrower. This can be done

Fig. 10.8. Harvestable sugarcane, Barbados (photo: J. Wienk).

after the stems are cut and laid in the rows; however, leaves are also burned from
the standing crop, after which the plants can be cut. The time between cutting
and processing the stems has to be as short as possible since delay results in loss
of sugar content. Processing the canes is usually carried out in mills near the
centres of cultivation. Before the juice can be pressed from the canes, they have
to be washed, cut into pieces, and shredded. Next, the shredded canes are
moved to crushers, which consist of large grooved rollers for macerating the
shredded canes. The crushed, macerated cane is subsequently moved to a
number of rollers with heavy hydraulic pressure for obtaining the sap.

At present, the major countries that produce cane sugar are Brazil, India and
China. The average world cane sugar yield is about 6.5 t/ha. In 2004, the total
world cane sugar production was almost 100 million t.

11 Vegetables

Allium Crops

Common onion – *Allium cepa*, Garlic – *Allium sativum*, Leek – *Allium porrum*,
Shallot – *Allium cepa* – *Aggregatum* group, Chive – *Allium schoenoprasum*;
Lily family – *Liliaceae*

Origin, history and spread

Common onion is believed to originate from central Asia, however true wild
plants do not occur. It is believed that the domestication started in Tadzhikistan,
Afghanistan and Iran. How and when *Allium cepa* spread is not clear but the
ancient Egyptians used onions in around 3000 BC; they have been found in the
coffins of mummies. Onions were also mentioned in the Old Testament as well
as in classical Greek writings. Onion was introduced into western and northern
Europe by the Romans around AD 300, and by the Middle Ages it was a popular
food. The Spanish introduced onion into the West Indies; from there it spread
to all parts of the Americas.

Garlic is believed to be derived from a wild ancestor, which grows in central
Asia. It spread to the Mediterranean region in ancient times. It was already a
popular food in Egypt around 3000 BC. It is an ancient crop in India and China
as well. European colonists introduced it into the New World.

Leek is derived from the wild *Allium ampeloprasum* and its probable first
cultivation was in the eastern Mediterranean area. Leek was mentioned in the
Bible. Since the Middle Ages it has grown in northern Europe.

At present, the three species are grown all over the world, but garlic more
in warmer regions and leek more in colder regions, because leek has a greater
cold tolerance.

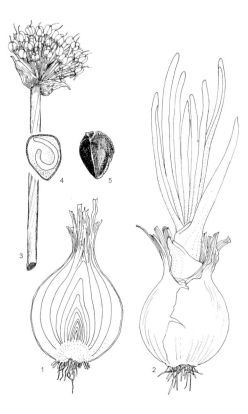

Fig. 11.1. Common onion: 1, longitudinal section of a bulb; 2, regrowth of a bulb (second growing season); 3, infructescence; 4, cross-section of the seed; 5, seed.

Botany

Common onion, garlic and leek are all biennial herbs, but mostly grown as annuals. Only when seed setting is required must they be grown for 2 years, because flowering occurs in the second year. They have in common that, in the first growing season, the overlapping more or less enlarged leaf-bases, including axillary buds, form a bulb or a stalk on a very short stem. The bulb of the common onion (Fig. 11.1) is formed of concentric, enlarged fleshy leaf-bases, which are called 'scales'. The outer leaf-bases dry and the inner leaf-bases thicken when the bulb is maturing. The bulbs may be flattened globose, globose or oval in shape and brown, white or red in colour. The stem is short and disc-like. The three to eight leaf-blades are alternate and cylindrical, hollow when mature, glaucous, up to 50 cm long with an acute top. The inflorescence is a compound umbel, formed on an elongated, hollow and swollen stem, arising from the centre of the bulb. At first the umbel is enclosed within a papery spathe. It may contain up to 2000 greenish-white flowers (Fig. 11.2). There are six stamens. The fruit is a capsule, which contains small wrinkled black seeds. The 1000-seed weight is 3–4 g. The adventitious roots, which arise from the short stem, are fibrous and shallow, spreading just beneath the ground.

Garlic has flat and slender leaves, up to 1 m long. The bulb is made of individual cloves, which are the expanded axillary buds. The cloves are enclosed by a number of leaf-sheaths (scales), which can be white or pinkish. Because of

Fig. 11.2. Inflorescence of common onion just before anthesis (left); inflorescences of chives (right).

the long time of selection and cultivation of garlic, this species has lost the ability to produce fertile seeds. Some cultivars do not flower at all, or flowers are aborted and replaced by bulbils.

Leek has three to ten flat leaves, alternating in two opposite rows, folded sharply, and up to 75 cm long. The overlapping leaf-bases form a long, thick cylindrical stalk or elongated bulb. The inflorescence of leek resembles that of common onion, the flowers can be white, purple or pink.

Uses, constituents and species

The main characteristic of *Allium* crops is their pungency. The volatile oils of *Allium* include disulphides and trisulphides. Fresh common onions contain traces of essential oil, which consists mostly of sulphur compounds. Garlic oil contains a great number of compounds; the characteristic flavour and aroma are found only after cutting or crushing of the cell membranes, which allows the enzyme alliinase to produce the disulphide of the characteristic garlic odour.

Common onion bulbs are used for salads, cooking (e.g. in soups), frying (e.g. with meat), and pickling. Not only the bulb, also the green onion leaves are eaten as vegetables and as culinary herbs. The leaves contain considerable amounts of vitamins A and C. Common onion is probably the main culinary herb in the world for enhancing the flavour of food.

Garlic is used mainly as a condiment for flavouring food, in fresh or dehydrated form. Garlic powder and oil are used in the food industry as flavouring in soups, sauces, sausages, cheeses, vinegars, pickles, and other products. Garlic is considered to be a medicinal plant for lowering blood sugar and cholesterol levels. It is also believed to protect against cancer.

Leek has a mild taste and is eaten raw or as a cooked vegetable, and used as an ingredient of soups and dishes.

As well as the crops described in this chapter, there are about 300 other *Allium* species, among them a number of ornamentals and minor crops. Chive, *Allium schoenoprasum*, is one such minor crop used for flavouring food. The perennial plant is native to Eurasia and widespread in the northern hemisphere. Chives grow in tufts, producing slender cylindrical leaves up to 25 cm long. The flowering stems bear globose heads of purple, pink or rose flowers (Fig. 11.2). The bulbs are lightly swollen and clustered in dense clumps. The mild-tasting leaves are used fresh or dried in egg dishes, soups, salads, cream cheese, and for garnishing. The chive plant is often grown as an ornamental.

Shallot, *Allium cepa – Aggregatum* cultivar group, is derived from common onion by selection among naturally occurring variants. Shallot is a biennial herb, mostly grown as an annual from bulbs. The difference with common onion is that the plant forms a cluster of three to 18 bulbs. The protective bulb-coat leaves are purple, brown or white. The bulbs are oblongoid, globular or oblate, up to 5 cm in diameter, variable in shape, size, colour and weight. The bulb is used as food and spice, for pickling, cooking and frying, but it is also used raw in salads.

Some members of the *Allium* genus can be troublesome weeds in pastures. This is not because of the competition with the grasses, but because of their sulphur compounds. Milk of cows grazing these pastures may have a pungent aroma, which makes it difficult to process the milk.

Ecology and agronomy

Allium crops thrive under a wide variety of climatic and soil conditions; however, they grow best at moderately cool temperatures and on fertile and well-drained clay or sandy loam, pH 6–7. Bulbing depends on temperature and day length.

Common onion grows best in cool climates; optimum temperature for growth ranges between 13 and 24°C. Onion is mostly sown directly at final spacing (Fig. 11.3). At present, in intensive agriculture, most of the wrinkled seed is pelleted, which enables sowing by precision seeding machines. Planting depth is 1–2 cm, with spacing of up to 10 cm. Onions can also be planted from sets, which are small dry onions, 2–3 cm in diameter. Sets have been grown from seed at very high plant densities. In tropical countries the seed is often sown in nurseries, and the seedlings are transplanted to the field.

Garlic growth is restricted to temperatures from 9 to 28°C. The many cultivars show a wide genetic variation in response to temperature and day length. Garlic is mostly grown from cloves. Plant depth is 3–5 cm, and spacing can be 10 cm in the rows and 20 cm between the rows.

The best temperature for growing leek is in the range of 20–25°C. Leek has a greater cold tolerance than common onion and can withstand freezing. The plant is mostly grown from seed. Seedlings are planted in trenches so that the soil can be earthed up to obtain a large, white, tender stalk.

In all *Allium* crops, organic and chemical fertilizers are generally applied, and because *Allium* crops have a shallow root system, a good supply of nutrients in the top of the soil is needed for good growth. N should be applied at the beginning of the growth period in order to stimulate vegetative growth before bulbing. Nutrient application can be based on soil fertility and the amount of

Fig. 11.3. Immature onion plants in the field (left); mature bulbs of onion (right).

nutrients removed by the crop, which for *Allium* species, per tonne of yield, is approximately 25 kg N, 35 kg P and 20–30 kg K. Because onion crops do not compete very well with weeds, weed control is essential. This can be done by mechanical or chemical weed control.

Common onions can be harvested when the necks are dry and about 50% of the tops of the plants bent over. Bulbs are mostly cut from the root system

Fig. 11.4. Leek: rows of plants in the field, to obtain large white tender stalks the plants are earthed-up (left); the plants as marketed (right).

mechanically; after lifting they are often left in the field to cure. After the curing process, the bulb will have a dry shrunken neck and dry outer scales. In 2005, the world total production of dry common onions was about 58 million t. China was by far the most important producer. Other important producers are India and the USA. India and The Netherlands are the largest exporters of common onion.

   Leek can be harvested any time when the stalk has the right size; the plant is harvested in the green state (Fig. 11.4). Garlic is harvested by digging up the underground bulbs when the leaves wither. In 2005, the world total production of garlic was approximately 15 million t; China, accounting for about 10 million t, was by far the most important producer.

Artichoke and Cardoon

Artichoke – *Cynara scolymus*, Cardoon – *Cynara cardunculus*; Daisy family – *Asteraceae*

Artichoke

Origin, history and spread

Artichoke or globe artichoke is probably one of the first ever cultivated vegetables. The plant is not known in the wild; it probably arose from a form of *Cynara cardunculus* (cardoon) on the shores of northern Africa. The ancient Egyptians, Greeks and Romans used artichoke as a food and a medicine. The Greeks and the Romans cultivated it. Theophrastus wrote in the 4th century BC that 'the head of *Scolymus* is most pleasant, being boiled or eaten raw, but chiefly when it is in flower, as also the inner substance of the heads (receptacle) is eaten'. Artichokes were an important vegetable on the menu at feasts in Rome. It was not until the 15th century that artichokes appeared in the rest of Europe. Colonists introduced it into the New World. At present artichoke is grown mainly in the Mediterranean region, Brazil, and the coastal parts of California.

Botany

Artichokes are perennial, silvery-green herbs, which may grow up to 2 m tall and spread outwards up to 2 m. In their appearance they are thistle-like (Fig. 11.5). The arching leaves are often more than 70 cm long and have deep pointed lobes. The leaves are not spiny. The flower stems grow erect and are terminated by a globular head of imbricate oval scales (bracts), which are purplish-green in colour. The scales have a fleshy base. The head may be 8–15 cm in diameter. The scales surround a great number of violet-blue florets in the centre (Fig. 11.6). The scales and florets are borne on the fleshy receptacle. Ray florets are absent. An unopened flower head resembles a green pine-cone (Fig. 11.7). Heads develop continuously on lateral stems. Suckers develop from the base of the plant. Artichokes are deep-rooted plants, the roots are tuberous. The above-

Fig. 11.5. A field of artichokes, Malta (photo: W.G. van den Beukel).

Fig. 11.6. Inflorescences of artichoke (left) and wild cardoon (right), Andalusia, Spain.

Fig. 11.7. Young inflorescences of artichoke: entire (left) and in longitudinal section (right).

ground portion of the plant dies down each year, but off-shoots rise from the rootstock next growing season.

Uses and constituents

The scales and the receptacle are eaten as a vegetable. It can be pickled, baked, fried, boiled or stuffed, and a part of the production is canned. In Italy dried receptacles are used in soups. Young artichokes can be eaten raw. The interior structures of peeled flower primordia can also be eaten. Artichoke is valued for its pleasant bitter taste that is caused by a number of phyto-chemicals such as cynaropicrin and cynarin, which can be found in the green parts of the plant. It contains about 3% protein, 0.2% fat, minerals and vitamins. The nutritive value of artichokes is low; one large head contains less than 50 kcal.

The leaves have been found to have medicinal value attributed to the pres-ence of caffeoylquinic acids and acid derivatives, cynarin and luteolin. It is be-lieved to heal all types of liver and gallbladder disorders and to prevent gallstones and liver diseases. Artichokes are also grown as ornamentals; dried flowering heads are used in floral arrangements.

Ecology and agronomy

Artichokes grow best in a frost-free coastal area with cool humid summers. The species needs a period of chilling, below 10°C, for vernalization and thus to produce flower heads. Hot summer temperatures will cause too rapid growth of flower heads, which may result in poor quality. In areas with hot dry summers, artichokes can best be grown in partial shade. Frost may cause damage to the crowns of the plants, which can be prevented by straw mulching or cutting down the stems and using the large leaves to cover the crowns. The best temperature for growing artichokes is around 24°C during the day and around 13°C during the night. Artichokes can be grown on a wide range of soils but produces best on deep, fertile, well-drained soils. The plants are moderately salt-tolerant.

Propagating by seed is possible although artichoke is usually propagated vegetatively because plants grown from seed lack uniformity. They are best planted by using root or crown divisions, which have to be set 15–20 cm deep in the soil. Planting suckers that are cut off from the base of the parent plants is an alternative. The soil has to be moist, so irrigation at planting may be necessary; however, artichokes will not tolerate waterlogging.

The plant spacing used depends on the production system and varies from 65 to 200 cm within the rows, and from 1.5 to 3.5 m between the rows. Fertilizer should be applied according to current soil fertility. The general recommendation may be 110–220 kg N/ha in the first growing season and 70–110 kg N/ha in each of the following growing periods. Furthermore, in each growing period, 25–50 kg P/ha and 30–100 kg K/ha are recommended. Artichokes require less fertilizer than most other vegetable crops to produce top yields. Artichokes require frequent water supply during the growing period, especially when the heads are forming, so irrigation may be necessary. Weed control is usually a combination of mechanical cultivation and herbicide application.

Artichokes can be harvested when the heads have achieved maximum size but just before the scales begin to spread open. The head is cut with 7–10 cm of stem remaining with the head. When the heads are over-mature, they will be loose, fibrous and inedible. The yield of heads can be up to 14 t/ha. After about 5 years of production, yields decline and replanting is required. In 2005, the world total production of artichoke was approximately 1.2 million t. Italy and Spain were the most important producers.

Cardoon

The closely related species cardoon resembles artichoke very much in habit and appearance, but it is smaller with spiny leaves and scales (Fig. 11.6). The leaves are deeply, once or twice pinnately cut into narrow spine-tipped lobes, with long yellow spines on the margin. The flower heads are 4–5 cm in diameter. Its origin is the Mediterranean area. It is now a minor vegetable crop mainly in the Mediterranean area. The plant is grown for its blanched leaf-stalks, which can be eaten as a cooked vegetable, and used in salads and soups. Cardoons are usually propagated by seed. When the plants are vegetatively full-grown the leaves have to be tied together, wrapped with paper and surrounded by a layer of hay or straw. In about a month the leaf-stalks will be blanched. Cardoon is not grown as a commercial crop.

Asparagus

Asparagus – *Asparagus officinalis*; Lily family – *Liliaceae*

Origin, history and spread

The origin of asparagus is probably the eastern Mediterranean area, especially along coasts and rivers. However, it also grows wild in other parts of Europe, the Caucasian area and western Siberia. Asparagus was already known in ancient Egypt. Painted bunches of asparagus on a sarcophagus suggest that it was cultivated in Egypt. The ancient Greeks used asparagus mainly for medicinal purposes; however, they used a different species, *Asparagus acutifolius*, which grew wild in Greece and still does today. Even today, people collect their young shoots. The word 'asparagus' comes from the Greek word 'asparagos', which means shoot or sprout. In ancient Rome, asparagus was considered to be a table delicacy. The Romans cultivated the crop, as stated by Cato and Pliny. After the Roman period the cultivation of asparagus disappeared more or less from Europe, although it reappeared in the 15th to 16th century, when the Arabs reintroduced it into France. Early settlers brought it into America.

Until the 19th century asparagus was grown for the young, green shoots; blanching started in the 19th century. Green asparagus was in many countries subsequently replaced by white asparagus. Today, asparagus is grown widely in temperate and subtropical regions.

Botany

Asparagus (Fig. 11.8) is a monocotyledonous, perennial, and erect herb, up to 2 m tall. The young stem is fleshy when still underground or up to 25 cm above the ground. The cylindrical stem bears triangular brownish scales, increasingly numerous towards the apex where they are very close together. The scales are the reduced true leaves of the plant. Older and longer stems are tough, strongly branched, with fine, needle-like foliage. In the axils of the scales are three to six subterete, green, needle-like, thin, 1–3 cm long branchlets called 'cladodes', functioning as leaves. Asparagus is dioecious, so male and female flowers are borne on different plants; however, occasionally hermaphrodite flowers are produced. The flowers are solitary or in pairs in the leaf axils. They are small, tubular-campanulate, pendulous, and yellowish or pale green. The petals of the male flowers are 6–8 mm long and those of the female flowers 4–6 mm. The fruit is a globose red berry containing one to six black seeds (Fig. 11.9). The average 1000-seed weight is 40 g. In the subsoil the plant forms a robust woody rhizome comprising a number of bud clusters and many unbranched fleshy storage roots that can penetrate the soil 1–1.5 m deep.

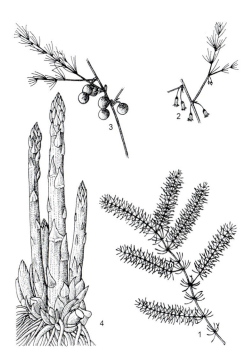

Fig. 11.8. Asparagus: 1, leafy shoot; 2, detail of a flowering shoot; 3, detail of a fruiting shoot; 4, crown with spears (line drawing: PROSEA volume 8).

Furthermore, there are numerous fibrous roots. The rhizomes and the buds make up the crown and together with the fleshy roots it forms the perennial portion of the plant.

Uses and constituents

Asparagus is considered to be a luxury vegetable; it is usually poached in water and consumed with butter and sometimes a little nutmeg. The part eaten is the

Fig. 11.9. Details of flowering (left) (photo: D. van der Linde) and fruiting (right) (photo: D.L. Schuiling) asparagus shoots.

Fig. 11.10. Green (left) (photo: D.L. Schuiling) and white (right) asparagus spears as marketed.

young, about 22-cm-long shoot that is called the 'spear'. The spear arises from a bud on the rhizome.

Most of the asparagus crop is sold to the fresh market, although a considerable portion is processed, i.e. canned, frozen, or dried for soups. There are green and white spears, both coming from the same species (Fig. 11.10). The difference is that white spears do not produce chlorophyll as they are grown in the dark. Spears have to be succulent and tender, toughness is not desired. Development of toughness is caused by the formation of lignin-containing fibre cells in the spear. There are always more fibre cells at the base of the spear than at the tip. White spears generally have higher fibre content than green spears.

The constituents of the green and white spears differ. Green asparagus contains 92% water, 3% protein, 2% carbohydrates and 0.2% fat, and vitamins A, B_1, B_2 and C, and the minerals Fe, Ca and P. White asparagus differs from green asparagus in protein content (2%), and also the contents of vitamins and minerals are lower.

On a small scale, asparagus is still used as a medicinal plant; it is known to have diuretic and laxative properties.

There are over 150 different wild species within the genus *Asparagus*. Only one species is cultivated as a human food. Several other species are cultivated as ornamental pot plants for their attractive foliage such as *Asparagus plumosus* and *Asparagus asparagoides*, both climbing vines native to South Africa.

Ecology and agronomy

The optimum temperature for good growth of asparagus is around 25°C, although the optimum temperature for the accumulation of carbohydrate reserves in the roots is slightly lower. As the gene centre of asparagus is in a subtropical area, the species requires long-day conditions and high light intensity but not too high temperature. Asparagus grows best on well-drained, moist, sandy loams, pH 5.9–6.5. The soil has to be free from compact layers to at least 1 m deep, preferably deeper, which allows the roots to penetrate deeply. It has been proved that crops with deep-penetrating roots produce considerably higher yields. The soil should be permeable and free of stones. The humus content of the soil has to be 2–3%.

Asparagus is a perennial plant; a planting may remain productive for 10–15 years. It can be established either from seed or vegetatively, but when established from seed, it may take up to 3 years to come into production. Vegetative propagation is by division of crowns, by meristem cultures or by layering. Growers are usually provided with seedlings or 1-year-old crowns from nurseries. Seedlings are often raised in containers or plastic trays. At present many seedlings are all-male hybrids, which provide the advantages of higher uniformity of the spears, higher yields and, because the plants do not produce seeds, no seedlings as weed plants.

The crowns meant for producing white spears are often planted in light sandy soils because of favourable temperature and working condition such as cultivation, ridge building, and harvest. For producing green spears, the soil may be somewhat heavier.

Asparagus should not be planted in any field in which asparagus has been cultivated in the past because of soil-borne diseases.

The crowns are planted in furrows about 20 cm deep. The furrow should have a W-shape; the crown is placed on top of the turned V-shaped ridge in the centre of the W-shaped furrow, and the roots are spread on both sides of that ridge. Furrows are not completely filled in at planting; the crowns should be covered by only about 8 cm of soil. The rest of the soil is gradually moved into the furrows with cultivation over the first growing season. The spacing in the row is 25–35 cm with 160–170 cm between the rows. The latter depends on mechanization. For achieving long and white spears, soil is mounded over the row in ridges to exclude light from the developing spears so that they do not produce chlorophyll (Fig. 11.11).

The recommended nutrient application for medium fertile soils for new plantings can be 100 kg N, 55 kg P and 150 kg K per hectare per year. When the field is in full production it can be 100 kg N, 45 kg P and 100 kg K per hectare per year.

Weed control has to start prior to planting or spear emergence in spring by tillage and eventually by herbicides. Perennial weeds, especially when they develop rhizomes, may cause nuisance when harvesting the spears. During the harvest period it is not desirable to use herbicides, weed control has to be done by cultivation. After the harvest period, herbicides can be used again. Although asparagus tolerates drought, it responds positively to irrigation. During the growth period the crop requires 400–500 mm of water. Water requirement is highest when the spears begin to develop.

If 1-year-old crowns are planted, spears can be harvested in the spring of the second year after planting. White spears are mostly cut by hand with a

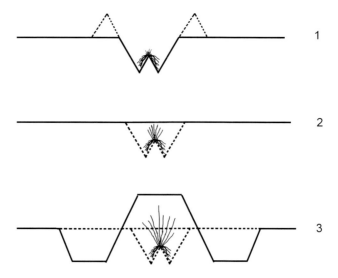

Fig. 11.11. A method (in three successive stages) to plant asparagus plants to obtain white spears.

special knife, just before they break through the surface. The length of the spears is then mostly about 25 cm. Green spears are mostly cut or snapped by hand, although mechanical harvesting is possible. Harvesting has to be done daily, often several times per day. The length of the harvest period depends on temperature but harvest should stop when the diameter and the size of the spears decrease. For an asparagus field that is in full production, the period can be about 8 weeks (Fig. 11.12). At present, some white asparagus is produced under black plastic tunnels for early harvesting.

Yields vary considerably: a field that is in full production may yield up to 10 t/ha for white spears and 5 t/ha for green spears. After harvesting, spears should be washed in cold water and cooled as soon as possible. After that the spears should be trimmed to a uniform length and graded by diameter. Spears for the fresh market can be stored for up to 2 weeks at low temperature (2–3°C) and high relative humidity (95%).

After the final harvest, a number of stems are allowed to grow in order to accumulate enough carbohydrate reserves for the next season via photosynthesis. At the end of the growing season the stems die back and can be removed.

In 2005, the world total production of fresh asparagus spears was approximately 6.6 million t. With almost 6 million t, China was by far the most important producer. A relatively small quantity (0.2 million t) enters international trade.

Fig. 11.12. Fields with asparagus: ridges in which white asparagus spears are produced, darker patches showing where spears were harvested (left); field of asparagus plants, growing from rhizomes from which earlier the spears were harvested (right) (photos: D.L. Schuiling).

Aubergine

Aubergine – *Solanum melongena*; Nightshade family – *Solanaceae*

Origin, history and spread

Aubergine or eggplant is a native of tropical Asia. It is believed to have originated in the Indian subcontinent with China as a secondary centre. The plant was first mentioned in a Chinese book from the 5th century; the next oldest record is from the Arabic world in the 9th century. The scientific name '*Melongena*' was an Arabic name for one type of aubergine. Aubergine reached Spain via North Africa in the 13th century and from there it spread throughout Europe. The Spaniards introduced it into the New World in the 17th century.

Eggplant is so-called because of the white or colourful egg-shaped fruits of some varieties. At first Europeans did not really know how to value the fruit, as proved by the different names they gave the fruit such as 'mad apple' or 'apple of love'. Some thought that eating the fruit would drive a person mad; others thought the fruit had aphrodisiac characteristics.

Now aubergine has spread throughout the (sub)tropics and the warm temperate regions. In cool temperate regions it is also grown in greenhouses.

Fig. 11.13. Aubergine: flowering and fruiting shoot (line drawing: PROSEA volume 8).

Botany

Solanum melongena (Fig. 11.13) is a much-branched, grey-green perennial, but is usually grown as an annual, up to 1.5 m high (Fig. 11.14). The stem is covered with fine hairs. The simple leaves are alternate, with petioles up to 10 cm. Leaf-blades are more or less oval, up to 25 cm long and 15 cm wide, the lower side covered with woolly hairs; base rounded or cordate, often unequal; margin deeply waved. The flowers are 3–5 cm in diameter, solitary or in two- or three-flowered cymes, rarely more, and they develop opposite the leaves (Fig. 11.15). The flowers are hermaphrodite or male. The calyx is tubular-campanulate, about 2 cm long, five- to seven-lobed, woolly and often bearing prickles. The corolla is violet in colour, deeply five- to six-lobed and radiating. The five or six stamens form a cone-like tube around the style. The fruit is a large, pendent, glossy, firm-fleshed berry, containing numerous seeds. Fruit shape ranges from globular to elongate and may be egg- or pear-shaped. Size can be up to 40 cm long and 20 cm in diameter (Fig. 11.16). The colour can be white, yellow, green, purple, black, or mixed in colour. Both shape and colour are very variable. Aubergine has a deeply penetrating taproot and a fibrous root system.

Cultivars, uses and constituents

Important cultivars are 'Black Beauty', 'Long Purple' and 'Florida Market'; numerous local cultivars are grown. The immature fruits are eaten as a cooked vegetable, sliced and fried, or eaten in curries and other dishes. Two classic recipes are 'ratatouille' from France, a vegetable stew, and 'iman bayeldi' from

Fig. 11.14. Habit of fruiting and flowering aubergine plant.

Fig. 11.15. Aubergine flower.

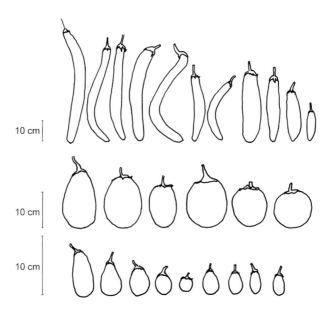

10 cm

10 cm

10 cm

Fig. 11.16. Diversity in shape and size of full-grown aubergine fruits.

Turkey, stuffed fruit boiled in olive oil. Aubergine is also widely used as a traditional medicine. Besides the culinary aubergine there is also an ornamental white aubergine (*Solanum ovigerum*), which is smaller and bears white fruits, resembling a hen's egg in shape and size. The fruit contains 90–93% water, 1–2% protein, 4–6% carbohydrate, about 1% fibre, and furthermore β-carotene and vitamin C.

Ecology and agronomy

Aubergine grows best at day temperatures between 21 and 35°C and night temperatures between 15 and 25°C. The species is day-length neutral. The best environmental conditions are found in lowland areas with little variation in temperature. It grows best on well-drained, sandy soil, high in organic matter, with optimum soil pH 6.0–6.8.

Growing aubergines usually starts from seed in a greenhouse, germination taking about 2 weeks. About 4–6 weeks after sowing the young plants can be transplanted in the field on raised beds or ridges. Watering, such as drip irrigation, is often necessary at the time of transplanting because transplants are very sensitive to water stress. Plant density can be about 25,000/ha, for ridge planting 80 cm between the rows and 50 cm between the plants is often used. To obtain a good production, heavy fertilizing is often necessary; depending on local conditions, 130 kg N/ha, 80 kg P/ha and 110 kg K/ha in split applications is recommended.

Before planting the field, farmyard manure can be used followed by additional fertilizer during the growing period. During dry periods irrigation is often required. Weed control can be done chemically or mechanically. Mechanical weed control should be shallow to avoid damage to the root systems. When plants are established the terminal growing point may be removed to stimulate lateral branching.

Flowering starts 6–8 weeks after transplanting, and the first fruits can be harvested about 5–6 weeks after flowering. Fruits should be harvested before they reach full maturity, because mature fruits have tough and spongy flesh with hard seeds and a bitter flavour. After having produced eight to 14 fruits, the economic life of the plant is over.

In 2005, the world total production of fresh aubergine fruits was approximately 31 million t. China, with 17 million t, is by far the most important producer. Other important producing countries are India and Egypt.

European Brassicas

European Brassicas – *Brassica oleracea*, Kale and collards – *Brassica oleracea* var. *acephala*, Cauliflower and broccoli – *Brassica oleracea* var. *botrytus*, Cabbages – *Brassica oleracea* var. *capitata*, Brussels sprouts – *Brassica oleracea* var. *gemmifera*, Kohlrabi – *Brassica oleracea* var. *gongylodes*; Wallflower family – *Brassicaceae*

Origin, history and spread

Brassica oleracea originates from the coastal areas of southern and south-western Europe and northern Africa, where the wild species can still be found. The plant has given rise to a remarkable number of vegetables with diverse vegetative appearance, such as kale, cabbage, cauliflower, broccoli, Brussels sprouts and kohlrabi. This is caused mainly by selection for different harvestable parts of the plant during many centuries. A leafy *B. oleracea* was probably first cultivated by the Greeks about 2500 years ago. They must have selected first for a low content of bitter-tasting glucosinolates because the wild species has a high content, making the plant inedible. The ancient Romans grew early types of cabbage and possibly precursors to broccoli and kohlrabi. In the following centuries kale and cabbage-like varieties spread over Europe. The Saxons and the Celts grew them in northern Europe. Cabbage is referred to in German writings dating 1000 years ago. Cabbage can often be seen on Dutch paintings from the 16th and 17th centuries.

Cauliflower and broccoli are thought to have been selected in the 15th or 16th century, cauliflower in northern Europe and broccoli in the eastern Mediterranean region. Kohlrabi was first described in 1554. Kohlrabi is probably derived from the so-called 'Pompeian kale' of the ancient Romans. Most of the varieties of *B. oleracea* probably arose in several different areas. Brussels sprouts first appeared as a spontaneous mutation in Belgium in the 18th century. Colonists introduced *B. oleracea* varieties into the New World. At present, *B. oleracea* varieties are grown throughout the temperate zone and on a small scale in the tropics and subtropics.

Botany

Cultivated forms of *B. oleracea* vary greatly in their vegetative appearance but are more or less similar in the generative stage (Fig. 11.17). They are annual or biennial erect plants, up to 2 m tall, glabrous, often much-branched in the upper part. The inflorescence is a raceme, bearing numerous flowers, usually yellow, sometimes white in colour. Flower buds are raised far above the ex-panded flowers. The flower consists of four erect sepals; four petals, about twice as long as the sepals; six stamens, two relatively short outer ones and four longer inner ones. The ovary is superior and consists of two united carpels divided by a false septum, and a single, lobed stigma. The fruit is a siliqua, 5–10 cm long, with numerous ovules, arranged in rows. Seed is globose, 2–4 mm in diameter, and grey-black to red-brown in colour. Seeds have no endosperm and consist mainly of two cotyledons, which are rich in oil. The 1000-seed weight is 2.5–4 g. The root system consists of a taproot with laterals.

Uses, constituents and varieties

Hundreds of cultivars exist, grown for various parts of the plant such as the leaves, stem, inflorescence, axillary buds, or a combination of these. Most of

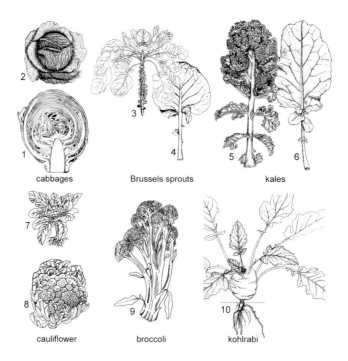

cabbages   Brussels sprouts   kales

cauliflower   broccoli   kohlrabi

Fig. 11.17. European Brassicas: 1, longitudinal section of red cabbage; 2, Savoy cabbage; 3, plant habit of Brussels sprouts; 4, leaf of Brussels sprouts; 5, leaf of curly kale; 6, leaf of marrow stem kale; 7, plant habit of cauliflower; 8, inflorescence of cauliflower; 9, inflorescence of fasciated type of broccoli; 10, plant habit of kohlrabi.

them are used as vegetables, mainly cooked or raw in salads, and highly valued because of the high vitamin C content. The Tartars introduced pickled cabbage, known as 'sauerkraut', into Europe. Pickling and cooking processes do not destroy the vitamin C. Pickled cabbage was once eaten by Dutch and English sailors during their long voyages of discovery as a defence against scurvy.

B. oleracea contains 80–90% water, 2–4% protein, 5–7% carbohydrates, some fibre, carotene, Ca, Fe, vitamins A, B_1 and B_2, and very high levels of antioxidant and anticancer compounds. One of the biochemical characteristics of *Brassica* species is the presence of sulphur-containing compounds such as glucosides and glucosinolates (see Rapeseed, Chapter 6). The sulphur compounds are responsible for the typical flavours of *Brassica* vegetables and cause also the odour produced during cooking. Some cultivars are grown as feed for livestock or as ornamentals. A part of the production is processed by quick-freezing, canning, pickling or drying. *B. oleracea* varieties are organized into groups, of which the following five have more than regional importance.

1. Variety *acephala*, including kale and collards.
2. Variety *botrytis*, including cauliflower and broccoli.
3. Variey *capitata*, including green, red and Savoy cabbage.
4. Variety *gongylodes*, including kohlrabi.
5. Variety *gemmifera*, including Brussels sprouts.

Kale and collards

The consumed parts of the plant are the leaves. Curly kales (Fig. 11.18) are hardy winter greens; they can withstand frost very well. A cold period is even desired because it improves the flavour by increasing the sugar content of the leaves. They are biennial varieties, but mostly grown as annuals. The plants usually have a simple, erect stem, bearing large, oblong leaves. The leaves are loosely arranged alternately up the stem. The colour of the leaves may be green or brownish-purple. Variety *acephala* most closely resembles the wild *B. oleracea*. Different cultivars vary greatly in height, 0.5–1.5 m, and the appearance of the leaves, from exposed to extremely curly and crimpy. The exposed forms can be eaten by humans but are often used for feeding livestock; the curly forms are more popular for human consumption. Another form called 'marrow stem kale' is grown for the enlarged stem, up to 2 m tall, and big leaves. The crop is cultivated in a lower plant density per hectare than the other forms to allow stem enlargement. Stem diameter can be about 10 cm. This form is used only as feed for livestock. It takes 3–4 months from sowing to harvesting. The whole plant can be cut, or the leaves can be stripped off so that the plant continues to produce new leaves. The leaves can be used fresh, or processed by quick-freezing or drying.

Collards are grown mainly in the USA. They are valued for their smooth, rather thick and tender leaves, which are used as greens. The plant produces a rosette of leaves and the leaf surface is generally exposed. In Western Europe, collards are grown as forage.

Fig. 11.18. Brussels sprouts (left); curly kale (right).

Kales with purple or yellow-green leaves are becoming more and more popular as ornamental garden plants. Because they are very winter-hardy, they can be used to maintain colour in winter gardens.

Cauliflower and broccoli

The consumed part of cauliflower is the 'head' (Fig. 11.19), consisting of a mass of swollen, proliferated floral meristems, also known as 'curd'. In the curd, no individual flower buds can be distinguished. The branches of the inflorescence are likewise greatly swollen. The colour of the curd is mostly white, sometimes purple. The white curd may become discoloured and develop an undesirable flavour when exposed to sunlight. To avoid that, a number of the larger leaves are tied loosely together over the head about 10 days before harvesting. This blocks the sunlight and prevents the formation of chlorophyll to keep the curd white. Some cultivars have leaves that are wrapped over the heads by themselves, so tying is unnecessary. Fifteen to 25 large oblong leaves, with a fleshy, white midrib, envelop the head. The leaves are almost sessile, glabrous, covered with a layer of wax, with smooth or curly edges, and blue-green in colour. Axillary buds usually do not develop shoots. The plant is 50–80 cm tall in the mature, vegetative stage.

Cauliflowers mature in 60–180 days, depending on cultivar and planting season. They can be harvested when the head is still white, compact and firm,

Fig. 11.19. European Brassicas: 1, cauliflower; 2, broccoli; 3, kohlrabi.

with a diameter of 15–25 cm. The heads are cut with some trimmed leaves still attached to protect the heads during transport.

The consumed part of broccoli, also known as 'calabrese', is the immature inflorescence consisting of green to purple buds and thick enlarged stems (Fig. 11.19). The name 'broccoli' is derived from the Italian word 'brocco', which means 'branch'. The flowering stems are often fasciated (some parallel stems grow together). Two types of broccoli can be distinguished: the single-headed type and sprouting broccoli. The single-headed type has a more or less compact, rather big main head, consisting of clusters of fully differentiated flower buds. Sprouting broccoli bears many small flower heads. The leaves are petiolate, divided and blue-green in colour.

Harvesting must be done before the buds open and the yellow petals become visible, which may be 50–85 days after transplanting. Broccoli heads are harvested with 10–15 cm of stem without leaves. When the terminal head is cut, shoots from axillary buds develop new, smaller heads.

At present, many cultivars exist with intermediate characteristics of cauliflower and broccoli, as hybrids between the two forms. Especially in Italy there is a great diversity of cauliflower- and broccoli-like vegetables. Some cauliflower cultivars have evolved in forms with peculiar head formations and different colours. This often causes confusion regarding the classification of these vegetables.

Cabbages

The consumed part of cabbage is the greatly enlarged terminal bud, consisting of a great number of tightly packed, enrolled leaves, borne on a short stem. The 'head' is encircled by a number of loose, basal leaves, spreading away from the head. They protect the head during growth. The basal leaves can be large, 35 cm × 35 cm; they are rather coarse and not suitable for human consumption. At harvest, or in preparing the head for consumption, they are usually removed. The different cabbage cultivars vary in colour, shape and leaf texture. There are hundreds of cultivars of cabbage; they can be grouped into smooth green, red, and Savoy cabbage.

The shape of the smooth green cabbage can be round, flattened or conical, and the colour may be blue-green or yellow-green. The head size can vary widely; the large heads are used for pickling ('sauerkraut' in Europe and 'kimchi' in Korea). Red cabbages are round or oblong, reddish-purple, and have very tight heads. Savoy cabbages are round, the leaves are crinkled and puckered, the outer leaves are blue-green, the inner leaves yellow-green. Savoy has a looser head than green and red cabbage (Fig. 11.20).

Cabbages can be harvested as soon as the heads are fairly firm, 80–120 days after germination. When they are relatively young they are tender and have a good flavour; over-mature cabbage can burst and become dull and fibrous. Cabbages can be stored for a rather long time in controlled atmosphere storage. Quality in cabbage depends upon rapid growth, so it needs sufficient nutrients and water supply. Cabbage is a cool-season crop which tolerates frost but not heat.

As well as cabbages for consumption, there are ornamental cabbages. Colours range from white, to pink, purple or red. The heads are mostly looser and the leaves are much fancier than 'normal' cabbages. They are valued in winter gardens due to their bright colours. Ornamentals are extremely cold-tolerant; light and moderate frost will intensify the colouring of the plants.

Fig. 11.20. Cabbages: 1, smooth green cabbage; 2, red cabbage; 3, Savoy cabbage.

Brussels sprouts

The consumed parts of Brussels sprouts are the enlarged buds in the axils of the petiolate leaves, known as 'sprouts' (Fig. 11.18). They resemble miniature cabbages but are much smaller, diameter 3–5 cm. The leaves are petiolate and the leaf-blade is rather small and subcircular. The first sprouts are formed at the base of the stem, working upwards in a tight spiral pattern. On top of the stem is a rosette of leaves. When the sprouts grow to a large size, the leaves will drop off and the stem becomes completely covered with tightly packed sprouts. About 3 weeks before harvest, the plant may be topped to remove the growing point. This stimulates the growth of the sprouts at the top of the plant and enhances uniform development, which facilitates once-over mechanical harvesting.

Plant densities in the field can manipulate the size of the plant and the sprouts. Increasing plant density results in reduced sprout size, increased stem length and sprouts spaced further apart on the stem, delayed maturity, and greater uniformity of sprout development.

Sprouts must be harvested when they are the right size (3–5 cm), when they are still firm and before they turn yellow. Harvesting can start 80–110 days after transplanting. The sprouts begin maturing from the bottom upwards. If harvesting is done by hand, the mature sprouts are picked first. The remaining sprouts can be picked later, repeatedly.

For harvesting the whole stalk at once mechanically, harvest time can be delayed. The best-quality sprouts are produced when days are sunny and the nights are cold. Brussels sprouts grow slowly and require about 100 days from transplanting for the first sprouts to mature. Sprouts can be marketed fresh, canned or frozen. Occasionally the tops of the plants are used as greens. At present, most of the cultivars are F_1 hybrids, valued for their uniformity and high yield.

Kohlrabi

The consumed part of kohlrabi is the enlarged, turnip-shaped stem that develops just above the ground, and from which the leaves develop (Fig. 11.19). The enlarged stem can be white-green or purple, with creamy white flesh; the shape can be globose or flattened. The inside consists mainly of much-enlarged pith surrounded by a ring of vascular bundles. The leaves are glaucous with slender petioles, arranged in a compressed spiral on the swollen stem. When the leaves fall off from the basal part, conspicuous leaf scars mark the surface. The plant is a biennial, but grown as an annual. The name 'kohlrabi' is German and means 'cabbage-turnip', which is a good description because the plant looks somewhat like a turnip but is botanically more related to the cabbage. The plant has to grow rapidly to ensure tender and succulent flesh, although the weather has to be cool, because hot weather may cause an inferior product. Kohlrabi can be harvested about 60 days after transplanting, when the stem is about 8 cm in diameter and the flesh is still tender. Over-mature kohlrabi will become ligneous and fibrous. The leaves of young plants can be used as mustard greens or cooked like spinach.

Kohlrabi can be eaten raw or as a cooked vegetable, but can also be added to soups and stews. The flavour is similar to that of turnip but milder and sweeter. In some European countries, it is also used as feed for livestock. When the leaves are removed, kohlrabi can be kept in good condition for 2–3 months, if stored under controlled atmosphere conditions.

Summary on the edible parts of cabbage or kale crops

- Kale and collards: exposed or curly and crimpy leaves.
- Cauliflower: mass of swollen, white or purple, proliferated floral meristems or curd.
- Broccoli: immature inflorescences consisting of green to purple buds and thick, enlarged stems.
- Cabbage: greatly enlarged terminal bud, consisting of a great number of tightly packed enrolled leaves, borne on a short stem, leaves smooth, green, red, or crinkled, green and yellow.
- Brussels sprouts: enlarged buds in the axils of the petiolate leaves.
- Kohlrabi: enlarged, turnip-shaped stem that develops just above the ground.

Ecology and agronomy

B. oleracea crops are generally temperate-region crops, although some varieties are more or less adapted to higher temperatures. Kales and Brussels sprouts are the most winter-hardy crops. They can withstand temperatures of −10°C, but also tolerate high summer temperatures. Cabbages, cauliflower and broccoli produce best at average daily temperatures of 15–20°C and a diurnal variation of at least 5°C, such conditions can also be found in tropical countries at elevations above 800 m. Some cauliflower cultivars, developed in India, are heat-tolerant, but when grown under tropical lowland conditions the curd quality is usually inferior. Low temperatures contribute to early flowering and completing floral development. Kohlrabi prefers temperatures of 15–20°C but it is resistant to frost. Cabbage crop varieties thrive best on medium heavy, well-drained and fertile soils (loam), having good moisture-retaining capacity and high organic matter content. Optimum pH is 6–7.

All cabbage crops are grown from seed, often sown on seedbeds, in paper pots, or peat blocks. Seedlings may grow up in a greenhouse, heated or non-heated, for an early crop or outdoors for later crops. Seedlings can be transplanted when they have seven to nine true leaves. Seedlings are less cold-tolerant than older plants. Seed can also be planted directly in the field, by a precision seeder, 1–2 cm deep. Crop rotation is very important because of soil-borne diseases such as club root. No cruciferous crops should have been grown in the field for 4 years. Plant density per hectare for the different crops is (×1000): kales and collards 30–40, cauliflower and broccoli 20–40, cabbages 25–60, Brussels sprouts 30–40, kohlrabi 200–270. A regular supply of water is required throughout the growing season. Most cabbage crops have a high N requirement, although fertilizer application should ensure adequate levels of all nutrients. The biennial varieties need vernalization during winter for flower initiation. Yield variation of the different crops in temperate regions is (t/ha): kales and collards 10–25, cauliflower 12–30, broccoli 4–10, cabbage 40–60, Brussels sprouts 5–10, kohlrabi 15–40. In 2005, the world total production of cabbages was about 70 million t and of cauliflower about 16 million t.

Oriental Brassicas

Chinese cabbage – *Brassica rapa* var. *pekinensis*, Pak choi – *Brassica rapa* var. *chinensis*, Chinese kale – *Brassica oleracea* var. *alboglabra*; Wallflower family – *Brassicaceae*

Chinese cabbage

Origin, history and spread

Chinese cabbage originated in China. It probably arose as a natural hybrid between turnip from the north of China and 'pak choi' from the south of China, both plants being varieties from *Brassica rapa*. Chinese cabbage is not known in the wild state. It was first mentioned in Chinese literature from the 5th century AD. It was not until the 19th century that it reached Japan and North America. It was introduced into Europe in the 20th century. At present it is one of the most important vegetables in East Asia and is grown all over the world.

Botany

Chinese cabbage (Fig. 11.21) is a biennial herb, but mostly grown as an annual. The plant is 20–50 cm tall in the vegetative stage, growing up to 1.5 m in the generative stage (for generative stage see Rapeseed, Chapter 6). During the vegetative stage leaves are arranged in an enlarged rosette, forming a more or

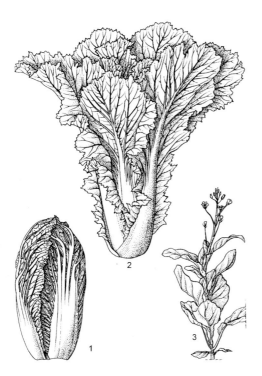

Fig. 11.21. Chinese cabbage (*Brassica rapa* var. *pekinensis*): 1, head; 2, habit; 3, flowering shoot (line drawing: PROSEA volume 8).

less compact head (Fig. 11.22). There are three main types: (i) the tall cylindrical type; and (ii) the hearted type, both headed cabbages; and (iii) the loose-headed type. The tall cylindrical type has long erect leaves, forming a compact but tapering head without wrapping leaves over the top. It grows up to 50 cm tall and 10–15 cm wide. Hearted types have large compact heads, with the leaves closely wrapped, usually over the top. It grows up to 25 cm tall and up to 20 cm wide. Loose-headed types have funnel-shaped, rosetted heads, but there is a lot of variation within this group. It includes a range from very loose to much denser types. The different types of Chinese cabbage cross easily so breeders have made many crosses that have characteristics in between the three types. The colour of the outer leaves ranges from dark green to pale green, sometimes almost yellow. The inner leaves are always paler. The shape of the leaves varies with growth stage.

Outer leaves can be narrowly ovate with long petioles, inner leaves broad, subcircular, and the surface can be smooth to wrinkled, sometimes hairy. The leaves are prominently veined. The bases of the leaf midribs are mostly white and swollen, often overlapping each other. The size of the leaves may be 20–90 cm × 15–35 cm. In the mature vegetative stage Chinese cabbage may form 120 leaves. The root system consists of a taproot and laterals.

Uses and constituents

The flavour of Chinese cabbage is very delicate. With the crispy texture of the leaves it is very suitable for salads. It can also be eaten as a cooked vegetable, but it needs light cooking otherwise the flavour is destroyed. It is very suitable for stir-frying. In Asia large quantities of Chinese cabbage are used to make

Fig. 11.22. Cabbages as marketed: 1, head of pak choi (*Brassica rapa* var. *chinensis*); 2, small head of pak choi variety marketed as 'baby bokchoy'; 3, head of Chinese cabbage (*Brassica rapa* var. *pekinensis*); 4, longitudinal section of head of Chinese cabbage.

salted and fermented pickles such as the famous Korean 'kimchi'. Leaves are sometimes dried and used in soups. The raw midribs of the leaves can be used in salads. Fresh leaves contain approximately 95% water, 2.2% carbohydrates, 1.2% protein, 0.2% fat and 0.5% fibre, as well as Ca, Fe, and vitamins A and C. It is low in energy value. Usually the mature heads are harvested; there are early- and late-maturing cultivars. In Asia it is also grown as a seedling crop (for harvesting very young plants), for semi-mature heads, or for the sweet flowering shoots. Seedling leaves are often rather hairy, and are best eaten cooked rather than raw.

Ecology and agronomy

Chinese cabbage grows best in climates with temperatures in the range of 13–22°C; it withstands light frost, but temperatures above 35°C can cause abnormal development of the heads. However, cultivars have been developed in the last decades that are more adapted to tropical climatic conditions. They prefer deep, rich, well-drained soils with high organic matter content. Very light, very heavy and very poor soils should be avoided. The soil pH should be between 6 and 7.5.

Irrigation will often be necessary because drought stress in the heading stage will prevent head formation; furthermore, Chinese cabbage has a high water requirement. To avoid damage by soil-borne diseases, it should be rotated with unrelated crops over at least 4 years. Seed, usually achieved from hybrids, is used to propagate Chinese cabbage. Seed can be sown directly as a seedling crop, although it is usually sown in a nursery and then transplanted. The 1000-seed weight is 2.5–3.5 g; 500 g of seed per hectare is needed for sowing. In transplanting, 50,000 plants/ha are used whereas for broadcasting approximately 80,000 plants/ha are used. The spacing of the plants in the row is often 40 cm. Application of fertilizers depends on type of soil, soil fertility and the nutrient depletion by the crop. The yield of Chinese cabbage can be 50 t/ha. To achieve this yield the crop removes about 150 kg N, 70 kg P and 330 kg K from the soil. Weed control can be carried out by weeding, cultivation and herbicides. Time from sowing to harvest varies from about 55 to 100 days, harvesting is mostly done by hand. Heads can be harvested when they are compact, which means that they do not collapse easily when pressed firmly with both hands. The average crop yields range from 10 to 60 t/ha, depending on cultivar, soil, climate and season.

Pak choi

Pak choi is probably older than Chinese cabbage. It arose in south China and is known to have been cultivated in China since the 5th century AD. It has now spread throughout Asia, especially in southern and central China and Taiwan. It is also grown in Europe and North America. Breeders have developed varieties that are adapted to tropical climatic conditions, so on a small scale pak choi is also grown in Africa and South America. However, it is less important commercially than Chinese cabbage.

Pak choi is an erect biennial herb, although mostly grown as an annual, in the mature vegetative stage 15–60 cm tall (Fig. 11.22). The 15–30 leaves are arranged spirally, are glabrous and spreading, and have a dull green colour. The petioles are enlarged, terete or flattened, 1.5–4 cm wide and 0.5–1 cm thick, white or greenish white, and growing upright forming a sub-cylindrical bundle. Petioles broaden at the base into a spoon-like shape. Petioles vary in length from 10 to 30 cm. Leaf-blades are orbicular to obovate, 7–20 cm × 7–20 cm. There are many forms varying in height, colour, shape of the leaf-blades, and length, thickness and shape of the petioles. One variety called 'rosette pak choi' is believed to be of very ancient origin. It forms a loose rosette of often puckered leaves with short petioles. This variety grows 15–25 cm high with a diameter up to 50 cm. Pak choi has a shallow, finely branched root system.

The crop is mostly eaten as a cooked vegetable; all of the above-ground parts are edible. The succulent petioles are highly valued. It is used in soups and stir-fried dishes. It is seldom eaten raw or pickled and fermented. In Asia pak choi is eaten in several stages of maturity: the large seedling stage, which means within 2 weeks of sowing; the more or less fully developed vegetative stage; and the early generative stage, for its young flowering shoots. In some parts of Asia, there is a long tradition of drying pak choi leaves for use in the winter months. It contains approximately 93% water, 3% carbohydrates, 1.7% protein, 0.7% fibre and 0.8% ash. It is a good source of vitamins and minerals such as β-carotene, vitamin C, Ca, P and Fe.

Most of the pak choi varieties are naturally cool-season crops and grow best at temperatures of 15–20°C; some varieties are adapted to higher temperatures. Most varieties tolerate some frost. It has to be grown in fertile, moisture-retentive soil at a pH range of 6–7.5. A lack of moisture leads to premature bolting and poor-quality plants. When the crop is sown early in the year, it may run to seed too rapidly. Pak choi resembles Chinese cabbage in its requirements and is grown in much the same way (Fig. 11.23). Pak choi should be harvested when the leaves look fresh. It can be harvested by either picking a few outer leaves at a time or cutting the whole plant above ground level. Picking a few leaves at a time stimulates the plant to re-sprout vigorously, which may give several pickings over a long period. Pak choi is best used fresh, as it does not store well. Yields vary from 10 to 30 t/ha.

Chinese kale

The consumed parts of Chinese kale, also called 'Chinese broccoli', are the young flowering stems with flower buds and small leaves (Fig. 11.24). When they are very young, the whole plant can be eaten. It is a close relative of European *B. oleracea*, possibly sharing a common ancestor with European broccoli. Botanically, it is very close to Portuguese cabbage (*Couve tronchuda*, syn. 'Braganza cabbage'). The Portuguese probably introduced it into Asia. In China and South-east Asia, including Japan, *Brassica* vegetables are very popular and numerous varieties exist. Most of them are derived from *Brassica rapa* and *Brassica juncea*.

Fig. 11.23. Intercropping of pak choi (*Brassica rapa* var. *chinensis*) and leek. The pak choi is about 1 month old, Bali, Indonesia (photo: D.L. Schuiling).

Fig. 11.24. Chinese kale (*Brassica oleracea* var. *alboglabra*) with flowering shoot above.

Chinese kale is an annual herb, up to 40 cm tall in the vegetative stage. All parts of the plant are glabrous and glaucous. The stem is single, narrow branching in the upper part, the flower stems are chunky, succulent and very smooth, on average 1–2 cm in diameter. Leaves alternate, they are petiolate, thick and firm. The leaf-blades are ovate to orbicular-ovate; they can be dull or glossy, smooth or wrinkled. Flowering starts 55–80 days after sowing, depending on cultivar and production system. The flowering shoot is less compact than the European broccoli shoots. Flowers are larger than most *Brassica* flowers and white or yellow.

Harvesting can be done at any growth stage, from 3 weeks after sowing up to the early flowering stage, but just before the first flowers open. After harvest, the plants can only be kept fresh for some days. Chinese kale is a vigorous, fast-growing plant, producing secondary shoots over a long period after the main flowering shoot has been cut. The secondary shoots can be harvested repeatedly. This process is called 'ratooning'. It is usually eaten as a cooked or fried vegetable, sometimes raw. The plant differs from most other *B. oleracea* varieties regarding climatic requirements; it can withstand some frost and also tolerates high temperatures, although it grows best at temperatures between 8 and 28°C.

The agronomy of Chinese kale is more or less similar to that of Chinese cabbage. There are numerous landraces of Chinese kale in China and Southeast Asia. The closely related Portuguese cabbage is a traditional cabbage from Portugal with a loose or open head. The whole plant can be eaten, but often young leaves and stalks with young, compact florets are selected; the leaves and ribs are very sweet. In Portugal, many landraces of Portuguese cabbage still exist. Yields of Chinese kale range from 6 to 12 t/ha.

Capsicum peppers

Sweet pepper – *Capsicum annuum*, Red pepper – *Capsicum frutescens*; Nightshade family – *Solanaceae*

Origin, history and spread

Capsicum peppers are native to tropical and subtropical America. Cultivated *Capsicum* peppers are probably derived from the wild peppers, among others from the 'wild bird pepper' (*Capsicum annuum* var. *aviculare*), because many cultivated types and wild bird pepper are genetically closely related. *Capsicum* was already consumed by man at least 7500 years ago. Domestication happened about 5500 years ago. Native Americans took wild peppers and selected various types, ranging from mild types to hot, very pungent types. Wild bird pepper can be found in the southern part of the USA, Mexico, the Caribbean and northern South America. Wild peppers are much smaller than cultivated peppers; they are the size of a pea and are extremely hot. Aztecs called peppers 'chilli', the same word that is used today. Indians used *Capsicum* species in cooking and as medicines. For example, the Mayas and Aztecs used the species to treat throat and lung ailments and to relieve toothaches.

When Columbus came to the New World in the 16th century, he first thought mistakenly that he had arrived in India. Therefore he called the population 'Indians' and their most important condiment 'red pepper', due to the same pungency as from black pepper (*Piper nigrum*), a totally unrelated species from India.

Spanish and Portuguese explorers introduced *Capsicum* pepper into Europe and Asia in the 16th century. After introduction, *Capsicum* pepper partly replaced black pepper as the most important condiment. However, using the name 'pepper' for two different species was confusing. Influenced by Dutch traders of black pepper, Spanish merchants tried to introduce the name 'pimento' for *Capsicum* pepper. They succeeded only partly because the name 'pimento' is used only for one form of sweet pepper, used mainly in Spain.

The name '*Capsicum*' is probably derived from the Greek word 'kapto', which means 'to bite', referring to the pungent taste of many *Capsicum* fruits. However, others say that the name is derived from the Latin word 'capsa', which means 'box', referring to the resemblance of the fruit to a box in which the seeds embedded.

At present, *Capsicum* pepper is widespread throughout the world, the sweet types mainly in temperate regions such as Europe, North America and Australia, often grown in greenhouses, for example in The Netherlands. The pungent red pepper types are grown mainly in Central and South America, Asia and Africa.

Botany

C. annuum (Fig. 11.25) is an erect, annual herb or sub-shrub with many branches and is up to 1.5 m tall. The stem is irregularly angular to subterete and about 1 cm in diameter, sometimes with a purple spot near the nodes. The leaves alternate and are simple, oval to lanceolate. Leaves are variable in shape and size:

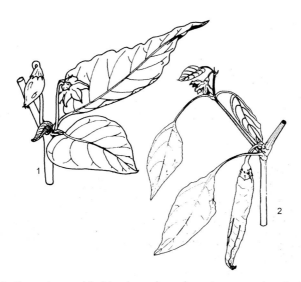

Fig. 11.25. Flowering and fruiting branches of capsicum species: 1, *Capsicum annuum*; 2, *Capsicum frutescens*.

they may be small but also up to 10–16 cm × 5–8 cm. Petiole may be up to 10 cm long. Leaves are mainly glabrous. The flowers are usually solitary and appear in the leaf axils but, because of growth habit, are in fact terminal. The calyx is bell-shaped with five teeth, formed by five fused sepals, and the corolla wheel- to bell-shaped with five or six fused petals. The calyx is green and the corolla is usually white in colour. There are five or six stamens inserted near the base of the corolla. The fruits are extremely variable in relation to length, width, colour and pungency (Fig. 11.26). The colour is often red, yellow or brown. Botanically, the fruit is a berry but is often called a 'pod' (Fig. 11.27). The shape of the fruit may be conical, spherical or box-like, and the fruit is filled with air. The seeds are attached to the interior ribs of the fruit and are flat, yellowish-white and kidney-shaped. The 1000-seed weight is 4–8 g. The species has a strong taproot and numerous lateral roots.

Capsicum frutescens (Fig. 11.25) is a perennial herb or sub-shrub, up to 2 m tall, mostly grown as an annual. Stem, leaves, flowers and roots are almost similar to those of *C. annuum*. The flowers are borne in clusters of two or more in the leaf axils. The colour of the flowers is purple or greenish white. Generally, the fruits are smaller than the fruits of *C. annuum*. The colour of the fruit is bright red, size and shape range widely.

Cultivars, uses and constituents

There are more than 20 *Capsicum* species; however, all of the cultivated varieties belong to five species: *C. annuum*, *C. frutescens*, *Capsicum chinense*, *Capsicum baccatum* and *Capsicum pubescens*.

C. annuum, *C. chinense* and *C. frutescens* have worldwide importance. The origin and evolution of *C. annuum*, *C. frutescens* and *C. chinense* are so

Fig. 11.26. Array of fruits of hot and sweet capsicum cultivars, seeds, and a green *Capsicum annuum* fruit in longitudinal section and in cross-section (photo: J. van Zee).

Fig. 11.27. Flower (left) and fruit (right) on the plant of sweet pepper (*Capsicum annuum*).

closely interwoven with each other that the three species of this complex are often difficult to distinguish. *C. baccatum* and *C. pubescens* have only regional importance, especially in Central and South America.

C. annuum, the non-pungent form being called 'bell pepper' or 'sweet pepper', and *C. frutescens*, also called 'red pepper', 'cayenne pepper' or 'chilli', are the main species. In general, it can be said that most of the cultivated *C. annuum* cultivars taste mild and sweet, and most of the *C. frutescens* cultivars taste pungent and hot; however, within both species exists a whole range in pungency.

Sweet peppers are used fresh or dried. Fresh peppers are used as a vegetable or in salads, they can be cooked and used in stews and other dishes, and they are sometimes sliced and pickled. In the south of Europe, pieces of the sweet pepper fruits are used to stuff olives. The fruits are also used in making the famous Hungarian 'goulash'. They are an important ingredient of the well-known French dish 'ratatouille'.

Dried sweet pepper is mostly ground and sold as a powdered product known as 'paprika', the Hungarian word for *Capsicum*. Paprika in Hungary is rather pungent; paprika in international trade is almost always non-pungent. Paprika is used to flavour or colour many foodstuffs such as meat, sauces, snack food, salads and sausages, but it is also used in poultry feed and pharmaceutical products.

Red peppers are also used fresh or dried (Fig. 11.28). They are mainly used to flavour dishes in mostly tropical countries. Red peppers, especially the very pungent ones, are called 'chillies'. Some chillies are so hot that they may irritate the skin at picking or processing the fruits. That is why the very hot peppers are picked only with gloves. Small-sized, pointed and very pungent red peppers are often called 'bird's eye' because these fruits resemble the pupil of a bird. Red

Fig. 11.28. Sun-drying of harvested hot pepper fruits (*Capsicum frutescens*), Java, Indonesia (photo: S. Bot).

peppers form an essential ingredient of curry powder, Tabasco sauce and chilli sauce. The powder that is made by grinding dried chillies is called 'cayenne pepper'. Very hot peppers are used in producing 'pepper spray', which can be used as a defensive weapon.

Capsicum species are also used medicinally, for example as a carminative, stimulant, digestive and pain reliever.

The pungency of the *Capsicum* species depends on the concentration of capsaicin in the fruit. Within the fruit, the concentration is highest in the seeds and the inner membranes. Capsaicin is a lipophilic chemical (methyl vanilly nonenamide) which can give a burning sensation in the mouth. *C. chinense* has a high content of capsaicin: one of the cultivars, 'red savina habanero', is considered to be one of the hottest chillies on earth. The pungency of *Capsicum* species is measured in 'Scoville units', introduced in 1912 by Wilbur Scoville. In this system, pungency is measured by multiples of 100, with sweet pepper at 0 and the red savina habanero at over 300,000 Scoville units. The principle of this method is based on dilution by *Capsicum* pepper extracts and organoleptic evaluation by human testers. Nowadays, pungency is also tested by means of high-performance liquid chromatography.

Owing to the fact that some *Capsicum* species or varieties are difficult to determine, the concentration of capsaicinoids (the combination of capsaicin and related compounds) is often used for classification.

Sweet pepper contains 92% water, 5.5% carbohydrates, 1.2% protein and 0.35% fat. Red pepper contains 86% water, 9% carbohydrates, 2% protein and 2% fat. Both types of *Capsicum* peppers contain Fe and Ca, and have a very high content of vitamins A and C.

In many countries, *Capsicum* pepper is the most popular condiment. However, consuming red peppers can be addictive, because it stimulates the production of endorphin in the brain, which may cause a euphoric feeling. Some cultivars are grown as ornamentals, for their unusual fruit shapes and their wide gamut of fruit colours.

Ecology and agronomy

Capsicum peppers can be grown at a wide range of altitudes, and annual precipitation may vary from 600 to 1200 mm. Optimal temperatures for fruit production are between 20 and 30°C. Because *Capsicum* peppers are frost-sensitive, only annual varieties can be grown in temperate regions. *Capsicum* peppers grow best on well-drained, sandy or silt-loam soil, pH 5.5–6.8. *Capsicum* peppers are propagated by seed and germination is best at about 25°C. The crop is often planted out in the field from pre-established seedlings. Transplanting of seedlings into the field takes place about 35 days after sowing, when the plants have reached the eight true leaves stage. Spacing is often wide, 0.6–1.2 m. However, plant densities may vary considerably, from 10,000 to 130,000 plants/ha. Many cultivars require staking to avoid lodging. Flowering begins 2–3 months after planting.

Fig. 11.29. Flowering capsicum pepper plants, staked, and on raised beds mulched with polythene sheet (left) (photo: D.L. Schuiling). Habit of fruiting capsicum pepper plant (right) (photo: J.D. Ferwerda).

For good growth the crop requires an ample supply of nutrients, 130 kg N, 80 kg P and 110 kg K per hectare is recommended, in split applications. *Capsicum* responds well to the application of green manure and farmyard manure.

Weed control has to start as pre-emergence treatment by cultivating or by using herbicides. After establishment, weeding can be done mechanically, although in the tropics it is often manually. An effective method of weed control is covering the soil with organic or plastic mulch (Fig. 11.29).

Picking the fruits may start 3–4 months after planting. The harvest time depends on the desired product: fruits are sometimes picked when they are unripe and green, and others are picked ripe, when the fruits are red, yellow, orange or brown. The fruits are generally hand-picked. In 2005 the world total production of fresh fruits of chillies and peppers was about 25 million t. China, Mexico, Turkey, Indonesia and Spain are the most important producers.

Carrot

Carrot – *Daucus carota*; Parsley family – *Apiaceae*

Origin, history and spread

The wild carrot is widely distributed in the temperate regions of Europe, North America and Asia, but also in northern Africa. The cultivated carrot was derived from the wild carrot in Afghanistan and possibly countries in the eastern part of the Mediterranean. It is believed that the first cultivated carrot had a purple-coloured taproot. A yellow form arose in the same area. It is assumed that carrot was first domesticated in Afghanistan and adjoining regions and spread from there to the Mediterranean. Carrots were known to the ancient Greeks and Romans. The Romans may have spread carrot into Europe but this is not certain. It is known from written sources that carrots had spread into Western Europe and China in the 14th century and into Japan in the 17th century. In his *Cruydeboeck* (1554), the Belgian doctor Dodonaeus (or Dodoens) described three different carrots: the yellow and the red carrots with enlarged taproots, and the wild carrot with thin ligneous roots. European colonists brought carrot into the Americas and Australia. White and orange carrots arose in Europe in the 18th century and were described for the first time in The Netherlands. Some Dutch landraces were the basis of most of the cultivars of orange carrots that are grown. Today, carrots are grown worldwide, mainly in the temperate regions. The orange carrot is by far the most important form, although white, yellow and red carrots are still grown on a small scale.

Botany

The cultivated carrot (Fig. 11.30) is a biennial herb but mostly grown as an annual. The plant is 20–50 cm tall at the vegetative stage in the first year and

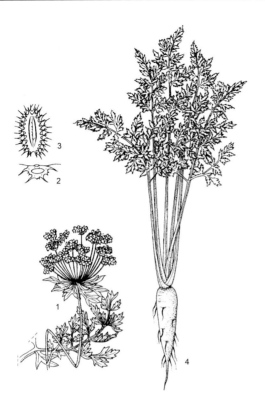

Fig. 11.30. Carrot: 1, flowering and fruiting shoot; 2, seed (mericarp); 3, fruit (schizocarp); 4, habit (line drawing: PROSEA volume 8).

grows up to 120–150 cm at the generative stage in the second year. In the first year, the plant develops eight to 12 leaves, growing in a rosette. The leaves are glabrous, have a long petiole, often with a sheath at the base, and a finely divided two- or three-pinnated leaf-blade. In the generative stage the plant develops branched, hollow, flowering stalks, bearing terminal inflorescences, which are compound umbels (Fig. 11.31). Each umbel consists of about 50 umbellets that subsequently consist of about 50 flowers. Each flower has five white, sometimes pinkish petals, two carpels, two styles and a diameter of about 2 mm. The flowers are mostly bisexual; however, male flowers occur. The fruit is an oblong-ovoid schizocarp, 2–4 mm long, at maturity splitting into two mericarps, with ridges that partly bear spines (spines cause problems at sowing, so they are often mechanically removed). The 1000-seed weight is 0.9–1.2 g. The taproot is swollen and fleshy in the first year; the colour is mostly orange but can also be red, yellow or white (Fig. 11.32). The shape can be more or less globular, cylindrical or conical. The length of the swollen part of the root depends on time of harvest and varies from 4 to 25 cm. Carrots have only a few lateral roots.

Varieties, uses and constituents

There are many varieties due to the different harvest times and different utilizations of the harvested products. Roughly two main groups can be

Fig. 11.31. Inflorescence of carrot (photo: D.L. Schuiling).

distinguished: (i) the anthocyanin-containing eastern carrots with branched roots that are mostly yellow to purple in colour, and having slightly dissected, greyish-green pubescent leaves; and (ii) the carotene-containing western carrot with unbranched roots that are mostly orange, yellow or red in colour, having strongly dissected, bright green and sparsely hairy leaves. The eastern form flowers in the first growing season, whereas the western form usually flowers in the second growing season. At present the western carrot, especially the orange form, is by far the most widely grown. The eastern carrot is grown only in Asia on a small scale (Fig. 11.32).

Carrot is one of the most popular vegetables worldwide. It is eaten raw or cooked. It is often used in mixed vegetable combinations. Carrots are often canned, dehydrated, or quick-frozen. Large quantities are processed into infant food, juice, as an ingredient of soups, sauces and stews. A part of the production is used as fodder. Especially horses are very fond of carrots.

Fresh carrots contain approximately 90% water, 1% protein, 7% carbohydrates (mainly sugar) and 1% fibre, furthermore Ca, Fe and vitamin C; the orange roots contain β-carotene, which is a prolific source of provitamin A. The scent of carrot-seed oil is appreciated in perfumes and liqueurs. Carrots are much used in folk medicine, for example as a carminative and a diuretic.

Ecology and agronomy

Although carrot is sometimes grown in tropical countries at higher altitudes, it is mainly a cool-season crop with optimum growth at 16–20°C. A regular supply

Fig. 11.32. Carrots on the market in East Java, Indonesia (left) (photo: D.L. Schuiling); colour varieties of the taproots of carrots (right).

of water is necessary to obtain well-formed and saleable roots. Vernalization at low temperatures is mostly required to induce flowering. Carrots prefer loamy sand and organic soils that should be well-drained and deep crumbly. Heavy clay soils may cause malformed and twisted roots. The optimum pH is 6–6.5.

Carrots are propagated from seed. Because of the seed shape and the presence of spines, seeds are often pelleted. Seed is mostly drilled with a precision seeder to a depth of 1–2 cm. Seed requirements depend on the desired grading concerning the weight of the roots at time of harvest. Next to the small baby carrots, the following grades occur: A = 50 g or less; B = 50–200 g; C = 200–400 g; D = more than 400 g. This is obviously linked with plant density, which can be 30–200 plants/m². Required amount of seed can be 1.0 to 4.5 kg (unpelleted) per hectare. Good seedbed preparations are essential because carrot seed germinates slowly and irregularly, and the seedling is rather weak. At germination, therefore, the soil has to crumble easily. The spacing in the row depends on the desired root size at time of harvest and varies from 3 to 8 cm; the rows may be 15–40 cm apart. Too dense spacing must be avoided because that often results in deformed roots. The minimum temperature for germination is 4°C. The first appearance of seedlings is 10–15 days after sowing. In the juvenile stage the plant develops a leaf rosette and starts thickening the taproot; the hypocotyl is drawn into the soil by root contraction.

In the seedling stage, high concentrations of nutrients in the soil may cause damage to the plants, so application of fertilizer can best be done before ploughing. During ploughing, the fertilizer will then get mixed well through the upper layer of the soil. Application of manure immediately before planting must be avoided, because it may cause roughness and branching of the taproots.

Application of fertilizer can be based on soil fertility and the removal of nutrients at harvest. Forty-five tonnes of harvested fresh carrot roots removes about 100 kg N, 15 kg P and 120 kg K from the soil. The uptake of nutrients per hectare varies; depending on yield it can be 140–150 kg N, 20–35 kg P, 265–330 kg K, 60–130 kg Ca and 12–21 kg Mg.

As carrots germinate slowly, weeds can cause a serious problem and therefore weed control is necessary. Mechanical control by shallow cultivation and chemical control are the options. To avoid problems in the juvenile stage of the plants, a pre-emergence treatment with herbicides is possible. That can be followed by a post-emergence treatment when the plants have five or six leaves. The rows of seedlings can be earthed-up to obtain maximum root length and to prevent green tops to the harvested roots.

Nowadays, many cultivars exist, which makes it possible to sow carrots in all seasons except the winter. When carrots can be harvested depends on cultivar, desired size and growing conditions. Full-grown carrots require up to 125 days to mature, baby carrots require 60–80 days. However, it is possible to harvest carrots any time when they have a usable size, maturing is not necessary.

Carrots are often harvested by pulling up the roots at the leaves, by hand or mechanically. Full-grown carrots are usually harvested in a different way: the foliage will be removed before harvest, after which the roots can be lifted mechanically, like sugarbeets.

A part of the carrot production for the fresh market is bunched with leaves, although most carrots are topped, which means that the leaves and the top of the 'root' are removed. Topped carrots will remain fresh for up to 5 months when they are stored at 1–4°C and high relative humidity of about 98%. Yields vary, depending on grading of the carrots, from 30 to 120 t/ha. Seed production in the temperate regions requires two growing seasons. Carrots have to be vernalized at low temperatures to induce flowering. However, during the winter the plants must be covered to prevent frost damage.

As wild carrots are widely distributed in the temperate regions, they can become a real threat where cultivated carrot seed is produced. The wild plants hybridize easily with the cultivated plants, resulting in a crop with valueless seed.

In 2005, the world total production of fresh carrots was approximately 24 million t. China, the Russian Federation and the USA are the most important producers. Poland, Ukraine, the UK, Italy and Japan are also substantial producers.

Chicory and Endive

Chicory – *Cichorium intybus*, Endive – *Cichorium endivia*; Daisy family – *Asteraceae*

Chicory

Origin, history and spread

Chicory probably originates from the Mediterranean region, but can now be found in the wild throughout Europe, North Africa and some parts of Asia. The ancient Egyptians, Greeks and Romans collected the green leaves of the wild plant to use them as a salad vegetable and they used the roots as a medicine. In France, it was discovered in the 18th century that roasted and ground chicory taproots can be used as an adulterant of, or substitute for, coffee. The use of chicory as a vegetable started in about 1850 in Belgium, more or less coincidentally. In a botanical garden someone threw some chicory roots into a dark shed and did not look at it for several weeks. After about 3 weeks, blanched heads had grown on the roots. The blanched leaves were less bitter and tastier than the green leaves. After this discovery, the cultivation of forced chicory became most important. At first the forced head was loose, called 'barbe de capucin'; later, breeders achieved compressed and compact slender heads, also called 'chicons', by selection. Chicory is also known as 'Belgian endive'. Chicory is grown mainly in Europe.

Botany

Chicory (Fig. 11.33) is a perennial herb that can reproduce by seed and by adventitious buds of the root. In the first growing season it produces alternating leaves in a more or less loose rosette on a very short stem. The leaves of the rosette are up to 50 cm long and up to 15 cm broad, often long-petioled, toothed to lobed, glabrous or hairy, with wide midribs and sometimes with reddish markings. The leaves resemble those of dandelion (*Taraxacum officinalis*). However, some varieties have entire or more rounded leaves.

In the second growing season, following a period of cold exposure, the species produces branched flowering stems, up to 2 m high, bearing small leaves, up to 7 cm long, entire and dentate, oblong-lanceolate, clasping the stem. The stem also bears numerous clusters of axillary, sessile flower heads, up to 5 cm in diameter. Flower heads consist of strap-shaped ray flowers, mostly blue, sometimes white (Fig. 11.34). The flower heads are surrounded by two rows of bracts. The fruits are naked achenes, 2–3 mm long, obovate, more or less angled, yellow to dark brown, with a very short pappus. The 1000-seed weight is about 1.8 g. All parts of chicory contain a bitter, milk-like juice.

Varieties, uses and constituents

Most of the chicory varieties are grown for forcing. Chicory plants grown for harvesting the roots are dug up after the first cultivation period, and the roots are trimmed to about 18 cm in length. The leaves are trimmed back to about

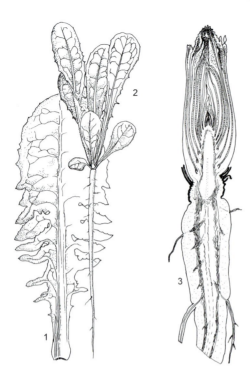

Fig. 11.33. Chicory: 1, leaf of a rosette; 2, juvenile plant in the first growing season; 3, longitudinal section of a taproot and head at the beginning of the second growing season.

3 cm, taking care to avoid damage to the growing point. After a period of cold storage the roots are planted in dark forcing chambers, in soil or water. In about 3–4 weeks, the growing point develops into a firm head of etiolated blanched leaves called the 'chicon'. A chicon should have a length of about 15 cm; the colour has to be white and yellow or white and red (Fig. 11.34). When the chicons are full-grown, they are cut off the roots and can be brought to the fresh market. The remaining roots can be used as cattle feed. Thus for producing

Fig. 11.34. 'Red' and 'white' heads of chicory (left); inflorescences of chicory (right) (photos: D.L. Schuiling).

chicons, two successive cultivation periods are needed: one in the field and one in a forcing chamber.

However, a number of varieties are destined for producing vegetables after the first cultivation period. These varieties are grown for field production of the edible leaves, such as 'Sugar Loaf' and 'Rossa di Verona'. There is great variety in these so-called 'non-forcing' chicory types, including green and red types, rounded or elongated heading types and rosette types. Some Italian non-forcing varieties are called 'radicchio'; they have reddish-purple, often more or less rounded leaves, with white veins. There are two major groups of radicchio: leaf types and head types, the latter resembling a small red cabbage. Chicory is used raw in salads or eaten as a cooked vegetable.

Special varieties of chicory that have thick fleshy roots are grown for making coffee substitute. After the roots have been harvested, they are cleaned, chopped, dried, roasted and ground. During roasting the inulin caramelizes, which gives the final product a dark brown colour and a special flavour. As well as a substitute for coffee it is also used as an adulterant, thus in a mixture with coffee.

Chicory leaves contain about 94% water, 3% carbohydrates, 1% protein, the minerals Na, K, Ca, Fe and P, and vitamins B and C. Fresh chicory roots contain approximately 5% protein, 4% sugar and 0.85% fat. The deep-growing taproot contains the carbohydrate inulin, which is a combination of pentose, levulose and dextrose. Recently, there has been increasing interest in chicory roots as a natural source of inulin, which can be used to produce fructose oligosaccharide, a low-calorie sweetener. This is an important development for the soft drink industry, for example.

Ecology and agronomy

Chicory is a cool-weather crop. It prefers moderate temperatures and a well-distributed rainfall. It needs about 130 frost-free days to produce roots that are suitable for forcing chicons. Chicory grows best on loose, friable, fertile loams or sandy loams, with good water-holding capacity and good internal drainage. Humus content should be below 2%, $CaCO_3$ content 2% or more and pH 6.5–7. For good root development, concerning shape and thickness, the soil must have a loose, undisturbed top layer of preferably 1 m depth.

Chicory is propagated by seed, which is often pelletized to improve precision sowing. Seed should be drilled at a depth of about 1–1.5 cm in rows. For growing an early crop, plants can be started in paper pots in a greenhouse and transplanted to the field. Germination and development of the seedling are slow, which means that, especially during this period, weed control is necessary. Growing chicory can be done on a flat field or on ridges. On a flat field the rows are spaced 37.5–50 cm apart and on ridges 50–75 cm apart, depending on the number of rows on the ridge. The spacing within the row should be finally 20–25 cm; thinning of the juvenile plants may be necessary. The number of plants per hectare, depending on variety and cropping system, may vary from 65,000 to 200,000. For forcing chicons it is important that the diameter of the roots is about 2.5–3 cm, because the thickness of the chicon is largely determined by the diameter of the root.

Chicory is a heavy feeder. However, when harvesting the roots for forcing chicons or producing surrogate coffee or fructose is the aim of the culture, fertilizing should be done very carefully, especially N. Applying too much N results in heavy heads at the expense of good roots. A good chicory yield of about 33 t of fresh roots per hectare removes from the soil approximately 150–180 kg N, 25–30 kg P and 200–250 kg K. Weed control can be done by cultivation and earthing-up and should be carried out carefully and shallow, to avoid damaging the taproots. Chemical weed control is also possible as herbicides are available.

Harvesting of the roots should be done as late in autumn as possible because there is an increase in size and weight of the taproots during sunny, cool weather. A part of the root production is taken to a factory for processing into surrogate coffee. However, most of the production is brought into cold store and subsequently used for forcing heads of chicory, often in special, dark, growing rooms with a hydroponic system. It takes about 3 weeks to develop harvestable heads or chicons from the growing points on top of the taproot.

The yield of roots depends among other things mainly on the number of plants per hectare and time of harvest, and may vary from 18 to 40 t/ha. In 2005, the world total production of chicory roots was about 0.9 million t, Belgium being the most important producer. France and The Netherlands are the largest producers of the chicon form and Italy of the green and red non-forcing forms.

Endive

The origin of endive is not clear: different authors mention Egypt or India, northern China or Sicily, although the wild *Cichorium endivia* ssp. *divaricatum* can still be found in the Mediterranean region. Endive was most likely brought into cultivation in this area. The ancient Egyptians, Greeks and Romans used it as a salad vegetable. It reached northern Europe probably by the 16th century and was first described in the USA in the 19th century. It is now grown throughout the temperate regions; Italy, France and The Netherlands are the largest producers.

Endive is a close relative of chicory. It is an annual or biannual herb that produces a compact rosette of mostly glabrous, sometimes thinly pubescent leaves on a shortened stem when young. Rosette leaves alternate and are sessile. The leaves can be broad with toothed and wavy margins, 10–25 cm × 8–15 cm. The rosette can be more or less open with a diameter of about 35 cm; this type is called 'escarolle'. The leaves can also be narrow, much divided and curled; this type is called 'frisée' (Fig. 11.35). The plant generally forms a loose head. The outside leaves of an endive head are green and bitter, while the inner leaves are greenish-yellow to creamy white and much milder flavoured, which is why endive leaves are sometimes bound together at the end of the growing period to achieve more blanched leaves. The generative development and generative habit of endive are very similar to those of chicory.

Endive leaves are used in salad mixtures or as a cooked vegetable. Most of the production is meant for the fresh market, a part is canned or frozen. The edible part of endive contains about 95% water, about 2% protein, 1.5% carbohydrates and 1% fibre, the minerals Ca, K, P, Fe and Mg, and the vitamins

Fig. 11.35. Endive: variety in shape of the leaves (photos: above, B. Rijk; below, PPO-PAGV Lelystad, The Netherlands).

A, B and C. Endive also contains inulin and intybin, which cause the bitter taste of the leaves.

The desired daily optimum temperature for good development of endive is 15–18°C. It prefers a permeable fertile soil with a pH of 6.5–7.8. Cultivation of endive is roughly similar to the cultivation of lettuce.

Cucurbits

Cucumber – *Cucumis sativus*, Gherkin – *Cucumis anguria*, Melon – *Cucumis melo*, Pumpkin and squash – *Cucurbita* spp., Watermelon – *Citrullus lunatus*, Bitter gourd – *Momordica charantia*; Pumpkin family – *Cucurbitaceae*

Introduction

Cucurbits include, among other species, cucumber, gherkin, melon, pumpkin and squash, watermelon and gourd. Most of the cucurbits have comparable morphology and are rather similar in their requirements for growth and development.

Cucurbits are annual herbs with prostrate, stretched or climbing vines, with spirally coiled tendrils. Tendrils arise in leaf axils. The often very long vines or stems are mostly roughly haired. The leaves alternate, they are large, simple and lobed. The campanulate flowers are borne in the axils of the leaves, the calyx is five-lobed, the five sepals are more or less grown together, stamens three, ovary usually inferior and containing numerous ovules. Male or female flowers can be discriminated by observing the presence or absence of the inferior ovary. The inferior ovary (the young fruit) of the female flower is already present at a very early stage. The colour of the flowers is mostly yellow or orange. Cucurbits are often monoecious, flowers are usually unisexual, thus with separated male or female flowers, although hermaphrodite flowers occur. The fruit is botanically called a 'false berry', also 'pepo'. Fruits are very variable in size, shape, colour, thickness and smoothness of the rind. Fruits mostly contain many large seeds, which are flat and ovate-oblong in outline. Seeds contain no endosperm. The root system is large, mostly distributed in the top 30 cm of the soil, some roots up to 1 m deep. Breeders have succeeded in developing parthenocarpic cultivars of several species. Parthenocarpic means that the fruit develops without fertilization and thus without seed development. This is possible only when pollination by insects like bees and flies can be prevented, explaining why parthenocarpic plants are grown only in glasshouses.

Cucurbits are usually cultivated for their large fruits, which are often eaten fresh, or used in salads, sometimes cooked and eaten as vegetables or as soup, and some species can be pickled. However, a number of cucurbits are not cultivated to provide food. For example, the fibrous vascular system of the fruit of the species loofah (*Luffa aegyptiaca*) is often used as a bath sponge or as a filter; some cucurbits have a very hard exocarp, which is exploited in some cultures by using the dried fruit to transport and store liquids or use of half the fruit as a drinking bowl; furthermore, many cucurbits are cultivated as ornamental gourds.

After removing the shells from the seeds, they can be roasted and/or salted, and eaten just like peanuts. The oil content of the seed can reach up to 50%. The species *Cucurbita pepo* is even cultivated for oil production; it is a highly valued salad oil.

As well as the traditional long-vined and wide-spreading cucurbits, breeders have developed bush, dwarf and short-vined cultivars. Seeds of ripe cucurbits have been used in medicines to treat prostate and bladder diseases.

Ecology and agronomy

Cucurbits are primarily warm-season crops, although breeders have succeeded in developing cultivars from several species that can be cultivated more or less successfully in cooler regions. However, the optimum temperature range is 18–30°C and the plants do not stand frost at all. In cool temperate regions, cucurbits are therefore often grown in glasshouses, mostly year-round. Cucurbits thrive best on deep, well-drained and thoroughly cultivated fertile loamy soils, pH 6–7. For good growth, it is recommended to add green manure or animal manure to the soil before planting. Especially heavy soils can be improved by adding organic matter. Photoperiod and temperature can influence the ratio of male to female flowers; for example, long days and high temperatures promote male sex expression of pumpkins. Most cucurbits require much water but they are sensitive to waterlogging. High air humidity is mostly harmful because it may stimulate leaf diseases by fungi.

Cucurbits are propagated by seed, sometimes by cuttings. Seeds require a warm soil of at least 16°C to germinate. If the soil does not have the right temperature, it is recommended to let the seeds germinate first in a glasshouse and later transplant the juvenile plants into the field when they have developed two 'true' leaves, which will be about 2 weeks after sowing. A method to warm the soil before sowing is by mounding or hilling the soil, which results in more effective exposure of the soil to sunlight. Hills are made by mounding the soil up to 25 cm high. Covering the soil with plastic mulch also favours the warming-up process.

Seeds have to be planted 2–3 cm deep. If plants are grown on a hill, they can be grown singly or in groups of two or three plants per hill. Spacing between single plants or groups depends on species and cultivar, and ranges from 1.5 to 3 m. Cucurbits rapidly produce a dense vegetative cover, vines may grow up to 5 m long. Except for cucurbits in glasshouses, plants are usually allowed to grow freely over the surface of the soil. In glasshouses, the plants are mostly trailed on to vertical structures or wires. The fast growth of the plants and the spreading over the surface of the soil make weeds seldom a problem. Some hand-weeding or hoeing at the start of the cultivation will mostly be sufficient. Moreover, covering the soil with plastic mulch, or straw, will effectively control weeds.

Cucurbits require low N and high P and K levels for good fruit production. Application of complete NPK fertilizers is based roughly on the removal of nutrients in a harvest which, per tonne of fruits, can be 3–6 kg N, 0.5–1 kg P and 5–6 kg K. The removal of nutrients by the various cucurbit species may differ slightly. Over-fertilizing has to be avoided because it promotes the growth of vines and leaves, and maleness, and it impedes fruit development.

As cucurbits require a fair amount of water, irrigation is often necessary.

The moment of harvesting varies widely because the fruits of the different species are picked in a range of maturity stages, from very young to ripe and dried.

Cucumber and gherkin

Cucumber probably originated in southern Asia. It is supposedly derived from the wild species *Cucumis hardwickii*, a species with small, bitter fruits. Cucumber has been grown in India for many centuries. The Indians passed it on to the ancient Egyptians, Greeks and Romans. They even knew how to pickle the fruits. (As far as cucurbits are concerned, ancient writings have to be interpreted carefully, because it is not always clear which cucurbit species is meant.) In the Middle Ages, cucumber was probably cultivated in the greater part of Europe because it was one of the cultivated plants on Charlemagne's list *Capitulare de Villis*. Nowadays it is grown throughout the world, outside, in plastic tunnels, or in glasshouses.

The vines or stems of cucumber are four-angled and covered with stiff hairs. The leaves are entire and more or less triangular in shape, irregularly toothed, and up to 20 cm in length. They have long petioles. The flowers are yellow, either male or female, and 2–4 cm in diameter. Originally, the fruits were globular in shape. Most modern cultivars have cylindrical, short to long, white, yellow or green fruits, often with a somewhat wrinkled or ribbed surface (Fig. 11.36). The flesh is pale green to white. The fruits have many seeds or are seedless. The flat seeds are white; 1000-seed weight is 20–35 g.

Immature cucumbers are usually eaten raw, in salads or sliced and served with vinegar or dressing, sometimes boiled, used in soups, or made into pickles. In South-east Asia, young shoots are eaten raw or steamed. The fruit contains 96% water, 2.2% carbohydrates and 0.3% fat, as well as minerals and vitamins. Cucumbers contain a special amino acid that stimulates the skin to retain good moisture content. That is why in beauty cures, sliced cucumber is often used to cover the face (cucumber mask).

Cucumbers can be divided into two main groups:

1. Cultivars with long fruits, so-called 'slicing cucumbers', 15–25 cm long, for outdoor cultivation, and longer cucumbers, up to 40 cm long, grown mainly in glasshouses. The latter are parthenocarpic; breeders have even developed cultivars that produce only female flowers. In glasshouses, cucumbers always grow on trellises or wires not only because of the limited space, but also because the long fruits bend when they lie on the ground.
2. The so-called 'pickling or ridge cucumbers', which are much shorter, less than 12 cm long, for outdoor cultivation. The fruits of this group are sometimes mistakenly called 'gherkins'. The real gherkin is *Cucumis anguria*, originally a weed in the Americas. The real gherkin is now grown mainly in the USA; its immature fruits are used for pickling.

Fig. 11.36. Cucumber (*Cucumis sativus*) (left to right): close-up of male flower (photo: D.L. Schuiling); details of two cucumber plants with fruits in various stages of development; leaves and tendrils.

There are many cucumber cultivars; some cultivars are suitable for harvesting the fruits several times during the growing season. From other cultivars, the fruits ripen all at once, although cucumbers are mostly harvested while immature, for pickling sometimes very young.

In 2005, the world total production of cucumbers and gherkins was about 42 million t. China, accounting for about 27 million t, was by far the most important producer.

Melon

Melons originated in Asia Minor and Africa, where wild melons still occur. It is supposed that they were domesticated about 4000 years ago. The ancient Romans cultivated melons; Pliny mentioned them in his writings as 'pepones'. Furthermore, melon seeds were found in the stomach of a 2000-year-old Chinese mummy, and the species was described in Charlemagne's *Capitulare de Villis*. Columbus brought it to America. Melon is now cultivated throughout the world, in the open field or in glasshouses.

The species has hairy, ridged stems. The leaves are large, 15–20 cm across; the shape may be orbicular, ovate or reniform. Leaves often have five to seven lobes. Petioles are up to 15 cm long. One plant may bear separate male and female flowers, although the combination of hermaphrodite and male flowers on the same plant also occurs. The male flowers occur in small groups, female and hermaphrodite flowers mostly solitary. The colour of the flowers is yellow.

Melon produces large fruits, up to 2 kg (rarely up to 5 kg). The shapes of the fruits vary; they can be globose or oblong, sometimes slender. The texture of the rind is smooth, or rough, netted, or wrinkled; the colour of the rind may be green, yellow or orange, or the rind may have green and yellow stripes; the colour of the flesh may be orange, green or whitish (Fig. 11.37). The cavity of the fruit is filled with many flattened, white seeds.

Melons can be divided into two groups, each of which consists of a number of different types.

1. Sweet melon types, which are the most important types, and mostly eaten fresh as fruit. For example: (i) Musk melon, globular or oblong fruit, up to 2 kg, rind mostly netted with a raised network that is lighter in colour than the rest of the fruit, rind can also be smooth, sometimes with broad ribs, rind colour is yellowish-green, flesh colour is orange or light green; (ii) Cantaloupe, globular to slightly ovoid fruit, up to 1.8 kg in weight, the rind may be scurfy or covered by warts, often with deep grooves, colour of the rind is greyish-green or brownish-yellow, flesh colour is orange, rarely green; (iii) Winter melon, ovoid fruit up to 2.5 kg in weight, the rind may be smooth or wrinkled, striped or spotted, rind colour is grey, green or yellow, flesh colour is white or greenish white; (iv) Ogen melon, named after a kibbutz in Israel where it was developed, the fruit is globular and relatively small, about 15–20 cm in diameter, the rind is ribbed and orange-yellow with green stripes, the flesh colour is green.

2. Non-sweet types, which are minor crops, and often eaten as a vegetable. For

Fig. 11.37. Whole fruits of the melon (*Cucumis melo*) cultivars 'Canteloupe' (left), 'Galia' (right) and 'Piel de Sapo' (back) (photo: D.L. Schuiling).

example: (i) Oriental pickling melon, the fruits are elongated when they are young and oval-cylindrical when they are mature, the rind is smooth and the colour is yellow with white longitudinal stripes. Mature fruits can be very big, weighing up to 5 kg, used in processing candy or the pulp of the fruit can be consumed with ice and sugar. However, most fruits are harvested when they are young and eaten as cucumber, or they are pickled; (ii) Snake melon, the fruit is long and slender, the rind is smooth, and the fruits are harvested when they are young and eaten as cucumber.

Although the musk melon and the cantaloupe melon are the most well-known types, there are many more melon types in addition to the ones described here.

An attempt to find a precise and complete botanical classification of all melon types and varieties is rather pointless; it is changing all the time because all melon varieties interbreed very easily.

Melons are low in protein and rich in sugars, vitamins and minerals. They contain about 90% water, 5–15% carbohydrates, about 1% protein, vitamins A, B and C, and minerals K, Ca, Fe, P and Mg.

Melon fruits mature 90–120 days after sowing. At maturity, most melons develop a crack at the point where the fruit is affixed to the vine; this enables the fruit to be separated easily from the vine. Another maturity indication is the scent of the fruit, which becomes stronger when the fruit is ripe. Harvesting melons is usually done manually. The main melon-producing countries are China, Turkey, India and Spain.

Pumpkin and squash

All *Cucurbita* species originated in Mexico and South America, where archaeological research indicates that they were probably already well known some 7000 years ago. After the exploratory expeditions by Columbus, *Cucurbita* species were introduced into the Old World. Nowadays, they are cultivated throughout the world.

As far as food production is concerned, *Cucurbita* is probably the most important genus in the *Cucurbitaceae* family. There are four major species in the *Cucurbita* genus (*Cucurbita maxima, Cucurbita mixta, Cucurbita moschata, Cucurbita pepo*), each species with many varieties. All together, there are more than 150 known varieties and interbreeding occurs very easily. Even more so than for melons, the attempt to fit all species and varieties of pumpkin and squash into a precise botanical classification system is almost impossible. Some varieties develop vines, sometimes up to 15 m long, others develop bushes. The flowers of *Cucurbita* spp. differ from other cucurbits because they are larger and bright yellow to orange in colour. It takes usually 3–4 months for the fruits to mature. Annual yields of mature fruits can be 20–25 t/ha. Weight of the individual mature fruits varies considerably and may range from 3 kg up to 20 kg. The very heavy fruits sometimes have an ornamental purpose: hollowed out and used as a lantern, for contests, or for showing at feasts, fairs and exhibitions.

Identifying the different species by their fruits is difficult because cultivars of the same species may have totally different fruits, whereas cultivars of different species may have very similar fruits. The fruits are, in contrast with melons and cucumbers, mostly not consumed fresh but cooked. Immature and mature fruits, also leaves and flowers, are used a vegetable. Seeds can be roasted and consumed as a snack food.

Cucurbita fruits with coarse-grained and strong-flavoured flesh are often used for pies, soups, chutney, jams and fodder, and often called 'pumpkin'. The ones with fine-grained flesh and mild flavour are usually called 'squash'. The latter are mostly used as a cooked, fried or steamed vegetable. The use of vernacular names of the different varieties is very confusing because different names are used for the same varieties. For example, pumpkin or squash can also be named 'gourd' or 'marrow'.

Some characteristics of the four major *Cucurbita* species (Fig. 11.38) are as follows:

- *C. maxima*: the stem is soft and round; leaves have almost a circular outline, and are not lobed or only lightly lobed; the fruit stalk is soft, spongy and swollen; the mature fruit is soft- or hard-shelled.
- *C. mixta*: the stem is hard, grooved and five-angled; leaves are large, cordate, glabrous or softly hairy, lightly lobed; the fruit stalk is hard and strongly thickened by cork; the mature fruit is soft-shelled.
- *C. moschata*: the stem is hard and angular; leaves are large, softly hairy and lightly lobed; the fruit stalk is hard and smoothly grooved; the mature fruit is soft-shelled.
- *C. pepo*: the stem is hard and angular; leaves are broadly triangular, stiff, prickly and deeply lobed; the fruit stalk is hard, deeply furrowed and five-

Fig. 11.38. Fruits belonging to different pumpkin/squash species: *Cucurbita moschata* and *Cucurbita pepo*, *C. moschata* 'Butternut' (butternut squash) (above, left to right); fruits of various cultivars of mainly *Cucurbita maxima* (below) (photos: D.L. Schuiling).

angled; it is an extremely variable species, in appearance and in the size, shape, colour and flavour of the fruits. *C. pepo* is the most cold-tolerant of the four species. Mature fruits of special cultivars are grown for the hull-less seeds from which oil can be obtained that is valued as edible oil. Oil percentage can be up to 50%. Seed yield is 500–1500 kg/ha. Some cultivars of this species are grown as ornamentals. Another variety of *C. pepo* is the French marrow 'courgette'. The fruits of these varieties are picked when they are still immature and about 10–25 cm long. They are cooked or fried whole or sliced.

Harvest time depends on whether immature or mature fruits are in demand. Since flowering is more or less a continuous process, picking immature fruits can be done several times during the growing period, the first time about 50 days after planting. The mature fruits of pumpkin and squash are usually harvested after all of the leaves have died away. If harvested carefully and stored adequately, the mature fruits can be kept for up to 6 months. The fruits of special cultivars are used as fodder.

In 2005, the world total fruit production of pumpkin, squash and gourd was about 20 million t. China, India and Ukraine were the main producers.

Watermelon

Watermelon originated in Africa, with a probable second centre of diversity in east India. It was cultivated in ancient Egypt, and known in southern Europe and western Asia before the Christian era. It reached China in the 10th century. The spread of watermelon in Europe is mainly attributed to the Turkish conquerors. African slaves and early settlers introduced it into the Americas. Nowadays it is cultivated throughout all tropical and subtropical regions.

Watermelon has thin and hairy stems, up to 5 m long. The leaves are large, up to 12 cm × 20 cm. The genus *Citrullus* differs clearly from other genera in this family by having simple leaves with three or four pairs of pinnated lobes, which themselves are further divided. This is in contrast with other genera, which have entire or non-pinnated, lobed leaves.

The fruits of watermelon are variable: shape can be round, oval, oblong or cylindrical; fruit weight 3–23 kg; the colour of the rind may be light, greyish or dark green, with or without dark or light green stripes, sometimes variegated; the rind is usually smooth; the colour of the flesh can be bright or deep red, orange-red, pink, yellow or white; the seeds can be large or small, the seed colour can be brown, black, green or red, sometimes mottled (Fig. 11.39).

Breeders have developed seedless cultivars, although the fruits mostly have a few seeds left. Watermelon has a strongly developed but rather shallow root system.

The fleshy, juicy fruits are mostly eaten fresh, but in some countries young fruits and rind are sometimes pickled, used in curries, or used in producing juices and beer. Watermelon is sometimes grown as fodder for livestock. The oil- and protein-rich seeds of large-seeded cultivars are sometimes consumed after they have been roasted and salted. In folk medicine, seeds have been used in producing medicines for treating fever, catarrhal afflictions, and disorders of urinary passages. They are regarded as having diuretic and curative properties.

The nutritional value of the flesh of watermelon is low. Next to 90% water, it contains about 9% carbohydrates, vitamins A, B and C, and minerals Ca, Fe, P and Mg.

Fig. 11.39. Main use of most watermelon (*Citrullus lanatus*) cultivars is to be eaten fresh (left) (photo: D.L. Schuiling); view from above and cross-section of watermelon (right).

Watermelon thrives best in sandy soils in hot, dry climates. However, irrigation is usually needed.

Watermelons are often sown directly in the field, but as watermelons require a long growing season, transplants can be used. The longer the growing season, the larger the fruit will be. Plant density varies from 5000 to 9000/ha.

Watermelons are mostly harvested when they are ripe, which is usually 4–5 months after sowing. Determination of ripeness can be done by a number of indicators, such as: tendrils become dry and turn brown; the glossy surface of the fruit turns dull; the underside of the fruit that is in contact with the soil turns pale yellow; and the sound of a hollow thump after knocking on the surface of the fruit. When harvesting, the fruit stalks should be cut as far from the fruit as possible, which may extend the storage life of the melon. Harvesting the fruits has to be done carefully because they are fragile and can be bruised or broken easily. Fruits can be stored for 2–3 weeks at 10°C.

Watermelons are usually grown for the fresh market. They are sold either whole or in halves or quarters. The most important producers of watermelon are the Russian Federation, China, Turkey, India and Iran.

Bitter gourd

Bitter gourd, also called 'balsam pear' or 'bitter melon', is cultivated mainly in India and East Asia. The species has ridged and pubescent stems. The first leaves are unlobed and heart-shaped with broad teeth along the margins. Older leaves are deeply palmately five- to seven-lobed, and 10–15 cm wide. The fruit is obovoid or oblong-cylindrical, bumpy, coarsely ridged, 10–20 cm long and pointed. The colour of the mature fruit is dark yellow to orange (Fig. 11.40). When the fruit is ripe it splits open to disclose the black seeds.

Fig. 11.40. Bitter gourd (*Momordica charantia*): male flower (left) (photo: D.L. Schuiling); near mature fruit on the plant (right).

The fruit has a bitter taste caused by the bitter substance momordicin. Peeling the fruits and brining removes the bitter taste. Young fruits are consumed as a vegetable; they are also used in pickles and curries.

Lettuce

Lettuce – *Lactuca sativa*; Daisy family – *Asteraceae*

Origin, history and spread

The origin of lettuce is uncertain but it is thought to be the Mediterranean region and the Near East. Lettuce is not known in the wild. The progenitor of many forms of lettuce was probably *Lactuca serriola*. Many types of *L. serriola* can still be found in the Mediterranean region and the Near East. It is an erect, branched plant, containing milky latex and bearing bitter-tasting leaves. The cultivation may have started in these regions but that is not recorded. The cultivation of lettuce in ancient Egypt, about 4500 years ago, is well known and was suggested by tomb paintings. The Greeks cultivated lettuce in 450 BC. Green and red forms of lettuce were eaten by the ancient Romans in the 1st century AD. Influenced by the Romans, lettuce spread throughout western Europe. Traders probably introduced it into China, where it was known in the 5th century. Various forms of lettuce were described in European literature in the 15th century. The oldest types of lettuce were the leafy types with loose rosettes; the bolted types have been known since the 14th century. European colonists introduced it into the New World. At present, lettuce is grown throughout the temperate and subtropical regions.

Botany

Lettuce is a variable, glabrous, lactiferous, annual herb. It usually forms a more or less dense basal rosette first, and later a tall, branched, flowering stem, up to 1 m tall. In the vegetative stage the leaves of the rosette are arranged variously, but often spirally, in compact or loose heads. Shape, size and colour of the rosette leaves differ with cultivar. They may be undivided to runcinate-pinnatifid, smooth to curly and fringed, dark green to yellowish-green, sometimes with red anthocyanin pigment. The leaves of the flowering stem become progressively shorter, ovate to orbicular in outline, entire and sessile. The inflorescence is a dense flat-topped panicle with flowers arranged in small heads. The head contains ten to 30 hermaphrodite, yellow flowers (Fig. 11.41). The fruit is a narrowly obovate achene, 3–8 mm long, compressed, tip surrounded by a white pappus of hairs. The 1000-seed weight is 0.8–1.2 g. The root system is a taproot and laterals, the taproot can reach 1.5 m depth. However, the root system is weak compared with the mass of leaves, especially in the vegetative stage.

Fig. 11.41. Close-up of part of lettuce inflorescence with two open flower heads (left) (photo: N. Keijzer); stem lettuce (right) (photo: Li Jin Long).

Varieties, uses and constituents

There are hundreds of cultivars of lettuce, although six varieties or cultivar groups can be distinguished.

1. Butterhead lettuce (syn. head lettuce): soft loose head of overlapping, succulent leaves, inner leaves yellowish and buttery in texture (Fig. 11.42). Most of the leaves are partially exposed. Usually eaten fresh and grown mainly in cool temperate areas.
2. Cos lettuce (syn. romaine lettuce): long narrow leaves, with thick midribs, in a tall, loose, upright cylindrical to oval head, up to 25 cm in height (Fig. 11.42). To achieve tender inner leaves, the leaves can be bound together. The edible part can fully be eaten fresh or cooked, or the inner leaves raw and the outer leaves cooked. It is suitable for subtropical conditions.
3. Latin lettuce (syn. grasse): forming a loose head with thick leathery leaves. Latin lettuce is tolerant of high temperatures; it is usually eaten fresh, and grown mainly in France.
4. Stem lettuce (syn. asparagus or stalk lettuce): grown for the fleshy 30–50 cm long, 3–6 cm thick stem, which has a crisp texture. The stem bears many leaves with a rosette at the apex; the leaves are edible when young (Fig. 11.41). It is grown mainly in Egypt, China and South-east Asia.
5. Crisp lettuce (syn. iceberg lettuce): firm heading, slightly heading and non-heading types occur. The firm heading types have closely packed leaves, the outer wrapper leaves covering the part usually consumed. Crisp lettuce has thick and crispy leaves with prominent flabellate veins and midribs (Fig. 11.42). The size of the head is variable, in some cultivars no larger than a tennis ball. This

variety is mostly eaten fresh. It tolerates warm temperatures. Crisp lettuce is grown mainly in Europe and the USA.

6. Bunching lettuce (syn. curled lettuce): thin, broad, smooth or curled or crinkled, green or reddish leaves in a loose rosette or on a short stem (Fig. 11.42). Bunching lettuce is eaten fresh, in mixed salads, and is also often used for garnishing. It tolerates warm temperatures.

Lettuce is grown mostly for its delicate leaves that can be eaten fresh, usually as a salad with a dressing of vinegar and oil, or different sauces. The peeled, fleshy stem of stem lettuce can be eaten raw but is mostly cooked. The leaves of lettuce, especially the curly and red ones, are often used as garnishing for all kinds of dishes. The oil extracted from the lettuce seed is used on a small scale as an aphrodisiac (Egypt). In the past, lettuce was used as a remedy for indigestion; the milky juice is a mild narcotic and sleep-inducer.

The edible part of butterhead lettuce contains about 96% water, 2% protein, 1% carbohydrate and 0.5% fibre. It also contains vitamins B_1, B_2 and C, carotene, and the minerals Ca, K, P and Fe. However, there are considerable differences in nutritional properties among the different lettuce varieties.

Ecology and agronomy

Most lettuce grows best at moderate day temperatures of 15–20°C. Higher temperatures can cause loose heads or no heads at all, and early bolting. Cool

Fig. 11.42. Lettuce cultivar groups: 1, green butterhead lettuce; 2, red butterhead lettuce; 3, crisp lettuce (or iceberg lettuce); 4, cos lettuce; 5, green bunching lettuce (or curled lettuce) cultivar 'Lolobionda'; 6, red bunching lettuce (or curled lettuce) cultivar 'Lollorossa' (photos: PPO-PAGV Lelystad, The Netherlands).

nights of about 10°C are essential for good development. As lettuce is a quick-growing crop having a poor root system, it requires much water. Lettuce grows on a wide range of soils as long as they are well-drained, having a good structure with a good water-holding capacity, and having a high fertility and pH 6–7. Lettuce has a high requirement for light. The red forms will attain a deep colour only in full sun. Most of the lettuce is produced in open fields; there is a limited production in greenhouses and plastic tunnels.

Lettuce is propagated by seed, mostly first grown in trays or pots of peat or paper in nurseries. Next, the seedlings are transplanted in the field (Fig. 11.43); however, direct seeding in the field is also an important method of planting, by drilling coated or pelleted seeds by precision seeders. The plants grow in rows that can be 45 cm apart; the in-row spacing varies depending on variety, ranging from 10 to 25 cm. The desired number of plants per hectare depends on variety and cropping system and can be 50,000–100,000. Before planting, the structure and water-holding capacity of the soil can be improved by applying about 15 t of farmyard manure per hectare, or growing a green-manure crop. Applying artificial fertilizers depends on soil fertility and has to be done very carefully to prevent high nitrate content of the plants (see Spinach, this chapter). Per hectare, the different lettuce varieties require between 100 and 190 kg N, P and K in balance. By harvesting 75,000 heads of lettuce

Fig. 11.43. Bunching lettuce (or curled lettuce) cultivation on ridges covered with plastic sheet to drain off excess rain during the rainy season, Bali, Indonesia (photo: D.L. Schuiling).

per hectare, approximately 75 kg N, 10–15 kg P and 115-150 kg K are removed from the soil. Young lettuce plants are susceptible to competition from weeds. Weed control is mostly a combination of herbicides and cultivation. Cultivation should be done just below the surface to avoid damage to the shallow root systems of lettuce. As lettuce needs plenty of water in the soil for good development, irrigation may be necessary.

Lettuce can be harvested 45–90 days after planting, depending on variety and climate. It can be done mechanically, by hand, or a combination of both methods. Harvesting should be done in a way that the lower leaves, which are often yellowing and partly covered by sand, remain in the field. It is done by cutting the plants at or slightly below the soil surface; the lower leaves then have to be removed before packing.

The yield depends on variety and cropping system and varies from 10 to 30 t/ha, fresh weight. Most of the lettuce is for the fresh market, the whole heads or shredded and packed.

Lettuce can be stored cold for a limited time, depending on variety, for a period of up to 4 weeks. Today, the shelf-life and appearance of lettuces at retailers can be improved by better packaging materials and special in-package atmospheres.

In 2005, the world total lettuce production was 22 million t. China, accounting for 11 million t, was by far the most important producer. The USA, counting for 5 million t, was the second most important producer.

Mushrooms

Button mushroom – *Agaricus bisporus,* Shii-take – *Lentinula edodes,* Oyster mushroom – *Pleurotus ostreatus*

Introduction

Although strictly speaking fungi cannot be classified as 'cultivated plants', their role in food supply is so important that adding a few species to the selection of this book is justified. Fungi can be components of foods and alcoholic beverages or play a role in processing them; for example fungi in blue cheeses, and yeast in bread, wine and beer. Some fungi are served as delicacies in restaurants; others form the basis of medicines. However, next to highly valued fungi, many feared fungi exist, which cause unwanted decay of several materials, or all kinds of disorders in humans, animals and plants.

Mushrooms belong to the fungi; they are very old, older than plants. That mushrooms were well known and used by early civilizations is proved by ancient Egyptian, Greek, Roman, Chinese and Mexican writings and/or images. Furthermore, they appear in archaeological findings, for example a fresco in Pompeii.

Throughout olden times mushrooms were mistrusted, and many people connected them with disease, decay and death. There is a reason of course: many mushrooms are edible, but some are highly toxic. Mushrooms are also called 'toadstools', which again suggests a negative image. In medieval times, the toad was a symbol of evil, associated with the devil, witchcraft and black magic. Mexican Indians consumed special mushrooms for their hallucinogenic qualities. Other people gave them a sacred status and used mushrooms in religious rituals.

When people began to use fungi as food is not clear. Prehistoric man may have collected mushrooms to use them as food; however, the first records of consumption of wild mushrooms date from classical Greek and Roman times. Around AD 600 the Chinese started to cultivate mushrooms.

Nowadays, mushrooms are so much valued that they are cultivated throughout the world. The three most important cultivated edible mushrooms account for about 70% of the world's supply: button mushroom about 30%, shii-take about 25%, and oyster mushroom about 15%. They have in common that they all consist of a cap, gills on the underside of the cap and a stem, which may be very short in oyster mushroom. There are many more, though less important cultivated mushrooms, such as jelly ear (*Auricularia* spp.), paddy straw mushrooms (*Volvariella volvaceae*) and velvet foot (*Flammulina velutipes*).

In the taxonomic arrangement of living organisms, mushrooms are placed in the Kingdom Fungi, which includes amongst others the division Basidiomycota. Characteristic for the basidiomycetes is that these fungi form club-shaped structures, called 'basidium', usually bearing four basidiospores. The most important cultivated mushrooms belong to the larger basidiomycetes.

As photosynthesis is not possible for fungi, they absorb their nutrition from fermented and/or decomposed living or dead organic material. During and after decomposition, a mass of very fine, thread-like, white mycelium develops throughout the decomposing or decomposed organic material. The mycelium can be seen as the vegetative part of the fungus. When circumstances are right, the mycelium will produce reproductive fruiting bodies, which can have a large number of different appearances. The fruiting bodies can be edible mushrooms.

The food value of different mushroom species varies; however, they are all low in energy content. Expressed as a percentage of dry weight, mushrooms contain 10–40% protein that is easily digested, 3–28% carbohydrate, 3–30% fibre and a little oil; moreover, mushrooms are particularly rich in a number of B-complex vitamins and moderate in vitamin C. A relatively new product called 'quorn' is made from the fungus *Fusarium graminearum*. It has high protein content, comparable with meat.

In nature, fungi can be seen as the decomposers of the biosphere: by their action, CO_2 is released into the atmosphere and nitrogenous compounds are returned to the soil. However, the ability of fungi to effect decomposition is often in direct conflict with human interests, because fungi grow on all kinds of foods and organic durable articles.

Some of the most highly valued culinary fungi are black truffle, morel and chanterelle; they are difficult to grow commercially. However, truffles are cultivated, although more or less indirectly. This involves choosing an outside site with suitable soil aspects and planting trees which most favour the production of truffles. Before planting the trees, they are inoculated with the truffle mycelium. In fact, it is no more than creating a desired ecosystem for the growth of truffles.

Mushrooms have been used medicinally; for example, they can be an ingredient of preparations used to prevent high cholesterol. Antitumour activity has been found in the polysaccharide fractions of the mushroom. The high Fe content can have a positive effect on the blood.

Button mushroom

The cultivation of button mushroom probably began in about 1650. Gardeners around Paris discovered that special mushrooms grew spontaneously on the decomposed manure that was left over after growing melons. This discovery was probably the beginning of commercial cultivation of button mushrooms.

Agaricus (Fig. 11.44) is a saprophytic mushroom with a ring called an 'annulus' around the stalk. The cap can be up to 15 cm in diameter, convex at first, to nearly plane when mature. The colour varies from light brown to cream and white. The surface of the cap can be smooth or have small scales; the colour can be white, cream or brown. The flesh is firm and white. Unlike other basidiomycetes, button mushroom forms two-spored basidia (hence the species name *bisporus*), covering the gills. The gills are free from the stem, pinkish to brown at first, dark brown to almost black when mature. The spore print is brown. The stem can be 2–8 cm long and 1–3 cm thick, smooth or with small scales.

Growing button mushrooms is different from growing shii-take and oyster mushrooms, because button mushroom is a secondary decomposer, whereas the others are primary decomposers. This means that the growing substrate of button mushrooms, which is usually straw mixed with stable litter and minerals, first has to be composted by bacteria and other fungi before button mushrooms can grow on it. The other mushrooms can decompose the raw material themselves, which can be substrates such as straw, wood logs, sawdust and wood shavings.

Fig. 11.44. Button mushroom (photos: PPO Paddestoelen, Horst, The Netherlands).

The mycelium of button mushroom is usually propagated vegetatively on sterilized cereal grain; it is called 'spawn'. Spawn and compost are mixed, and then kept at 24°C and high relative humidity and high CO_2. The spawn will grow and produce an extensive network of mycelia in the compost. To colonize the compost entirely takes up to about 20 days. The compost has to be covered with a layer of casing material, which can be a mixture of peat and limestone. The layer stimulates fruit body formation. About 6 weeks after the beginning of the cultivation, the first mushrooms may appear and the first crop can be harvested. Just before that, air temperature has to be reduced to 18°C. A growing mushroom doubles its weight every 24 h. After the first harvest, the crop grows in repeating 5-day cycles called 'flushes', although successive harvests show decreasing yields. Three to five crops are harvested before the substrate and crop residues have to be removed. The main portion of button mushroom is grown for the fresh market, but a considerable portion is canned or used for the manufacture of dried or canned soups and sauces. As button mushrooms do not need light for growing, they are usually grown in special dark mushroom houses, sometimes in caves.

Most of the mushrooms are harvested and sold immature, when the cap is still closed and more or less attached to the stalk, diameter 3–5 cm. A smaller part is harvested when they are more mature, which means that the cap is opened to expose the gills, when the diameter can be up to 15 cm. The latter is called 'portabella'.

Shii-take

The name 'shii-take' is derived from the name of a Japanese tree called 'shii', on which shii-take grows, and the Japanese name for fungus, which is 'take'. It is also called 'black forest mushroom'. This mushroom has been well known in China and Japan for about 2000 years, and was probably first cultivated in China about 1000 years ago. Nowadays, shii-take is the most popular mushroom in Japan and China, and is particularly valued for its excellent taste and flavour. Shii-take (Fig. 11.45) is a saprophytic mushroom. The cap is usually 5–12 cm in diameter, but can be up to 20 cm. When young, the cap is concave and it will flatten in maturing. The cap is open to expose the gills. The colour of the cap is brownish-grey to reddish-brown, often darker in the centre. On top of the cap, small scales may occur. The colour of the gills is yellowish-white when young, brownish when older. The stem is covered with brown scales. The flesh is firm and juicy, though not watery, white to brownish in colour, and has an intense mushroom flavour.

Shii-take is grown both indoors and outdoors, on natural or so-called synthetic logs. Natural logs may be cut from various hardwoods like oak, shii trees, beech and chestnut. The most efficient diameter of the logs is between 10 and 15 cm. The logs are inoculated with spawn, which is propagated on wood pieces, cereals, or sawdust. A row of holes is made in the log, which are filled with the spawn. Next, the holes are covered with wax, plastic or other material, to prevent drying of the spawn. After that, mycelium growth may start.

Fig. 11.45. Shii-take (photos: PPO Paddestoelen, Horst, The Netherlands).

Synthetic logs are polypropylene bags filled with about 2.5 kg of a mixture composed of sawdust or wood shavings, straw, maize cobs and cereal bran. After sterilizing the mixture, the spawn is added. When the growth of the mycelium inside the bag is completed, the bag may be removed before the fruiting bodies develop. Fruiting requires a drop in temperature, increased humidity, and light.

The harvest time depends on the desired product: for the fresh mushroom, harvesting is done when 50–60% of the cap is open; for dry mushrooms, harvesting is done when they are more mature and 70–80% of the cap is open.

The biggest portion of the mushrooms grown is preserved, mainly by drying or by canning or pickling in vinegar. However, a considerable portion is for the fresh market.

Oyster mushroom

Oyster mushroom can be found in the wild, mainly in the temperate zones and in the cooler seasons in subtropical areas. It grows in particular on dead trees but also on living trees as a parasite. The name 'oyster mushroom' comes from the shell-like appearance of the cap.

Oyster mushrooms (Fig. 11.46) grow in clusters. The colour of the cap varies with age, the intensity of light, and what time of the year the mushroom fruits. The colour can be from whitish, pale brown to dark bluish-grey. The cap is first convex, becoming plane in maturing, from kidney- to almost circular-shaped, and has a diameter of 4–15 cm, sometimes up to 30 cm. The margin of the cap is in-rolled. The gills are widely spaced and yellowish-white in colour, the spore print is lilac. The stem is nearly absent or very short when the mushroom is growing from the side of a log, tree, or another substrate. Growing from the sides of the substrate also means that the stem growth is strongly eccentric compared with the cap. When the mushroom is growing on top of a log, it may develop a distinctive and thick stem, which is hairy at the base and

Fig. 11.46. Oyster mushroom (photos: PPO Paddestoelen, Horst, The Netherlands).

positioned more or less in the centre of the cap. The gills merge into the stem. The taste of oyster mushroom may vary widely; the texture of the flesh can be soft to chewy. *Pleurotus ostreatus* has an extremely high carbohydrate content of 57%, based on dry weight.

Within the genus *Pleurotus*, many species exist that are often difficult to distinguish from one another, and what is generally considered to be 'oyster mushrooms' may involve different species.

Oyster mushroom is easy to cultivate on a variety of substrates, such as natural logs, straw, sawdust and hulls; almost all substrates that contain cellulose can be used, including a wide range of plant wastes. The substrate has to be inoculated with spawn that developed on grain or straw. Inoculation occurs by mixing the spawn with the substrate or attaching it to the substrate. The advantage of oyster mushroom culture is cheapness of the culture, which is partly the reason why the species is cultivated worldwide. A disadvantage is the tremendous number of spores that are released by the oyster mushrooms during the cultivation. When the mushrooms are grown in a mushroom house, the spores may cause allergies and respiratory problems among the workers in the house. The optimal temperature for growth of the mycelium is about 25–28°C. The optimal temperature for the development of fruiting bodies depends on species and ranges from 10 to 28°C. The fruiting body of oyster mushroom cannot endure high CO_2 concentrations in the growing house; the formation of the fruiting body requires light.

When the mushrooms can be harvested depends on the desired product. Different products require different cap diameters, which can range from 2 to 10 cm. The mushrooms are sold fresh, or preserved by drying, canning or brining.

Okra, Kangkong and Ceylon Spinach

Okra – *Abelmoschus esculentus* (Mallow family – *Malvaceae*), Kangkong – *Ipomoea aquatica* (Morning glory family – *Convolvulaceae*); Ceylon spinach – *Basella alba* (Madeira-vine family – *Basellaceae*)

Okra

Origin, history and spread

Okra, also called 'lady's finger' or 'gumbo', probably originated somewhere around Ethiopia. It was cultivated by the Egyptians around the 12th century BC. Some authors believe that the centre of origin was in South-east Asia. However, little is known about early history, domestication and distribution. The plant has been well known in India for many ages, and reached America via the African slave trade route. The name 'okra' means 'lady fingers' in Igbo, a language spoken in Nigeria. Although okra is now cultivated in many tropical and sub-tropical countries, at present it is mostly cultivated in India and Nigeria.

Botany

Okra (Fig. 11.47) is an annual or perennial, stout, erect, few-branching, herbaceous plant, growing to a height of 4 m. The leaves are spirally arranged; the

Fig. 11.47. Kangkong: 1, detail of stem with leaves. Ceylon spinach: 2, detail of stem with leaves. Okra: 3, leaf; 4, full-grown fruit; 5, young fruit.

leaf-blade is palmate with five to seven lobes and up to 50 cm in diameter, with few spines; petiole up to 50 cm long. The flowers arise solitary from the leaf axils; the calyx is spathaceous, 2–6 cm long, splitting on one side during the expansion of the corolla; the corolla consists of five obovate petals, 3–7 cm long and white to yellow in colour. The flowers are 4–8 cm in diameter and have a dark purple centre. The fruit is a cylindrical to pyramidal capsule, usually ribbed, spineless in cultivars, 5–20 cm long, 1–5 cm in diameter, greenish-purple when young and brownish-cream when mature (Fig. 11.48). The fruit contains numerous globose, black seeds, which are 3–6 mm in diameter. The 1000-seed weight is about 60 g.

Okra has a strong taproot, which grows up to 50 cm deep; on four sides of the taproot about 24 to 35 laterals originate, spreading up to 45 cm.

Cultivars, uses and constituents

There are several okra cultivars differing in height, colour of the capsule, and prominence of the ribs of the capsule; dwarf forms are about 1 m tall. Important cultivars are 'Clemson Spineless' and 'Pusa Sawani'.

The fruits have to be picked when they are young, about 8–10 cm long, when the fruit is tender and the seeds are still white. When the fruits are too mature, they become too woody and cannot be eaten. Okra is eaten in many ways; it can be raw, boiled, steamed, fried and stir-fried. Okra is eaten as a vegetable, as a component of many dishes: 'gumbo', stews and casseroles. Gumbo is a thick soup, containing vegetables including okra, and meat, originally from

Fig. 11.48. Okra: close-up of okra fruits in various developmental stages on the plant (left); detail of flowering plant (right) (photos: J. Wienk).

Africa. Because of the mucilaginous properties of the plant, the fruits are also used as a thickening agent in soups and in gumbo. Mature seeds yield edible oil, which is used as a salad oil. The seeds can be baked and ground into meal for use as a coffee substitute. The leaves are sometimes used as a vegetable or as feed for cattle. The fibre of mature okra stems is sometimes processed into rope and paper. To preserve okra, it is dried, canned, powdered, pickled, or frozen. The edible part of the fruit contains approximately 90% water, 2% protein, 7% carbohydrates and 1% fibre. It is a good source of the vitamins A, B-complex and C, and minerals, especially Fe and Ca.

Ecology and agronomy

Okra is primarily a warm-season plant. Okra requires full sunlight and temperatures above 20°C for normal growth; optimal day temperature is 30–35°C. It grows well on well-drained, fertile, light or heavy soils, which have been worked at least 25 cm deep, pH 6–7.5. The plant does not tolerate chilling and frost. Once established, the species is one of most heat- and drought-tolerant cultivated plants in the world. Okra is a short-day plant; however, cultivars differ considerably in sensitivity to photoperiod.

Okra is propagated from seed; the seeds should be soaked for 24 h before planting. The seed has to be planted in rows, 1–2 cm deep. Rows can be spaced 60–100 cm apart and spacing within the row ranges from 10 cm to about 30 cm. Plant density depends on cultivar, varying widely from 20,000 to 150,000 plants/ha. Seedlings will emerge after 3–4 weeks when dry seed is used and after 1 week when the seed has been soaked in water. Before planting, manure or artificial fertilizer has to be mixed well into the top 10 cm of the soil. The uptake of minerals is rather high. When the crop yields 10 t of fresh fruits per hectare, the uptake is 100 kg N, 10 kg P, 60 kg K, 80 kg Ca and 40 kg Mg. Fertilizer can be given in three split applications. Weeding or cultivating is needed only during the first month after emergence of the plants, and is often combined with earthing-up the rows. Weed control can also be done chemically; herbicides are available.

Okra can grow very fast; flowering will start about 2 months after sowing, developing fruits should be harvested 7–8 days after flowering. The fruits are harvested when 8–10 cm long. As the plant continues to flower, fruits can be picked every 2 days. Picking fruits is usually done manually. For seed production, the whole mature crop can be harvested at once.

After harvesting, the fruits can be stored for only 3–5 days at about 10°C; lower temperatures may cause damage to the fruits. Okra should not be washed until shortly before use; washing makes the fruit slimy. Due to extensive cultivation, yields are often low, 2–10 t of fresh fruits per hectare. Yields above 30 t/ha can be realized under optimal conditions. In 2005, the world total okra production was about 5 million t of fresh fruits.

Kangkong

Origin, history and spread

Kangkong, also called 'water spinach', is probably native to India and South-east Asia. The species naturalizes easily and can now be found wild in South and South-east Asia, tropical Africa, and South and Central America. In many areas, it can be considered as a troublesome weed. Kangkong may cause environmental damage; for example, masses of plants can obstruct water flow in drainage systems. It is a fast-growing and aggressive plant, which can be a serious threat to waterways. However, the plant is also valued as a leafy vegetable. Kangkong is now cultivated and eaten as a vegetable in South-east Asia, Taiwan and southern China.

Botany

Kangkong (Fig. 11.47) is a perennial or occasionally annual, trailing, creeping, floating or erect herb, with milky sap. The stem is vine-like, hollow, and about 3 m long, sometimes longer. When growing in water, stems can be floating. The leaves are simple, alternate, petiolate; blades variable, cordate, triangular or lanceolate, 3–15 cm × 0.5–10 cm, heart-shaped or hastate at the base; leaf-blade is glabrous or rarely pilose, with pointed tip. The flowers are funnel-shaped, solitary or in few-flowered clusters at leaf axils. They consist of five free sepals; five united petals, up to 5 cm wide, pink to white in colour, often with a dark centre (Fig. 11.49). The fruit is an oval or spherical capsule, about 1 cm wide, woody at maturity, containing one to four greyish seeds. The 1000-seed weight is about 40 g. When growing in water, roots arise from the nodes of the stem.

Fig. 11.49. Stand of white-flowering broad-leaved kangkong seen from above (photo: J. van Zee).

Cultivars, uses and constituents

Two types can be determined, with many cultivars of each: (i) the red type with red-purple stems, dark-green leaves and pink to lilac flowers; and (ii) the green type with whitish-green stems, green leaves and white flowers.

Young tops and leaves are cooked like spinach, or fried in oil and eaten as a vegetable, or used in various dishes. A small portion of the production is canned. The vines are used as fodder for cattle and pigs. Fresh leaves and tops contain approximately 90% water, 3% protein, 5% carbohydrates, 1% fibre, 0.3% fat and 1.6% ash, the minerals Ca, Mg and Fe, and vitamin C and provitamin A.

Ecology and agronomy

Kangkong is primarily a plant of the humid tropical lowlands. It requires short-day conditions and a warm, wet climate to flourish; mean temperature has to be above approximately 25°C. The species thrives best in water or relatively moist soil. Soil may be clay-loam or loam. Optimum pH range is 5.3–6.0.

Kangkong is propagated from seed or stem cuttings, usually germinated and grown in nursery beds first. Two methods of cultivation are used: dry and wet. In both cases, large amounts of organic material and plenty of water are required to achieve adequate yields. Heavy fertilizing benefits leafy growth. Weed control is mostly done manually.

In dry cultivation, the plants are usually planted on raised beds, at a spacing of 15 cm, and often supported by trellises. In wet cultivation, usually about 30-cm-long cuttings are planted in mud. As the vines grow, the field is flooded and a continuous slow flow of water is required. In wet cultivation, weeds are controlled by the flooding. Harvesting may start 30–40 days after sowing. Yields vary greatly; in dry cultivation annual yields of around 40 t of fresh product per hectare are reported. In wet cultivation, yield can be up to 100 t/ha. World production statistics are not available.

Ceylon spinach

Origin, history and spread

Ceylon spinach, also called 'basella', 'Indian spinach' or 'climbing spinach', originated probably in India. In South-east Asia, China and India it has been grown since ancient times. As the species naturalizes easily, it can be found along roadsides and as weed on arable land. Basella is now cultivated in Africa, tropical America and tropical Asia, particularly in moist lowlands.

Botany

Ceylon spinach (Figs 11.47 and 11.50) is a fast-growing perennial, often grown as an annual, glabrous, succulent vine, up to 10 m in length. The stem is much-

Fig. 11.50. Ceylon spinach (left to right): plant habit showing its twining character; inflorescence with flower buds in the top and young fruits covered with fleshy perianth nearer the base; infructescence (photos: B.P. Schuiling).

branching, slender, smooth, sometimes almost leafless and greenish or reddish. The leaves alternate, and are ovate to heart-shaped, 5–15 cm × 5–13 cm, acute or acuminate, usually cordate at the base, and with short petioles. The inflorescence is an axillary, hanging, long-peduncled spike, up to 20 cm long. The flowers are sessile, 3–4 mm long, white, rose or purple; the ovary is rounded, with three styles and containing violet juice; the flower contains five stamens. The fruit is a depressed globose, fleshy, dark purple pseudo-berry, about 5 mm in diameter. The fruit contains one seed; 1000-seed weight is 30–40 g. Prostrate stems may develop new roots at the nodes.

Types, uses and constituents

Three types can be distinguished: (i) the most common type with dark green, ovate or almost round leaves; (ii) with red stems and reddish, ovate to almost round leaves, often planted as an ornamental; and (iii) with heart-shaped, dark green leaves. The three types are sometimes considered to be three distinct species: *Basella alba*, *Basella rubra* and *Basella cordifolia*. The thick fleshy leaves and young shoots are cooked, boiled or fried, and eaten as a slightly mucilaginous vegetable. It is used in stews and soups or as a green salad. Deep-freezing is a good method to preserve the leaves. The red juice of the fruit is used for dyeing purposes, in cosmetics, as ink, for colouring foods, and as dye for official seals. Naturalized basella is often collected in the wild to use it as a vegetable. For example, in Tanzania it is a component of 'mboga', a dish made of many different collected wild vegetables.

The edible part of fresh Ceylon spinach contains approximately 91% water, 2% protein, 4% carbohydrates, 0.3% fat and 1.3% fibre. It is a source of the vitamins A and C, and the minerals Ca and Fe.

Ecology and agronomy

Ceylon spinach grows best during warm rainy periods with a small amount of shade, with minimum day temperature of 15°C, optimum day temperature of 25–30°C and night temperature not below 10°C. It grows in a variety of soils but it prefers a well-drained, moisture-retentive sandy loam, rich in organic matter; it tolerates a pH in the range 4.3 to 7.0. It is a short-day plant.

Basella is propagated from seed or stem cuttings. Germination requires a minimum temperature of 18–21°C. It can be planted in a nursery or directly in the field. For commercial production, densities of 50,000 plants/ha are used. In home gardens, vines are often grown on trellises, but in commercial crops it is usually grown without support. The species thrives under conditions of moderate soil fertility and responds well to N.

Basella is a fast-growing plant, the first harvest of young shoots and leaves can be about 40 days after planting. Subsequently it can be harvested several times until flowering becomes abundant. A good crop can yield about 50 t of young shoots and leaves per hectare. World production data are not available.

Radish and Cress

Radish – *Raphanus sativus*, Cress – *Lepidium sativum*; Wallflower family – *Brassicaceae*

Radish

Origin, history and spread

The origin of radish is uncertain; however, it could be derived from the closely related wild species *Raphanus raphanistrum*, which can be found in the Mediterranean region, western Asia, western Europe and near the Black Sea. The two species can be crossed easily. When domestication took place is also uncertain, but radish was certainly grown in ancient Egypt, because it was a component of the diet of the slaves who built the Cheops pyramid (2700 BC). The earliest cultivated radishes were the ones with black, swollen roots. Radish was grown in China by about 500 BC. The ancient Romans cultivated radish and they probably brought the species to western and central Europe. In northwestern Europe, radish seeds are found in archaeological sites from the Roman period. Because radish seeds were sometimes found in a small bottle mixed with seeds of other species, the seeds possibly were considered medicinal. Radish was mentioned in Charlemagne's list of cultivated plants *Capitulare de Villis* (around AD 800), where it was called 'radices'. From China, the species spread to Japan around AD 700. From Europe, explorers brought it to the new World. White radishes were first grown in Europe around AD 1500; red radishes were bred in the 18th century. Nowadays, radish is grown throughout the temperate regions. However, in different regions, people value completely different types.

Botany

Raphanus sativus (Fig. 11.51) is an annual or biennial erect herb, more or less hairy, and 15–150 cm tall. Leaves alternate; lower leaves in a rosette, leaf-blades oblong, oblong-ovate to lyrate-pinnatifid, often parted, margins more or less crenated, 5–60 cm long; higher leaves are smaller, less pinnatifid, lanceolate-spathulate, margins often dentate. All leaves are petioled, lower leaves having longer petioles than higher ones. The inflorescence is a terminal, long, many-flowered raceme. The flowers are white or lilac and characteristic for the wallflower family: four sepals, four petals and six stamens. The diameter of the flowers is about 1.5 cm. The fruit is a cylindrical siliqua, which is a pod-like capsule, with a variable length of 5 to 60 cm, containing two to 12 seeds in one-seeded compartments. Compartments together have a sponge-like appearance. The fruit has a seedless beak, which can be very long. Seed is ovoid-globose in shape, 2–3 mm in diameter and yellowish-brown in colour; the 1000-seed weight is 10–19 g.

Parts of the plant can be swollen. What part of the plant will swell depends on variety; the swollen part or the storage organ is in some varieties composed of the upper part of the taproot and hypocotyl, in other varieties it is mainly hypocotyl. The so-called 'leaf radishes', which are vegetable, fodder or green manure varieties, often have no swollen parts. The colour of the storage organ is variable: black, white, red, pink, yellow, or combinations of red and white or green and white. The colour of the flesh is mostly white, although red-fleshed varieties occur. The shape of the storage organs is globose, ellipsoid, conical or cylindrical. The root system consists of a taproot and laterals.

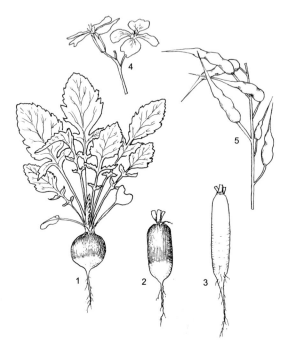

Fig. 11.51. Garden radish: 1, cultivar 'Sparkler'; 2, cultivar 'French breakfast'; 3, cultivar 'Icicle'; 4, flowers; 5, detail of a stem with fruits (siliqua).

Fig. 11.52. Radishes: 1, garden radish cultivar 'French breakfast'; 2, garden radish cultivar 'Scarlet Globe'; 3, elongated white type of winter radish ('rettich').

Varieties, uses and constituents

Within the species *R. sativus*, great variation in size, colour, shape and use of the storage organs occurs. In Western countries the small red, white, or red and white garden radish, diameter 2–3 cm, and the larger black or white winter radish, diameter up to 10 cm, mostly globose or ellipsoid in shape, are grown widely, as a field crop or in glasshouses (Figs 11.52 and 11.53). In Asia, large varieties are preferred. Nowadays, the large, white, cylindrical radish of Japanese origin, the so-called 'daikon' type, is the centre of interest. The daikon radish has a milder and sweeter taste than the small garden radish. The storage organs of the larger forms can attain a weight of 20 kg or more. In Japan, the giant, white 'Sakurajima radish' is grown, which can even attain a weight of 45 kg. Radish can be eaten fresh as a salad vegetable; it can also be cooked or preserved by pickling or drying. The fresh storage organ of the small radish is valued as a pungent appetizer or for garnishing. The larger varieties, which are commonly used in South-east Asia, are often cooked as vegetables or used in soups and sauces. As well as the storage organs, the leaves or fruits of a few varieties are also eaten. When leaves are eaten, they may be fresh in a salad or cooked and prepared as spinach. The very long fruits, up to 60 cm long, of the so-called 'rat-tailed' radish, are edible and consumed when they are still immature, either fresh, cooked or pickled. Some varieties with rapid leaf production are

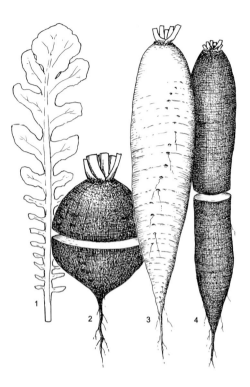

Fig. 11.53. Winter radish: 1, leaf; 2, round black type, cut in half to show the white flesh; 3, elongated white type; 4, elongated black type, cut in half to show the white flesh.

grown as fodder crop or as green manure. Some varieties are grown for the excellent salad oil that can be extracted from the seeds. On a small scale, seedlings of radish are grown for use as an appetizer or for garnishing.

The storage organ of radish contains approximately 94% water, 5% carbohydrates, 0.5% protein and 0.1% fat, furthermore Ca, P, Fe, and the vitamins A, B and C. Fresh radishes contain the enzyme diatase, which stimulates the digestion of starch, which is why radish combines well with starch foods like bread and rice. Eating radish is assumed to have a positive effect in preventing cancer. Radish juice is used to treat convulsive cough.

Ecology and agronomy

Radish is primarily a crop for the temperate regions or a cool-season crop. Cool conditions stimulate prosperous growth. Short day length stimulates root development; high temperature stimulates the development of inflorescences. The crop requires well-drained, light, sandy, deep, soils, with pH 6–6.5. Heavy soils may bring about misshapen roots.

Radish is propagated by seed. Seeding rate depends on variety, ranging from 10 to 60 kg/ha. The seed is often sown directly in rows in the field or sometimes in a glasshouse. For early cultivation, the seed can first be sown in paper pots in a nursery, and transplanted later into the field. Radish is often grown on prepared beds. The small, western, garden radish needs a spacing of 10–25 cm between the rows and about 3 cm between the plants in the row. The varieties with the large storage organs, for example the daikon radish, require a much wider spacing of 30 cm between the rows and 15–25 cm between the plants in the row. For giant radishes, spacing has to be even wider, depending on the required size of the storage organ.

To obtain good-quality storage organs, which are mild and tender, the plant must grow rapidly with enough fertilizer and moisture. Application of fertilizer can be in accordance with removal of nutrients from the soil, which is, per 50 t of storage organs, about 120 kg N, 20 kg P and 165 kg K. Irrigation or sprinkling may be necessary because stagnation of growth may cause hot-tasting and tough storage organs. Weed control can best start before sowing the radish. As radish usually germinates and grows rapidly, it may suppress weeds rather well. If necessary, mechanical weed control between the rows is possible; moreover, a few herbicides are available.

Radish is usually grown as a sole crop; however, intercropping occurs, for example with lettuce.

The required growing period depends on climate, variety and desired product. Small western radish can be harvested 3–5 weeks after sowing, whereas larger Asiatic varieties need often up to 9 weeks, sometimes even more, to 'mature'. Harvesting mechanically is possible; however, it is often done by hand. As soon as possible after harvesting, the product has to be stored at 1°C and high humidity. When the leaves are left on the storage organs, the storage life is about 1 week. When the leaves are removed, the storage life can be up to several months. Yields of fresh storage organs depend on climate, variety and cropping system, and vary considerably, ranging from 7 to 60 t/ha when conditions are optimal; even 90 t/ha is recorded. Production statistics are not available.

Cress

Apart from radish, there are also other cruciferous plants from which seedlings are mainly used as salad plants, especially as an appetizer or for garnishing. Garden cress, *Lepidium sativum* (Fig. 11.54), is an example. It is probably a native of south-western Asia. According to Vavilov, Ethiopia is the main centre of origin because of the wide variability in that region. The species has been cultivated for centuries because there is much written evidence that the ancient Egyptians, Greeks and Romans used it in spicy salads. The species is valued because of the pleasantly pungent flavour; it is also an appreciated source of vitamins. The cotyledons of garden cress seedlings are deeply three-lobed, which is in contrast with other seedlings that are also used as salad plants. The mature leaves are pinnate or bipinnate. Some varieties have crisped leaves. Garden cress grows very rapidly in temperate conditions. The seedlings are ready for consumption about 2 weeks after sowing. At that stage, the cotyledons have developed. The plants are grown on prepared beds in the field or in greenhouses. Garden cress seedlings are also traded ready for use, in plastic trays filled with peat or synthetic material with a nutrient solution (Fig. 11.55).

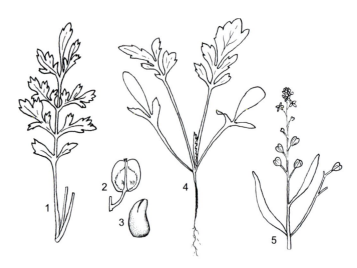

Fig. 11.54. Garden cress: 1, leaf; 2, fruit (siliqua); 3, seed; 4, seedling; 5, inflorescence with flowers and fruits.

Fig. 11.55. Garden cress as marketed (left) and cress sprouts in detail (right) (photo: The Greenery, The Netherlands).

Spinach

Spinach – *Spinacia oleracea*; Goosefoot family – *Chenopodiaceae*

Origin, history and spread

The wild form of spinach is not known. It was already cultivated over 2000 years ago in the Middle East, probably in Iran. It was known by the ancient

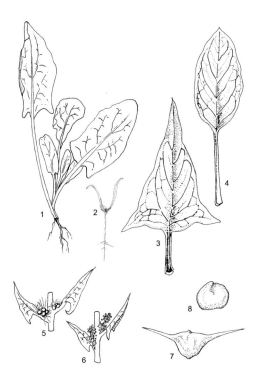

Fig. 11.56. Spinach: 1, habit of a vegetative plant; 2, seedling; 3 and 4, variety in shape of the leaves; 5, detail of a stem with female flowers in the leaf axils; 6, detail of a stem with male flowers in the leaf axils; 7, seed with spines; 8, globular seed.

Greeks and Romans. Traders brought spinach into China and India in the 7th century. The Arabs introduced spinach into Spain in the 12th century, and from there it spread throughout Europe. European settlers brought it to the New World. The name 'spinach' is derived from the Persian word 'ispanai', which means 'green hand'. At present, spinach is widely cultivated in the temperate regions of the world.

Botany

Spinach (Fig. 11.56) is an annual herb that forms a rosette of alternate and simple leaves on a very short stem. The leaves are glabrous and tender and may differ from smooth to wrinkled, ovate, rounded or broad arrow-shaped. The leaves are usually long-petioled and variable in size, leaf-blade 9–30 cm × 7–20 cm, depending on soil fertility and plant spacing. Leaves are lobed at the base and sometimes with lobed margins. A vegetative crop is up to 25 cm high; plants of most varieties produce 22–26 leaves. Spinach quickly runs to seed, producing a stem that bears triangular, narrow, pointed leaves, about 75 cm high. Spinach is usually dioecious, plants generally producing either male or female flowers, although monoecious plants occur. The flowers are small and green, male flowers borne in clusters on long, terminal spikes, female flowers mostly axillary. Flowers resemble those of sugarbeet. Male plants bolt and flower earlier than female plants and die soon after flowering (Fig. 11.57). The 1000-seed weight is 10–15 g. Spinach

Fig. 11.57. Detail of a spinach stem with leaves and fruits.

has a deep taproot with laterals, although most of the root system is rather shallow.

Varieties, uses and constituents

There are two main groups of varieties: (i) the smooth-seeded summer spinach; and (ii) the prickly-seeded winter spinach. Most commercial varieties are now smooth-seeded. Varieties may also be divided into two different groups by leaf type (Fig. 11.58): (i) wrinkled leaves, mostly for the fresh market; and (ii) smooth leaves, mostly for processing, primarily freezing, drying or canning.

Spinach is primarily a pot herb or green, the leaves are normally consumed as a cooked vegetable; however it is becoming moderately popular as a salad vegetable. As the leaves give a green colouring, spinach can also be used for colouring food, liqueurs, etc. Spinach is also much used as infant food.

Spinach tends to accumulate N in the form of nitrate, particularly after heavy fertilizing for achieving dark green and succulent leaves. In the digestive tract, nitrate is converted to nitrite, which oxidizes haemoglobin, forming methaemoglobin. It is assumed that this substance can cause a disorder called methaemoglobinaemia, which may cause a health problem in infants who consume large amounts of spinach.

Spinach is a nutritious vegetable and contains large amounts of minerals and vitamins: Ca, P, Fe, K, and vitamins A, B and C. Furthermore it contains roughly 92% water, 2.5% protein, 0.6% carbohydrate and 0.3% fat. Leaves of spinach often have high oxalic acid content. Although it is not toxic, oxalic acid does lock

Fig. 11.58. Two types of spinach leaves: flat (left) and curled and bumpy (right).

up minerals, especially Ca, in food and makes it unavailable for absorption by the human body. Cooking reduces the oxalic acid content.

A closely related member of the same family is garden beet (*Beta vulgaris*). Two varieties are grown as vegetables: spinach beet and chard. The varieties are grown solely for their leaves and used like spinach. The flavour also resembles that of spinach but is slightly milder. Both varieties resemble spinach in their botanical characteristics. Chard differs somewhat from spinach beet by having broader, whiter leaf stalks, which can be eaten as a separate vegetable. The varieties are commercially not important; they are grown mainly in home gardens.

Ecology and agronomy

Spinach is essentially a hardy, cool-season crop; young plants tolerate low temperatures down to −10°C. The plant grows well on a wide range of soils, but it yields best on a moist, sandy loam, high in organic matter. Spinach prefers neutral to slightly alkaline soils, with an optimum pH of 6.2–6.9. The heaviest crops can be produced on N-rich soils. The plant grows well when exposed to full sunlight but prefers light shade when the temperature is high. Spinach requires much water in the growing period; drought can cause the plants to run to seed very quickly. In general, hot and dry summers speed up the change-over to the generative stage.

In several regions spinach can be sown almost throughout the year, although it is important to choose the right varieties for the different growing periods in winter, spring, summer and autumn. This is because spinach becomes reproductive in response to the day length and different varieties do so at different day lengths.

Spinach is propagated by seed. How much seed is required depends on production system and period of the year, ranging from 40 to 150 kg/ha. Seeds should be planted 1–3 cm deep. The seed can be broadcast or sown in rows, sometimes on beds. The spacing between the plants in the row varies from 5 to 15 cm; the distance between the rows depends on production system, but may vary from 12.5 to 30 cm. The seeds may germinate within 10 days and the leaves can, when conditions are favourable, be harvested about 5–8 weeks after planting, depending on variety. To achieve a good yield, the plants need to have access to sufficient fertilizer. How much fertilizer has to be applied depends on soil analysis and the removal of nutrients by the crop; 20 t of harvested leaves removes about 70 kg N, 10 kg P and 100 kg K per hectare. Weed control is necessary because spinach does not compete well with weeds. It can be done by two to four tillage operations during the growing season or by using herbicides. Tillage should be shallow to avoid damage to the shallow root system. Irrigation is often needed during dry periods; 25 mm of water every week is generally sufficient.

The plants should be harvested while young and tender and before the development of the seed stalk. Harvesting can be done by mechanically cutting the entire rosettes at the soil surface. After that, regrowth is not possible because the apical meristems are removed. Cutting about 3 cm above the soil surface allows the plants to regrow and to obtain a second harvest. Spinach meant for canning, drying or freezing should be processed immediately after harvesting. Spinach for the fresh market can be stored only for 10–14 days at 0°C and 95–100% relative humidity. The leaves must remain fully turgid to be marketable. The yield varies widely; from 8 to 40 t of fresh leaves per hectare. In 2005, the world total production of fresh spinach leaves was approximately 13 million t; China (11 million t) was by far the most important producer.

Tomato

Tomato – *Lycopersicon esculentum*; Nightshade family – *Solanaceae*

Origin, history and spread

The genus *Lycopersicon* is native to western South America. Tomato (*Lycopersicon esculentum*) originated in Peru from the wild cherry form (*L. lycopersicum* var. *cerasiforme*), which has a fruit diameter of about 3 cm. Tomato was probably first domesticated in the Vera Cruz–Puebla area in Mexico. Other sources suggest that domestication first occurred in Peru.

After the discovery of America, the plant was brought to Europe by Columbus after his second exploratory expedition. The species was only slowly accepted in Europe because, as a relative of the nightshade family, it was suspected that the tomato was poisonous like related species such as deadly nightshade (*Atropa bella-donna*) and thorn-apple (*Datura stramonium*). A second reason to reject the plant was the unfamiliar, unpleasant smell. When tomato was introduced in Europe, it was known as 'golden apple' ('pomo d'oro', from which the Italian name 'pomodoro' is derived) or 'Peruvian apple', but also 'devil's wolf apple'. All together this delayed the utilization of the plant for many centuries. The popularity of the fruit increased only just shortly after World War II. At present the species is much appreciated and is grown all over the world, in temperate regions often in greenhouses.

Botany

Tomato (Fig. 11.59) is a weak-stemmed herb with a lot of axillary branches. Under natural conditions it forms a spreading, straggling bush. Cultivated tomatoes in greenhouses are mostly trained into a single stem by pruning axillary branches. The stem is solid and coarsely hairy. The leaves are spirally arranged; compound, pinnate or bipinnate. The leaflets are lobed or toothed. Flowers, usually occurring in cymes or branched cymes, are about 2 cm in diameter. They have a five-lobed green calyx, five yellow petals and six stamens. In many cultivars the bright yellow anthers form a cone, enclosing the style (Fig. 11.60).

The fruit is a fleshy, juicy berry, usually red, also dark red or yellow and sometimes striped. Tomatoes can be either bilocular (cherry tomato) or mult-locular (most cultivars). The seeds, located inside the locular cavities, are flattened ovoid, 1.3–5 mm × 2–4 mm and hairy; 1000-seed weight is 2.5–3.5 g. The shape of the fruits can be sub-globose, egg-shaped, angular or irregularly globose with bulges (Fig. 11.61). On the surface of the green parts of the plant are yellow glands, which cause the characteristic 'tomato smell' when touched.

Fig. 11.59. Tomato: 1, flowering and fruiting shoot; 2 and 3, fruit types; 4, seed (line drawing: PROSEA volume 8).

Fig. 11.60. Details of tomato plants with leaves and infructescences; and a flower (inset).

Tomato plants grown from seed have a taproot system; when grown from stem cuttings, the plants form fibrous root systems.

Cultivars, uses and constituents

Tomatoes are grown for the fresh market or for processing. A great number of cultivars are selected or bred for both fields and greenhouses. Tomatoes grown in greenhouses are practically always for the fresh market. Size, colour, shape, flavour and productivity are important factors in selection. What variety is grown is determined by the final use; for example, growing tomatoes in the USA and in Europe requires different cultivars because tomatoes of a larger size are preferred in the USA. Most field cultivars do not perform well in greenhouses. Many wild relatives of the tomato are used for cross-breeding to obtain effective resistances to various tomato diseases.

Tomato is used as an ingredient of salads, is eaten fresh, fried or baked, and is used in soups and sauces. Large quantities are processed into ketchups, tomato juice and purée. A part of the production is canned or sun-dried. Fresh tomatoes contain approximately 94% water, 1% protein, 0.2% fat, 3–4% carbohydrates and also Fe, Mg, vitamins A, B and C. The

Fig. 11.61. Tomato: various fruit types from a garden collection (left) and commercially grown fruits (right).

green parts of the plant contain tomatine, a toxic glycoalkaloid; the seeds contain 24% edible oil.

Ecology and agronomy

For high yields, tomato requires sunny environments. The optimum temperature range is 21–24°C. Night temperatures should be above 10°C; it does not withstand frost. Well-distributed rainfall is required. In a cool climate, the crop needs about 20 mm/week during the growing period; in a hot climate, about 70 mm/week. Additional irrigation is often practised. The crop is sensitive to waterlogging. Tomatoes grow on many soil types, ranging from sandy loam to clay-loam soils, preferably rich in organic matter and a soil pH of 6–6.5.

Tomatoes, both for fresh market and processing, are usually started as transplants. For raising transplants, up to 100 g of seed is sown in a seedbed and this provides enough seedlings for 1 ha. Plant spacing ranges from 50 to 150 cm, depending on cropping system. After transplant into the open field, they are usually staked and later on pruned. Application of fertilizers depends on soil fertility, cultivar and aim of cultivation. Per hectare, 180 kg N, 40 kg P and 150 kg K is recommended; N and K in split applications. As tomatoes do not compete very well with weeds, especially in the humid tropics, weed control is essential. Before transplanting the seedlings, a pre-emergence treatment by cultivation or by herbicides has to be carried out. After transplanting, weed control has to be maintained. Covering the soil with a layer of organic mulch, or black plastic, is also practised. Mulching will additionally reduce the loss of water.

In greenhouses, tomatoes are often planted in artificial substrates with plant-driven water movement and fertilizer application. The stems are trained up a string, which is tied to an overhead wire, while all lateral shoots are pruned off. The apex of the stem is removed when the plant has produced the required number of fruit bunches. In greenhouses in the temperate regions, artificial heat, illumination and CO_2 are often needed.

The first fruits can be harvested 10–14 weeks after sowing. Fresh market tomatoes are often harvested at the mature-green stage and ripened in storage. Several pickings can be taken and harvesting may last up to 2 months. Tomatoes grown for processing are usually harvested by machines only once when fully ripe.

The world average yield of fresh fruit, grown in an open-field system, varies from 8 to 38 t/ha. The yield in a greenhouse system can be extremely high and up to 420 t/ha. In 2005, the world total production of fresh tomatoes was about 125 million t. China, the USA, Turkey, India, Egypt and Italy were the largest producers.

Bibliography

Abbott, D.L. (1984) *The Apple Tree: Physiology and Management*. Grower Books, London.

Aguilar, N.O., Pham Truong, T.T. and Oyen, L.P.A. (1999) *Ocimum basilicum* L. In: De Guzman, C.C. and Siemonsma, J.S. (eds) *Plant Resources of South-East Asia No. 13: Spices*. Backhuys Publishers, Leiden, The Netherlands, pp. 151–156.

Akehurst, B.C. (1981) *Tobacco*. Longman, London.

Alim, A. (1978) *A Handbook of Bangladesh Jute*. Shamsul Islam Associated Printers Limited, Dhaka.

Ambrosi, H., Dettweiler, E., Ruhl, E.H., Schmid, J. and Schumann, F. (1994) *Farbatlas Rebsorten*. Ulmer, Stuttgart, Germany,

Andrews, D.J., Rajewski, J.F. and Kumar, K.A. (1993) Pearl millet: new feed grain crop. In: Janick, J. and Simon, J.E. (eds) *New Crops*. Wiley, New York, pp. 198–208.

Anggoro, Permadi, H. and Van der Meer, Q.P. (1993) *Allium cepa* L. cv. group *Aggregatum*. In: Siemonsma, J.S. and Kasem Piluek (eds) *Plant Resources of South-East Asia No. 8: Vegetables*. Pudoc Scientific Publishers, Wageningen, The Netherlands, pp. 68–71.

Anon. (2000) Tobacco. *The Columbia Electronic Encyclopedia © 1994*. Columbia University Press, Columbia, New York.

Anon. (2002) An apple a day keeps the doctor away. *Vegetarians in Paradise* web magazine; available at http://www.vegparadise.com/highestperch39.html (accessed 18 July 2006).

Atherton, J.G. and Rudich, J. (1986) *The Tomato Crop*. Chapman & Hall, London.

Austin, D.F. (1997) *Convolvulaceae* (Morning Glory Family). Florida Atlantic University, Boca Raton, Florida; available at http://www.fau.edu/divdept/biology/people/convolv.htm (accessed 8 August 2006).

Banga, O. (1963) *Main Types of Western Carotene Carrot and Their Origin*. Tjeenk Willink, Zwolle, The Netherlands.

Barnes, A.C. (1974) *The Sugar Cane*. Leonard Hill Books, Aylesbury, UK.

Barnes, D.K., Goplen, B.P. and Baylor, J.E. (1988) Highlights in the USA and Canada. In: Hanson, A.A., Barnes, K. and Hill, R.R. Jr (eds) *Alfalfa and Alfalfa Improvement*. Agronomy Monograph No. 29. American Society of Agronomy/Crop Science Society of America/Soil Science Society of America, Madison, Wisconsin, pp. 1–24.

Baudoin, J.P. (1989) *Phaseolus lunatus* L. In: Van der Maessen, L.J.G. and Sadikin, S. (eds) *Plant*

Resources of South-East Asia No. 1: Pulses. Pudoc Scientific Publishers, Wageningen, The Netherlands, pp. 57–60.

BC Ministry of Agriculture, Fisheries and Food (1993) *Greenhouse Vegetable Production Guide for Commercial Growers, 1993–1994 Edition.* British Columbia Ministry of Agriculture, Fisheries and Food, Victoria, British Columbia, Canada.

Beukema, H.P. and Van der Zaag, D.E. (1990) *Introduction to Potato Production.* Centre for Agricultural Publishing and Documentation (Pudoc), Wageningen, The Netherlands.

Bhardwaj, H.L., Rangappa, M. and Hamama, A.A. (1999) Chickpea, faba bean, lupin, mungbean, and pigeon pea: potential new crops for the Mid-Atlantic Region of the United States. In: Janick, J. (ed.) *Perspectives on New Crops and New Uses.* ASHS Press, Alexandria, Virginia, p. 202.

Blackburn, F. (1984) *Sugar Cane.* Longman, London.

Blasse, W. (1987) *Sauerkirschen* (*Sour Cherries*). VEB Deutscher Landwirtschaftsverlag, Berlin.

Blazdell, P. (2000) The mighty cashew. *Interdisciplinary Science Reviews* 25(3), 220–226.

Bonthuis, H. and Donner, D.A. (2002) *The 77th Recommended List of Varieties, Field Crops of The Netherlands.* Plant Research International, Wageningen, The Netherlands.

Bonthuis, H., Donner, D.A. and Van Vliegen, A. (2004) *The 79th Recommended List of Varieties, Field Crops of The Netherlands.* Centre for Genetic Resources, Wageningen, The Netherlands.

Bosland, P.W. (1996) Capsicums: innovative uses of an ancient crop. In: Janick, J. (ed.) *Progress in New Crops.* ASHS Press, Arlington, Virginia, pp. 479–487.

Brigham, R.D. (1993) Castor: return of an old crop. In: Janick, J. and Simon, J.E. (eds) *New Crops.* Wiley, New York, pp. 380–383.

Brouwer, W. (1972) *Handbuch des Speziellen Pflanzenbaues: Roggen.* Paul Parey, Berlin.

Brown, D.T. (ed.) (1998) *Cannabis: The Genus Cannabis.* Medicinal and Aromatic Plants Book from C.H.I.P.S. Culinary and Hospitality Industry Publications Services, Weimar, Texas.

Brown, K. (2002) Agave sisalana *Perrine.* University of Florida, Center for Aquatic and Invasive Plants, Gainesville, Florida.

Bultitude, J. (1983) *Apples, a Guide to the Identification of International Varieties.* McMillan Press, London.

Caliskan, S., Arslan, M., Arioglu, H. and Isler, N. (2004) Effect of planting method and plant population on growth and yield of sesame (*Sesamum indicum* L.) in a Mediterranean type of environment. *Asian Journal of Plant Sciences* 3(5), 610–613.

Chandra, S. (1984) *Edible Aroids.* Australian National University, Canberra/Clarendon Press, Oxford, UK.

Chang, S.T. and Hayes, W.A. (eds) (1978) *The Biology and Cultivation of Edible Mushrooms.* Academic Press, New York.

Chang, S.T. and Miles, P.G. (2004) *Mushrooms, Cultivation, Nutritional Value, Medicinal Effect, and Environmental Impact,* 2nd edn. CRC Press, Boca Raton, Florida.

Chapman, G.P. (1992) *Grass Evolution and Domestication.* Cambridge University Press, Cambridge, UK.

Chapot, H. (1975) *Citrus.* Technical Monograph No. 4. Ciba-Geigy Ltd, Basle, Switzerland.

Chatt, E.M. (1953) *Cocoa, Cultivation, Processing Analyses.* Interscience Publishers, New York.

Choopong Sukumalanandana and Verheij, E.W.M. (1992) *Fragaria×ananassa* (Duchesne) Guédes. In: Verheij, E.W.M. and Coronel, R.E. (eds) *Plant Resources of South-East Asia No. 2: Edible Fruits and Nuts.* Pudoc, Wageningen, The Netherlands, pp. 171–175.

Christman, S. (2000) *Brassica oleracea* var. capitata. Floridata.com LC, Tallahassee, Florida; available at http://www.floridata.com/ref/b/brass_ole.cfm (accessed 8 August 2006).

Cobley, L.S. (1956) *The Botany of Tropical Crops.* Longmans, Green and Co. Ltd, London.

Cobley, L.S. (1963) *An Introduction to the Botany of Tropical Crops.* Longmans, Green and Co. Ltd, London.

Cobley, L.S. and Steele, W.L. (1976) *An Introduction to the Botany of Tropical Crops.* Longman, London.

Collins, J.L. (1960) *The Pineapple: Botany, Cultivation and Utilisation.* Interscience Publishers, New York.

Cook, B.G. (1992) *Arachis pintoi* Krap. & Greg., nom. Nud. In: 't Mannetje, L. and Jones, R.M. (eds) *Plant Resources of South-East Asia No. 4: Forages.* Pudoc Scientific Publishers, Wageningen, The Netherlands, pp. 48–50.

Cook, C.D.K. (1990) Origin, autecology, and spread of some of the world's most troublesome aquatic weeds. In: Pieterse, A.H. and Murphy, K.J. (eds) *Aquatic Weeds: the Ecology and Management of Nuisance Aquatic Vegetation.* Oxford University Press, New York, pp. 31–38.

Cooke, D.A. and Scott, R.K. (1993) *The Sugar Beet Crop.* Chapman & Hall, London.

Cornelis, J., Nugteren, J.A. and Westphal, E. (1985) Kangkong (*Ipomoea aquatica* Forss.): an important leaf vegetable in South-East Asia. *Abstracts on Tropical Agriculture* 10(4), 9–21.

Coursy, D.G. (1967) *Yams. An Account of the Nature, Cultivation and Utilisation of the Useful Members of the Dioscoreaceae.* Longmans, London.

Crandall, P.C. (1995) *Bramble Production: the Management and Marketing of Raspberries and Blackberries.* Food Products Press, New York.

Crane, J.C. and Iwakiri, B.T. (1981) Morphology and reproduction in pistachio. *Horticultural Reviews* 3, 376–393.

CRFG (1996) Avocado, *Persea* species. California Rare Fruit Growers, Inc., Fullerton, California; available at http://www.crfg.org/pubs/ff/avocado.html (accessed 10 August 2006).

CRFG (1996) Currants, *Ribes* spp. California Rare Fruit Growers, Inc., Fullerton, California; available at http://www.crfg.org/pubs/ff/currants.html (accessed 10 August 2006).

CRFG (1996–2001) Fruit Facts. California Rare Fruit Growers, Inc., Fullerton, California; available at http://www.crfg.org/pubs/frtfacts.html (accessed 10 August 2006).

Dahal, K.R., Utomo, B.I. and Brink, M. (2003) *Agave sisalana* Perrine. In: Brink, M. and Escobin, R.P. (eds) *Plant Resources of South-East Asia No. 17: Fibre Plants.* Backhuys Publishers, Leiden, The Netherlands, pp. 68–75.

Dajue, L.I. and Mündel, H. (1996) *Safflower* Carthamus tinctorius *L.* Promoting the Conservation and Use of Underutilized and Neglected Crops No. 7. International Plant Genetic Resources Institute, Rome.

Darwinkel, A. (1996) *Secale cereale* L. In: Grubben, G.J.H. and Soetjiptopartohardjono (eds) *Plant Resources of South-East Asia No. 10: Cereals.* Backhuys Publishers, Leiden, The Netherlands, pp. 123–127.

Davies, D.R. (1989) *Pisum sativum* L. In: Van der Maesen, L.J.G. and Somaatmadja, S. (eds) *Plant Resources of South-East Asia No. 1: Pulses.* Pudoc, Wageningen, The Netherlands, pp. 63–64.

Davies, P. and Gibbons, B. (1993) *Field Guide to Wild Flowers of Southern Europe.* Crowood Press, Marlborough, UK.

DCCD (2006) *About Cashew.* Directorate of Cashewnut & Cocoa Development, Cochin, India; available at http://dacnet.nic.in/cashewcocoa/tech.htm (accessed 20 July 2006).

De Guzman, C.C. (1999) *Rosmarinus officinalis* L. In: De Guzman, C.C. and Siemonsma, J.S. (eds) *Plant Resources of South-East Asia No. 13: Spices.* Backhuys Publishers, Leiden, The Netherlands, pp. 194–197.

De Jong, P.M., Wind, K. and Neuteboom, J.H. (1991) Tropical grassland B – MSc course in Animal Science and Aquaculture; Practical course in Botany of tropical grasses and legumes. Department of Field Crops and Grassland Science, Wageningen Agricultural University, Wageningen, The Netherlands.

De Kraker, J. (1991) *De teelt van spinazie (The Growing of Spinach).* Teelthandleiding nr. 38. Proefstation voor de Akkerbouw en de Groententeelt in de Vollegrond, Lelystad, The Netherlands.

De Kraker, J. (1994) *Teelt van sla in al haar soorten* (*Growing All Varieties of Lettuce*). Teelthandleiding nr. 63. Proefstation voor de Akkerbouw en de Groenteteelt in de Vollegrond, Lelystad, The Netherlands.

De Laroussilhe, F. (1980) *Techniques agricoles et productions tropicales XXIX*. Le Manguier, Paris.

De Rougemont, G.M. (1989) *A Field Guide to the Crops of Britain and Europe*. Collins, London.

De Waard, P.W.F. and Anunciado, I.S. (1999) *Piper nigrum* L. In: De Guzman, C.C. and Siemonsma, J.S. (eds) *Plant Resources of South-East Asia No. 13: Spices*. Backhuys Publishers, Leiden, The Netherlands, pp. 189–194.

Deanon, J.R. (1967) Eggplant, tomato and pepper. In: Knott, J.E. and Deanon, J.R. (eds) *Vegetable Production in South-East Asia*. University of the Philippines Press, Los Baños, Philippines, pp. 97–137.

Degras, L. (1993) *The Yam, a Tropical Root Crop*. McMillan Press, London.

Dickinson, C. and Lucas, J. (1979) *The Encyclopedia of Mushrooms*. Orbis Publishing, London.

Diederichsen, A. and Rugayah (1999) *Coriandrum sativum* L. In: De Guzman, C.C. and Siemondsma, J.S. (eds) *Plant Resources of South-East Asia No. 13: Spices*. Backhuys Publishers, Leiden, The Netherlands, pp. 104–108.

Duke, J.A. (1981) *Handbook of Legumes of World Economic Importance*. Plenum Press, New York, New York.

Duke, J.A. (1983) *Anacardium occidentale* L. In: Handbook of Energy Crops (unpublished). Purdue University, Center for New Crops & Plants Products, West Lafayette, Indiana.

Duke, J.A. (1983) *Avena sativa* L. In: Handbook of Energy Crops (unpublished). Purdue University, Center for New Crops & Plants Products, West Lafayette, Indiana.

Duke, J.A. (1983) *Brassica juncea* (L) Czern. In: Handbook of Energy Crops (unpublished). Purdue University, Center for New Crops & Plants Products, West Lafayette, Indiana.

Duke, J.A. (1983) *Brassica nigra* (L) Koch. In: Handbook of Energy Crops (unpublished). Purdue University, Center for New Crops & Plants Products, West Lafayette, Indiana.

Duke, J.A. (1983) *Camellia sinensis* (L.) Kuntze. In: Handbook of Energy Crops (unpublished). Purdue University, Center for New Crops & Plants Products, West Lafayette, Indiana.

Duke, J.A. (1983) *Cannabis sativa* L. In: Handbook of Energy Crops (unpublished). Purdue University, Center for New Crops & Plants Products, West Lafayette, Indiana.

Duke, J.A. (1983) *Carica papaya* L. In: Handbook of Energy Crops (unpublished). Purdue University, Center for New Crops & Plants Products, West Lafayette, Indiana.

Duke, J.A. (1983) *Carthamus tinctorius*. In: Handbook of Energy Crops (unpublished). Purdue University, Center for New Crops & Plants Products, West Lafayette, Indiana.

Duke, J.A. (1983) *Cichorium intybus*. In: Handbook of Energy Crops (unpublished). Purdue University, Center for New Crops & Plants Products, West Lafayette, Indiana.

Duke, J.A. (1983) *Cocos nucifera* L. In: Handbook of Energy Crops (unpublished). Purdue University, Center for New Crops & Plants Products, West Lafayette, Indiana.

Duke, J.A. (1983) *Coffea arabica* L. In: Handbook of Energy Crops (unpublished). Purdue University, Center for New Crops & Plants Products, West Lafayette, Indiana.

Duke, J.A. (1983) *Corchoris olitorius* L. In: Handbook of Energy Crops (unpublished). Purdue University, Center for New Crops & Plants Products, West Lafayette, Indiana.

Duke, J.A. (1983) *Elaeis guineensis* Jacq. In: Handbook of Energy Crops (unpublished). Purdue University, Center for New Crops & Plants Products, West Lafayette, Indiana.

Duke, J.A. (1983) *Eleusine coracana* (L.) Gaertn. In: Handbook of Energy Crops (unpublished). Purdue University, Center for New Crops & Plants Products, West Lafayette, Indiana.

Duke, J.A. (1983) *Glycine max* (L.) Merr. In: Handbook of Energy Crops (unpublished). Purdue University, Center for New Crops & Plants Products, West Lafayette, Indiana.

Duke, J.A. (1983) *Gossypium hirsutum* L. In: Handbook of Energy Crops (unpublished). Purdue University, Center for New Crops & Plants Products, West Lafayette, Indiana.

Duke, J.A. (1983) *Helianthus annuus* L. In: Handbook of Energy Crops (unpublished). Purdue University, Center for New Crops & Plants Products, West Lafayette, Indiana.

Duke, J.A. (1983) *Hevea brasiliensis* (Willd.) Muell.-Arg. In: Handbook of Energy Crops (unpublished). Purdue University, Center for New Crops & Plants Products, West Lafayette, Indiana.

Duke, J.A. (1983) *Juglans regia* L. In: Handbook of energy crops (unpublished). Purdue University, Center for New Crops & Plants Products, West Lafayette, Indiana.

Duke, J.A. (1983) *Medicago sativa* L. In: Handbook of Energy Crops (unpublished). Purdue University, Center for New Crops & Plants Products, West Lafayette, Indiana.

Duke, J.A. (1983) *Phaseolus vulgaris* L. In: Handbook of Energy Crops (unpublished). Purdue University, Center for New Crops & Plants Products, West Lafayette, Indiana.

Duke, J.A. (1983) *Ricinis communis* L. In: Handbook of Energy Crops (unpublished). Purdue University, Centre for New Crops & Plants Products, West Lafayette, Indiana.

Duke, J.A. (1983) *Saccharum officinarum* L. In: Handbook of Energy Crops (unpublished). Purdue University, Center for New Crops & Plants Products, West Lafayette, Indiana.

Duke, J.A. (1983) *Sinapis alba* L. In: Handbook of Energy Crops (unpublished). Purdue University, Center for New Crops & Plants Products, West Lafayette, Indiana.

Duke, J.A. (1983) *Vicia faba* L. In: Handbook of Energy Crops (unpublished). Purdue University, Center for New Crops & Plants Products, West Lafayette, Indiana.

Duke, J.A. (1983) *Vigna unguiculata* (L.) Walp. ssp. *Unguiculata*. In: Handbook of Energy Crops (unpublished). Purdue University, Center for New Crops & Plants Products, West Lafayette, Indiana.

Duke, J.A. (1983) *Vitis vinifera* L. In: Handbook of Energy Crops (unpublished). Purdue University, Center for New Crops & Plants Products, West Lafayette, Indiana.

Duke, J.A. (1983) *Zea mays* L. In: Handbook of Energy Crops (unpublished). Purdue University, Center for New Crops & Plants Products, West Lafayette, Indiana.

Duke, J.A. (2001) *Handbook of Nuts*. CRC Press, Boca Raton, Florida.

Duke, J.A. and DuCellier, J.L. (1993) *CRC Handbook of Alternative Cash Crops*. CRC Press, Boca Raton, Florida.

Eastmond, A. and Robert, M.L. (2000) Henequen and the challenge of sustainable development in Yucatan, Mexico. *Biotechnology and Development Monitor* 41, 11–15.

Ebskamp, A.G.M. and Bonthuis, H. (2000) *75th Recommended List of Varieties, Field Crops of The Netherlands*. Centre for Plant Breeding and Reproduction Research, Wageningen, the Netherlands.

Eden Seeds (2005) Asian Vegetables – All. Eden Seeds, Lower Beechmont, Queensland, Australia; available at http://edenseeds.com.au/content/seeds.asp?section=5&letter=ALL (accessed 8 August 2006).

Ellis, R.T. (1995) Tea, *Camellina sinensis* (*Camelliaceae*). In: Smart, J. and Simmonds, N.W. (eds) *Evolution of Crop Plants*, 2nd edn. Longman Scientific & Technical, Harlow, UK, pp. 22–27.

Elzebroek, A.T.G. and Wind, K. (2000) *Manual – Knowledge of cultivated plants*. Group Crop and Weed Ecology, Wageningen University and Research Centre, Wageningen, The Netherlands.

Eshbaugh, W.H. (1993) History and exploitation of a serendipitous new crop discovery. In: Janick, J. and Simon, J.E. (eds) *New Crops*. Wiley, New York, pp. 132–139.

Espino, R.R.C., Jamaluddin, S.H., Bechamas, S. and Nasution, R.E. (1992) *Musa* L. (edible cultivars). In: Verheij, E.W.M. and Coronel, R.E. (eds) *Plant Resources of South-East Asia No. 2: Edible Fruits and Nuts*. Pudoc, Wageningen, The Netherlands, pp. 225–233.

Evensen, S.K. and Standal, B.R. (1984) *Use of Tropical Vegetables to Improve Diets in the Pacific Region*. Research Series 028. College of Tropical Agriculture and Human Resources, University of Hawaii, Manoa, Hawaii.

FAO (1993) International Trade in NWFPs. Corporate Document Repository – Forestry Department,

Food and Agriculture Organization of the United Nations, Rome; available at http://www.fao.org/docrep/v9631e/v9631e04.htm (accessed 15 August 2006).

Feher, E. (1992) *Asparagus*. Akademiai Kiado, Budapest.

Ferguson, L. and Arpaia, M. (1990) New subtropical tree crops in California. In: Janick, J. and Simon, J.E. (eds) *Advances in New Crops*. Timber Press, Portland, Oregon, pp. 331–337.

Ferraris, R. (1973) *Pearl Millet*. Review Series No. 1/1973. Commonwealth Agricultural Bureaux, Farnham Royal, UK.

Fery, F.L. (2002) New opportunities in *Vigna*. In: Janick, J. and Whipkey, A. (eds) *Trends in New Crops and New Uses*. ASHS Press, Alexandria, Virginia, pp. 424–428.

Finn, C. (1999) Temperate berry crops. In: Janick, J. (ed.) *Perspectives on New Crops and New Uses*. ASHS Press, Alexandria, Virginia, pp. 324–334.

Flach, M. (1997) *Sago palm* Metroxylon sagu *Rottb*. Promoting the Conservation and Use of Underutilized and Neglected Crops No. 13. Institute of Plant Genetics and Crop Plant Research, Gatersleben, Germany/International Plant Resources Institute, Rome.

Flach, M. and Siemonsma, J.S. (1999) *Cinnamomum verum* J.S. Presl. In: De Guzman, C.C. and Siemonsma, J.S. (eds) *Plant Resources of South-East Asia No. 13: Spices*. Backhuys Publishers, Leiden, The Netherlands, pp. 99–104.

Flach, M. and Tjeenk Willink, M. (1999) *Myristica fragrans*. In: De Guzman, C.C. and Siemonsma, J.S. (eds) *Plant Resources of South East Asia No. 13: Spices*. Backhuys Publishers, Leiden, The Netherlands, pp. 143–148.

Fleming, J.E. and Galwey, N.W. (1995) Quinoa (*Chenopodium quinoa*). In: Williams, J.T. (ed.) *Cereals and Pseudocereals*. Underutilized Crops Series. Chapman & Hall, London, pp. 71–83.

Fleming, M.P. and Clarke, R.C. (1998) Physical evidence for the antiquity of *Cannabis sativa* L. (*Cannabaceae*). *Journal of the International Hemp Association* 5(2), 80–92.

Fondio, L. and Grubben, G.J.H. (2004) *Corchorus olitorius* L. In: Grubben, G.J.H. and Denton, O.H. (eds) *Plant Resources of Tropical Africa: 2 Vegetables*. PROTA Foundation, Wageningen, The Netherlands/Backhuys Publishers, Leiden, The Netherlands/Technical Centre for Agricultural and Rural Cooperation, Wageningen, The Netherlands, pp. 221–223.

Foster, A.E. (1981) Barley, *Hordeum vulgare*. In: McClure, T.A. and Lipinsky, E.S. (eds) *Barley, Hordeum vulgare*. CRC Press, Boca Raton, Florida.

Freeling, M. and Walbot, V. (eds) (1994) *The Maize Handbook*. Springer Verlag, New York.

Friedrich, G. and Schuricht, W. (1990) *Nusse und Quitten*. Neuman Verlag, Leipzig-Radebeul, Germany.

Fryxell, P.A. (1978) *The Natural History of the Cotton Tribe (Malvaceae, tribe Gossypideae)*. Texas A&M University Press, College Station, Texas.

Gane, A.J. (1985) The pea crop – agricultural progress, past, present and future. In: Hebblethwaite, P.D., Heath, M.C. and Dawkins, T.C.K. (eds) *The Pea Crop: a Basis for Improvement*. Butterworths, London, pp. 3–15.

Gangolly, S.R. (1957) *The Mango*. Indian Council of Agricultural Research, New Delhi.

Gentry, H.S. (1982) *Agaves of Continental North America*. University of Arizona Press, Tuscon, Arizona.

Giesen, B.J. (1989) *Kersen: teelt en rassenoverzicht (Cherries: Cropping and List of Cultivars)*. Uitgeverij H. Veenman en Zonen, Wageningen, The Netherlands.

Gildemacher, B.H. and Jansen, G.J. (1993) *Cucumis sativus* L. In: Siemonsma, J.S. and Kasem Piluek (eds) *Plant Resources of South-East Asia No. 8: Vegetables*. Pudoc Scientific Publishers, Wageningen, The Netherlands, pp. 157–160.

Gooding, M.J. and Davies, W.P. (1997) *Wheat Production and Utilization, Systems, Quality and the Environment*. CAB International, Wallingford, UK.

Grace, M.R. (1977) *Cassava Processing*. FAO Plant Production and Protection Series No. 3. Food and Agriculture Organization of the United Nations, Rome.

Grafton, G. (1995) Apples and Cider in the United Kingdom. University of Birmingham, Birmingham, UK; available at http://www.bham.ac.uk/GraftonG/cider/history1.htm (accessed 15 May 2004).

Grashoff, C. (1992) Variability in yield of faba beans (*Vivia faba* L.). Thesis, Wageningen Agricultural University, Wageningen, The Netherlands.

Grieve, M. (1931) Thyme, garden. *A Modern Herbal* [electronic version, 1995]. Botanical.com; available at http://www.botanical.com/botanical/mgmh/t/thygar16.html (accessed 1 August 2006).

Grubben, G.J.H. and Siemonsma, J.S. (1996) *Fagopyrum esculentum* Moench. In: Grubben, G.J.H. and Soetjiptopartohardjono (eds) *Plant Resources of South-East Asia No. 10: Cereals.* Backhuys Publishers, Leiden, The Netherlands, pp. 95–99.

Grubben, G.J.H. and Sukprakarn, S. (1993) *Lactuca sativa* L. In: Siemonsma, J.S. and Kasem Piluek (eds) *Plant Resources of South-East Asia No. 8: Vegetables.* Pudoc Scientific Publishers, Wageningen, The Netherlands, pp. 186–190.

Hacker, J.B. (1992) *Desmodium uncinatum* (Jacq.) D.C. In: 't Mannetje, L. and Jones, R.M. (eds) *Plant Resources of South-East Asia No. 4: Forages.* Pudoc Scientific Publishers, Wageningen, The Netherlands, pp. 116–118.

Hacker, J.B. (1992) *Setaria sphacelata.* In: 't Mannetje, L. and Jones, R.M. (eds) *Plant Resources of South-East Asia No. 4: Forages.* Pudoc Scientific Publishers, Wageningen, The Netherlands, pp. 201–203.

Häfliger, E. and Scholz, H. (1980) *Grass Weeds*, Vols 1 and 2. Documenta, Ciba-Geigy, Basle, Switzerland.

Halevy, A.H. (ed.) (1985) *Handbook of Flowering*, Vol. II. CRC Press, Boca Raton, Florida.

Hancock, J.F. (1992) *Plant Evolution and the Origin of Crop Species.* Department of Horticulture, Michigan State University, East Lancing, Michigan/Prentice Hall, Englewood Cliffs, New Jersey.

Hardon, J.J., Rajanaidu, N. and Van der Vossen, H.A.M. (2001) *Elaeis guineensis* Jacq. In: Van der Vossen, H.A.M. and Umali, B.E. (eds) *Plant Resources of South-East Asia No. 14: Vegetable Oils and Fats.* Backhuys Publishers, Leiden, The Netherlands, pp. 85–93.

Harris, P.M. (1992) *The Potato Crop.* Chapman & Hall, London.

Hartana, I. and Vermeulen, H. (2000) *Nicotiana tabacum* L. In: Van der Vossen, H.A.M. and Wessel, M. (eds) *Plant Resources of South-East Asia No. 16: Stimulants.* Backhuys Publishers, Leiden, The Netherlands, pp. 93–99.

Hill, A. (1952) *Economic Botany.* Harvard University, Cambridge, Massachusetts, pp. 356–357.

Hubbard, C.E. (1968) *Grasses.* Penguin Books Ltd, Middlesex, UK.

Huke, R.E. and Huke, E.H. (1990) *Rice: Then and Now.* International Rice Research Institute, Manila, the Philippines.

Hulse, J.H. (1994) Nature, composition and utilization of food legumes. In: Muehlbauer, F.J. and Kaiser, W.J. (eds) *Expanding the Production and Use of Cool Season Food Legumes.* Kluwer Academic Publishers, Dordrecht, The Netherlands, pp. 77–97.

Humanes, G.J. and López-Villalta, M.C. (1992) *Produción de aceita de oliva de calidad. Influencia del cultivo.* Junta de Andalucia, Seville, Spain.

Hummer, K.E. (1999) *Corylus Genetic Resources, Hazelnut.* US Department of Agriculture, Agricultural Research Service National Clonal Germplasm Repository, Corvallis, Oregon.

Hymowitz, T. (1990) Soybeans: the success story. In: Janick, J. and Simon, J.A. (eds) *Advances in New Crops.* Timber Press, Portland, Oregon, pp. 159–163.

ICRISAT/FAO (1996) *The World Sorghum and Millets Economies. Facts, Trends and Outlook.* International Crops Research Institute for the Semi-Arid Tropics, Patancheru, India/Food and Agriculture Organization of the United Nations, Rome.

Ipor, I.B. and Oyen, L.P.A. (1999) *Laurus nobilis* L. In: De Guzman, C.C. and Siemonsma, J.S. (eds) *Plant Resources of South-East Asia No. 13: Spices.* Backhuys Publishers, Leiden, The

Netherlands, pp. 134–137.

Jackson, D.I. (1986) *Temperate and Subtropical Fruit Production.* Butterworths Horticultural Books, Wellington.

Jain, S.K. and Sutarno, H. (1996) *Amaranthus* L. (grain amaranth). In: Grubben, G.J.H. and Soetjiptopartohardjono (eds) *Plant Resources of South-East Asia No. 10: Cereals.* Backhuys Publishers, Leiden, The Netherlands, pp. 75–79.

Janick, J. (undated) Reading 4–3. The Origin of Fruits and Fruit Growing. Purdue University, Department of Horticulture & Landscape Architecture, West Lafayette, Indiana; available at http://www.hort.purdue.edu/newcrop/history/lecture04/r_4-3l.html (accessed 19 July 2006).

Janick, J. and Moore, J.N. (eds) (1975) *Advances in Fruit Breeding.* Purdue University Press, West Lafayette, Indiana.

Janick, J., Blasé, M.G., Johnson, D.L., Jolliff, G.D. and Myers, R.L. (1996) Diversifying US crop production. In: Janick, J. (ed.) *Progress in New Crops.* ASHS Press, Alexandria, Virginia, pp. 98–109.

Jansen, P.C.M. (1989) *Lens culinaris* Medikus. In: Van der Maesen, L.J.G. and Somaatmadja, S. (eds) *Plant Resources of South-East Asia No. 1: Pulses.* Pudoc, Wageningen, The Netherlands, pp. 51–53.

Jansen, P.C.M. (1989) *Vicia faba* L. In: Van der Maesen, L.J.G. and Somaatmadja, S. (eds) *Plant Resources of South-East Asia No. 1: Pulses.* Pudoc, Wageningen, The Netherlands, pp. 64–66.

Jansen, P.C.M. (1999) *Cuminum cyminum* L. In: De Guzman, C.C. and Siemondsma, J.S. (eds) *Plant Resources of South-East Asia No. 13: Spices.* Backhuys Publishers, Leiden, The Netherlands, pp. 108–111.

Jansen, P.C.M. and Premchand, V. (1996) Xanthosoma. In: Flach, M. and Rumawas, F. (eds) *Plant Resources of South-East Asia No. 9: Plants Yielding Non-seed Carbohydrates.* Backhuys Publishers, Leiden, The Netherlands, pp. 159–164.

Jansen, P.C.M., Siemonsma, J.S. and Narciso, J.O. (1993) *Brassica oleracea* L. In: Siemonsma, J.S. and Kasem Piluek (eds) *Plant Resources of South-East Asia No. 8: Vegetables.* Pudoc Scientific Publishers, Wageningen, The Netherlands, pp. 108–111.

Jennings, D.L. (1988) *Raspberries and Blackberries: their Breeding, Diseases and Growth.* Academic Press, London.

Jones, J.K. (1976) Strawberry, *Fragaria ananassa* (*Rosaceae*). In: Simmonds, N.W. (ed.) *Evolution of Crop Plants.* Longman, London, pp. 237–242.

Jones, R.J., Brewbaker, J.L. and Sorensson, C.T. (1992) *Leucaena leucocephala* (Lamk) de Wit. In: 't Mannetje, L. and Jones, R.M. (eds) *Plant Resources of South-East Asia No. 4: Forages.* Pudoc Scientific Publishers, Wageningen, The Netherlands, pp. 150–154.

Jong, F.S. (1995) Research for the development of sago palm cultivation in Serawak, Malaysia. Thesis, Wageningen Agricultural University, Wageningen, The Netherlands.

Kalkman, C., Nauta, M.M. and Van der Meijden, R. (2003) *Planten voor dagelijks gebruik, botanische achtergronden en toepassingen* (*Plants for Daily Use, Botanical Backgrounds and Applications*). KNNV Uitgeverij, Utrecht, The Netherlands.

Kaprovickas, A. (1973) Evolution of the genus *Arachis.* In: Moav, R. (ed.) *Agricultural Genetics.* National Council for Research and Development, Jerusalem.

Kasem Piluek and Beltran, M.M. (1993) *Raphanus sativus* L. In: Siemonsma, J.S. and Kasem Piluek (eds) *Plant Resources of South-East Asia No. 8: Vegetables.* Pudoc Scientific Publishers, Wageningen, The Netherlands, pp. 233–237.

Katzer, G. (2004) Peppermint (*Mentha piperita* L.). Gernot Katzer's Spice Pages, Graz, Austria; available at http://www.uni-graz.at/~katzer/engl/Ment_pip.html (accessed 1 August 2006).

Katzer, G. (2006) Pepper (*Piper nigrum* L.). Gernot Katzer's Spice Pages, Graz, Austria; available at http://www.uni-graz.at/~katzer/engl/Pipe_nig.html (accessed 1 August 2006).

Kauffman, C.S. and Weber, L.E. (1990) Grain amaranth. In: Janick, J. and Simon, J.E. (eds) *Advances in New Crops.* Timber Press, Portland, Oregon, pp. 127–139.

Kerkhoven, G.J. and Mutsaers, H.J.W. (2003) *Gossypium* L. In: Brink, M. and Escobin, R.P. (eds) *Plant Resources of South-East Asia No. 17: Fibre Plants.* Backhuys Publishers, Leiden, The Netherlands, pp. 139–150.

Khandakar, A.L. and Van der Vossen, H.A.M. (2003) *Corchorus* L. In: Brink, M. and Escobin, R.P. (eds) *Plant Resources of South-East Asia No. 17: Fibre Plants.* Backhuys Publishers, Leiden, The Netherlands, pp. 106–114.

King Orchards (2003) Tart is smart. King Orchards, Central Lake, Michigan; available at http://www.mi-cherries.com/variety2.htm (accessed 19 July 2005).

Koopmans, A., Ten Have, H. and Subandi (1996) *Zea mays* L. In: Grubben, G.J.H. and Soetjiptopartohardjono (eds) *Plant Resources of South-East Asia No. 10: Cereals.* Backhuys Publishers, Leiden, The Netherlands, pp. 143–149.

Krishnan, K.B., Doraiswamy and Chellamani, K.P. (2005) Jute. In: Franck, R.R. (ed.) *Bast and Other Plant Fibres.* Woodhead Publishing Limited, Abington, UK in association with The Textile Institute, Cambridge, UK, pp. 94–208.

Krochmal, A. and Krochmal, C. (1982) *Uncultivated Nuts of the United States.* Agriculture Information Bulletin No. 450. US Department of Agriculture, Forest Service, Washington, DC.

Kuntohartono, T. and Thijsse, J.P. (1996) *Saccharum officinarum* L. In: Flach, M. and Rumawas, F. (eds) *Plant Resources of South-East Asia No. 9: Plants Yielding Non-seed Carbohydrates.* Backhuys Publishers, Leiden, the Netherlands, pp. 143–148.

Kuo, C.G. and Toxopeus, H. (1993) *Brassica rapa* L. cv. group Chinese cabbage. In: Siemonsma, J.S. and Kasem Piluek (eds) *Plant Resources of South-East Asia No. 8: Vegetables.* Pudoc Scientific Publishers, Wageningen, The Netherlands, pp. 127–130.

La Dinh Moi (1999) *Mentha arvensis* L. In: De Guzman, C.C. and Siemonsma, J.S. (eds) *Plant Resources of South-East Asia No. 13: Spices.* Backhuys Publishers, Leiden, The Netherlands, pp. 344–349.

Langer, R.H.M. and Hill, G.D. (1982) *Agricultural Plants.* Cambridge University Press, Cambridge, UK.

Langham, D.R. and Wiemers, T. (2002) Progress in mechanizing sesame in the US through breeding. In: Janick, J. and Whipkey, A. (eds) *Trends in New Crops and New Uses.* ASHS Press, Alexandria, Virginia, pp. 157–173.

Lapinskas, P. (1979) The potential of *Phaseolus coccineus* and of hybrids with *P. vulgaris* as pulse crop for the UK. PhD thesis, University of Cambridge, Cambridge, UK.

Larkcom, J. (1991) *Oriental Vegetables. The Complete Guide for Garden and Kitchen.* John Murray (Publishers) Ltd, London.

Larue, J.H. and Johnson, R.S. (1989) *Peaches, Plums and Nectarines.* Cooperative Extension Publication No. 3331. University of California, Tulare, California.

Lelley, J. (1991) *Pilzanbau, Biotechnologie der Kulturspeisepilze.* Eugen Ulmer GmbH & Co., Stuttgart, Germany.

Lerner, B.R. and Dana, M.N. (2005) *Growing Cucumbers, Melons, Squash, Pumpkins and Gourds.* Department of Horticulture, Purdue University, West Lafayette, Indiana.

Leung, A. and Foster, S. (1996) *Encyclopedia of Common Natural Ingredients Used in Foods, Drugs and Cosmetics,* 2nd edn. John Wiley & Sons, New York.

Leuschner, K. and Manthe, C.S. (1996) *Drought-tolerant Crops for Southern Africa: Proceedings of the SADC/ICRISAT Regional Sorghum and Pearl Millet Workshop, 25–29 July 1994, Gabarone, Botswana.* International Crops Research Institute for the Semi-Arid Tropics, Patancheru, India.

Levins, H. (2004) Symbolism of the pineapple. Hoag Levins web site; available at http://www.levins.com/pineapple.html (accessed 20 July 2006).

Lisson, S.N. (2003) *Linum usitatissimum* L. In: Brink, M. and Escobin, R.P. (eds) *Plant Resources of South-East Asia No. 17: Fibre Plants.* Backhuys Publishers, Leiden, The Netherlands, pp. 172–179.

Litz, R.E. (1997) *The Mango – Botany, Production and Uses.* CAB International, Wallingford, UK.

Looney, N.E. and Jackson, D.I. (1999) Stonefruit. In: Jackson, D.I. and Looney, N.E. (eds) *Temperate and Subtropical Fruit Production.* CAB International, Wallingford, UK.

Loussert, R. and Brousse, G. (1978) *L'olivier.* Techniques agricoles et productions Méditerranéennes, Paris.

Mabberley, D.J. (1987) *The Plant Book. A Portable Dictionary of the Higher Plants.* Cambridge University Press, Cambridge, UK.

McIlroy, R.J. (1967) *An Introduction to Tropical Cash Crops.* Ibadan University Press, Ibadan, Nigeria.

McKay, J.W. (1969) Chestnuts. In: Jaynes, R.A. (ed.) *Handbook of North American Nut Trees.* The Northern Nut Growers Association, Knoxville, Tennessee, pp. 264–286.

MAFF (1981) *Bush Fruits.* Reference Book No. 4. Ministry of Agriculture, Fisheries and Food, London.

Maggioni, L., Janes, H., Hayes, A., Swinburne, T. and Lipman, E. (1998) *Report of a Working Group on Malus/Pyrus.* International Plant Genetic Resources Institute, Rome.

Magness, J.R., Markle, G.M. and Compson, C.C. (1971) *Food and Feed Crops of the United States.* Interregional Research Project IR-4, IR Bulletin No. 1/NJAES Bulletin No. 828. New Jersey Agricultural Experimental Station, New Brunswick, New Jersey.

Makasheva, R.Kh. (1983) *The Pea.* Published for the US Department of Agriculture and the National Science Foundation, Washington, DC. Amerind Publishing Co. Pvt. Ltd, New Delhi.

Mangelsdorf, P.C. (1974) *Corn, Its Origin, Evolution and Improvement.* The Belknap Press of Harvard University Press, Cambridge, Massachusetts.

't Mannetje, L. (1992) *Pennisetum purpureum.* In: 't Mannetje, L. and Jones, R.M. (eds) *Plant Resources of South-East Asia No. 4: Forages.* Pudoc Scientific Publishers, Wageningen, The Netherlands, pp. 191–192.

't Mannetje, L. (1992) *Stylosanthes guianensis* (Aublet) Swartz. In: 't Mannetje, L. and Jones, R.M. (eds) *Plant Resources of South-East Asia No. 4: Forages.* Pudoc Scientific Publishers, Wageningen, The Netherlands, pp. 211–213.

't Mannetje, L. and Kersten, S.M.M. (1992) *Chloris gayana.* In: 't Mannetje, L. and Jones, R.M. (eds) *Plant Resources of South-East Asia No. 4: Forages.* Pudoc Scientific Publishers, Wageningen, The Netherlands, pp. 90–92.

't Mannetje, L. and Kersten, S.M.M. (1992) *Paspalum dilatatum.* In: 't Mannetje, L. and Jones, R.M. (eds) *Plant Resources of South-East Asia No. 4: Forages.* Pudoc Scientific Publishers, Wageningen, The Netherlands, pp. 178–180.

't Mannetje, L. and Kersten, S.M.M. (1992) *Paspalum plicatulum.* In: 't Mannetje, L. and Jones, R.M. (eds) *Plant Resources of South-East Asia No. 4: Forages.* Pudoc Scientific Publishers, Wageningen, The Netherlands, pp. 183–185.

Marshall, G. (1988) *Flax: Breeding and Utilisation. Proceedings of the EEC Flax Workshop, held in Brussels, Belgium.* Kluwer Academic Publishers, Dordrecht, The Netherlands.

Marshall, H.G. and Pommeranz, Y. (1982) Buckwheat: description, breeding, production, and utilization. *Advances in Cereal Science and Technology* 4, 157–210.

Masefield, G.B., Wallis, M., Harrison, S.G. and Nicholson, B.E. (1969) *The Oxford Book of Food Plants.* Oxford University Press, London.

Mastebroek, H.D., Van Soest, L.J.M. and Siemonsma, J.S. (1996) *Chenopodium* L. (grain chenopod). In: Grubben, G.J.H. and Soetjiptopartohardjono (eds) *Plant Resources of South-East Asia No. 10: Cereals.* Backhuys Publishers, Leiden, The Netherlands, pp. 79–83.

Mears, P.T. (1992) *Pennisetum clandestinum.* In: 't Mannetje, L. and Jones, R.M. (eds) *Plant Resources of South-East Asia No. 4: Forages.* Pudoc Scientific Publishers, Wageningen, The Netherlands, pp. 187–189.

Meijers, P.G. (1962) *Onze cultuurgewassen in het verleden* (*Our Cultivated Crops in the Past*). NV Uitgeversmaatschappij/W.E.J. Tjeenk Willink, Zwolle, The Netherlands.

Menninger, E.A. (1977) *Edible Nuts of the World.* Horticultural Books, Stuart, Florida.

Miège, J. and Lyonga, S.N. (1982) *Yams. Ignames.* Clarendon Press, Oxford, UK.

Mills, H.A. (2001) Spinach, *Spinacia oleracea.* University of Georgia, Department of Horticulture, Athens, Georgia; available at http://www.uga.edu/vegetable/spinach.html (accessed 9 August 2006).

Mitchell, A. and Wilkinson, J. (1982) *Bomengids, Kennen en herkennen (The Trees of Britain and Northern Europe).* BV Uitgeversmaatschappij Tirion, Baarn, The Netherlands.

Mohd Noor A. Ghani and Wessel, M. (2000) *Hevea brasiliensis* (Willd. ex Juss.) Müll. Arg. In: Boer, E. and Ella, A.B. (eds) *Plant Resources of South-East Asia No. 18: Plants Producing Exudates.* Backhuys Publishers, Leiden, The Netherlands, pp. 73–82.

Monk, K.A., De Fretes, Y. and Reksodiharjo-Lilley, G. (1997) *The Ecology of Indonesia Series.* Vol. V. *The Ecology of Nusa Tenggara and Maluku.* Oxford University Press, Oxford, UK.

Morton, J.F. (1987) *Fruits of Warm Climates.* Julia F. Morton, Miami, Florida.

Morton, J.F. (1987) Avocado. *Fruits of Warm Climates.* Julia F. Morton, Miami, Florida, pp. 91–102.

Morton, J.F. (1987) Papaya. *Fruits of Warm Climates.* Julia F. Morton, Miami, Florida, pp. 336–346.

Morton, J.F. (1987) Pineapple. *Fruits of Warm Climates.* Julia F. Morton, Miami, Florida, pp. 18–28.

Muehlbauer, F.J. (1993) Food and grain legumes. In: Janick, J. and Simon, J.E. (eds) *New Crops.* Wiley, New York, pp. 256–265.

Muehlbauer, F.J., Short, R.W., Summerfield, R.J., Morrison, K.J. and Swan, D.G. (1981) *Description and Culture of Lentils.* Cooperative Extension Publication EB 0957. College of Agriculture, Washington State University, Pullman, Washington.

Muelbauer, F.J., Short, R.W. and Kaiser, W.J. (1982) *Description and Culture of Garbanzo Beans.* Cooperative Extension Publication EB 1112. College of Agriculture, Washington State University, Pullman, Washington.

Nakasone, H.Y. and Paull, R.E. (1998) *Tropical Fruits.* CAB International, Wallingford, UK.

Neuvel, J.J. (1991) *Teelt van Tuinbonen (Cultivation of Faba Beans).* Proefstation voor de Akkerbouw en de Groenteteelt in de Vollegrond, Lelystad, The Netherlands.

Neve, R.A. (1991) *Hops.* Chapman and Hall, London.

NewCROP™ New Crop Resource Online Program (1995) Homepage. Purdue University, Center for New Crops & Plants Products, West Lafayette, Indiana; available at http://www.hort.purdue.edu/newcrop/ (accessed 2 August 2006).

NewCROP™ New Crop Resource Online Program (1999) Corn, Maize, *Gramineae Zea mays* L. Purdue University, Center for New Crops & Plants Products, West Lafayette, Indiana; available at http://www.hort.purdue.edu/newcrop/Crops/Corn.html (accessed 2 August 2006).

Nguyen, K.D., Tran, H. and Siemonsma, J.S. (1999) *Cinnamomum* Schaeffer. In: De Guzman, C.C. and Siemonsma, J.S. (eds) *Plant Resources of South-East Asia No. 13: Spices.* Backhuys Publishers, Leiden, The Netherlands, pp. 94–99.

Nguyen, T.T., De Guzman, C.C. and Jansen, P.C.M. (1999) *Anethum graveolens* L. In: De Guzman, C.C. and Siemondsma, J.S. (eds) *Plant Resources of South-East Asia No. 13: Spices.* Backhuys Publishers, Leiden, The Netherlands, pp. 71–74.

Nichols, M.A. (1990) Asparagus – the world scene. *Acta Horticulturae* 271, 25–31.

Nichols, M.A. (1993) *Asparagus officinales* L. In: Siemonsma, J.S. and Kasem Piluek (eds) *Plant Resources of South-East Asia No. 8: Vegetables.* Pudoc Scientific Publishers, Wageningen, The Netherlands, pp. 91–93.

Nicolson, D.H. and Wiersema, J.H. (2004) Proposal to conserve *Sesamum indicum* against *Sesamum orientale (Pedaliaceae). Taxon* 53, 210–211.

Norman, M.J.T., Pearson, C.J. and Searle, P.G.E. (1995) *The Ecology of Tropical Food Crops.* Cambridge University Press, Cambridge, UK.

Oelke, E.A., Oplinger, E.S., Bahri, H., Durgan, D.R., Putnam, D.H., Doll, J.D. and Kelling, K.A. (1990) Rye, Alternative Field Crops Manual. University of Wisconsin, Madison,

Wisconsin/University of Minnesota, St Paul, Minnesota; available at http://www.hort.purdue.edu/newcrop/afcm/rye.html (accessed 25 July 2006).

Ohler, J.G. and Magat, S.S. (2001) *Cocos nucifera* L. In: van der Vossen, H.A.M. and Umali, B.E. (eds) *Plant Resources of South-East Asia No. 14: Vegetable Oils and Fats.* Backhuys Publishers, Leiden, The Netherlands, pp. 76–84.

Onwueme, I.C. (1978) *The Tropical Tuber Crops.* University of Ife, Ile-Ife, Nigeria/John Wiley and Sons, Chichester, UK.

Onwueme, I.C. (1996) *Dioscorea.* In: Flach, M. and Rumawas, F. (eds) *Plant Resources of South-East Asia No. 9: Plants Yielding Non-seed Carbohydrates.* Backhuys Publishers, Leiden, The Netherlands, pp. 85–97.

Opeña, R.T. (1993) *Brassica juncea* (L.) Czernjaew. In: Siemonsma, J.S. and Kasem Piluek (eds) *Plant Resources of South-East Asia No. 8: Vegetables.* Pudoc Scientific Publishers, Wageningen, The Netherlands, pp. 104–108.

Opeña, R.T. and Van der Vossen, H.A.M. (1993) *Lycopersicon esculentum* Miller. In: Siemonsma, J.S. and Kasem Piluek (eds) *Plant Resources of South-East Asia No. 8: Vegetables.* Pudoc Scientific Publishers, Wageningen, The Netherlands, pp. 199–205.

Oplinger, E.S., Oelke, E.A., Kaminski, A.R., Combs, S.M., Doll, J.D. and Schuler, R.T. (1990) Castorbeans, Alternative Field Crops Manual. University of Wisconsin, Madison, Wisconsin/University of Minnesota, St Paul, Minnesota; available at http://www.hort.purdue.edu/newcrop/afcm/castor.html (accessed 25 July 2006).

Oplinger, E.S., Putnam, D.H., Kaminski, A.R., Hanson, C.V., Oelke, E.A., Schulte, E.E. and Doll, J.D. (1990) Sesame, Alternative Field Crops Manual. University of Wisconsin, Madison, Wisconsin/University of Minnesota, St Paul, Minnesota; available at http://www.hort.purdue.edu/newcrop/afcm/sesame.html (accessed 25 July 2006).

Orkwar, C.G., Asiedo, R. and Ekanayake, I.J. (1998) *Food Yams, Advances in Research.* International Institute of Tropical Agriculture, Ibadan, Nigeria.

OSU (2002) Brussels sprouts, *Brassica oleracea* (*Gemmifera* group). Oregon State University, Corvallis, Oregon; available at http://www.orst.edu/Dept/NWREC/brussprt.html (accessed 8 August 2006).

OSU (2002) Globe artichoke, *Cynara scolymus.* Oregon State University, Corvallis, Oregon; available at http://www.orst.edu/Dept/NWREC/artichgl.html (accessed 8 August 2006).

OSU (2004) Sugarcane & Sugar Beets. Oregon State University, Corvallis, Oregon; available at http://oregonstate.edu/instruct/css/330/seven/index.htm (accessed 7 August 2006).

OSU (2006) Kale. Oregon State University, Corvallis, Oregon; available at http://food.oregonstate.edu/v/kale.html (accessed 8 August 2006).

Oyen, L.P.A. (1991) *Pyrus pyrifolia* (NL. Burman) Nakai. In: Verheij, E.W.M. and Coronel, R.E. (eds) *Plant Resources of South-East Asia No. 2: Edible Fruits and Nuts.* Pudoc, Wageningen, The Netherlands, pp. 272–276.

Oyen, L.P.A. (1993) *Cichorium endivia* L. In: Siemonsma, J.S. and Kasem Piluek (eds) *Plant Resources of South-East Asia No. 8: Vegetables.* Pudoc Scientific Publishers, Wageningen, The Netherlands, pp. 142–144.

Oyen, L.P.A. and Andrews, D.J. (1996) *Pennisetum glaucum* (L.) R. Br. In: Grubben, G.J.H. and Soetjiptopartohardjono (eds) *Plant Resources of South-East Asia No. 10: Cereals.* Backhuys Publishers, Leiden, The Netherlands, pp. 119–123.

Oyen, L.P.A. and Soenoeadji (1993) *Allium fistulosum* L. In: Siemonsma, J.S. and Kasem Piluek (eds) *Plant Resources of South-East Asia No. 8: Vegetables.* Pudoc Scientific Publishers, Wageningen, The Netherlands, pp. 73–77.

Oyen, L.P.A. and Umali, B.E. (2001) *Carthamus tinctorius* L. In: van der Vossen, H.A.M. and Umali, B.E. (eds) *Plant Resources of South-East Asia No. 14: Vegetable Oils and Fats.* Backhuys Publishers, Leiden, The Netherlands, pp. 70–76.

PAGV (1989) *Witlof, teelt van de wortel, produktie van lof* (*Chicory, Growing the Root, Producing the Leaves*). Teelthandleiding nr. 12. Proefstation en Consulentschap in Algemene Dienst voor de Akkerbouw en de Groenteteelt in de Vollegrond, Lelystad, The Netherlands.

Paje, M.M. and Van der Vossen, H.A.M. (1993) *Citrullus lanatus* (Thunberg) Matsum & Nakai. In: Siemonsma, J.S. and Kasem Piluek (eds) *Plant Resources of South-East Asia No. 8: Vegetables*. Pudoc Scientific Publishers, Wageningen, The Netherlands, pp. 144–148.

Paje, M.M. and Van der Vossen, H.A.M. (1993) *Cucumis melo* L. In: Siemonsma, J.S. and Kasem Piluek (eds) *Plant Resources of South-East Asia No. 8: Vegetables*. Pudoc Scientific Publishers, Wageningen, The Netherlands, pp. 153–157.

Pandey, R.K. and Westphal, E. (1989) *Vigna unguiculata* (L.) Walp. In: Van der Maesen, L.J.G. and Somaatmadja, S. (eds) *Plant Resources of South-East Asia No. 1: Pulses*. Pudoc, Wageningen, The Netherlands, pp. 77–81.

Parry, J.W. (1962) *Spices: their Morphology, Histology, and Chemistry*. Chemical Publishing Co., New York.

Pastor, M., Navarro, C. and Vega, V. (1995) *Poda de formación del olivar*. Departemento de Olivicultura de la Direccion General de Investigacion Agrarian, Junta de Andalucia, Seville, Spain.

Perry, L.P. (1991) *Growing Hops in New England*. University of Vermont Extension System, Department of Plant and Soil Science, Morrisville, Vermont.

Peterson, R.F. (1965) *Wheat*. Grampian Press, London.

Petzold, H. (1984) *Birnensorten*. Neumann Verlag, Leipzig-Radebeul, Germany.

Phillips, R.L. (1994) *The Coconut*. Fact Sheet HS-40. Horticultural Sciences Department, University of Florida, Gainesville, Florida.

Poll, J.T.K. (1991) *Teelt van witte asperge* (*Growing White Asparagus*). Proefstation voor de Akkerbouw en Groenteteelt in de Vollegrond, Lelystad, The Netherlands.

Poll, J.T.K. (1992) *Teelt van groene asperge* (*Growing Green Asparagus*). Proefstation voor de Akkerbouw en de Groenteteelt in de Vollegrond, Lelystad, The Netherlands.

Polunin, O. (1969) *Flowers of Europe*. Oxford University Press, London.

Poulos, J.M. (1993) *Capsicum* L. In: Siemonsma, J.S. and Kasem Piluek (eds) *Plant Resources of South-East Asia No. 8: Vegetables*. Pudoc Scientific Publishers, Wageningen, The Netherlands, pp. 136–140.

Powell, A.A. (2001–2005) Asian Pear Culture in Alabama. Auburn University, Auburn, Alabama; available at http://www.aces.edu/dept/peaches/pearasiancult.html (accessed 18 July 2006).

Prentice, A.N. (1972) *Cotton, With Special Reference to Africa*. Tropical Agriculture Series. Longman Group Limited, London.

Prinsley, R.T. and Tucker, G. (1987) *Mangoes – A Review*. Commonwealth Science Council, London.

Purseglove, J.W. (1968) *Tropical Crops, Dicotyledons 1*. Longman, Green and Co. Ltd, London, UK.

Purseglove, J.W. (1968) *Tropical Crops, Dicotyledons 2*. Longman, Green and Co. Ltd, London.

Purseglove, J.W. (1972) *Tropical Crops, Monocotyledons 1*. Longman Group Ltd, London.

Purseglove, J.W. (1972) *Tropical Crops, Monocotyledons 2*. Longman Group Ltd, London.

Purseglove, J.W., Brown, E.G., Green, C.L. and Robbins, S.R.J. (1981) *Spices*, Vol. 1. Longman, Harlow, UK.

Purwaningsih, H. and Brink, M. (1999) *Foeniculum vulgare* Miller. In: De Guzman, C.C. and Siemondsma, J.S. (eds) *Plant Resources of South-East Asia No. 13: Spices*. Backhuys Publishers, Leiden, The Netherlands, pp. 126–130.

Rachie, K.O. and Peters, L.V. (1977) *The Eleusines. A Review of the World Literature*. International Crops Research Institute for the Semi-Arid Tropics, Hyderabad, India.

Rahayu, M. and Jansen, P.C.M. (1996) *Setaria italica* (L.) P. Beauvois cv. group Foxtail millet. In: Grubben, G.J.H. and Soetjiptopartohardjono (eds) *Plant Resources of South-East Asia No. 10: Cereals*. Backhuys Publishers, Leiden, The Netherlands, pp. 127–130.

Rahmansyah, M. (1993) *Basella alba* L. In: Siemonsma, J.S. and Kasem Piluek (eds) *Plant Resources* _of South-East Asia No. 8: Vegetables._ Pudoc Scientific Publishers, Wageningen, The Netherlands, pp. 93–95.

Raintree (2006) Artichoke (_Cynara scolymus_). Raintree Nutrition, Carson City, Nevada; available at http://www.rain-tree.com/artichoke.htm (accessed 8 August 2006).

Rasmusson, D.C. (1985) _Barley._ Agronomy Monograph No. 26. American Society of Agronomy/Crop Science Society of America/Soil Science Society of America, Madison, Wisconsin.

Raven, P.H., Evert, R.F. and Curtis, H. (1981) _Biology of Plants._ Worth Publishers, New York.

Reddy, P.S. (1988) _Groundnut._ Indian Council of Agricultural Research, New Delhi.

Reich, L. (1991) Red currants, White currants. Rain Public Internet Broadcasting; available at http://www.rain.org/greennet/docs/exoticveggies/html/redcurrantswhitecurrants.htm (accessed 10 August 2006).

Reich, L. (1991) _Uncommon Fruits Worthy of Attention._ Addison-Wesley, Reading, Massachusetts.

Rhodes, D. (2006) HORT410 – Vegetable Crops. Purdue University, Department of Horticulture & Landscape Architecture, West Lafayette, Indiana; available at http://www.hort.purdue.edu/rhodcv/hort410/cole/co00001.htm (accessed 8 August 2006).

Ricard, J.M. (2003) _La truffe. Guide technique de trufficulture._ Ctifl – Centre technique interprofessionel des fruits et legumes, Paris.

Rieger, M. (2004) Cashew – _Anacardium occidentale._ Mark's Fruit Crops web site, University of Georgia; available at http://www.uga.edu/fruit/cashew.html (accessed 20 July 2006).

Rieger, M. (2004) Pistachio – _Pistacia vera._ Mark's Fruit Crops web site, University of Georgia; available at http://www.uga.edu/fruit/pistacio.html (accessed 15 August 2006).

Rieger, M. (2005) Apricot – _Prunus armeniaca._ Mark's Fruit Crops web site, University of Georgia; available at http://www.uga.edu/fruit/apricot.html (accessed 18 July 2006).

Rieger, M. (2005) Pineapple – _Ananas comosus._ Mark's Fruit Crops web site, University of Georgia; available at http://www.uga.edu/fruit/pinapple.html (accessed 20 July 2006).

Rivas, L. and Holmann, F. (2000) _Early Adoption of_ Arachis pintoi _in the Humid Tropics: The Case of Dual-purpose Livestock Systems in Caqueta, Colombia._ Centro Internacional de Agricultura Tropical, Cali, Colombia.

Robinson, J.C. (1996) _Bananas and Plantains._ Crop Production Science in Horticulture Series No. 5. CAB International, Wallingford, UK.

Rogers, D.J. (1963) Studies of _Manihot esculenta_ Crantz and related species. _Bulletin of the Torrey Botanical Club_ 90, 43–54.

Ronchi, C.P. and Silva, A.A. (2004) Weed control in young coffee plantations through post emergence herbicide application onto total area. _Planta Daninha_ 22(4), 607–615.

Roxas, V.P. (1993) _Cucurbita ficifolia_ Bouché. In: Siemonsma, J.S. and Kasem Piluek (eds) _Plant Resources of South-East Asia No. 8: Vegetables._ Pudoc Scientific Publishers, Wageningen, The Netherlands, pp. 165–167.

Rumbaugh, M.D. (1990) Special purpose forage legumes. In: Janick, J. and Simon, E. (eds) _Advances in New Crops._ Timber Press, Portland, Oregon, pp. 183–190.

Runham, S.R. (1996) An Updated Review of the Potential Uses of Plants Grown for Extracts including Essential Oils and Factors Affecting their Yield and Composition. MAFF Ref. ST0105. Ministry of Agriculture, Fisheries and Food, London; available at http://www.defra.gov.uk/farm/crops/industrial/research/reports/RDREP03.pdf (accessed 12 September 2007).

Ryder, E.J. (1980) _Leafy Salad Vegetables._ AVI Publishing Company, Westport, Connecticut.

Ryder, E.J. (1999) _Lettuce, Endive and Chicory._ Crop Production Science in Horticulture No. 9. CAB International, Wallingford, UK.

Ryder, E.J. (2002) The new salad crop revolution. In: Janick, J. and Whipkey, A. (eds) _Trends in New_

Crops and New Uses. ASHS Press, Alexandria, Virginia, pp. 408–412.

Ryder, E.J. and Waycott, W. (1993) New directions in salad crops: new forms, new tools, and old philosophy. In: Janick, J. and Simon, J.E. (eds) *New Crops.* Wiley, New York, pp. 528–532.

Sagwansupyakorn, C. (1993) *Brassica oleracea* L. cv. group Chinese kale. In: Siemonsma, J.S. and Kasem Piluek (eds) *Plant Resources of South-East Asia No. 8: Vegetables.* Pudoc Scientific Publishers, Wageningen, The Netherlands, pp. 115–117.

Saichol Kesta and Verheij, E.W.M. (1991) *Vitis vinifera* L. In: Verheij, E.W.M. and Coronel, R.E. (eds) *Plant Resources of South-East Asia No. 2: Edible Fruits and Nuts.* Pudoc, Wageningen, The Netherlands, pp. 304–310.

Samson, J.A. (1992) *Citrus sinensis.* In: Verheij, E.W.M. and Coronel, R.E. (eds) *Plant Resources of South-East Asia No. 2: Edible Fruits and Nuts.* Pudoc, Wageningen, The Netherlands, pp. 138–141.

Samson, J.A. (1992) *Citrus×paradisi.* In: Verheij, E.W.M. and Coronel, R.E. (eds) *Plant Resources of South-East Asia No. 2: Edible Fruits and Nuts.* Pudoc, Wageningen, The Netherlands, pp. 133–135.

Sandsted, R.F., Wilcox, D.A., Zitter, T.A. and Muka, A.A. (1985) *Asparagus Information Bulletin 202.* Cornell University, Ithaca, New York.

Sauer, J.D. (1993) *Fragaria* – strawberries. In: *Historical Geography of Crop Plants: A Select Roster.* CRC Press, Boca Raton, Florida, pp. 127–130.

Sauer, J.D. (1993) *Historical Geography of Crop Plants: A Select Roster.* CRC Press, Boca Raton, Florida.

Schoneveld, J.A. (1991) *Teelt van Peen (Growing Carrots).* Teelthandleiding nr. 36. Proefstation voor de Akkerbouw en de Groententeelt in de Vollegrond, Lelystad, The Netherlands.

Schoorel, E.F. and Van der Vossen, H.A.M. (2000) *Camellia sinensis* (L.) Kuntze. In: Van der Vossen, H.A.M. and Wessel, M. (eds) *Plant Resources of South-East Asia No. 16: Stimulants.* Backhuys Publishers, Leiden, The Netherlands, pp. 55–63.

Schrader, W.L. and Mayberry, K.S. (1997) *Artichoke Production in California.* Publication No. 7221. University of California, Division of Agriculture and Natural Resources, Oakland, California.

Schuiling, D.L. and Flach, M. (1985) *Guidelines for the Cultivation of Sago Palm.* Department of Tropical Crop Science, Wageningen Agricultural University, Wageningen, The Netherlands.

Schuiling, D.L. and Jong, F.S. (1996) *Metroxylon sagu* Rottb. In: Flach, M. and Rumawas, F. (eds) *Plant Resources of South-East Asia No. 9: Plants Yielding Non-seed Carbohydrates.* Backhuys Publishers, Leiden, The Netherlands, pp. 121–126.

Schulze-Kraft, H. and Teitzel, J.K. (1992) *Brachiaria decumbens.* In: 't Mannetje, L. and Jones, R.M. (eds) *Plant Resources of South-East Asia No. 4: Forages.* Pudoc Scientific Publishers, Wageningen, The Netherlands, pp. 58–59.

Schulze-Kraft, H. and Teitzel, J.K. (1992) *Brachiaria ruzizienzis.* In: 't Mannetje, L. and Jones, R.M. (eds) *Plant Resources of South-East Asia No. 4: Forages.* Pudoc Scientific Publishers, Wageningen, The Netherlands, pp. 65–67.

Schuster, W.H. (1992) *Ölplanzen in Europa.* DLG-Verlag-GmbH, Frankfurt am Main, Germany.

Scott, R.K., Harper, F., Wood, D.W. and Jaggard, K.W. (1974) Effect of seed size on growth, development and yield of monogerm sugar beet. *Journal of Agricultural Science, Cambridge* 82, 517–530.

Scott, W.O. and Aldrich, S.R. (1970) *Modern Soybean Production.* S & A Publications, Champaign, Illinois.

Seegeler, C.J.P. (1983) *Oil Plants in Ethiopia, Their Taxonomy and Agricultural Significance.* Agricultural Research Report No. 921. Pudoc, Wageningen, The Netherlands.

Seegeler, C.J.P. and Oyen, L.P.A. (2001) *Ricinis communis* L. In: Van der Vossen, H.A.M. and Umali, B.E. (eds) *Plant Resources of South-East Asia No. 14: Vegetable Oils and Fats.*

Backhuys Publishers, Leiden, The Netherlands, pp. 115–120.

Serr, E.F. (1969) Persian walnuts in the western states. In: Jaynes, R.A. (ed.) *Handbook of North American Nut Trees*. The Northern Nut Growers Association, Knoxville, Tennessee, pp. 240–263.

Shanmugasundaram, S. and Sumarno (1989) *Glycine max* (L.) Merr. In: Van der Maesen, L.J.G. and Somaatmadja, S. (eds) *Plant Resources of South-East Asia No. 1: Pulses*. Pudoc, Wageningen, The Netherlands, pp. 43–47.

Shoemaker, J.S. (1955) *Small-Fruit Culture*. McGraw-Hill, New York.

Shorter, R. and Patatnothai, A. (1989) *Arachis hypogaeae*. In: Van der Maesen, L.J.G. and Somaatmadja, S. (eds) *Plant Resources of South-East Asia No. 1: Pulses*. Backhuys Publishers, Leiden, The Netherlands, pp. 35–39.

Sibma, L., Grashoff, C. and Klein Hulze, J.A. (1989) *Ontwikkeling en groei van veldbonen Vicia faba onder Nederlandse omstandigheden* (*Development and Growth of Broad Beans Vicia faba in The Netherlands*). Gewassenreeks 3. Pudoc, Wageningen, The Netherlands.

Siemonsma, J.S. (1993) *Abelmoschus esculentus* (L.) Moench. In: Siemonsma, J.S. and Kasem Piluek (eds) *Plant Resources of South-East Asia No. 8: Vegetables*. Pudoc Scientific Publishers, Wageningen, The Netherlands, pp. 57–60.

Siemonsma, J.S. and Na Lampang, A. (1989) *Vigna radiate* (L.) Wilczek. In: Van der Maesen, L.J.G. and Somaatmadja, S. (eds) *Plant Resources of South-East Asia No. 1: Pulses*. Pudoc, Wageningen, The Netherlands, pp. 71–74.

Simmons, A.F. (1972) *Growing Unusual Fruit*. David and Charles, Newton Abbot, UK.

Simon, J.E., Chadwick, A.F. and Craker, L.E. (1984) *Herbs: an Indexed Bibliography 1971–1980. The Scientific Literature on Selected Herbs, and Aromatic and Medicinal Plants of the Temperate Zone*. Archon Books, Hamden, Connecticut.

Simon, J.E., Quinn, J. and Murray, R.G. (1990) Basil: a source of essential oils. In: Janick, J. and Simons J.E. (eds) *Advances in New Crops*. Timber Press, Portland, Oregon, pp. 484–489.

Simon, J.E., Morales, M.R. and Charles, D.J. (1993) Specialty melons for the fresh market. In: Janick, J. and Simon, J.E. (eds) *New Crops*. Wiley, New York, pp. 547–553.

Simon, P.W. (2003) Carrot Facts. Department of Horticulture, University of Wisconsin, Madison, Wisconsin; available at http://www.hort.wisc.edu/usdavcru/simon/carrot_facts.html (accessed 8 August 2006).

Singh, R.K., Singh, U.S. and Khush, G.S. (2000) *Aromatic Rices*. Science Publishers, Enfield, New Hampshire.

Singh, S. (2004) *Advances in Citriculture*. Jagmander Book Agency, New Delhi.

Skerman, P.J. and Riveros, F. (1990) *Tropical Grasses*. Food and Agricultural Organization of the United Nations, Rome.

Smartt, J. (1989) *Phaseolus coccineus* L. In: Van der Maessen, L.J.G. and Somaatmadja, S. (eds) *Plant Resources of South-East Asia No. 1: Pulses*. Pudoc Scientific Publishers, Wageningen, The Netherlands, pp. 56–57.

Smartt, J. (1989) *Phaseolus vulgaris* L. In: Van der Maesen, L.J.G. and Somaatmadja, S. (eds) *Plant Resources of South-East Asia No. 1: Pulses*. Pudoc Scientific Publishers, Wageningen, The Netherlands, pp. 60–63.

Smartt, J. (1994) *The Groundnut Crop*. Chapman & Hall, London.

Smith, B.W. (1950) *Arachis hypogaea*: aerial flower and subterranean fruit. *American Journal of Botany* 37, 802–815.

Smith, K. (1997) Cocos nucifera. *Ethnobotanical Leaflets*. Southern Illinois University, Carbondale, Illinois.

Sneep, J. (1982) The domestication of spinach and the breeding history of its varieties. *Euphytica* 1982, Suppl. 2.

Somchai Sukonthasing, Montri Wongrakpanich and Verheij, E.W.M. (1992) *Mangifera indica* L. In: Verheij, E.W.M. and Coronel, R.E. (eds) *Plant Resources of South-East Asia No. 2: Edible Fruits and Nuts.* Pudoc, Wageningen, The Netherlands, pp. 211–216.

Sosef, M.S.M. and Boer, E. (2000) *Coffea liberica* Bull ex Hiern. In: Van der Vossen, H.A.M. and Wessel, M. (eds) *Plant Resources of South-East Asia No. 16: Stimulants.* Backhuys Publishers, Leiden, The Netherlands, pp. 74–78.

Stanton, T.R. (1955) *Oat Identification and Classification.* Technical Bulletin No. 1100. US Department of Agriculture, Washington, DC.

Stenhouse, J.W. and Tippayaruk, J.L. (1996) *Sorghum bicolor* (L.) Moench. In: Grubben, G.J.H. and Soetjiptopartohardjono (eds) *Plant Resources of South-East Asia No. 10: Cereals.* Backhuys Publishers, Leiden, The Netherlands, pp. 130–136.

Stephens, J.M. (2003) *Kangkong* – Ipomoea aquatica *Forssk., also* Ipomoea reptans *Poir.* HS618. Horticultural Sciences Department, Institute of Food and Agricultural Sciences, University of Florida, Gainesville, Florida.

Stoll, K. (1988) *Geschichte der Pomologie in Europa.* Druck Stutz & Co. AG, Wadenswil, Switzerland.

Stone, P.J. and Savin, R. (1999) Grain quality and its physiological determinants. In: Satorre, E.H. and Slafer, G.A. (eds) *Wheat, Ecology and Physiology of Yield Determination.* Haworth Press, New York, pp. 13–133.

Struik, P.C. and Wiersema, S.G. (1999) *Seed Potato Technology.* Wageningen Pers, Wageningen, The Netherlands.

Sulistiorini, D. and Van der Meer, Q.P. (1993) *Allium ampeloprasum* L. cv. group Leek. In: Siemonsma, J.S. and Kasem Piluek (eds) *Plant Resources of South-East Asia No. 8: Vegetables.* Pudoc Scientific Publishers, Wageningen, The Netherlands, pp. 62–64.

Sumeru Ashari (1992) *Citrus reticulata.* In: Verheij, E.W.M. and Coronel, R.E. (eds) *Plant Resources of South-East Asia No. 2: Edible Fruits and Nuts.* Pudoc, Wageningen, The Netherlands, pp. 135–138.

Summerfield, R.J. and Roberts, E.H. (1985) *Grain Legume Crops.* William Collins Sons, London.

Suranant Subhadrabandhu (1992) *Prunus* L. In: Verheij, E.W.M. and Coronel, R.E. (eds) *Plant Resources of South-East Asia No. 2: Edible Fruits and Nuts.* Pudoc, Wageningen, The Netherlands, pp. 262–266.

Sutarno, H., Danimihardja, S. and Grubben, G.J.H. (1993) *Solanum melongena* L. In: Siemonsma, J.S. and Kasem Piluek (eds) *Plant Resources of South-East Asia No. 8: Vegetables.* Pudoc Scientific Publishers, Wageningen, The Netherlands, pp. 255–258.

Sutarno, H., Hadad, E.A. and Brink, M. (1999) *Zingiber officinale* Roscoe. In: De Guzman, C.C. and Siemonsma, J.S. (eds) *Plant Resources of South-East Asia No. 13: Spices.* Backhuys Publishers, Leiden, The Netherlands, pp. 238–244.

Suttie, J.M. (2000) *Hay and Straw Conservation for Small-scale Farming and Pastoral Conditions.* FAO Plant Production and Protection Series No. 29. Food and Agriculture Organization of the United Nations, Rome.

Swiader, J.M., Ware, G.W. and McCollum, J.P. (1992) *Producing Vegetable Crops.* Interstate Publishers, Danville, Illinois.

Takagi, H., Kuo, C.G. and Sakamoto, S. (1996) *Ipomoea batatas* (L.) Lamk. In: Flach, M. and Rumawas, F. (eds) *Plant Resources of South-East Asia No. 9: Plants Yielding Non-seed Carbohydrates.* Backhuys Publishers, Leiden, The Netherlands, pp. 102–107.

Tamaguchi, M. (1990) Asian vegetables. In: Janick, J. and Simon, J.E. (eds) *Advances in New Crops.* Timber Press, Portland, Oregon, pp. 387–390.

Tay, D.C.S. and Toxopeus, H. (1993) *Brassica rapa* L. cv. group pak choi. In: Siemondsma, J.S. and Kasem Piluek (eds) *Plant Resources of South-East Asia No. 8: Vegetables.* Pudoc Scientific Publishers, Wageningen, The Netherlands, pp. 130–134.

Teskey, B.J.E. and Shoemaker, J.S. (1972) *Tree Fruit Production.* AVI Publishing Company, Westport, Connecticut.

Thames, S.F. and Schuman, T.P. (1996) New crops or new uses for old crops: where should the emphasis be? In: Janick, J. (ed.) *Progress in New Crops*. ASHS Press, Alexandria, Virginia, pp. 8–18.

Thomas, H. and Jones, I.T. (1976) Origins and identification of weed species of *Avena*. In: Price Jones, D. (ed.) *Wild Oats in World Agriculture*. Agricultural Research Council, London, pp. 1–57.

Thompson, H.C. and Kelly, W.C. (1957) *Vegetable Crops*. McGraw-Hill Book Company, New York.

Thomson, M.M., Lagerstedt, H.B. and Mehlenbacher, S.A. (1996) Hazelnuts. In: Janick, J. and Moor, J. (eds) *Fruit Breeding*, Vol. III. *Nuts*. John Wiley & Sons, New York, pp. 294–301.

Tindall, H.D. (1983) *Vegetables in the Tropics*. McMillan, London.

Tous, J. and Ferguson, L. (1996) Mediterranean fruits. In: Janick, J. (ed.) *Progress in New Crops*. ASHS Press, Alexandria, Virginia, pp. 416–430.

Toxopeus, H. (2001) *Brassica*. In: Van der Vossen, H.A.M. and Umali, B.E. (eds) *Plant Resources of South-East Asia No. 14: Vegetable Oils and Fats*. Backhuys Publishers, Leiden, The Netherlands, pp. 65–70.

Toxopeus, H. and Lubberts, J.H. (1999) *Carum carvi* L. In: De Guzman, C.C. and Siemondsma, J.S. (eds) *Plant Resources of South-East Asia No. 13: Spices*. Backhuys Publishers, Leiden, The Netherlands, pp. 91–94.

Toxopeus, H. and Lubberts, J.H. (1999) *Sinapis alba* L. In: De Guzman, C.C. and Siemonsma, J.S. (eds) *Plant Resources of South-East Asia No. 13: Spices*. Backhuys Publishers, Leiden, The Netherlands, pp. 204–207.

Toxopeus, H. and Utomo, I. (1999) *Brassica nigra* (L.) W.D.J. Koch. In: De Guzman, C.C. and Siemonsma, J.S. (eds) *Plant Resources of South-East Asia No. 13: Spices*. Backhuys Publishers, Leiden, The Netherlands, pp. 85–88.

Tsegaye, A. (2002) On indigenous production, genetic diversity and crop ecology of enset (*Ensete ventricosum*). PhD Thesis, Wageningen University, Wageningen, The Netherlands.

Turner, D. and Muir, K. (1985) *The Handbook of Soft Fruit Growing*. Croom Helm, Beckenham, UK.

UGA (2005) HORT 3020, Introduction to Fruit Crops. University of Georgia, Department of Horticulture, Athens, Georgia; available at http://www.uga.edu/fruit/courseinfo.html (accessed 10 August 2006).

Upjohn, B., Dear, B. and Sandral, G. (2004) Subterranean clover, *Trifolium subterranean*. *Agnote DPI-268 (third edition)*. NSW Department of Primary Industries, Orange, New South Wales, Australia; available at http://www.agric.nsw.gov.au/reader/1462 (accessed 24 July 2006).

USDA (1998) *Savory Herbs: Culture and Use*. Farmer's Bulletin No. 1977. US Department of Agriculture, Washington, DC.

USDA/NRCS (2004) The PLANTS Database, Version 3.5. US Department of Agriculture, Washington, DC/National Plant Data Center, Baton Rouge, Louisiana; available at http://plants.usda.gov (accessed 3 August 2006).

Utami, D. and Jansen, P.C.M. (1999) *Piper* L. In: De Guzman, C.C. and Siemonsma, J.S. (eds) *Plant Resources of South-East Asia No. 13: Spices*. Backhuys Publishers, Leiden, The Netherlands, pp. 183–188.

Utami, N.W. and Brink, M. (1999) *Myristica*. In: De Guzman, C.C. and Siemonsma, J.S. (eds) *Plant Resources of South-East Asia No. 13: Spices*. Backhuys Publishers, Leiden, The Netherlands, pp. 139–143.

Utomo, W.B.I. and Rahayu, E. (2000) *Nicotiana rustica* L. In: Van der Vossen, H.A.M. and Wessel, M. (eds) *Plant Resources of South-East Asia No. 16: Stimulants*. Backhuys Publishers, Leiden, The Netherlands, pp. 91–93.

Vainio-Mattila, K. (2000) Wild vegetables used by the Sambaa in the Usambara Mountains, NE Tanzania. *Annals of Botany Fennici* 37, 57–67.

Valmayor, R.V. and Wagih, M.E. (1996) *Musa* (plantain and cooking banana). In: Flach, M. and Rumawas, F. (eds) *Plant Resources of South-East Asia No. 9: Plants Yielding Non-seed Carbohydrates.* Backhuys Publishers, Leiden, The Netherlands, pp. 126–131.

Van der Hoek, H.N. and Jansen, P.C.M. (1996) *Panicum miliaceum* L. cv. group Proso millet. In: Grubben, G.J.H. and Soetjiptopartohardjono (eds) *Plant Resources of South-East Asia No. 10: Cereals.* Backhuys Publishers, Leiden, The Netherlands, pp. 115–119.

Van der Maesen, L.J.G. (1986) Cajanus *DC and* Alylosia *W. & A. (Leguminosae).* Paper No. 85-4. Wageningen Agricultural University, Wageningen, The Netherlands.

Van der Maesen, L.J.G. (1989) *Cajanus cajan* (L) Millsp. In: Van der Maesen, L.J.G. and Somaatmadja, S. (eds) *Plant Resources of South-East Asia No. 1: Pulses.* Pudoc, Wageningen, The Netherlands, pp. 39–42.

Van der Maesen, L.J.G. (1990) Pigeonpea: origin, history, evolution and taxonomy. In: Nene, Y.L., Hill, S.H. and Sheila, V.K. (eds) *The Pigeonpea.* CAB International, Wallingford, UK, pp. 15–46.

Van der Meer, Q.P. (1993) *Allium tuberosum* Rottler ex Sprengel. In: Siemonsma, J.S. and Kasem Piluek (eds) *Plant Resources of South-East Asia No. 8: Vegetables.* Pudoc Scientific Publishers, Wageningen, The Netherlands, pp. 80–82.

Van der Meer, Q.P. and Agustina, L. (1993) *Allium chinense* G. Don. In: Siemonsma, J.S. and Kasem Piluek (eds) *Plant Resources of South-East Asia No. 8: Vegetables.* Pudoc Scientific Publishers, Wageningen, The Netherlands, pp. 71–73.

Van der Meer, Q.P. and Leong, A.C. (1993) *Allium cepa* L. cv. group Common Onion. In: Siemonsma, J.S. and Kasem Piluek (eds) *Plant Resources of South-East Asia No. 8: Vegetables.* Pudoc Scientific Publishers, Wageningen, The Netherlands, pp. 68–71.

Van der Meer, Q.P., Anggoro and Permadi, H. (1993) *Allium sativum* L. In: Siemonsma, J.S. and Kasem Piluek (eds) *Plant Resources of South-East Asia No. 8: Vegetables.* Pudoc Scientific Publishers, Wageningen, The Netherlands, pp. 77–80.

Van der Vossen, H.A.M. (1993) *Brassica oleracea* L. cv. groups cauliflower & broccoli. In: Siemonsma, J.S. and Kasem Piluek (eds) *Plant Resources of South-East Asia No. 8: Vegetables.* Pudoc Scientific Publishers, Wageningen, The Netherlands, pp. 111–115.

Van der Vossen, H.A.M. (1993) *Spinacia oleracea* L. In: Siemonsma, J.S. and Kasem Piluek (eds) *Plant Resources of South-East Asia No. 8: Vegetables.* Pudoc Scientific Publishers, Wageningen, The Netherlands, pp. 266–268.

Van der Vossen, H.A.M. and Duriyaprapan, S. (2001) *Helianthus annuus* L. In: Van der Vossen, H.A.M. and Umali, B.E. (eds) *Plant Resources of South-East Asia No. 14: Vegetable Oils and Fats.* Backhuys Publishers, Leiden, The Netherlands, pp. 101–107.

Van der Vossen, H.A.M. and Sambas, E.N. (1993) *Daucus carota* L. In: Siemonsma, J.S. and Kasem Piluek (eds) *Plant Resources of South-East Asia No. 8: Vegetables.* Pudoc Scientific Publishers, Wageningen, The Netherlands, pp. 167–171.

Van der Vossen, H.A.M., Soenaryo and Mawardi, S. (2000) *Coffea* L. In: Van der Vossen, H.A.M. and Wessel, M. (ed.) *Plant Resources of South-East Asia No. 16: Stimulants.* Backhuys Publishers, Leiden, The Netherlands, pp. 66–74.

Van der Werf, H. (1994) Crop physiology of fibre hemp (*Cannabis sativa* L.). PhD thesis, Wageningen Agricultural University, Wageningen, The Netherlands.

Van der Zwet, T. and Childers, N.F. (1982) *The Pear – Cultivars to Marketing.* Horticultural Publications, Gainesville, Florida.

Van Eijnatten, C.L.M. (1991) *Anacardium occidentale* L. In: Verheij, E.W.M. and Coronel, R.E. (eds) *Plant Resources of South-East Asia No. 2: Edible Fruits and Nuts.* Pudoc, Wageningen, The Netherlands, pp. 60–64.

Van Ginkel, M. and Villareal, R.L. (1996) *Triticum* L. In: Grubben, G.J.H. and Soetjiptopartohardjono (eds) *Plant Resources of South-East Asia No. 10: Cereals.* Backhuys Publishers, Leiden, The Netherlands, pp. 137–143.

Van Wijk, C. (1989) *Korte teeltbeschrijving Chinese Kool (Short Description of Cultivating Chinese*

Cabbage). Teeltbeschrijving nr. 8. Proefstation en Consulentschap in Algemene Dienst voor de Akkerbouw en de Groenteteelt in de Vollegrond, Lelystad, The Netherlands.

Van Wijk, C.A.P. and Zwanepol, S. (1992) *Teelt van Rammenas (Cultivation of Chinese Radish)*. Teelthandleiding nr. 44. Proefstation voor de Akkerbouw en de Groenteteelt in de Vollegrond, Lelystad, The Netherlands.

Vaughan, J.G. and Geisler, C.A. (1997) *The New Oxford Book of Food Plants*. Oxford University Press, Oxford, UK.

Vavilov, N.I. (1951) *The Origin, Variation, Immunity and Breeding of Cultivated Plants*. Ronald Press, New York.

Vavilov, N.I. (1992) *Origin and Geography of Cultivated Plants*. Cambridge University Press, Cambridge, UK.

Veltkamp, H.J. and De Bruijn, G.H. (1996) *Manihot esculenta*. In: Flach, M. and Rumawas, F. (eds) *Plant Resources of South-East Asia No. 9: Plants Yielding Non-seed Carbohydrates*. Backhuys Publishers, Leiden, The Netherlands, pp. 107–113.

Vergara, B.S. and De Datta, S.K. (1996) *Oryza sativa*. In: Grubben, G.J.H. and Soetjiptopartohardjono (eds) *Plant Resources of South-East Asia No. 10: Cereals*. Backhuys Publishers, Leiden, The Netherlands, pp. 106–115.

Verheij, E.W.M. and Stone, B.C. (1992) Citrus. In: Verheij, E.W.M. and Coronel, R.E. (eds) *Plant Resources of South-East Asia No. 2: Edible Fruits and Nuts*. Pudoc, Wageningen, The Netherlands, pp. 119–126.

Verheij, E.W.M. and Snijders, C.H.A. (1999) *Syzygium aromaticum*. In: De Guzman, C.C. and Siemonsma, J.S. (eds) *Plant Resources of South-East Asia No. 13: Spices*. Backhuys Publishers, Leiden, The Netherlands, pp. 211–218.

Villegas, V.N. (1991) *Carica papaya* L. In: Verheij, E.W.M. and Coronel, R.E. (eds) *Plant Resources of South-East Asia No. 2: Edible Fruits and Nuts*. Pudoc, Wageningen, The Netherlands, pp. 108–112.

Vogel, S. and Graham, M. (1979) Sorghum and millet: food production and use. *Report of Workshop in Nairobi, Kenya*, 4–7 July 1978. International Development Research Centre, Ottawa, Ontario, Canada.

Volk, T.J. and Ivors, K. (2001) *Tom Volk's Fungus of the Month for April 2001*. University of Wisconsin–La Crosse, Wisconsin; available at http://botit.botany.wisc.edu/toms_fungi/apr2001.html (accessed 9 August 2006).

Wagih, M.E. and Wiersema, S.G. (1996) *Solanum tuberosum* L. In: Flach, M. and Rumawas, F. (eds) *Plant Resources of South-East Asia No. 9: Plants Yielding Non-seed Carbohydrates*. Backhuys Publishers, Leiden, The Netherlands, pp. 148–154.

Wagner, W.L., Herbst, D.R. and Sohmer, S.H. (1999) *Manual of the Flowering Plants of Hawaii*, revised edition. University of Hawaii Press, Honolulu, Hawaii.

Wang, J.-K. (1983) *Taro, A Review of* Colocasia *and Its Potentials*. University of Hawaii Press, Honolulu, Hawaii.

Ware, G.W. and McCollum, J.P. (1968) *Producing Vegetable Crops*. Interstate Printers and Publishers, Danville, Illinois.

Warren, D. (1987) *Brazil and the Struggle for Rubber*. Cambridge University Press, Cambridge, UK.

Webster, C.C. and Baulkwill, W.J. (1989) *Rubber*. Longman Scientific & Technical, Harlow, UK.

Wee, Y.C. and Charuphant Thongtham, M.L. (1991) *Ananas comosus* (L.) Merr. In: Verheij, E.W.M. and Coronel, R.E. (eds) *Plant Resources of South-East Asia No. 2: Edible Fruits and Nuts*. Pudoc, Wageningen, The Netherlands, pp. 66–71.

Weiss, E.A. (1971) *Castor, Sesame, and Safflower*. Leonard Hill, London.

Weiss, E.A. (1983) *Oilseed Crops*. Tropical Agriculture Series. Longman Group, London.

Weiss, E.A. (1997) *Essential Oil Crops*. CAB International, Wallingford, UK.

Weiss, E.A. (2000) *Oilseed Crops*. World Agriculture Series. Blackwell Science, Oxford, UK.

Weiss, E.A. and De la Cruz, Q.D. (2001) *Sesamum orientale* L. In: Van der Vossen, H.A.M. and

Umali, B.E. (eds) *Plant Resources of South-East Asia No. 14: Vegetable Oils and Fats.* Backhuys Publishers, Leiden, The Netherlands, pp. 123–128.

Welch, R.W. (1995) *The Oat Crop, Production and Utilization.* Chapman & Hall, London.

Wendel, J.F. (1995) Cotton. In: Smartt, J. and Simmonds, N.W. (eds) *Evolution of Crop Plants.* Longman, Harlow, UK, pp. 358–366.

Wessel, M. and Toxopeus, H. (2000) *Theobroma cacao.* In: Van der Vossen, H.A.M. and Wessel, M. (eds) *Plant Resources of South-East Asia No. 16: Stimulants.* Backhuys Publishers, Leiden, The Netherlands, pp. 113–121.

Westphal, E. (1985) *Cultures vivrières tropicales.* Pudoc, Wageningen, The Netherlands.

Westphal, E. (1993) *Ipomoea aquatica* Forsskal. In: Siemonsma, J.S. and Kasem Piluek (eds) *Plant Resources of South-East Asia No. 8: Vegetables.* Pudoc Scientific Publishers, Wageningen, The Netherlands, pp.181–184.

Whiley, A.W. (1991) *Persea americana* Miller. In: Verheij, E.W.M. and Coronel, R.E. (eds) *Plant Resources of South-East Asia No. 2: Edible Fruits and Nuts.* Pudoc, Wageningen, The Netherlands, pp. 249–254.

Whitaker, T.W. and Davies, G.N. (1962) *Cucurbits.* Leonard Hill, London.

Widjaja, E.A. and Sukprakarn, S. (1993) *Cucurbita* L. In: Siemonsma, J.S. and Kasem Piluek (eds) *Plant Resources of South-East Asia No. 8: Vegetables.* Pudoc Scientific Publishers, Wageningen, The Netherlands, pp. 160–165.

Widodo, S.H. (1999) *Thymus vulgaris* L. In: De Guzman, C.C. and Siemonsma, J.S. (eds) *Plant Resources of South-East Asia No. 13: Spices.* Backhuys Publishers, Leiden, The Netherlands, pp. 220–223.

Wilbert, J. (1987) *Tobacco and Shamanism in South America.* Yale University Press, New Haven, Connecticut.

Wilson, J.E. and Siemonsma, J.S. (1996) *Colocasia esculenta.* In: Flach, M. and Rumawas, F. (eds) *Plant Resources of South-East Asia No. 9: Plants Yielding Non-seed Carbohydrates.* Backhuys Publishers, Leiden, The Netherlands, pp. 69–72.

Wind, K. and Elzebroek, A.T.G. (1989) *Graslandplanten (Grassland Species).* Praktijkreeks Veehouderij, Uitgeversmaatschappij C. Misset BV, Doetinchem, The Netherlands.

Wit, F. (1969) The clove tree. In: Ferwerda, F.P. and Wit, F. (eds) *Outlines of Perennial Crop Breeding in the Tropics.* H. Veenman and Zonen NV, Wageningen, The Netherlands, pp. 163–174.

Wood, G.A.R. (1975) *Cocoa.* Tropical Agriculture Series. Longman Group, London.

Wrigley, G. (1988) *Coffee.* Tropical Agriculture Series. Longman Scientific & Technical, Harlow, UK.

Yadava, U.L., Burris, J.A. and McCrary, D. (1990) Papaya: a potential annual crop under middle Georgia conditions. In: Janick, J. and Simon, J.E. (eds) *Advances in New Crops.* Timber Press, Portland, Oregon, pp. 364–366.

Yu, C. (2005) Sisal. In: Franck, R.R. (ed.) *Bast and Other Plant Fibres.* Woodhead Publishing, Abingdon, UK in association with The Textile Institute, Cambridge, UK, pp. 228–273.

Zarger, T.G. (1969) Black walnuts – as nut trees. In: Jaynes, R.A. (ed.) *Handbook of North American Nut Trees.* The Northern Nut Growers Association, Knoxville, Tennessee, pp. 203–211.

Zeven, A.C. (1972) The partial and complete domestication of the oil palm (*Elaeis guineensis*). *Economic Botany* 26(3), 274–280.

Zohary, D. and Hopf, M. (1993) *Domestication of Plants in The Old World – The Origin and Spread of Cultivated Plants in West Asia, Europe, and the Nile Valley.* Clarendon Press, Oxford, UK.

PROSEA volumes which have been used frequently

Boer, E. and Ella, A.B. (eds) (2000) *Plant Resources of South-East Asia No. 18: Plants Producing Exudates.* Backhuys Publishers, Leiden, The Netherlands, 189 pp.

Brink, M. and Escobin, R.P. (eds) (2003) *Plant Resources of South-East Asia No. 17: Fibre Plants.* Backhuys Publishers, Leiden, The Netherlands, 456 pp.

De Guzman, C.C. and Siemonsma, J.S. (eds) (1999) *Plant Resources of South-East Asia No. 13: Spices.* Backhuys Publishers, Leiden, The Netherlands, 400 pp.

Flach, M. and Rumawas, F. (eds) (1996) *Plant Resources of South-East Asia No. 9: Plants Yielding Non-seed Carbohydrates.* Backhuys Publishers, Leiden, The Netherlands, 237 pp.

Grubben, G.J.H. and Soetjiptopartohardjono (eds) (1996) *Plant Resources of South-East Asia No. 10: Cereals.* Backhuys Publishers, Leiden, The Netherlands, 199 pp.

't Mannetje, L. and Jones, R.M. (eds) (1992) *Plant Resources of South-East Asia No. 4: Forages.* Pudoc Scientific Publishers, Wageningen, The Netherlands, 300 pp.

Siemonsma, J.S. and Kasem Piluek (eds) (1993) *Plant Resources of South-East Asia No. 8: Vegetables.* Pudoc Scientific Publishers, Wageningen, The Netherlands, 412 pp.

Van der Maessen, L.J.G. and Somaatmadja, S. (eds) (1989) *Plant Resources of South-East Asia No. 1: Pulses.* Pudoc, Wageningen, The Netherlands, 105 pp.

Van der Vossen, H.A.M. and Umali, B.E. (eds) (2001) *Plant Resources of South-East Asia No. 14: Vegetable Oils and Fats.* Backhuys Publishers, Leiden, The Netherlands, 299 pp.

Van der Vossen, H.A.M. and Wessel, M. (eds) (2000) *Plant Resources of South-East Asia No. 16: Stimulants.* Backhuys Publishers, Leiden, The Netherlands, 201 pp.

Verheij, E.W.M. and Coronel, R.E. (eds) (1991) *Plant Resources of South-East Asia No. 2: Edible Fruits and Nuts.* Pudoc, Wageningen, The Netherlands, 445 pp.

Glossary of Botanical Terms

Abaxial	(i) Applied to an embryo which is out of the axis of the seed by the one-sided thickness of the albumen; (ii) the side of a lateral organ away from the axis.
Achene	A small, hard, dry, indehiscent fruit, strictly of one free carpel as in buttercup; occasionally consisting of more than one carpel.
Acuminate	Having a gradually diminishing point.
Acute	Distinctly and sharply pointed, but not drawn out.
Adaxial	The side or face next to the axis.
Adventitious	(Buds) produced abnormally from the stem instead of the axils of the leaves; (roots) which do not arise from the radicle but from another part.
Aerenchyma	Term for thin-walled cells and large intercellular spaces found in the stems of some marsh plants, serving for aeration or floating.
Alternate	Arranged spirally or alternately, not in whorls or opposite pairs.
Androecium	The male element; the stamens as a unit of the flower.
Angled	Two planes forming an edge.
Angular	When organs show a determinate number of angles, e.g. quadrangular.
Annual	Completing the life cycle in 1 year.
Anther	Pollen-bearing part of a stamen.
Anthesis	The time the flower is expanded, or the time when pollination takes place.
Apetalous	Without petals or with a single perianth.
Apex	The growing point of a stem or root.
Aril	An extra covering to the seed in some plants.
Articulate	Jointed, or with places where separation takes place naturally.
Ascending	Curving upwards.
Auricle	A small lobe or ear as appendage to the leaf.
Awn	A bristle-like projection from the tip or back of the glumes of some grasses.
Axil	The angle formed between the axis and any organ which arises from it, especially of a leaf.
Basidiomycete	Fungi producing spores on basidia.
Basidium	The spore mother-cells of some fungi.
Bast	Phloem or fibrous tissues serving for mechanical support.

Beak	Long point at the end of a fruit.
Berry	A pulpy fruit with immersed seeds.
Biennial	A plant which flowers, fruits and dies in its second year or season.
Bipinnate	Double pinnate, with the primary divisions again divided.
Bisexual	Having both sexes present and functional in the same flower.
Blade	Part of the leaf above the sheath, also known as lamina, often flat, sometimes bristle-like.
Bloom	(i) Blossom; (ii) the white waxy covering on some fruits and leaves.
Bract	A modified leaf beneath a flower or part of an inflorescence.
Bracteole	A small or secondary bract.
Bristle	Stiff hair or a very fine straight awn, also applied to the upper part of the awn.
Bud	The nascent state of a flower or branch.
Bulb	A modified bud, usually underground.
Bulbil	A (small) bulb, usually axillary.
Burr	A rough or prickly covering around seeds, fruits or spikelets.
Calyx	The outer whorl of floral parts (sepals), which may be free or united.
Campanulate	Bell-shaped.
Canopy	The uppermost leafy layer of a tree or a forest.
Cap	A term for the husk of a nut.
Capitulum	A close head of sessile flowers, e.g. cauliflower.
Capsule	A dry dehiscent fruit of two or more carpels, splitting when ripe into valves, or opening by slits or pores.
Carpel	Unit of an ovary or fruit.
Caruncle	A wart or protuberance near the hilum of a seed.
Caryopsis	Grain or grass-fruit in which the seed coat is united with the ovary wall.
Catkin	A deciduous spike consisting of unisexual apetalous flowers.
Cauline	Belonging to the stem or arising from it.
Cherelle	Young cacao fruit.
Chupon	Orthotropic subterminal shoot.
Cladode	A branch of a single internode simulating a leaf, e.g. in *Asparagus*.
Cob	The spike of maize.
Coenocarpium	The collective fruit of an entire inflorescence, e.g. a pineapple.
Coleoptile	The first leaf (sheath) in germination of monocotyledons.
Concave	Hollow, as the inside of a saucer.
Cordate	Heart-shaped.
Coriaceous	Leathery texture.
Corm	The base of a stem swollen with reserve materials into a bulbous shape.
Corolla	The second or inner whorl of a flower, consisting of petals, which can be free or united.
Cortex	The bark or rind.
Cotyledon	The first leaf of the embryo (seed-lobe).
Crenate	With blunt teeth.
Crown	(i) Plant section with buds near soil level; (ii) the canopy of a tree; (iii) corolla.
Culm	A stalk, the peculiar hollow stem of grasses.
Cultivar	Cultivated variety.
Cupola	Nearly hemispherical, like an acorn cup.
Cymose	Repeatedly branching inflorescence with the oldest flower at the end of each branch.
Deciduous	Falling off at maturity or end of life.

Dentate	Margin toothed with the pointed teeth directed outwards.
Dichasium	A false dichotomy in which two lateral shoots arise from the primary axis below the flower, which terminate the apex.
Digitate	A compound leaf in which all the leaflets are borne on the apex of the petiole; also applied to an inflorescence.
Dimorph	Occurring under two forms.
Dioecious	Unisexual, the male and female elements in different plants.
Disc floret	A flower in the centre of a flower-head (e.g. sunflower).
Distal	Remote from the place of attachment; the converse of proximal.
Distichous	Disposed in two vertical ranks, as the florets in many grasses.
Drupe	A fleshy indehiscent fruit with a stone usually containing one seed (e.g. a plum).
Ellipsoid	A solid object which is elliptical in section.
Embryo	The rudimentary plant within a seed.
Endocarp	The innermost layer of the pericarp or fruit wall.
Endosperm	Part of a seed containing most of the reserves.
Endotrophic	Applied to mycorrhiza when the fungus attacks the cells of a root itself.
Epicalyx	An involucre of bracts below the flower, resembling an extra calyx.
Exocarp	The outer layer of a pericarp.
Fasciate	Used of the condition of a stem when several have grown together.
Fascicle	A compact cluster.
Filiform	Thread-shaped.
Floret	An individual flower in a dense inflorescence.
Flush	A brief period of rapid shoot growth, with unfolding of the primordia which had accumulated during the previous quiescent period.
Foliage	The leafy covering, especially of trees.
Glabrous	Not hairy.
Gland	A definite secreting structure on the surface, embedded or ending a hair.
Glandular	Having or bearing secreting structures or glands.
Glaucous	Pale bluish-green, or with a whitish bloom which can be rubbed off.
Globose	Spherical or nearly so.
Globular	Spherical in shape.
Globule	A small globe or spherical particle.
Glomerule	A condensed head of almost sessile florets; a cluster of heads in a common involucre.
Glumes	Basal bracts in grass spikelets.
Grain	Naked seed of grasses.
Gynoecium	The female portion of a flower.
Gynophore	A stalk supporting the gynoecium, formed by an elongated receptacle.
Hastate	Spear-shaped, with the basal lobes turned outwards.
Herbaceous	With the texture, colour and properties of a herb.
Hermaphrodite	With both male and female organs.
Hesperidium	A polycarpellary, syncarpous berry, pulpy within, with a tough external rind (like orange).
Hilum	The scar left on the seed indicating its point of attachment.
Hirsute	Covered with long, moderately stiff and not interwoven hairs.
Hull	A widely used term that includes both the palea and lemma of a seed.
Husk	Likewise, a widely used term that includes the palea and lemma of a seed.
Hypanthium	A cup-like receptacle usually derived from the fusion of the floral envelopes and androecium on which are seemingly borne calyx, corolla and stamens.
Hypocotyl	The young stem below the cotyledons.

Indehiscent	Not opening by valves or along regular lines.
Inflorescence	The flowering region or the mode of flowering of a plant.
Infructescence	A ripened inflorescence in the fruiting stage.
Internode	Portion between two successive nodes of the culm, flower-head or spikelet.
Involucre	A whorl of bracts beneath a flower or flower cluster.
Jorquette	Successive whorl of plagiotropic branches.
Keel	Lower fused petals ridged like the keel of a boat, as in the pea family.
Kernel	The nucellus of an ovule or of a seed, the whole body within the coats.
Lacerate(d)	Torn, or irregularly cleft.
Lactiferous	Latex-bearing.
Lamina	Blade, the expanded part of a leaf or petal.
Lanceolate	Lance-shaped; narrow and tapering towards the tip.
Lateral	Fixed on or near the side of an organ.
Leaflet	One part of a compound leaf.
Lemma	Lower of two bracts enclosing the flower.
Lenticular	Shaped like a doubly convex lens.
Ligule	Outgrowth of the inner junction of the leaf-sheath and blade, often membranous, sometimes as hairs.
Linear	Long and narrow, with parallel sides.
Lobe	Any division of an organ or specially rounded division.
Locular	Divided by internal partitions into compartments as in anthers and ovaries.
Loculicidal	The cavity of a pericarp dehiscent by the back, the dorsal suture.
Lodicule	A small scale outside the stamens in the flower of grasses.
Longitudinal	In the direction of the length.
Lyrate	Pinnatifid with the terminal lobe large and rounded, the lower lobes small (lyre-shaped).
Mesocarp	The middle layer of a pericarp.
Monocarp	Only fruiting once.
Monocotyledon	Having one cotyledon or seed-lobe.
Monoecious	The stamens and pistils in separate flowers.
Mucilaginous	Slimy.
Multiloculares	Compound spores.
Mycelium	The vegetative portion of the thallus of fungi.
Mycorrhiza	The symbiotic union of fungi and roots of plants.
Nectariferous	Nectar-bearing.
Node	A joint on the stem where a leaf is (was) attached.
Nodule	A small knot or rounded body.
Nucellus	(i) The body of the ovule containing the embryo sac; (ii) the kernel of an ovule.
Nut	A hard and indehiscent one-seeded fruit.
Oblong	Longer than broad, with the sides parallel or almost so.
Obovate	Reverse of egg-shaped (ovate).
Obovoid	An obovate solid.
Obtuse	Blunt or rounded at the end.
Ocrea	A tubular stipule, or pair of opposite stipules so combined.
Opposite	When two leaves or two branches are borne at the same node on opposite sides of the stem.
Orbicular	Flat with a more or less circular outline.
Orthotropic	Having a more or less vertical direction of growth.
Ovary	The female part of a flower, enclosing the ovules.
Ovate	Egg-shaped.

Ovoid	A solid object which is egg-shaped in section.
Ovule	The immature seed in the ovary before fertilization.
Palea	Upper of two bracts enclosing the flower.
Palmate	With three or more lobes or leaflets radiating like fingers from the palm of a hand.
Panicle	A branched inflorescence.
Paniculate	Resembling a panicle.
Papilionaceous	Type of flower characteristic of the pea family.
Pappus	Modified calyx of the *Compositae* commonly either membranous or in the form of a parasol of hairs (e.g. dandelion).
Parenchyma	Tissue composed of cells more or less isodiametric, especially such tissue as the pith.
Paripinnate	A pinnate leaf without the odd terminal leaflet.
Parthenocarp	Producing fruit without true fertilization.
Pedicel	Applied to the stalk of the spikelet.
Peduncle	In grasses, the stalk of a spikelet-cluster or inflorescence.
Pendulous	Drooping; hanging down.
Perennial	Of more than 2 years' duration.
Perianth	The sepals and petals of a flower.
Pericarp	The wall of a fructified ovary.
Periderm	The outer bark.
Petal	A flower leaf of the corolla.
Petiole	Leaf-stalk.
Phloem	The bast elements of a vascular bundle.
Pinnate	A compound leaf with more than three leaflets arranged in two rows on a single common stalk.
Pinnatifid	Pinnately lobed, but completely divided into leaflets.
Pistil	The female organ of the flower, usually consisting of ovary, style(s) and stigma(s).
Pith	The spongy centre of an exogenous stem, chiefly consisting of parenchyma.
Plagiotropic	Having an oblique or horizontal direction of growth.
Plumose	Feathered, as the pappus of thistles.
Pod	A dry and many-seeded dehiscent fruit, a legume or silique.
Polycarpellary	Of many carpels, free or united.
Pome	An inferior fruit of many cells, of which the apple is the type.
Poricidal	Applied to anthers which open by pores.
Precocious	Flowering and fruiting at an early stage.
Primordium	A member or organ in its earliest stage.
Prostrate	Lying flat on the ground.
Pseudocarp	A fruit with its accompanying parts as a strawberry.
Pseudothallus	The axis of a crowded inflorescence as a glomerule or umbel.
Pubescent	Covered with short, soft hairs.
Pulvinus	The swollen base of a petiole.
Punctate	Marked with dots or translucent glands.
Pyriform	With the shape of a pear.
Raceme	An unbranched inflorescence with the individual flowers stalked.
Racemose	Raceme-like.
Rachis	Main axis of an inflorescence.
Radicle	The rudimentary root of the embryo.
Ratoon growth	Regrowth after the first harvest.
Receptacle	The end of the flower stalk on which the parts of the flower are borne.

Reniform	Kidney-shaped.
Resin	A group of oxidized hydrocarbons, solidified or hardened turpentine, and insoluble in water.
Reticulate	Having the veins disposed like the threads of a net.
Rhizomes	More or less swollen stems, usually underground.
Rhomboid	Quadrangular, with the lateral angles obtuse.
Rosette	A cluster of leaves or other organs in a circular form.
Rugose	Wrinkled.
Runcinate	Saw-toothed or sharply incised margin.
Runner	A stolon, an elongated lateral shoot, rooting at intervals, the intermediate part apt to perish, and thus new plants arise (e.g. strawberry).
Sagittate	Enlarged at the base into two acute straight lobes, like the barbed head of an arrow.
Saprophytic	Living upon dead organic matter.
Scabrous	Rough to the touch.
Scale	Miniature leaf without blade, found at the base of stems, or any thin, membranous, usually non-green body.
Schizocarp	A pericarp which splits into one-seeded portions, split-fruits.
Sepal	One of the individual leaves of a calyx.
Septum	A partition.
Serrate	With minute, forward-pointing teeth.
Sessile	Not stalked.
Sheath	Lower part of the leaf, the part that encloses the stem.
Shell	The hard envelope of a nut.
Siliqua	A fruit characteristic of the wallflower family, elongated and pod-like, but with a central partition and opening from below by two valves.
Simple	Not compound, as in leaves with a single blade.
Solitary	Single, only one from the same place.
Spadix	A spike bearing flowers sometimes sunken, enclosed in a spathe.
Spathe	Large sheathing bract.
Spathulate	Expanded rather suddenly towards the apex, so as to resemble the outline of a spoon; however, the organ is rather flat.
Spike	An unbranched, elongated flower-head, bearing stalkless flowers.
Spikelet	A unit of a grass flower-head, usually with two glumes and one or more florets.
Spine	A stiff, sharp-pointed projection or tip.
Spur	A hollow and slender extension of some part of the flower, usually nectariferous as the corolla of the violet.
Stalk	Any lengthened support of an organ.
Stamen	Pollen-bearing part of the flower.
Staminate	Applied to flowers which are wholly male.
Staminode	A sterile or abortive stamen.
Standard	The broad, usually large upper petal of a flower of the pea family.
Stigma	The part of the female organ of the flower, which receives the pollen.
Stipule	A scaly or leaf-like outgrowth at the base of a petiole.
Stolon	Prostate or creeping stem, rooting at the nodes and there giving rise to vegetative shoots and culms.
Style	The connecting portion between stigma and ovary.
Succulent	Juicy, fleshy.
Sucker	A shoot of subterranean origin.

Sympodial	Of a stem in which the growing point either terminates in an inflorescence or dies, growth being continued by a subtending lateral growing point.
Syncarpous	Composed of two or more united carpels.
Taproot	The primary descending root, forming a direct continuation of the radicle.
Tassel	The male inflorescence, a terminal panicle of maize.
Tendril	A filiform production, cauline or foliar, by which a plant may secure itself in its position.
Terete	Cylindrical; circular in transverse section.
Testa	The outer coat of the seed.
Tetrafoliate	With four leaflets.
Thorn	A woody sharp-pointed structure formed from a modified branch.
Thyrse	A compound inflorescence composed of a panicle (indeterminate axis) with the secondary and ultimate axes cymose (determinate).
Torus	The receptacle of a flower; that portion of the axis on which the parts of the flower are inserted.
Transverse	Across; broader than long.
Trifoliate	Three-leaved.
Trifoliolate	With three leaflets.
Trilobed	With three lobes.
Truncate	Ending abruptly as through cut off, very blunt.
Tuber	Swollen, underground part of a stem or root.
Tubercule	A spherical or ovoid swelling.
Tubular	Trumpet-shaped.
Umbel	An inflorescence with branches radiating like the ribs of an umbrella.
Umbellet	A small umbel or a simple one.
Umbilicus	Hilum, a scar left on the seed indicating its point of attachment.
Undulate	Wavy, said for example of a leaf margin if the waves run in a plane at right angles to the plane of the leaf blade.
Unisexual	Of one sex, having stamens or pistils only.
Utricle	A small bladdery pericarp, or any bladder-shaped appendage.
Valve	Segment of a dehiscent fruit.
Vein	A strand of vascular tissue in a flat organ, as a leaf.
Vine	Any trailing or climbing stem or runner; or the plant which bears grapes.
Whorl	A ring of similar parts radiating from a node.
Wings	Lateral petals, characteristic of flowers of the pea family.
Xylem	The wood elements of a vascular bundle.

Index of Plant Names

The main entry for each species is listed in **bold**.

Scientific name	Common name(s)
Abelmoschus esculentus (L.) Moench.	Okra **475**
Aegilops speltoides Tausch	Goatgrass 357
Aegilops squarrosa L.	Goatgrass 358
Agaricus bisporus (Lange) Imbach	Mushroom, button **469**
Agave americana L.	Century plant or maguey 158
Agave cantala Roxb.	Maguey or cantala 158
Agave fourcroydes Lem.	Hennequen 158
Agave funkiana K. Koch & C.D. Bouche	Ixtle de Jaumave 158
Agave lecheguilla Torr.	Lechuguilla 158
Agave letonae F.W. Taylor ex Trel.	Hennequen, Salvadorian or letona fibre 158
Agave sisalana Perrine	Sisal **155**
Agave tequilana F.A.C. Weber	Agave, Weber blue 158
Allium cepa L.	Onion, common **409**
Allium cepa L. *Aggregatum* group	Shallot **409**
Allium porrum L.	Leek **409**
Allium sativum L.	Garlic **409**
Allium schoenoprasum L.	Chives **409**
Alocasia macrorrhiza (L.) Scott	Taro, wild 393
Amaranthus spp.	Amaranth **340**
Anacardium occidentale L.	Cashew **112**
Ananas comosus L. Merill	Pineapple **83**
Anethum graveolens L.	Dill **306**
Apios americana Medic.	Groundnut 196
Apium graveolens L.	Celery and celeriac **306**
Arachis batizocoi Krapov & W.C. Greg.	Groundnut, wild diploid 192
Arachis cardenasii Krapov & W.C. Greg.	Groundnut, wild diploid 192
Arachis hypogaea L.	Groundnut **192**
Arachis monticola Krapov. & Rig	Peanut, mountain 192